PROCEEDINGS OF SYMPOSIA
IN PURE MATHEMATICS
Volume XXIII

PARTIAL DIFFERENTIAL EQUATIONS

Symposium in Pure Mathematics, University of California at Berkeley, 1971.

AMERICAN MATHEMATICAL SOCIETY
PROVIDENCE, RHODE ISLAND
1973

PROCEEDINGS OF THE SYMPOSIUM IN PURE MATHEMATICS OF THE AMERICAN MATHEMATICAL SOCIETY

HELD AT THE UNIVERSITY OF CALIFORNIA
BERKELEY, CALIFORNIA
AUGUST 9-27, 1971

Edited by
D. C. SPENCER

Prepared by the American Mathematical Society under National Science Foundation Grant GP-28200

Library of Congress Cataloging in Publication Data

Symposium in Pure Mathematics, University of California at Berkeley, 1971.
Partial differential equations.

(Proceedings of symposia in pure mathematics, v. 23)
"An outgrowth of lectures delivered at the eighteenth Summer Research Institute of the American Mathematical Society ... held ... from August 9 to August 27, 1971."
Includes bibliographical references.
1. Differential equations, Partial--Congresses.
I. Spencer, Donald Clayton, 1912- ed.
II. American Mathematical Society. III. Title.
IV. Series.
QA374.S93 1971 515'.353 72-4071
ISBN 0-8218-1423-0

AMS (MOS) subject classifications (1970). Primary 35-XX

Printed in the United States of America

CONTENTS

Seminar on Nonlinear Problems

Seminar on Geometry

Seminar on Mathematical Physics

Seminar on Singular Integral Operators

PREFACE

The papers in these Proceedings are an outgrowth of lectures delivered at the eighteenth Summer Research Institute of the American Mathematical Society. The topic of the institute was partial differential equations, and it was held at the University of California at Berkeley from August 9 to August 27, 1971. The institute was financed by the National Science Foundation.

Notes of lectures were distributed during the conference (and remaining notes shortly afterward) to the participants, and many of the papers appearing in this volume are revised versions of the informal notes. Although some of the papers are expositions of known material, many contain new results. The papers are arranged under the headings of the five seminars of the conference: linear problems, nonlinear problems, geometry, mathematical physics, and singular integral operators.

The organizing committee for the institute consisted of: Alberto P. Calderón, Lars Hörmander, Charles B. Morrey, Jr., Louis Nirenberg (Chairman), James B. Serrin, Isadore M. Singer and Donald C. Spencer.

The editor would like to thank the many persons who cooperated to make the institute and this volume possible. Of special direct help were Lillian R. Casey, Beth Clarke, Hope Daly (conference secretary), Carole Kohanski and Margaret Reynolds.

D. C. SPENCER

EXISTENCE AND REGULARITY OF HYPERSURFACES OF R^n WITH PRESCRIBED MEAN CURVATURE

MARIO MIRANDA

1. **Existence and regularity of hypersurfaces of R^n with prescribed mean curvature.**

1.1. Let R^n be the Euclidean n-dimensional space and $\Omega \subset R^n$ an open set with Lipschitz boundary $\partial\Omega$. Let Γ be a Borel subset of $\partial\Omega$ and A a function in $L^1(\Omega)$. On the class $\mathscr{B} = \mathscr{B}(\Omega)$ of Borel subsets of Ω we can consider the functional

$$(1.1)\quad \mathscr{F}(E) = \int_\Omega A(x)\phi_E(x)\,dx + |D\phi_E|(\Omega) + \int_{\partial\Omega} |\phi_E - \phi_\Gamma|\,dH_{n-1}, \qquad E \in \mathscr{B},$$

where ϕ_E and ϕ_Γ are the characteristic functions of E and Γ, and where $|D\phi_E|(\Omega)$ is the total variation on Ω of the gradient of ϕ_E, i.e.,

$$(1.2)\qquad |D\phi_E|(\Omega) = \sup\left\{\int_\Omega \phi_E \operatorname{div} G\,dx;\ G \in [C_0^1(\Omega)]^n,\ |G(x)| \leqq 1 \quad \forall x\right\}.$$

We recall that (see [**1**]) when $|D\phi_E|(\Omega) < +\infty$ then ϕ_E has a trace on $\partial\Omega$ belonging to $L^1_{\mathrm{loc}}(\partial\Omega)$; in the second integral of (1.1), ϕ_E denotes this trace. When $|D\phi_E|(\Omega) = +\infty$ we mean $\mathscr{F}(E) = +\infty$. H_{n-1} is the $(n-1)$-dimensional Hausdorff measure on R^n.

We will prove the following existence theorem.

THEOREM 1. *For any choice of Ω, Γ, A there exists $E_0 \in \mathscr{B}(\Omega)$ such that*

$$(1.3)\qquad \mathscr{F}(E_0) = \inf \mathscr{F}(\mathscr{B}) = \inf_{E \in \mathscr{B}(\Omega)} \mathscr{F}(E).$$

PROOF. It is obvious that

$$(1.4)\qquad \inf \mathscr{F}(\mathscr{B}) \geqq -\int_\Omega |A(x)|\,dx > -\infty,$$

so we can assume that

AMS 1970 *subject classifications.* Primary 35J20.

$$\inf \mathscr{F}(\mathscr{B}) = \gamma \in R. \tag{1.5}$$

We will denote by $\{E_h\}_h$ a sequence of sets of $\mathscr{B}(\Omega)$ such that

$$\lim_{h\to\infty} \mathscr{F}(E_h) = \gamma. \tag{1.6}$$

We recall now that is is possible (see [**2**]) to extend ϕ_Γ to a function $f \in L^1_{\mathrm{loc}}(R^n)$ with $|Df| \in L^1_{\mathrm{loc}}(R^n)$. For every set $E \in \mathscr{B}(\Omega)$, let us denote by f_E the function of $L^1_{\mathrm{loc}}(R^n)$ which is equal to ϕ_E on Ω and equal to f on $R^n - \Omega$. We can then write the functional $\mathscr{F}$ in the following way (see [**1**]):

$$\mathscr{F}(E) = \int_\Omega A \cdot \phi_E \, dx + \sup_\rho \left\{ |Df_E|(B_\rho) - \int_{B_\rho - \Omega} |Df(x)| \, dx \right\}; \tag{1.7}$$
$$B_\rho = \{x \in R^n, |x| < \rho\}.$$

From (1.6) we have

$$\sup_h |Df_{E_h}|(B_\rho) < +\infty, \qquad \forall \rho > 0. \tag{1.8}$$

Then the sequence $\{f_{E_h}\}_h$ is compact in $L^1_{\mathrm{loc}}(R^n)$ (see [**3**]), and, for simplicity, we will assume that $\{f_{E_h}\}_h$ converges in $L^1_{\mathrm{loc}}(R^n)$ towards a function f_{E_0}. So (see [**3**])

$$|Df_{E_0}|(B_\rho) \leqq \liminf_{h\to\infty} |Df_{E_h}|(B_\rho) < +\infty, \qquad \forall \rho > 0. \tag{1.9}$$

From (1.8) we have $\mathscr{F}(E_0) < +\infty$ and $\forall \varepsilon > 0 \ \exists \rho(\varepsilon) > 0$ s.t.

$$\begin{aligned} &\mathscr{F}(E_0) - \mathscr{F}(E_h) \\ &\leqq \varepsilon + \int_\Omega (\phi_{E_0} - \phi_{E_h}) A \, dx + |D\phi_{E_0}|(B_{\rho(\varepsilon)}) - |D\phi_{E_h}|(B_{\rho(\varepsilon)}), \qquad \forall h. \end{aligned} \tag{1.10}$$

At the limit, we therefore have, recalling (1.9),

$$\liminf_{h\to\infty} \{\mathscr{F}(E_0) - \mathscr{F}(E_h)\} \leqq \varepsilon. \tag{1.11}$$

Since $\varepsilon > 0$ is arbitrary and (1.6) holds, we have

$$\mathscr{F}(E_0) = \inf \mathscr{F}(\mathscr{B}). \qquad \text{Q.E.D.} \tag{1.12}$$

REMARK 1. Theorem 1 remains valid if instead of considering the class $\mathscr{B}(\Omega)$ of all Borel subsets of Ω, we consider the class $\mathscr{B}_M(\Omega)$ of the sets containing a fixed $M \in \mathscr{B}(\Omega)$.

1.2. We now want to say a few words about the geometric properties of the set E_0 found in Theorem 1. Let B be an open ball contained with its boundary in Ω and centered at a point $x_0 \in \partial E_0$. Let us assume $\partial E_0 \cap B$ to be the graph of a C^2-function of $(n-1)$ variables, i.e., for simplicity,

$$E_0 \cap B = \{x \in B; (x_1, \ldots, x_{n-1}) \in K, x_n < g(x_1, \ldots, x_{n-1})\}, \tag{2.1}$$

where K is an open set of R^{n-1} and $g \in C^2(K)$. We can write $\mathscr{F}(E_0)$ in the following way (see [**1**]):

$$\mathscr{F}(E_0) = \int_{\Omega - B} A\phi_{E_0}\,dx + |D\phi_{E_0}|(\Omega - \bar{B}) + \int_{\partial\Omega} |\phi_E - \phi_\Gamma|\,dH_{n-1}$$

$$+ \int_{\partial B} |\phi_{E_0}^+ - \phi_{E_0}^-|\,dH_{n-1} + \int_K \left\{(1 + |Dg|^2)^{1/2} + \int^g A\,dx_n\right\} dx_1 \cdots dx_{n-1}, \tag{2.2}$$

where $\phi_{E_0}^+$ and $\phi_{E_0}^-$ are the exterior and interior traces of ϕ_{E_0} on ∂B. As a consequence of the minimum property of E_0, we will have

$$\frac{d}{dt}\int_K \left\{(1 + |D(g + t\psi)|^2)^{1/2} + \int^{g+t\psi} A\,dx_n\right\} dx_1 \cdots dx_{n-1}\Big|_{t=0} = 0, \quad \forall \psi \in C_0^1(K). \tag{2.3}$$

Then, if A is continuous, we obtain that

$$\int_K \left\{\sum_{i=1}^{n-1} D_i g \cdot D_i\psi/(1 + |Dg|^2)^{1/2} + \psi A(x_1, \ldots, x_{n-1}, g)\right\} dx_1 \cdots dx_{n-1} = 0, \quad \forall \psi \in C_0^1(K), \tag{2.4}$$

and so, by integrating by parts and recalling the arbitrariness of ψ, we get the following pointwise equality:

$$\sum_{i=1}^{n-1} D_i\left(\frac{D_i g}{(1 + |Dg|^2)^{1/2}}\right) = A(x_1, \ldots, x_{n-1}, g), \qquad \forall (x_1, \ldots, x_{n-1}) \in K,$$

i.e., A is the sum of the principal curvatures of ∂E_0, so ∂E_0 is a minimal surface when $A \equiv 0$.

1.3. In this subsection we will discuss the regularity of ∂E_0. It is possible to prove the following theorem.

THEOREM 2. *If Ω, Γ, A verify the hypothesis of* §1 *and moreover $|A(x)| \leqq A_0 < +\infty$, $\forall x \in \Omega$, $|D\varphi_{E_0}|(\Omega) < +\infty$, then there exists $\Omega_0 \subset \Omega$ open such that $\partial E_0 \cap \Omega_0$ is an $(n-1)$-dimensional manifold of class $C^{1,\alpha}$ (for some $\alpha, 0 < \alpha < 1$) and $H_s(\Omega - \Omega_0) = 0$, $\forall s > n - 8$ (for $n \leqq 7$, $\Omega - \Omega_0 = \varnothing$).*

The proof of this result is contained in [**4**]. Before giving some indications on the proof we want to give a little history of the result. First DeGiorgi in 1960 had considered the case $A \equiv 0$ and proved the result with $s = n - 1$. However, DeGiorgi's proof has never been presented to a regular journal for publication. I wrote this proof again in [**5**]. Massari's proof is an improvement of mine. Almgren [**6**] and Simons [**7**] proved that $\Omega - \Omega_0 = \varnothing$ up to the dimension 7. They also studied the problem of minimal cones, to which the study of singularities had been reduced by a technique of Fleming-Triscari-DeGiorgi. In 1969 Bombieri, DeGiorgi and Giusti [**11**] proved that $\Omega - \Omega_0$ can be nonempty starting from $n = 8$ and with $A \equiv 0$. In 1970 Federer [**12**], still in the case $A \equiv 0$, proved that $H_s(\Omega - \Omega_0) = 0$, $\forall s > n - 8$. Massari repeats Federer's proof for any bounded $A(x)$ in his paper. Results of the type of Theorem 2 have been recently obtained by Allard [**13**]. Allard considers varifolds, in the sense of Almgren, of any dimensions, and he requires only an integral bound on the mean curvature. So Allard's results are more general, but in the special case of codimension 1 less precise than Massari's.

Now we want to give some indications about the proof of Theorem 2. The fundamental lemma is the following:

LEMMA 1 (DEGIORGI). $\forall n \in N$, $n \geqq 2$, $\forall A_0 \in R$, $A_0 > 0$, $\forall \alpha \in R$, $0 < \alpha < 1$, *there exists* $\sigma = \sigma_{A_0}(n, \alpha) > 0$, $\sigma \in R$, *such that if* E_0 *minimizes the functional* $\mathcal{F}$ *in* $\Omega \subset R^n$, *if* $|A(x)| \leqq A_0$, *if* $\bar{B}_\rho(x) = \{y \in R^n; |y - x| \leqq \rho\} \subset \Omega$ *and*

$$(3.1) \qquad \Lambda(\rho, x) = |D\phi_{E_0}|(\bar{B}_\rho(x)) - |D\phi_{E_0}(B_\rho(x))| \leqq \varepsilon\rho^{n-1} \leqq \sigma_{A_0}(n, \alpha)\rho^{n-1}$$

with $\rho \leqq \varepsilon^2$, *then*

$$(3.2) \qquad \Lambda(\alpha\rho, x) \leqq \alpha^{1/2} \cdot \varepsilon \cdot (\alpha\rho)^{n-1}.$$

The proof of this lemma uses a large part of Massari's paper. A simple remark about Lemma 1 is that its application can be repeated. So if (3.1) is verified it is also true that

$$(3.3) \qquad \Lambda(\alpha^h\rho, x) \leqq \alpha^{h/2} \cdot \varepsilon \cdot (\alpha^h\rho)^{n-1}, \qquad \forall h \in N.$$

Let us now denote by $v(x)$ the vector

$$(3.4) \qquad \lim_{\rho \to 0+} \frac{D\phi_{E_0}(B_\rho(x))}{|D\phi_{E_0}|(B_\rho(x))},$$

which exists $|D\phi_{E_0}|$-almost everywhere in Ω (see the Vitali-Lebesgue theorem). Our way of studying the regularity of $\partial E_0 \cap \Omega$ consists of analyzing the behavior of $v(x)$. An easy consequence of Lemma 1 is the following:

LEMMA 2. *With the same hypothesis as Lemma* 1 *for* n, Ω, A, E_0, ρ *and* $x \in \partial E_0$ *we can conclude that* $v(x)$ *exists.*

PROOF. It is easy to check that if $B_1 \subset B_2 \subset \Omega$ is any couple of Borel subsets and if we denote $\phi = \phi_{E_0}$, then

$$(3.5) \qquad \left|\frac{D\phi(B_1)}{|D\phi|(B_1)} - \frac{D\phi(B_2)}{|D\phi|(B_2)}\right| \leqq 2\left(\frac{|D\phi|(B_2)}{|D\phi|(B_1)}\right)^{1/2} \cdot \left(\frac{|D\phi|(B_2) - |D\phi(B_2)|}{|D\phi|(B_2)}\right)^{1/2}.$$

From (3.3) and (3.5) and other general inequalities (see [**4**]) it follows that $\{D\phi(B_{\alpha^h\rho}(x))/|D\phi|(B_{\alpha^h\rho}(x))\}_h$ is a Cauchy sequence. It can be proved in fact that

$$\left|\frac{D\phi(B_{\alpha^{h+k}\rho}(x))}{|D\phi|(B_{\alpha^{h+k}\rho}(x))} - \frac{D\phi(B_{\alpha^h\rho}(x))}{|D\phi|(B_{\alpha^h\rho}(x))}\right| \leqq c(n, \alpha) \cdot \varepsilon^{1/2} \cdot \alpha^{h/4}. \tag{3.6}$$

So the limit (3.4) exists on the sequence $\{\alpha^h\rho\}_h$. By using (3.5) again it can be proved that the limit exists for $\rho \to 0+$ and $|\nu(x)| = 1$, and

$$\left|\nu(x) - \frac{D\phi(B_{\alpha^h\rho}(x))}{|D\phi|(B_{\alpha^h\rho}(x))}\right| \leqq \varepsilon^{1/2} \cdot c(n, \alpha) \cdot \alpha^{h/4} \quad \text{if } \rho < \frac{\omega_{n-1}}{2(n-1)\omega_n A_0} \tag{3.7}$$

($\omega_n = \operatorname{meas}_n\{x \in R^n;\ |x| < 1\}$, $\omega_{n-1} = \operatorname{meas}_{n-1}\{x \in R^{n-1};\ |x| < 1\}$).

It is then possible to prove the following:

LEMMA 3. *In the conditions of Lemma 2 there exists a $\delta > 0$ such that $\nu(y)$ exists $\forall y \in B_\delta(x) \cap \partial E_0$ and is Hölder continuous.*

Lemma 3 implies the existence of an open subset Ω_0 of Ω where ∂E_0 has a Hölder continuous tangent hyperplane. By blowing up the singular points, as is done in [**12**], the estimate of $\Omega - \Omega_0$ can be obtained.

REMARK 2. For the case considered in Remark 1 we have the following:

THEOREM 3. *If Ω, Γ, A verify the hypothesis of Theorem 2 and $M \subset \Omega$ has a C^1 boundary in Ω, then, for any set $E_0 \in \mathscr{B}_M(\Omega)$ minimizing $\mathscr{F}$ on $\mathscr{B}_M(\Omega)$, there exists $\Omega_1 \subset \Omega$ open, $\Omega_1 \supset \partial M \cap \Omega$, such that $\partial E_0 \cap \Omega_1$ is a C^1-manifold of dimension $n - 1$.*

The proof of Theorem 3 can be found in [**14**] for the case $A(x) \equiv 0$. The argument works also in the general case $|A(x)| \leqq A_0$ but it has not yet been published.

Let us observe too that, in the presence of obstacles M, $\partial E_0 \cap \Omega$ cannot be, in general, of class C^2, as can be easily seen by pushing a minimal hypersurface with a ball. The results at the end of this article show that if $\partial M \cap \Omega$ is a C^2-manifold of dimension $n - 1$ then ∂E_0 is a $C^{1,\alpha}$-manifold (for any α, $0 < \alpha < 1$) of the same dimension in a neighborhood of $\partial M \cap \Omega$.

REMARK 3. Emmer [**21**] has applied Massari's result to solve a problem in the theory of capillarity.

2. **Boundary behavior of minimal hypersurfaces in R^n and the Dirichlet problem for the minimal hypersurfaces equation.**

2.1. We now state conditions on Ω and Γ that are sufficient to assure $\phi_{E_0}|_{\partial\Omega} = \phi_\Gamma$. We restrict ourselves to the case $A \equiv 0$.

The new condition on Ω is the following: for $x \in \partial\Omega$ there exists an isometry i of R^n such that $ix = 0$ and there exist $\rho > 0$, an open set $G \subset R^{n-1}$, and a Lipschitz function ψ of $n - 1$ variables such that

$$i\Omega \cap B_\rho(0) = \{x \in B_\rho;\ (x_1, \dots, x_{n-1}) \in G, x_n > \psi(x_1, \dots, x_{n-1})\} \tag{1.1}$$

and

$$\int_G \sum_{i=1}^{n-1} \frac{D_i\psi \cdot D_i\alpha}{(1+|D\psi|^2)^{1/2}}\,dx_1 \cdots dx_{n-1} \geqq 0, \qquad \forall \alpha \in C_0^1(G),\ \alpha \leqq 0. \tag{1.2}$$

When these conditions are verified we will say that Ω is pseudoconvex at x. Obviously, every convex set is pseudoconvex at any of its boundary points. In the case $\psi \in C^2(G)$, the integral inequality (1.2) is equivalent to the following pointwise one

$$\sum_{i=1}^{n-1} D_i \frac{D_i\psi}{(1+|D\psi|^2)^{1/2}} \geqq 0, \quad \text{on } G. \tag{1.3}$$

So the sets whose boundaries are regular and have nonnegative mean curvature (see [**15**]) can be seen as the regular pseudoconvex sets.

We will prove the following:

THEOREM 4. *If $\Omega \subset R^n$, $n \geqq 2$, is pseudoconvex at $x \in \partial\Omega$, if no part of $\partial\Omega$ around x is a minimal hypersurface, if $\Gamma \subset \partial\Omega$ and E_0 minimizes the functional $\mathscr{F}$,*

$$\mathscr{F}(E) = |D\phi_E|(\Omega) + \int_{\partial\Omega} |\phi_E - \phi_\Gamma|\,dH_{n-1}, \tag{1.4}$$

in the class of all Borel subsets of Ω, and if $\mathscr{F}(E_0) < +\infty$, then

$$\text{if } x \in \mathring{\Gamma},\ \exists \rho > 0 : B_\rho(x) \cap E_0 = B_\rho(x) \cap \Omega, \tag{1.5}$$

$$\text{if } x \in \partial\Omega - \bar{\Gamma},\ \exists \rho > 0 : B_\rho(x) \cap E_0 = \varnothing. \tag{1.6}$$

The proof of Theorem 4 is based on the following lemma, which can be considered as an extension of the strong maximum principle for the solutions of minimal surface equations.

LEMMA 4. *If $\Omega \subset R^n$, $n \geqq 2$, is pseudoconvex at $x \in \partial\Omega$ and E_0 minimizes the functional $\mathscr{F}$ in $B_\rho(x)$ with $\phi_\Gamma = \phi_{E_0}|_{\partial B_\rho(x)}$ and if $E_0 \subset B_\rho(x) \cap \Omega$, $x \in \partial E_0$, then there exists $\rho' > 0$ such that*

$$B_{\rho'}(x) \cap E_0 = B_{\rho'}(x) \cap \Omega. \tag{1.7}$$

PROOF. Let us assume that the isometry relative to the pseudoconvexity at x is the identity. Let σ be a positive number such that $G \supset \{x \in R^{n-1};\ |x| < \sigma\}$ and let us denote by ψ_σ the solution of the equation

$$\sum_{i=1}^{n-1} D_i \frac{D_i\psi_\sigma(x)}{(1+|D\psi_\sigma|^2)^{1/2}} = 0, \qquad x \in R^{n-1},\ |x| < \sigma, \tag{1.8}$$

with the condition

$$\psi_\sigma(x) = \psi(x), \qquad x \in R^{n-1},\ |x| = \sigma. \tag{1.9}$$

Let us denote by Ω_σ the open set $\Omega \cap \{(x, x_n);\ x \in R^{n-1},\ |x| < \sigma,\ x_n > \psi_\sigma(x)\}$.

We now have (see [**16**])

$$(1.10) \qquad |D\phi_{E_0\cap\Omega_\sigma}|(B_\rho) + |D\phi_{E_0\cup\Omega_\sigma}|(B_\rho) \leqq |D\phi_{E_0}|(B_\rho) + |D\phi_{\Omega_\sigma}|(B_\rho),$$

and, for the minimum property of E_0 and Ω_σ,

$$(1.11) \qquad |D\phi_{E_0}|(B_\rho) \leqq |D\phi_{E_0\cap\Omega_\sigma}|(B_\rho),$$

$$(1.12) \qquad |D\phi_{\Omega_\sigma}|(B_\rho) \leqq |D\phi_{E_0\cup\Omega_\sigma}|(B_\rho).$$

So the relations (1.10), (1.11) and (1.12) have to be equalities. The equality in (1.12) implies $\Omega_\sigma = \Omega_\sigma \cup E_0$, that is, $E_0 \subset \Omega_\sigma$. Since $0 \in \partial E_0$, we must have $0 = \psi_\sigma(0)$ and so (see [**14**]) ∂E_0 must be regular at 0. Then, in a neighborhood of 0, ∂E_0 is the graph of an analytic function. Applying the classical strong maximum principle to this function and to ψ_σ we must have $\{(x, x_n);\ x \in R^{n-1},\ x_n = \psi_\sigma(x)\} \subset \partial E_0$. Since we can repeat this argument for any small σ, we have (1.7). Q.E.D.

Proof of Theorem 4. Let $x \in \partial\Omega - \bar{\Gamma}$. Then there exists $\rho > 0$ such that $B_\rho(x) \cap \Gamma = \varnothing$. Because of the property of E_0 and an easy remark (see [**16**]) we have that E_0 minimizes the functional $\mathscr{F}$ in $B_\rho(x)$. So x cannot belong to ∂E_0 (see Lemma 4); so there must exist a $\rho' > 0$ such that $B_{\rho'}(x)E_0 = \varnothing$. Q.E.D.

2.2. We will now consider the nonparametric case. We will prove the following:

Theorem 5. *For any $\Omega \subset R^n$, $n \geqq 2$, open, bounded and locally pseudoconvex, for any $g \in C(\partial\Omega)$, there exists a unique $f \in C(\bar{\Omega})$, analytic in Ω, such that $f|_{\Omega} = g$ and*

$$\sum_{i=1}^n D_i \frac{D_i f(x)}{(1 + |Df|^2)^{1/2}} = 0, \qquad \forall x \in \Omega.$$

This result is a consequence of the results contained in [**15**] and [**17**] in the case $\partial\Omega \in C^2$, and those contained in [**18**] and [**17**] in the case Ω is strictly convex. Theorem 5 contains these two previous results and its proof is contained in [**16**]. For the convenience of the reader we repeat this proof here.

Proof of Theorem 5. Let us denote by B a ball containing $\bar{\Omega}$ in its interior. Let us extend g over B in order to get $g \in C(B) \cap H^{1,1}(B)$ (see [**2**]). Let us denote by W the class of functions $w \in L^1(B)$ which are equal to g on $B - \Omega$ and let us consider on W the functional

$$(2.2) \qquad \begin{aligned} \mathscr{T}(w) = \sup&\left\{\int_\Omega \left(w \sum_{i=1}^n D_i\gamma + \gamma_{n-1}\right) dx;\ \gamma \in [C_0^1(\Omega)]^{n+1},\ |\gamma| \leqq 1\right\} \\ &+ \int_{\partial\Omega} |w^- - g|\, dH_{n-1}, \end{aligned}$$

where w^- is the interior trace of w on $\partial\Omega$. There exists $f \in W$ such that $\mathscr{T}(f) = \inf \mathscr{T}(W)$ (see [**16**]) and f is analytic on Ω. We now want to prove that $f \in C(\bar{\Omega})$ and $f|_{\partial\Omega} = g$. Let us first observe that the set $E_0 = \{(x, x_{n+1});\ x \in B,\ x_{n+1} \subset f(x)\}$

has the property

$$(2.3)\qquad |D\phi_{E_0}|(B \times R) \leqq |D\phi_M|(B \times R), \quad \forall M \subset R^{n+1}, (M - E_0) \cup (E_0 - M) \subset\subset \bar{\Omega} \times R.$$

It is sufficient now to prove that $\forall \{x_h\}_h \subset \Omega$, $x_h \to x \in \partial\Omega$, $f(x_h) \to l \Rightarrow l = g(x)$. This must be true because $(x_h, f(x_h)) \in \partial E_0 \cap (\Omega \times R)$; so $(x, l) \in \partial E_0$ and then l has to be equal to $g(x)$ (see Lemma 4). Q.E.D.

In the case of obstacles we have the following result.

THEOREM 6. *If $\Omega \subset R^n$, $n \geqq 2$, is bounded and locally pseudoconvex, if $g \in C(\partial\Omega)$ and $\psi \in C^1(\Omega) \cap C(\bar{\Omega})$ with $\psi|_{\partial\Omega} \leqq g$, then there exists a unique $f \in C^1(\Omega) \cap C(\bar{\Omega})$, $f \geqq \psi$, on Ω, $f|_{\partial\Omega} = g$ and*

$$(2.4)\qquad \int_\Omega (1 + |Df|^2)^{1/2}\, dx \leqq \int_\Omega (1 + |D\gamma|^2)^{1/2}\, dx, \quad \forall \gamma \in C^1(\Omega) \cap C(\bar{\Omega}), \gamma \geqq \psi, \gamma|_{\partial\Omega} = g.$$

REMARK 4. Using results contained in [**19**] or [**20**] the function f found in Theorem 6 is $C^{1,\alpha}(\Omega)$ (for any α, $0 < \alpha < 1$) if $\psi \in C^2(\Omega)$.

BIBLIOGRAPHY

1. M. Miranda, *Comportamento della successioni convergenti di frontiere minimali*, Rend. Sem. Mat. Univ. Padova **38** (1967), 238–257. MR **36** #5785.

2. E. Gagliardo, *Caratterizzazioni delle tracce sulla frontiera relative ad alcune classi di funzioni in n variabili*, Rend. Sem. Mat. Univ. Padova **27** (1957), 284–305. MR **21** #1525.

3. M. Miranda, *Distribuzioni aventi derivate misure insiemi di perimetro localmente finito*, Ann. Scuola Norm. Sup. Pisa (3) **18** (1964), 27–56. MR **29** #2364.

4. U. Massari, *Esistenza e regolarità ipersuperfici di curvatura media assegnata in R^n*, Arch. Rational Mech. Anal. (to appear).

5. M. Miranda, *Sul minimo dell'integrale del gradiente di una funzione*, Ann. Scuola Norm. Sup. Pisa (3) **19** (1965), 626–665. MR **32** #6271.

6. F. J. Almgren, Jr., *Some interior regularity theorems for minimal surfaces and an extension of Bernstein's theorem*, Ann. of Math. (2) **84** (1966), 277–292. MR **34** #702.

7. J. Simons, *Minimal varieties in riemannian manifolds*, Ann. of Math. (2) **88** (1968), 62–105. MR **38** #1617.

8. W. H. Fleming, *On the oriented plateau problem*, Rend. Circ. Mat. Palermo (2) **11** (1962), 69–90. MR **28** #499.

9. D. Triscari, *Sulle singolarità delle frontiere orientate di misura minima nello spazio euclideo a 4 dimensioni*, Matematiche (Catania) **18** (1963), 139–163. MR **29** #6357c.

10. E. DeGiorgi, *Una estensione del teorema di Bernstein*, Ann. Scuola Norm. Sup. Pisa (3) **19** (1965), 79–85. MR **31** #2643.

11. E. Bombieri, E. DeGiorgi and E. Giusti, *Minimal cones and the Bernstein problem*, Invent. Math. **7** (1969), 243–268. MR **40** #3445.

12. H. Federer, *The singular sets of area minimizing rectifiable currents with codimension one and of area minimizing flat chains modulo two with arbitrary codimension*, Bull. Amer. Math. Soc. **76** (1970), 767–771. MR **41** #5601.

13. W. K. Allard, *A regularity theorem for the first variation of the area integrand*, Bull. Amer. Math. Soc. **77** (1971), 772–776.

14. M. Miranda, *Frontiere minimali con ostacoli*, Ann. Univ. Ferrara **16** (1971), 29–37.

15. H. Jenkins and J. Serrin, *The Dirichlet problem for the minimal surface equation in higher dimensions*, J. Reine Angew. Math. **229** (1968), 170–187. MR **36** #5519.

16. M. Miranda, *Un principio di massimo forte per le frontiere minimali e una sua applicazione alla risoluzione del problema al contorno per l'equazione delle superfici di area minima*, Rend. Sem. Mat. Univ. Padova **45** (1971), 355–366.

17. E. Bombieri, E. DeGiorgi and M. Miranda, *Una maggiorazione a priori relativa alle ipersuperfici minimali non parametriche*, Arch. Rational Mech. Anal. **32** (1969), 255–267. MR **40** #1898.

18. M. Miranda, *Un teorema di esistenza e unicità per il problema dell'area minima un n variabili*, Ann. Scuola Norm. Sup. Pisa (3) **19** (1965), 233–249. MR **31** #6144.

19. H. Lewy and G. Stampacchia, *On existence and smoothness of solutions of some non-coercive variational inequalities*, Arch. Rational Mech. Anal. **41** (1971), 241–253.

20. M. Giaquinta and L. Pepe, *Esistenza e regolarita per il problema dell'area minima con ostacoli in n variabili*, Ann. Scuola Norm. Sup. Pisa **25** (1971), 481–507.

21. M. Emmer, *Esistenza, unicità e regolarita delle superfici di equilibrio nei capillari*, Ann. Univ. Ferrara **18** (1972).

22. E. Santi, *Sul problema al contorno per l'equazione delle superfici di area minima su domini limitati qualunque*, Ann. Univ. Ferrara **17** (1971), 13–26.

23. W. K. Allard, *On the first variation of a manifold*, Princeton University, Princeton, N.J. (preprint).

UNIVERSITA DI FERRARA, FERRARA, ITALY

RECENT APPLICATIONS OF INDEX THEORY FOR ELLIPTIC OPERATORS

I. M. SINGER

1. **Introduction.** In the past few years, index theory has been extended to situations where the index of an operator is no longer an integer. These new indices have interesting applications to geometry and topology, and what I would like to do in this article is to describe some of these applications. It is my impression that specialists in partial differential equations have shied away from index theory for elliptic operators because of the essential use of K-theory and characteristic classes. I have tried to choose examples which minimize the use of characteristic classes but are nevertheless revealing. In any case, Appendix I contains a brief description of characteristic classes, and how they are used in index formulae.

This article is mostly expository; for more details of the basic material see [**4**], [**5**], [**9**]–[**14**], [**27**]. The recent applications are taken directly from [**1**], [**7**] for the mod 2 index (§4), from [**8**], [**21**] for fixed point formulae (§5), and from [**23**] for the families index (§6). What *is* new is the discussion of transversal elliptic operators and their index. This is joint work with M. F. Atiyah. In Appendix II, I have included our proof that the transversal index is a distribution on G when G is compact. L. Hörmander has observed that using his notion of wave front sets [**20**] one can extend the result to noncompact Lie groups with a more elegant proof. We include his proof in Appendix II as well.

2. **Different indices.** Let $A: H_0 \to H_1$ be a continuous linear map. In general, the index of A is the difference between Ker A (the null space of A) and Cok A $= H_1/A(H_0)$. One must give some meaning to Ker A − Cok A.

(a) *The ordinary index.* When A is a Fredholm operator between Banach spaces, then Ker A and Cok A are finite dimensional vector spaces and their dimensions are the only invariants. So, ind A = dim Ker A − dim Cok A, and ind: $\mathscr{F} \to Z$ is continuous where $\mathscr{F}$ denotes the set of Fredholm operators in the norm topology.

AMS 1970 *subject classifications.* Primary 58G10.

This covers the case of an elliptic operator on a compact manifold where H_i, $i = 0, 1$, are appropriate Sobolev spaces. We remind the reader that elliptic complexes lead to Fredholm operators as well. That is, if $0 \to C^\infty(E_0) \xrightarrow{d_0} C^\infty(E_1) \xrightarrow{d_1} \cdots \to C^\infty(E_N) \to 0$ is a complex of pseudodifferential operators of order k on vector bundles with exactness on the symbol level:

$$[(\operatorname{Ker} \sigma(d_i)(\xi) = \sigma(d_{i-1})(\xi)(E_{i-1}) \text{ for } \xi \neq 0 \in T^*(M)].$$

then $\sum \bigoplus d_i + (d_{i-1})^*$ is elliptic on $\sum \bigoplus C^\infty(E_i)$. In particular, $\sum \bigoplus (d_{2i} + d_{2i-1}^*)$: $\sum C^\infty(E_{2i}) \to \sum C^\infty(E_{2i} + 1)$ is elliptic.

(b) *The G-index*. Suppose now H_i, $i = 0, 1$, are G-spaces where G is a group, i.e., we have two weakly continuous representations ρ_i of G as bounded operators on H_i. Suppose that A is a G-operator in $\mathscr{F}$, i.e., $\rho_1(g)A = A\rho_0(g)$, $g \in G$. Then $\operatorname{Ker} A$ and $\operatorname{Cok} A$ are invariant under G so are also G-spaces. Now $\operatorname{ind}_G A = \operatorname{Ker} A - \operatorname{Cok} A$ is to be interpreted as the formal difference of two G-spaces. When G is compact, a finite representation ρ is completely determined up to similarity by its character function χ_ρ defined by $\chi_\rho(g) = \operatorname{trace} \rho(g)$. We can therefore think of $\operatorname{ind}_G A$ as the element of $\chi_{\operatorname{Ker} A} - \chi_{\operatorname{Cok} A}$ of $R(G)$, the character ring of G, i.e., the linear span of characters. Note $\operatorname{ind}_G = \operatorname{ind}$ when $G = (e)$. Because the set of characters is discrete when G is compact, the usual proof of the stability of index under perturbations goes through for ind_G as well. Such is the situation when A is an elliptic operator, say the Laplacian on a compact manifold and G is the (compact) group of isometries. When G is not compact, then other assumptions, such as transversality, seem to be needed to get an index formula. More will be said about this when we discuss fixed point formulae.

(c) *The transversal index*. When a compact group G operates on a manifold M as a group of diffeomorphisms, the usual ellipticity assumptions can be weakened and one still obtains an index.

DEFINITION. A transversal cotangent vector at m is a cotangent vector φ at m which annihilates vectors at m tangent to the orbit of G at m. We will denote the space of transversal vectors by $T_G^*(M)$.

DEFINITION. A pseudodifferential operator A is transversal elliptic if

(1) A is a G operator, and

(2) $\sigma_A(\varphi)$ is an isomorphism for all nonzero $\varphi \in T_G^*(M)$, where σ_A denotes the principal symbol of A, i.e., $T_G^*(M) - 0$ are noncharacteristic.

In the above definition we have assumed A operates on functions, but the same definition makes sense for $A: C^\infty(E) \to C^\infty(F)$ with E and F G-bundles in the sense of [**4**], [**11**]. Note that transversal elliptic is more general than elliptic.

When A is transversal elliptic, one can define an index. Let $\{X_1, \ldots, X_r\}$ denote an orthonormal base of g, the Lie algebra of G relative to an inner product invariant under ad G. Let Δ denote the Laplacian $-\sum_{j=1}^r X_j^2$ on G. The action of G on M induces operators $\tilde{X}_j$ and $\tilde{\Delta}$ on M. Let λ be a nonnegative real number and suppose A is of order $k > 0$. Consider the sequence

$$0 \to C^\infty(E) \xrightarrow{(A,\tilde{\Delta}^{k/2} - \lambda^{k/2}I)} C^\infty(F) \oplus C^\infty(E) \xrightarrow{(\tilde{\Delta}^{k/2} - \lambda^{k/2}I) - A} C^\infty(F) \to 0.$$

This is an elliptic complex: At the first stage, suppose $\sigma_{\tilde{\Delta}^{k/2}}(\varphi) = 0$ and $\sigma_A(\varphi) = 0$. The first condition implies $\varphi \in T_G^*(M)$ and so the second condition implies $\varphi = 0$ by transversal ellipticity. Hence the symbol is injective at the first stage. At the second stage, suppose (f, e) satisfies $\sigma_{\tilde{\Delta}^{k/2}}(\varphi)f - \sigma_A(\varphi)e = 0$. Write $\varphi = \varphi_1 \oplus \varphi_2$ with $\varphi_2 \in T_G^*(M)$ and φ_1 orthogonal to $T_G^*(M)$ relative to an invariant inner product on M. If $\varphi_1 \neq 0$, let $e_1 = (\sigma_{\tilde{\Delta}^{k/2}}(\varphi))^{-1}e$ so that $(\sigma_A(\varphi), \sigma_{\tilde{\Delta}^{k/2}}(\varphi))e_1 = (\sigma_A(\varphi)e_1, e)$ $= (f, e)$ for $\tilde{\Delta}^{k/2}$ and A commute. If $\varphi_1 = 0$ then $\sigma_{\tilde{\Delta}^{k/2}}(\varphi) = 0$ so that $\sigma_A(\varphi)e = 0$ $= \sigma_A(\varphi_2)e = 0$. Hence either $e = 0$ or $\varphi = 0$; in either case, we have exactness at the second stage. At the last stage, given f, then the equation $\sigma_{\tilde{\Delta}^{k/2}}(\varphi_1)f_1 - \sigma_A(\varphi_2)e_1$ $= f$ can always be solved for f. (The above elliptic complex used the Laplacian Δ on G which induces $\tilde{\Delta}$ elliptic in the orbit directions. A more interesting elliptic complex can be formed using the rank G generators of the center of the enveloping algebra of G.)

Since the complex above is exact it follows that Ker $A \cap \operatorname{Ker}(\tilde{\Delta}^{k/2} - \lambda^{k/2}I)$ and Cok $A \cap \operatorname{Ker}(\tilde{\Delta}^{k/2} - \lambda^{k/2}I)$ are finite dimensional. But in an irreducible representation ρ, $\Delta = \lambda_\rho I$. Hence if we let $H_s^\rho(E)$ denote the closed subspace of $H_s(E)$ which transforms according to the representation ρ, we obtain

THEOREM. *$A: H_s^\rho(E) \to H_{s-k}^\rho(F)$ is a Fredholm operator.*

As a consequence we see that a transversal elliptic operator A has an index with values in $\operatorname{Hom}(R(G), Z)$; $\operatorname{ind}_T A: \chi_\rho \to \operatorname{ind}(A|_{H_s^\rho(E)})$. It turns out that $\operatorname{ind}_T A: R(G) \to Z$ is an invariant distribution on G, i.e., $R(G) \subset C^\infty(G)$ and $\operatorname{ind}_T(A)$ can be extended to a continuous G-map: $C^\infty(G) \to C$. (We include a proof in Appendix II.) When A is actually elliptic, then $\operatorname{ind}_T A(\chi) = \langle \operatorname{ind}_G A, \chi\rangle$ so that as a distribution, $\operatorname{ind}_T A = \operatorname{ind}_G A \cdot dg$ where dg is normalized Haar measure.

Finally, one might think of A as an operator on the orbit space M/G even though this need not be a manifold. It is not clear as yet what the applications of ind_T will be. When the orbits of G are of constant dimension, M/G is a rational homology manifold and perhaps ind_T will allow extensions of fixed point formulas to this category. For a semisimple group, the Cassimir operator is transversal elliptic relative to a maximal compact subgroup. So there may be applications to group representations.

(d) *The* mod 2 *index*. Let $\tilde{\mathscr{F}}$ denote the space of skew adjoint Fredholm operators on a real Hilbert space. If $A \in \tilde{\mathscr{F}}$, then the parity of dim Ker A is fixed under perturbations (in $\tilde{\mathscr{F}}$) and is denoted by $\operatorname{ind}_1 A$, an element of Z_2. The following heuristic argument can be made precise for elliptic operators of positive order. If $A_t \in \tilde{\mathscr{F}}$ for $t \in [0, 1]$, then dim Ker $A_t \leqq$ dim Ker A_0 for $t < \varepsilon$ and the inequality if it occurs is due to eigenvalues of A_t approaching 0 so that the corresponding eigenspaces in the limit lie in Ker A_0. Since A_t is skew adjoint on a real space, the eigenvalues are imaginary and occur in pairs $\pm i\lambda(t) \neq 0$. Hence $\lim \pm i\lambda(t) = 0$

implies $\mp i\lambda t \to 0$ so that the parity of Ker $A(t)$ is not changed as $t \to 0$. To be more precise, one can show that $\text{ind}_1 : \tilde{\mathscr{F}} \to Z_2$ is continuous [**9**].

One can rephrase $\text{ind}_1 A$ as determining whether or not Ker A admits complex linear structure. The nonzero "eigenspaces" do, using $A/|A| = A/(-A^2)^{1/2}$ as multiplication by i; in fact that is the heart of the invariance under perturbation proof. Viewed this way we can define a new index, ind_2, as a mod 2 index. Let H be a complex Hilbert space and let $\tilde{\tilde{\mathscr{F}}} = [A:H \to H$; A is conjugate linear, A is real Fredholm and $A^* = -A]$. Then $\text{ind}_2 : \tilde{\tilde{\mathscr{F}}} \to Z_2$ given by $\text{ind}_2 A = \dim_C \text{Ker}\, A$ mod 2 is continuous. For the nonzero "eigenspaces" of A are quaternion spaces since they are invariant under 1, i, and $j = A/|A|$. So any perturbation contributes even parity to the kernel. Again, one may say that ind_2 measures whether Ker A is quaternionic or not. The various ind $= \text{ind}_0$, ind_1, and ind_2 are part of an 8-periodic sequence in which the index ind_j measures whether a kernel is a Clifford algebra C_j or not, where $C_1 = \boldsymbol{C}$ and $C_2 =$ Quaternions. We shall say more about Clifford algebras in the next section.

(e) *The index of families.* Suppose X is a connected compact Hausdorff space and we have a continuous map $a: X \to \mathscr{F}$, the set of Fredholm operators; i.e., we have a family of Fredholm operators $\{a(x)\}_{x \in X}$ parameterized by X. Then $\{\text{Ker}\, a(x)\}_{x \in X}$ and $\{\text{Cok}\, a(x)\}_{x \in X}$ form a family of kernels and cokernels.

How can we make sense out of their difference? Here is where K-theory enters naturally into operator theory, for if K-theory did not exist, it would be invented in order to define a families index. It turns out in fact that the basic theorems in K-theory have a natural interpretation in operator theory [**14**], [**33**].

If the family of vector spaces $\{\text{Ker}\, a(x)\}_{x \in X}$ is of constant dimension, then the semicontinuity property of the index makes $\{\text{Ker}\, a(x)\}_{x \in X}$ and $\{\text{Cok}\, a(x)\}_{x \in X}$ vector bundles over A. [See the next paragraph.] The set of isomorphism classes of vector bundles over X form a semigroup under $\oplus$ with tensor product $\otimes$ as a multiplication. We complete, to make a ring, as in the construction of the integers from the nonnegative integers. The only complication is there is no cancellation law, so that we must add an equivalence relation, $E \sim F$ if there exists a vector bundle G with $E \oplus G$ isomorphic to $F \oplus G$. This ring is denoted by $K(X)$ and we define $\text{ind}_{\mathscr{F}}\, a$ to be the element in $K(X)$ given by the difference of the vector bundles, $\{\text{Ker}\, a(x)\}_{x \in X} - \{\text{Cok}\, a(x)\}_{x \in X}$.

To see that one obtains vector bundles, let us first suppose that $a(x)$ is surjective for all $x \in X$. Then $a(x): \text{Ker}\, a(x)^{\perp} \to H$ is an isomorphism. Let b denote the inverse. If y is sufficiently close to x, then $a(y): \text{Ker}\, a(x)^{\perp} \to H$ is an isomorphism with inverse $b(y)$ so that $\text{Ker}\, a(y)$ is complementary to $\text{Ker}\, a(x)^{\perp}$. If $u \in H$, then $u - b(y)a(y)u \in \text{Ker}\, a(y)$ so that $\text{Ker}\, a(y)$ is a full complement to $\text{Ker}\, a(x)^{\perp}$ and $P_0: \text{Ker}\, a(y) \to \text{Ker}\, a(x)$ is an isomorphism where P_0 is orthogonal projection on $\text{Ker}\, a(x)$. Thus the family of vector spaces $\{\text{Ker}\, a(x)\}_{x \in X}$ is locally a product and in fact a vector bundle E over X. Since the cokernel is trivial, in this case $\text{ind}_{\mathscr{F}} a = E$, a vector bundle.

When $a(x)$ is not surjective, we can find a subspace V of H of finite codimension, so that $P_V \circ a(x): H \to V$ is surjective for all $x \in X$. (At each point $x \in X$ we can find such a V_x. This V_x will suffice for a neighborhood of x. Take a finite cover and intersect the locally obtained V_{x_j} to obtain the desired V.) It is then reasonable to define $\text{ind}_{\mathcal{F}}\, a$ as the formal difference $\text{Ker}(P_V \circ a) - V^\perp \times X$, an element of $K(X)$. One must show that if another appropriate $\tilde{V}$ had been chosen then $\text{Ker}\, P_{\tilde{V}} \circ a - \tilde{V}^\perp \times X = P_V \circ a - V^\perp \times X$ in $K(X)$. Using $V \cap \tilde{V}$, it suffices to check the case $\tilde{V} \subset V$. Then $\text{Ker}\, P_{\tilde{V}} \circ a = \text{Ker}\, P_V \circ a + W$ where W is a vector bundle whose fiber W_x at x maps isomorphically onto $V - \tilde{V} = V \cap \tilde{V}^\perp$ under $a(x)$. Also $W_x \cap \text{Ker}\, P_V \circ a(x) = (0)$. Hence $\text{Ker}\, P_{\tilde{V}} \circ a - \text{Ker}\, P_V \circ a = W = \tilde{V}^\perp \times X - V^\perp \times X = V \times X - \tilde{V} \times X$ as elements of $K(X)$.

Finally, we remark that there is a natural ring homomorphism from $K(X)$ to rational cohomology $H^*(X, Q)$ given by the Chern character $\text{ch}: K(X) \to H^*(X, Q)$. See Appendix I for a brief discussion of this and characteristic classes.

3. **The standard examples.** Most applications of index theory to geometry and topology come from a few examples; from a sophisticated point of view really only one example. The new applications also come from these examples. Let me review them quickly. Before doing so I would like to point out for the *ordinary index* it appears that soon direct proofs of the index theorem will be forthcoming. See Patodi [**29**] for the Gauss-Bonnet and Riemann-Roch case and Gilkey [**17**] for the Gauss-Bonnet and the Hirzebruch signature case.[1]

The approach taken there is in terms of the heat equation (or zeta function). If $A: H_0 \to H_1$ is an elliptic operator of first order, then $A^*A = L_0$ on H_0 and $AA^* = L_1$ are nonnegative elliptic operators with eigenexpansions. If V_λ is the λth eigenspace for L_0, then $A(V_\lambda)$ is the λth eigenspace for L_1 when $\lambda \neq 0$. Hence

$$\text{(I)} \qquad \text{tr}(e^{tL_0} - e^{tL_1}) = \text{ind}\, A.$$

Now e^{tL_j} for $t < 0$ is an integral operator with kernel $p_j(t, m_1, m_2)$ which has an asymptotic expansion for t small, $j = 0, 1$.

$$p_j(t, m, n) = (c_{j,0} + c_{j,1}t + c_{j,2}t^2 + \cdots)/(4\pi t)^{d/2}.$$

The coefficients in the expansion can be computed formally in terms of the total symbol of L_j [**31**]. Because of (I), the nonzero powers must cancel and $c_{0,d/2} - c_{1,d/2}$ yields the index of A. When this procedure is applied to the natural operators below, one finds that $c_{0,d/2} - c_{1,d/2}$ depends on the curvature and its covariant derivatives. Patodi and Gilkey have shown in the cases mentioned that it depends only on the curvature and is in fact the appropriate combination of characteristic classes as occur in the various index formulae.

(i) The standard example of an elliptic complex is the de Rham complex $\ldots \to C^\infty(\Lambda^p) \xrightarrow{d} C^\infty(\Lambda^{p+1}) \to \ldots$ with d exterior derivative. The corresponding elliptic

[1] Added in Proof. Gilkey and Atiyah-Bott-Patodi have recently obtained such a proof by the heat equation method applied to the signature operator (ii) with coefficients in a vector bundle.

operator is $d + d^*: C^\infty(\Lambda_e) \to C_\infty(\Lambda_0)$ where $\Lambda_e = \sum \bigoplus \Lambda^{2j}$ and $\Lambda_0 = \sum \bigoplus \Lambda^{2j+1}$. By the Hodge theorem, $\operatorname{ind}(d + d^*) = \chi(M)$, the Euler characteristic of M for $\operatorname{Ker}(d + d^*)$, is the set of harmonic even forms and $\operatorname{Cok}(d + d^*)$ is the set of harmonic odd forms. Here d^* is the adjoint in terms of a Riemannian metric on M which induces an inner product on Λ^p. In fact, $d^* = \pm * d *$ where $\langle \omega, \tau \rangle \operatorname{vol}_M = \omega \wedge *\tau$. The index theorem gives the Gauss-Bonnet theorem.

(ii) When $\dim M = 4k$, then $*^2 = +1$. Let $\Lambda^\pm$ denote the ± 1 eigenforms of $*$. It is easy to check that $d + d^*: C^\infty(\Lambda^-)$ and is elliptic. We shall call this operator the signature operator for its index is $S(M)$ the signature of M. (The signature of M is the index of the quadratic form given by cup product on $H^{2k}(M)$. In terms of differential forms, the quadratic form is $q(\tau) = \int_M \tau \wedge \tau = \int_M \langle \tau, *\tau \rangle \operatorname{vol}_M$ so that its index is given by the number of harmonic forms in Λ^+ minus the number of harmonic forms in Λ^-.) The index theorem gives the Hirzebruch index theorem expressing $S(M)$ in terms of Pontrjagin classes, $S(M) = \int_M \prod_j x_j/\tanh x_j$. (See Appendix I.)

(iii) If M is a complex manifold, one has the elliptic complexes $\ldots \to C^\infty(\Lambda^{p,q}) \xrightarrow{\bar\partial_p} C^\infty(\Lambda^{p,q+1})$ where $\Lambda^{p,q}$ are the $p + q$ forms with p holomorphic differentials and q antiholomorphic ones locally. Now index $\bar\partial_p$ is the Euler characteristic of $H^*(M, \tilde\Lambda^{p,0})$ the cohomology of M with coefficients in the sheaf of germs of holomorphic p-forms $\tilde\Lambda^{p,0}$. For a Kahler manifold, $\operatorname{ind} \bar\partial_p = \sum (-1)^p b_{p,q}$ where $b_{p,q} = \dim H^{p,q}$, the harmonic forms of type (p, q). If one writes $\Lambda^{p,q} = \Lambda^{0,q} \otimes \Lambda^{p,0}$, one could replace $\Lambda^{p,0}$ by any holomorphic vector bundle W and get the elliptic complex

$$\to C^\infty(\Lambda^{0,q} \otimes W) \xrightarrow{\bar\partial_W} C^\infty(\Lambda^{0,q+1} \otimes W) \to .$$

The index theorem gives the Hirzebruch-Riemann-Roch theorem, $\operatorname{ind} \bar\partial_W = \int_M \operatorname{ch}(W)\tau(M)$ where $\tau(M)$ is the characteristic class $\prod_i y_i/(i - e^{-y_i})$. (See Appendix I.)

For later purposes we remark here that the Hodge index theorem says that $S(M) = \sum_p \operatorname{ind} \bar\partial_p$ for Kahler manifolds (and of course $\chi(M) = \sum_p (-1)^p \operatorname{ind} \bar\partial_p$).

(iv) Our final example uses spinors and Clifford algebras. We give a brief description. Following Dirac, we seek a differential operator D which is a square root of the Laplacian $\Delta = -\sum (\partial/\partial\chi_j)^2$. Write $D = \sum A_j \partial/\partial\chi_j$ and assume A_j constant matrices. Then $D^2 = \Delta$ is equivalent to $A_j^2 = -1$ and $A_j A_k = -A_k A_j$. Thus we seek matrices with these properties.

We abstract slightly. Let V be an inner product space of dimension n. The Clifford algebra $Q(V)$ is an associative algebra containing 1, V, and satisfying the relation $v^2 = -\langle v, v \rangle 1$. So $v_1 \cdot v_2 + v_2 \cdot v_1 = -2\langle v_1, v_2 \rangle 1$. In particular, if $\{e_1, \ldots, e_n\}$ is an orthonormal base, $e_i e_j = -e_j e_i$, $i \neq j$ (as in exterior multiplication) but $e_i^2 = -1$ (as opposed to $e_i \wedge e_i = 0$ in exterior multiplication).

There is a natural linear map φ from $C(V)$ to $\Lambda(V)$: Let $l_v: \Lambda(V) \to \Lambda(V)$ be left multiplication, $l_v(\omega) = v \wedge \omega$. Let $i_v = l_v^*$. It is easy to check that $(l_v - i_v)^2 = -\langle v, v \rangle 1$. Hence $v \mapsto l_v - i_v$ extends to an algebraic map

$$\Phi: C(V) \to \mathrm{Hom}(\Lambda(V), \Lambda(V)).$$

Finally, let $\varphi(u) = \Phi(u)(1)$ for any $u \in C(V)$ so that $\Phi(u)$ becomes left multiplication on $C(V)$ when $C(V)$ is identified with $\Lambda(V)$ *via* φ. Note that $l_v - i_v = \Phi(v)$ is the symbol of $(d + d^*)/i$ on v when $V = T^*(M, m)$. That is one reason why Clifford algebras are useful to elliptic operator theory. Since $\bigcup_{m \in M} \Lambda(T^*(M, m))$ forms a vector bundle over M, the linear isomorphism above makes

$$Cl = \bigcup_{m \in M} C(T^*(M, m))$$

a vector bundle and C^∞ differential forms can be viewed as C^∞ sections of Cl.

The algebras $C(V)$ are either simple or the direct sum of two simple algebras depending on dim V mod 8. Hence they are full matrix algebras over $\boldsymbol{R}$, $\boldsymbol{C}$, or $\boldsymbol{H}$ (the quaternions) or direct sum of two such (see [**6**]). They start with $\boldsymbol{R}, \boldsymbol{C}, \boldsymbol{H}, \boldsymbol{H} \otimes \boldsymbol{H}$ and the general ind_k of §2 involves whether a kernel which is a C_k module is actually a C_{k+1} module (where $C_k = C(R^k)$).

When dim $V = 2l$, then C_{2l} is simple and is represented as a full matrix algebra over $\boldsymbol{R}$ or $\boldsymbol{H}$ of real dimension 2^{2l}. So it operates on a vector space Δ of dimension 2^l over $\boldsymbol{R}$ or dimension 2^{l-1} over $\boldsymbol{H}$. The even elements C_{2l}^+ of C_{2l} form a subalgebra which has two irreducible pieces $\Delta^\pm$ in Δ called the half spin spaces. The odd elements C_{2l}^- interchange Δ^+ with Δ^-. In particular, $V \subset C_{2l}^-$ does.

Having done the algebra, we would now like to globalize. We have Cl whose fibre at each point m is $Cl(T^*(M, m))$. However the choice of the vector space Δ is not canonical and there is an obstruction to their global existence as a vector bundle. When this obstruction (the second Stiefel-Whitney class) vanishes, Δ, Δ^+ and Δ^- can be defined globally as vector bundles. Such manifolds are called *spin manifolds*, and a specific choice of Δ gives a *spin structure*. Such choices are in 1-1 correspondence with $H^1(M, Z_2)$. When one has a specific spin structure, one can define the desired *Dirac* operator $D: C^\infty(\Delta^+) \to C^\infty(\Delta^-)$.

To do so, note that in the constant coefficient case $\sigma_D(\xi) = \sum A_j \xi_j$ so one would now expect $\sigma_D(\xi): \Delta^+ \to \Delta^-$ as above with $\xi \in C_{2l}^-$. This defines σ_D with D a first order operator, and hence D using covariant differentials or a partition of unity. The index theorem gives ind D as a combination of Pontrjagin classes denoted by

$$\hat{A}(M) = \int_M \prod \frac{x_i}{e^{x_i/2} - e^{-x_i/2}}.$$

When M is a complex manifold of complex dimension l, it turns out that M is a spin manifold when the canonical line bundle $K = \Lambda^{l,0}$ has a holomorphic square root. Furthermore, the set of spin structures on M is in 1-1 correspondence with the $[L; L^2 = K, L$ a holomorphic line bundle$]$. Choosing a spin structure means choosing an L, in which case $D: \Delta^+ \to \Delta^-$ turns out to be the same as $\bar{\partial}: C^\infty(\Lambda^{0,\mathrm{even}} \otimes L) \to C^\infty(\Lambda^{0,\mathrm{odd}} \otimes L)$.

4. **Applications of the** mod 2 **index.** The classical example of a skew adjoint real operator with $\text{ind}_1 = 1$ is d/dx operating on real C^∞ functions of the circle. This is the operator $*d$ where d is the total differential and $*$ the usual algebraic operator sending p-forms to $n - p$ forms. More generally $d + \delta$ where $\delta = d^* = \pm * d *$ is selfadjoint, and maps $C^\infty(\Lambda_e) \to C^\infty(\Lambda_0)$ (see §3(I)). On odd dimensional manifolds $*(d + \delta) = + *d \pm d*$ is elliptic and maps $C^\infty(\Lambda_e) \to C^\infty(\Lambda_0)$. It is easy to verify that D on even forms defined by $D|_{\Lambda^{2p}} = (-1)^p d* + *d$ is skew adjoint elliptic when $\dim M = 4k + 1$ (selfadjoint when $4k + 3$). Then $\operatorname{Ker} D$ is the sum of the even harmonic forms so $\text{ind}_1 D \equiv \sum \beta_{2p} \equiv \frac{1}{2}\sum \beta_p$ (mod 2). The right-hand side is called the real Kervaire semicharacteristic and denoted by $k(M)$. So D is a generalization of d/dx and $\text{ind}_1 D$ is a known topological invariant. We now give a simple application of the fact that $\text{ind}_1 D$ is a perturbation invariant.

THEOREM [1]. *Let M be a compact oriented $4k + 1$ manifold.* (1) *Suppose M admits two independent vector fields. Then $k(M) = 0$.* (2) *If X is a nonvanishing vector field on M and X is homotopic to $-X$, then $k(M) = 0$.*

PROOF. In both cases we want to show that $\text{ind}_1 D = 0$. Though one can avoid Clifford algebras by direct computations, it is more illuminating to use Clifford multiplication.

LEMMA. *Suppose X is a nonvanishing vector field on M and r_x denotes right Clifford multiplication. Then $r_x^{-1}(d + \delta)r_x$ and $d + \delta$ have the same symbol. In particular they differ by a* 0*th order operator.*

PROOF. $\sigma_{r_x^{-1}(d+\delta)r_x(v)} = r_x^{-1}\sigma_{d+\delta}(v)r_x = ir_x^{-1}\Phi(v)r_x^{-1} = i\Phi(v) = \sigma_{d+\delta}(v)$ because $\Phi(v)$ is left multiplication.

Finally note that $D = r_\omega(d + \delta)$ where ω is the volume element or top dimensional form so that ω lies in the center of the Clifford algebra.

Suppose now x_1 and x_2 are two orthonormal vector fields so that $(x_1x_2)^2 = -x_1^2x_2^2 = -1$ in the Clifford sense. Then $r_{x_1x_2}^{-1}Dr_{x_1x_2}$ has the same symbol as D by the lemma above and is skew adjoint for $r_{x_1x_2}$ is unitary. Thus $\tilde{D} = \frac{1}{2}\{D + r_{x_1x_2}^{-1}Dr_{x_1x_2}\}$ has the same symbol as D and is skew adjoint. Hence $\text{ind}_1 D = \text{ind}_1 \tilde{D}$. But $\tilde{D}$ is invariant under $r_{x_1x_2}$ whose square is -1. Hence $\operatorname{Ker} \tilde{D}$ is invariant under $r_{x_1x_2}$ and is even dimensional so that $k(M) = 0$.

For the second part of the theorem, suppose x is a nonvanishing vector field normalized to have length 1. Then $(x \cdot \omega)^2 = +1$. Again $\tilde{D}_1 = \frac{1}{2}(D + r_{x\cdot\omega}^{-1}Dr_{x\cdot\omega})$ is skew and with the same symbol as D so that $k(M) \equiv \dim \operatorname{Ker} \tilde{D}_1 \bmod 2$. Now $\operatorname{Ker} \tilde{D}_1$ splits into ± 1 eigenspaces of $x \cdot \omega$ whose dimensions mod 2 ($a(x)$ and $b(x)$) are homotopy invariants of x. Since $r_{-x\cdot\omega} = -r_{x\cdot\omega}$, $a(-x) = b(x)$ so that if x is homotopic to $-x$ we have $a(x) = a(-x)$. But $k(M) = a(x) + b(x) = a(x) + a(-x) = 2a(x)$ when x is homotopic to $-x$.

REMARK. The first part of the theorem can be strengthened to give a beautiful generalization of the Hopf theorem which states that if x is a vector field with isolated zeros, then the sum of its singularities is the Euler characteristic. The generalization is this: Suppose M is of dimension $4k + 1$ and x_1, x_2 are two C^∞ vector fields with isolated singularities, i.e., the set of points where x_1 and x_2 are linearly dependent is finite. (Having only isolated singularities is not the generic situation for a pair x_1, x_2. There are obstructions.) One measures the local singularity as an element of $\pi_{4k}(V_{n,2}) = Z_2$ where $V_{n,2}$ is the Stiefel manifold of 2 frames in R_n. Then the sum of the singularities of (x_1, x_2) equals $k(M)$. See Atiyah and Dupont [7].

The theorem whose proof we have sketched above is a main step in the proof; it shows that no singularities implies $k(M) = 0$.

We could also have used the trick above to prove the part of Hopf's theorem that states: *If x is a nonvanishing vector field on M, then $\chi(M)$, the Euler characteristic, is* 0. For it is easy to see that $r_x^{-1} d + \delta|_{C^\infty(\Lambda_e)} r_x$ has the same symbol as

$$d + \delta|_{C^\infty(\Lambda_0)}.$$

Hence $\operatorname{ind}(d + \delta)|_{C^\infty(\Lambda_e)} = \operatorname{ind}(d + \delta)|_{C^\infty(\Lambda_0)}$ so that $\chi(M) = -\chi(M) = 0$.

The same idea can be used to prove divisibility properties of topological invariants in the presence of independent vector fields. We will give one illustration of this that involves the *signature* $S(M)$ of M as well as $\chi(M)$.

We now have $D^+ = d + \delta: C^\infty(\Lambda_e^+) \to C^\infty(\Lambda_0^-)$ and $D^-: C^\infty(\Lambda_e^-) \to C^\infty(\Lambda_0^-)$. Also $\operatorname{ind} D^+ + \operatorname{ind} D^- = \chi(M)$ while $\operatorname{ind} D^+ - \operatorname{ind} D^- = S(M)$ so that $\operatorname{ind} D^\pm = \frac{1}{2}(\chi \pm S)$.

THEOREM [1]. *Let M be a compact oriented manifold of dimension $4k$. Suppose M allows a tangent field of oriented 2-planes. Then $\chi(M) \equiv 0 \bmod 2$ and $\chi(M) \equiv S(M) \bmod 4$.*

PROOF. Let U be the oriented 2-plane as a section of Cl (normalized so that $U^2 = -1$). Then r_U leaves $\Lambda^\pm$ invariant for it commutes with $\Phi(\omega)$. It also preserves $\Lambda_{e,0}$ because U is even. Again $\tilde{D}_\pm = \frac{1}{2}\{D_\pm + r_U^{-1} D_\pm r_U\}$ has the same symbol as $D_\pm$. Since $U^2 = -1$, $\operatorname{Ker} \tilde{D}^\pm$ and $\operatorname{Ker}(D^\pm)^*$ are even dimensional for they are invariant under U. Hence $\frac{1}{2}(\chi(M) \pm S(M))$ are even dimensional.

The final application of mod 2 indices is to Riemann surface theory. Let M be a Riemann surface. Note that $\bar{\partial}: f \to (\partial f/\partial \bar{z})\, d\bar{z}$ is a basic elliptic operator: $C^\infty(1) \to C^\infty(\Lambda^{0,1})$. We can tensor with a holomorphic line bundle L to get: $\bar{\partial}: C^\infty(L) \to C^\infty(L \otimes \bar{K})$ where $K = \Lambda^{1,0}$. This operator is not skew adjoint; it does not even map sections of L to itself. But under certain hypotheses on L and with a hermitian inner product we obtain a skew adjoint operator. First, a hermitian metric on M and a hermitian inner product on a line bundle L gives a *conjugate* linear isomorphism $h: C^\infty(L) \to C^\infty(\bar{L})$. So $h \circ \bar{\partial}: C^\infty(L) \to C^\infty(\bar{L} \otimes K)$. Suppose finally that $L^2 = K$ so that $\bar{L} \otimes K = \bar{L} \otimes L \otimes L \simeq L$; hence $h \circ \bar{\partial}$ gives a conjugate linear map on $C^\infty(L)$. The operator $h \circ \bar{\partial}$ is skew adjoint for if $s_1, s_2 \in C^\infty(L)$, then $(\bar{\partial} s_1) \otimes s_2 + s_1 \otimes \bar{\partial} s_2 = \bar{\partial}(s_1 \otimes s_2) = d(s_1 \otimes s_2)$ for $s_1 \otimes s_2 \in L^2 = K$ and is a form of type (1, 0). So

$$\int_M \bar{\partial} s_1 \otimes s_2 = -\int_M s_1 \otimes \bar{\partial} s_2 \quad \text{or} \quad \langle h \circ \bar{\partial} s_1, s_2 \rangle = -\langle s_1, h \circ \bar{\partial} s_2 \rangle.$$

Hence $h \circ \bar{\partial}$ is a skew adjoint conjugate linear elliptic operator. From the discussion of the previous section $\mathrm{ind}_2(h \circ \bar{\partial}) = \dim_C \mathrm{Ker}\, h \circ \bar{\partial} \bmod 2$ is perturbation invariant. We obtain a theorem of Atiyah [**2**].

THEOREM. *Let M_t be a C^∞ family of compact Riemann surfaces, $t \in C$ and $|t| < 1$. Let L_t be a C^∞ family of holomorphic line bundles such that $L_t^2 = K_t$. Then*

$$\dim_C \Gamma(L_t) \bmod 2$$

is independent of t, where $\Gamma(L_t)$ denotes the holomorphic sections of L_t.

When interpreted in terms of divisor classes and complete linear systems, this is a classical result in Riemann surface theory.

One can go further. If $L_0^2 = K$, then all other square roots of K are of the form $L_0 \otimes R$ where $R^2 = 1$. Since $[R; R^2 = 1] \simeq H^1(M, Z_2)$, the set $\mathscr{S}(M)$ of square roots of K is a Z_2 affine space with translation group $H^1(M, Z_2)$. Using the *index formula* for ind_2 one obtains [**2**] the

THEOREM. *The function $\varphi: \mathscr{S}(M) \to Z_2$ given by $\varphi(L) = \dim_C \Gamma(L) \bmod 2$ is a quadratic function whose associated bilinear form is the cup product on $H^1(X, Z_2)$. The function φ has $2^{g-1}(2^g + 1)$ zeros.*

The discussion above is a special case of the Dirac operator on spin manifolds of dimension $8k + 2$. In that case ind_2 is a Z_2 valued function on the set of spin structures. See the end of §3.

5. **Applications of** ind_G. The most interesting applications of index theorems for ind_G are fixed point formulas. We have already used elliptic operator theory to prove the prototype: *Every one-parameter group of diffeomorphisms on a compact orientable manifold M with nonzero Euler characteristic has a fixed point*. For if X is the vector field which is the generator of the one-parameter group, then no fixed points implies X never vanishes which by the special case of Hopf's theorem implies $\chi(M) = 0$.

If E is an elliptic complex over the compact manifold M, let $H^j(M, E)$ be $\mathrm{Ker}\, d_j/\mathrm{image}\, d_{j-1}$, the jth cohomology. Suppose G acts on E. This means G acts on M as a group of diffeomorphisms and lying over this action are isomorphisms $\tilde{g}_j: g^* E_j \to E_j$ so that $\tilde{g}_j(m): E_j(gm) \to E_j(m)$. Suppose also G commutes with the operators d_j as operators on $C^\infty(E_k)$. Then $g \in G$ induces an operator $\check{g}_j$ on $H^j(M, E)$. Finally, let $L(E, g) = \sum (-1)^j \mathrm{tr}(\check{g}_j)$. If E is a two step elliptic complex giving an elliptic operator $d_0 = A$, then $L(E, g) = \mathrm{ind}_G(A)(g)$ so our present discussion is an extension of the G-index notion.

A fixed point formula is a formula for $L(E, g)$ in terms of the fixed point set Mg of g, the normal bundle to Mg and the principal symbol of the elliptic operator on Mg. There are two different hypotheses under which one gets a fixed point theorem. (i) G compact but no assumptions about the fixed point set [**11**]; (ii) no assumptions about G, but the fixed point set Mg must be a set of isolated points and the action of g simple at Mg, i.e., the graph of g is transversal to the diagonal in $M \times M$ [**4**], [**5**].

It would be interesting to find a common generalization, especially a formula for the noncompact group with a relaxation of the transversality condition. Both theorems use elliptic theory in essential but different ways. In the second theorem all trace (a pun!) of the operator disappears from the final formula. In the first theorem, in applications to standard operators, the operators also disappear from the final formula. As a result, many of the applications have alternative proofs which do not use elliptic theory.

In this article, I do not propose to sketch how elliptic operators are used to get fixed point formulas. Aside from the original papers [**4**], [**5**], [**9**], [**10**], [**11**], there have been some expositions [**3**], [**32**]. Instead I would like to describe the formulas and give some applications.

The second theorem is easy to state. If $m \in Mg$, then dg_m leaves $T(M, m)$ fixed and g leaves the fibre at m, $E_j(m)$ fixed. So $\tilde{g}_j(m)$ is a linear transformation on $E_j(m)$. Then

$$\text{(I)} \qquad L(E, g) = \sum_{m \in Mg} \sum_{j} (-1)^j \operatorname{tr}(\tilde{g}_j(m))/|\det(1 - dg_m)|.$$

For example, on a Riemann surface with $\bar{\partial}: 1 \to \Lambda^{0,1}$, the formula says $1 - \operatorname{trace}(dg|_{H^{0,1}}) = \sum_{m \in Mg} (1 - g'(m))^{-1} = \sum_{m \in Mg} \operatorname{Res}(z - g(z))^{-1}\, dz$, if z is a local coordinate system about the fixed point. In this case, Residue makes sense even when g is not transversal and gives the correct formula. I earlier commented that a relaxation of transversality is desired in the general case; perhaps a clue is the way that generalized residue theory will work in the higher dimensional complex case.

The first theorem is harder to state for it involves characteristic classes. Let N^g denote the normal bundle to Mg. Then the fixed point formula for $L(E, g)$ involves the character of the principal symbol of $E|_{Mg}$, characteristic classes of N^g characteristic classes of Mg. In more detail, for each $m \in Mg$, the vector space N^g_m is invariant under g, acting as a rotation. The eigenvalue $+1$ does not occur for N^g_m is normal to Mg. One can decompose N^g_m into the -1 eigenspace and two-dimensional spaces on which g acts as rotation through angle $\theta, 0 < \theta < \pi$. Globally, one can decompose N^g into $N^g(-1) \oplus \sum_{0<\theta<\pi} N^g(\theta)$ with $N^g(-1)$ a real bundle and the $N^g(\theta)$ complex vector bundles. Then $L(E, g)$ can be expressed in terms of the character of the principal symbol of $E|_{Mg}$, Pontrjagin classes of $N^g(-1)$ and of Mg, and Chern classes of the $N^g(\theta)$.

We shall now illustrate how this works out in special cases. First suppose M is a complex manifold and W a holomorphic vector bundle over M. The elliptic complex is $\dots \longrightarrow \Lambda^{0,q} \otimes W \xrightarrow{\bar\partial} \Lambda^{0,q+1} \otimes W \longrightarrow \dots$. Suppose g is a holomorphic map $g: M \to M$ with simple fixed points which has an extension to W. If $m \in Mg$, let $\tilde{g}_m$ denote the induced complex linear action of g on W_m. Then substitution in (I) gives $\sum_{m \in \text{fixed pts}} \text{trace } \tilde{g}_m/\det_C(1 - dg_m)$ as the Lefschetz number, where $\det_C$ means complex determinant of $1 - dg$ as a complex linear map of $T^{1,0}(M, m)$. The special case of $W = \Lambda^{p,0}$ is of particular interest. Let the Lefschetz number

$$\sum (-1)^q \operatorname{trace}(\breve{g} | H^{p,q}(M))$$

be denoted by $L(p, g)$. (When M is Kahler, $H^{p,q}(M)$ is (p, q) cohomology in the Hodge splitting of $H^{p+q}(M, \boldsymbol{C})$. In general $H^{p,q}(M) = H^q(M, \tilde{\Lambda}^{p,0})$, the qth cohomology with coefficients in the sheaf of germs of forms of type $(p, 0)$.

Suppose now X is a holomorphic vector field with simple isolated zeros, so that $g_t = e^{tx}$ is a one-parameter group of holomorphic transformations with simple points. Let $l_m(X)$ denote the linear transformation which X induces at the fixed point m, so that $dg_t(m) = e^{tl_m(X)}$. Then

$$L(p, g_t) = \sum_{m \in Mg} \text{trace } e_p^{tl_m(X)}/\det_C(1 - e^{tl_m(X)})$$

and $e_p^{tl_m(X)}$ is the induced action of $e^{tl_m(X)}$ on $\Lambda_m^{p,0}$. Suppose $\{\alpha_m\}$ are the eigenvalues of $l_m(X)$; then, $\operatorname{tr} e_p^{tl_m(X)} = \sigma^p(e^{\alpha_m t})$ and $L(p, g_t) = \sum_{m \in Mg} \sigma^p(e^{\alpha_m t})/\prod_{\alpha_m}(1 - e^{\alpha_m t})$ where σ^p is the pth elementary symmetric function.

Now let $\chi_y(M, g_t) = \sum_p L(p, g_t) y^p$ [**18**] so that $\chi_{-1}(M, g_t) = \chi(M)$ and $\chi_{+1}(M, g_t) = S(M)$, the signature of M, by the Hodge index theorem, when M is Kahler.

THEOREM [**21**]. *Let M be a compact complex manifold and X a holomorphic vector field with simple isolated zeros, then $\chi_y(M, g_t)$ is independent of t.*

PROOF. Note that if M is Kahler, the theorem is trivial because g_t, being connected to 1, induces the trivial map on $H^{p,q} \subset H^{p+q}(M, \boldsymbol{C})$. In general, let α_{kj}, $j = 1, \dots, n$, denote the eigenvalues of $l_{m_k}(s)$ at the kth fixed point m_k. Then

$$\chi_y(M, g_t) = \sum_{k,p} \frac{\sigma^p(e^{\alpha_{kj} t})}{\prod_j (1 - e^{\alpha_{kj} t})} y^p = \sum_k \prod_j \frac{1 + y e^{\alpha_{kj} t}}{1 - e^{\alpha_{kj} t}}.$$

From the definition of $\chi_y(M, g_t)$, we have $\chi_y(M, g_t) = \sum_r P_r(y) e^{c_r t}$ with $P_r(y)$ polynomials in y and the c_r are distinct in $\boldsymbol{C}$. Since $\chi_y(M, g_t)$ is holomorphic in t so is

$$P(t) = (d/dt)\chi_y(M, g_t) = \sum_{k,l} \frac{-\alpha_{kj}(1 + y)}{(1 + y e^{\alpha_{kl} t})(1 - e^{-\alpha_{kl} t})} \prod_j \frac{1 + y e^{\alpha_{kj} t}}{(1 - e^{\alpha_{kj} t})}.$$

Suppose we can show for all but a finite number of c's in $\boldsymbol{C}$ that $\lim_{t \to \infty} P(ct) = 0$. Then, since $P(t) = \sum_r P_r(y) c_r e^{c_r t}$, we conclude say by Phraegman Lindelof that $P(t) \equiv 0$ proving the theorem. But if the real part of $\alpha_{kj}/c \neq 0$, then $\lim_{t \to \infty} P(ct) = 0$. Q.E.D.

If we switch to a compact holomorphic vector field, i.e., one for which g_t lies in a *compact* group of diffeomorphisms with no assumptions on the fixed point set, even more is true.

THEOREM [**21**]. *Let X be a compact holomorphic vector field on a compact complex manifold. Then*

(a) $\chi_y(M, g_t)$ *is independent of t.*

(b) $\chi_y(M, 1) = \sum(-y)^{s(k)}\chi(M_k^{g_t}, 1)$ *where $(M_k^{g_t})$ are the components of M^{g_t} and $s(k)$ = the number of α_{kj} which are positive on the normal bundle to $M_k^{g_t}$.*

PROOF. Note that when $y = -1$, (ii) becomes the well-known

$$\chi(M) = \sum_k \chi(M_k^{g_t}).$$

When $y = +1$, one obtains $S(M) = \sum_k(-1)^{s(k)}S(M_k^{g_t}) = \sum_k S(M_k^{g_t})$ with another orientation. This is true for C^∞ manifolds as well, and can be proved using the formula for ind_G on the signature operator of §3(ii).

It suffices to consider the circle case, $t \in S^1$. The normal bundle $N_k^{S_1}$ of a connected component $M_k^{S_1}$ splits into $\sum N_k^{S_1}(r)$ with t acting by t^r on $N_k^{S_1}(r)$ with r a nonzero integer. Then the index formula gives

$$\chi_y(M, g_t) = \int_{M^{S_1}} \prod_{j,r} (1 + ye^{-x_j(r)}t^{-r})/(1 - e^{-x_j(r)}t^{-r})\tau(M^{S_1})$$

where the Chern classes of $N^{S_1}(r)$ are the elementary symmetric functions in $x_j(r)$, and $\tau(M^{S_1})$ is the Todd class of M^{S_1}. Thus $\chi_y(M, g_t)$ which by definition is of the form $\sum_{r=-s}^{s} p_r t^r$ has no pole at 0 or ∞ and hence is constant.

Part (b) of the theorem is obtained by evaluating the constant at $t = 1$.

The theorems above and methods of proof by Kosniowski and Lusztig suggested to Atiyah and Hirzebruch that they try the same idea on the Dirac operator for spin manifolds. They found

THEOREM [**8**]. *Let M be a compact connected smooth oriented spin manifold of* dim $4k$. *If $\hat{A}(M) \neq 0$, then any compact subgroup of the group of diffeomorphisms of M is finite.*

METHOD OF PROOF. One shows that if S^1 acts nontrivially on M, then $\hat{A}(M) = 0$. By using a double cover of S^1, if necessary, one can assume S^1 preserves a spin structure (and a Riemann metric) so that S^1 commutes with the Dirac operator D. By assumption, the fixed point set M^{S_1} is a proper subset of M so that the normal bundle $N^{S_1} = \bigoplus \sum N^{S_1}(r)$ is nontrivial. Then the ind_G formula gives

$$\mathrm{spin}(M, t) = \int_{N^{S_1}} \prod_{r,j} (t^{-1/2}e^{x_j/2(r)}/t^{-1}e^{x_j(r)})\hat{A}(N^{S_1}).$$

There is a problem of sign we will not worry about here. In any case it turns out that the right side is a well-defined rational function as in the complex case. But inspection shows that it vanishes at $t = 0$ and ∞ and hence is identically zero. So $\hat{A}(M) = \text{spin}(M, 1) = 0$.

This result is a striking example of how the elliptic operator theory disappears from the final statement. One immediately raises the question whether there is a proof strictly within the domain of differential topology. And one must raise the further question as to whether the result is purely topological. That is, the theorem makes sense with smooth replaced by PL and diffeomorphisms by PL homeomorphisms. Is it still true? One might also ask whether S^1 can act nontrivially (continuously but not smoothly) on a smooth spin manifold with $\hat{A} \neq 0$. By now, many results obtainable by fixed point formulas have alternate proofs using differential geometry and topology directly. To date, as far as I know, the above theorem has no such proof.

Before leaving the subject of spin manifolds one should mention Lichnerowicz's result [**22**] (a type of Kodaira vanishing theorem). He showed that the Laplacian associated to the Dirac operator can be written as a nonnegative operator plus the scalar curvature. Hence on a spin manifold with positive scalar curvature, there are no harmonic spinors; by the index theorem, $\hat{A}(M) = 0$. This result is the only restriction I know of for positively curved manifolds aside from the classical one that $\Pi_1(M)$ is finite. The simplest example of a four-dimensional simply connected spin manifold can be found in Milnor [**24**], [**25**]. One might speculate that there is a relation between positive curvature and nontrivial S^1 action since the only known examples of positively curved spaces are homogeneous. From the point of view of global nonlinear partial differential equations, the results above are puzzling. For specifying the curvature is solving a global nonlinear problem. It is hard to see how the $\hat{A}$-genus invariant would enter.

6. **Applications of the index for families.** Natural examples of families of Fredholm operators arise in the study of fibre bundles. Let $\pi: Z \to Y$ be a differentiable fibre bundle with fibre M. Then a "natural" elliptic operator on M will induce a family of elliptic operators on $\{\pi^{-1}(y)\}_{y \in Y}$ which will be Fredholm on appropriate Hilbert spaces. Occasionally the index of this family is nontrivial and has some interesting consequences. We discuss this situation for the signature operators which we defined in §3, and give an application due to Lusztig [**23**]. It involves the *Novikov higher signature.*

Let M be a compact connected $2k$ manifold. Suppose $b_1, \ldots, b_n$ is a basis of $H^1(M, Z)$, with $\xi_1, \ldots, \xi_n$ a dual basis of $\text{Hom}(H^1(M, Z), Z)$. Suppose $r = 2k - 4l$ and $i_1 < i_2 < \ldots < i_r$ is a subset of $\{1, 2, \ldots, n\}$. Then there exists a $4l$ manifold $M_{i_1 \ldots i_r}$ dual to the cohomology class $b_{i_1} \ldots b_{i_r}$ with trivial normal bundle; furthermore its signature is independent of the representative chosen. Consider the expression $\sum_r \sum_{i_1 < \cdots < i_r} 2^{r/2} S(M_{i_1 \cdots i_r}) \xi_{i_1} \wedge \ldots \wedge \xi_{i_r}$.

It is not hard to show that this is independent of a choice of basis for $H^1(M, Z)$ as an element of the Grassmann algebra over $\text{Hom}(H^1(M, Z), Z)$. This element of the Grassmann algebra is called the Novikov higher signature of M. Novikov conjectured it was a homotopy invariant of M [**26**]; a complete proof was obtained by Hsiang and Farrell [**19**]. Recently, Lusztig [**23**] has shown that the higher signature can be interpreted as the index of an appropriate family from which the homotopy invariance will follow. I would like to sketch his construction.

First observe that the Grassmann algebra of $\mathrm{Hom}(H^1(M, Z), Z)$ is in fact the cohomology of an n-torus T^n, so that the higher signature can be interpreted as an element of $H^*(T^n, Z)$. For Kahler manifolds, especially Riemann surfaces, tori appear naturally as Picard varieties. Let $\mathrm{Pic}(M)$ denote the set of isomorphism classes of holomorphic line bundles with first Chern class 0. Then $\mathrm{Pic}(M)$ is a complex torus which in the Riemann surface case can be identified with the Picard variety. Let $Z = \mathrm{Pic}(M) \times M$ and for each $L \in \mathrm{Pic}(M)$ consider the elliptic complex $0 \to \dots \to C^\infty(\sum_p \Lambda^{p,q} \otimes L) \xrightarrow{\bar\partial} C^\infty(\sum_p \Lambda^{p,q+1} \otimes L) \to \dots$. When L is the trivial line bundle, then the Hodge index theorem gives $\bar\partial_L = S(M)$. In general, $\{\bar\partial_L\}_{L \in \mathrm{Pic}(M)} = D$ gives a family of elliptic operators whose $\mathrm{ind}_{\mathscr{F}}$ lies in $K(\mathrm{Pic}(M))$. Either the Grothendieck-Riemann-Roch or the index theorem for families gives $\mathrm{ch}\,\mathrm{ind}_{\mathscr{F}} D \in H^*(\mathrm{Pic}(M))$ is the Novikov higher signature of M.

When M is not a Kahler manifold, one can still interpret the higher genus analytically. The operator $\bar\partial$ is replaced by the signature operator while $\mathrm{Pic}(M)$ is easily interpreted as the connected component of the character group of $\pi_1(M)$. So if $\chi \in \mathrm{Pic}(M)$, then associated to it is the flat line bundle L_χ coming from the character $\chi: \pi_1(M) \to S^1$. Again let $Z = \mathrm{Pic}(M) \times M$ and consider the family of elliptic operators $\{D^+ \otimes L_\chi\}_{\chi \in \mathrm{Pic}(M)} = D$. Then the index theorem for families shows that $\mathrm{ind}_{\mathscr{F}} D$ has Chern character equal to the Novikov higher signature.

Appendix I. Let E be a complex vector bundle of dimension n over a compact manifold M. Assume E is equipped with a hermitian metric and M with a Riemannian metric. Let Ω denote an associated curvature form. It is a 2-form with values in $\mathrm{Hom}(E, E)$. In terms of a local coordinate system $(x_1, \dots, x_m)$ and a local basis of E, $\Omega = R^\beta_{ij\alpha}$ which assigns to $\partial/\partial x_i, \partial/\partial x_j$ the matrix $R^\beta_{ij\alpha}$ as a linear transformation on E relative to the given base. Then $\det_E(I + R^\beta_{ij\alpha}/2\pi(-1)^{1/2})$ is a sum of even dimensional ordinary forms on M. The jth Chern class c_j is the $2j$-form occurring in this determinant. So $\sum c_j(E) = \det(I + \Omega/2\pi(-1)^{1/2})$. The forms c_j are closed and *via* the de Rham theorem represent cohomology classes independent of the choice of metrics.

We now introduce the following formalism (Hirzebruch [**18**]). In any power series in $y_1, \dots, y_n$ which is symmetric in $y_1, \dots, y_n$, we can replace the y_j's by the elementary symmetric functions of the y_j's and replace these in turn by c_j. Thus any symmetric power series represents a polynomial in the Chern classes.

Now define $\mathrm{ch}(E) = \sum_{j=1}^n e^{y_j}$. So for example when $\dim E = 1$, then $\mathrm{ch}(E) = e^{c_1(E)}$. When $\dim E = 2$ and $\dim M = 2$, $\mathrm{ch}(E) = 1 + y_1 + y_1^2/2! + 1 + y_2 + y_2^2/2! = 2 + (y_1 + y_2) + (y_1 + y_2)^2/2 - y_1 y_2 = 2 + c_1 + c_1^2 - c_2$. The

Chern map $\mathrm{ch}: E \to H^*(M, C)$ can be extended to a homomorphism $\mathrm{ch}: K(X) \to H^*(M, C)$ which in fact is characterized by (i) being functorial under mappings between spaces and (ii) $\mathrm{ch}(E) = e^{c_1(E)} = e^{\Omega/2\pi(-1)^{1/2}}$ when E is a line bundle.

At the risk of boring the reader, let us illustrate the formalism above in the Riemann-Roch theorem for a line bundle W over a Riemann surface. Then

$$\mathrm{ind}\, \bar{\partial}_W = \int_M \mathrm{ch}(W)\tau(M) \quad \text{with } \mathrm{ch}(W) = e^{c_1(W)} = 1 + c_1(W)$$

and

$$\tau(M) = y/(1 - e^{-y}) = 1/(1 - y/2 + y^2/6) = 1 + y/2 = 1 + c(M)/2$$

where $c(M)$ means c_1 of the tangent bundle. So

$$\mathrm{ch}(W)\tau(M) = (1 + c(W))(1 + c(M)/2) = 1 + c(W) + c(M)/2.$$

Finally $\int_M$ means integrate the form of highest degree, so we get $\mathrm{ind}\, \bar{\partial}_W = \int_M c(W) + \int_M c(M)/2$, and this is easily interpreted as the classical Riemann-Roch theorem.

If E is a real vector bundle, the Pontrjagin classes are defined in terms of Chern classes of $E \otimes C$, the complexification of E by $p_j(E) = (-1)^j c_{2j}(E \otimes C)$. The formalism for real bundles is this. If f is a formal power series in $x_1^2, \ldots, x_n^2$ which is symmetric in x_j^2, f can be expressed in terms of the elementary symmetric functions in x_j^2. For the kth elementary symmetric function, substitute p_k so that f represents a polynomial in the Pontrjagin classes.

So for example on a 4-manifold, there is only one Pontrjagin class p_1. The signature $S(M)$ equals $\alpha \int_M p_1$ where α is the coefficient of x^2 in the power series expansion $x/\tanh x$ and $\hat{A}(M) = \beta \int_M p_1$ where β is the coefficient of x^2 in the expansion of $x/(e^{x/2} - e^{-x/2})$ [$\alpha = 1/3$ and $\beta = -1/24$; see [**18**, §1.5]].

Appendix II.

THEOREM. *The transversal elliptic index is a distribution.*

PROOF. Let $A: C^\infty(E) \to C^\infty(F)$, E, F complex vector bundles over M be transversal elliptic relative to a compact Lie group action G. Let Δ denote the natural Laplacian on G and let $\tilde{\Delta}$ denote the induced operators on $C^\infty(E)$ and $C^\infty(F)$. That is, $\Delta = -\sum X_r^2$ relative to an orthonormal base of an invariant inner product on g, the Lie algebra of G. Then $\tilde{\Delta} = -\sum \tilde{X}_r^2$ where $\tilde{X}_r$ is the corresponding "vector field" operator G induces on $C^\infty(E)$ and $C^\infty(F)$. Since A is transversal elliptic and $\tilde{\Delta}$ is elliptic along the orbits, as shown in §2(c), the operator $B: C^\infty(E) \to C^\infty(F) \oplus C^\infty(E)$ given by $Bu = (Au, \tilde{\Delta}^r u)$ has injective symbol. [For convenience, assume A is of order $2r$ which we can accomplish by the usual normalization via a Laplacian on M.] Hence B^*B is elliptic and there exists an operator $\tilde{Q}$ (of order $-4r$) $\ni \tilde{Q}B^*B = I + C$, C of order $-\infty$. Let $Q = \tilde{Q}B^*$ (of order $-2r$) so that $Q = P \oplus R: C^\infty(F) \oplus C^\infty(E) \to C^\infty(E)$ and $QB = I + C$, i.e.,

$$PA + R\tilde{\Delta}^r = I + C. \qquad \text{(I)}$$

We will assume that r is sufficiently large ($r > \dim M/4$) so that an operator of order $-2r$ is of trace class, for example, is an integral operator with continuous kernel (all on $H_0 = L_2$).

Let Π denote orthogonal projection on the closure of Ker A. Then

(II) $$\Pi R\tilde{\Delta}^r\Pi = \Pi + C_1 \text{ on } \Pi(H_0) \text{ so, on } \Pi(H_0),$$
$$\Pi R\tilde{\Delta}^r = I + K,\ K \text{ of trace class.}$$

Now let T be the "inverse" of $\tilde{\Delta}^r$ on $\Pi(H_0)$: on each nonzero eigenspace of $\tilde{\Delta}^r$ in Ker A, T is the inverse, and $T = 0$ on Ker $A \cap$ Ker $\tilde{\Delta}$. So T is a bounded operator and

(III) $$\tilde{\Delta}^r T = T\tilde{\Delta}^r = I - p_0 \text{ where } p_0 \text{ is projection on Ker } A \cap \text{Ker } \tilde{\Delta};$$
a finite dimensional space by §2(c).

Fundamental fact. T is of trace class.

PROOF. Since ΠR is of trace class (for trace class operators is an ideal (not closed!) in all bounded operators), it suffices to show that $\alpha = \Pi R - T$ is of trace class. But

$$\alpha = \alpha(I - p_0) + \alpha p_0 = \alpha\tilde{\Delta}^r T + \alpha p_0 = (\Pi R - T)\tilde{\Delta}^r T + \alpha p_0$$
$$= (I + K - (I - p_0))T + \alpha p_0$$

which is of trace class.

What this fact means is that each irreducible representation occurs in Ker A with multiplicity that does not grow too fast. That is, if $\lambda_j \neq 0$ is an eigenvalue of $\tilde{\Delta}$ which occurs with dimension n_j in Ker A, then

(IV) $$\sum_{\lambda_j \neq 0} n_j^2/\lambda_j^{2r} = \operatorname{tr}(T^2) < \infty.$$

Let $\lambda(\omega)$ be the eigenvalue of Δ on ω, an irreducible representation on V_ω of dim $n(\omega)$. If $m(\omega)$ is the multiplicity of V_ω in Ker A, then (IV) says

(V) $$\sum_{\omega \neq \omega_0} n^2(\omega)m^2(\omega)/\lambda^{2r}(\omega) < \infty, \qquad \omega_0 \text{ the trivial representation.}$$

The transversal index is an element of Hom$(R(G), Z)$ which is the difference of two such, one obtained from Ker A and the other from Cok A. The first maps $\omega \to m(\omega)$; we now show it extends to a distribution on $C^\infty(G)$. By taking adjoints, a similar argument applies to Cok A.

For $f \in C^\infty(G)$, let $\hat{f}$ be its Fourier transform:

$$\hat{f}(\omega) = \int_G \bar{\omega}(g)f(g)\,dg.$$

The Plancherel formula says that the map $f \to \hat{f}$ extends to $L_2(G)$ and for f_1,

$f_2 \in L_2(G)$, $\int_G f_1 \bar{f}_2 = \sum_\omega \operatorname{tr}(\hat{f}_1(\omega)(\hat{f}_2(\omega))^*)n(\omega)$. If ρ is the representation of G on Ker A, consider the map $\varphi: C^\infty(G) \to \boldsymbol{C}$ given by

$$\varphi(f) = \operatorname{tr} \int_G \bar{\rho}(g) f(y)\, dg.$$

Note that when $f = \chi_\omega$ the character of ω, then $\varphi(\chi_\omega) = m(\omega)$. So φ is our desired functional provided we can show it is well defined and a distribution. In fact, we show that $\varphi \in H_{-2r}(G)$ by showing that $\Delta^{-r}\varphi \in L_2(G)$. ($\Delta^{-r}$ means 0 on constants and $(\Delta^{-1})^r$ on the orthogonal complement of constants.)

Now $\varphi(f) = \sum_\omega \operatorname{tr}(\hat{f}(\omega))m(\omega) = \langle \hat{f}, \hat{\varphi} \rangle$ where $\hat{\varphi}(\omega) = (m(\omega)/n(\omega))I_\omega$ and I_ω is the identity transformation on V_ω. So

$$\widehat{\Delta^{-r}\varphi}(\omega) = (m(\omega)/\lambda^r(\omega)n(\omega))I_\omega, \qquad \omega \neq \omega_0.$$

So

$$\|\hat{\Delta}^{-r}\varphi\|^2 = \sum_{\omega \neq \omega_0} \frac{m^2(\omega)}{\lambda^{2r}(\omega)n^2(\omega)} \operatorname{tr}(I_\omega)n(\omega) = \sum_{\omega \neq \omega_0} m^2(\omega)/\lambda^{2r}(\omega)$$

which is finite by (V).

We conclude with a discussion of the index for a transversal elliptic operator when G is not compact. That the index is a distribution is due to L. Hörmander (unpublished) and I thank him for permission to include the result here. Suppose E is a complex over M:

$$0 \to C^\infty(E_0) \xrightarrow{d_0} C^\infty(E_1) \xrightarrow{d_1} \cdots \to C^\infty(E_N) \to 0$$

with the d_j pseudodifferential operators of order $r > 0$ and $d_j d_{j-1} = 0$. Let $H^j(M, E) = \operatorname{Ker} d_j/\operatorname{image} d_{j-1}$. Let $C_j \subset T^*(M) - 0$ be the characteristic set at the jth place:

$$C_j = [\xi \in T^*(M) - 0; \sigma_{d_{j-1}}(\xi)(E_{j-1}) \subsetneqq \sigma_{d_j}(\xi)^{-1}(0)].$$

Suppose the Lie group G acts on E as in §5. E is *transversal elliptic* if $T^*_G(M) - 0 \cap \bigcup_j C_j$ is empty. [This is the natural extension of the two step case; note that because G is not compact, the device of using adjoints and reducing to the two step complex does not work.]

We can no longer define the transversal index in terms of characters of G because a priori the representation of G on the cohomology of E is not a discrete direct sum. We want to define ind_T directly as a linear functional on $C_0^\infty(G)$. The action of G on E gives a representation ρ_j of G on $H^j(M, E)$. One would like to define for $f \in C_0^\infty(G)$, the operator ${}_jT_f$ on $H^j(M, E)$ as $\int_G \bar{\rho}_j(g) f(g)\, dg$ and then put $(\operatorname{ind}_T E)(f) = \sum (-1)^j \operatorname{tr}({}_jT_f)$. But $H^j(M, E)$ is not finite dimensional and not even a Hilbert space. So we must go to Hilbert space completions and prove that the corresponding ${}_jT_f$ are of trace class.

Choose hermitian metrics on E_j and a volume form on M (or density if M is not orientable). Let $\tilde{d}_j: L_2(E_j) \to L_2(E_{j+1})$ be the closed unbounded operator whose graph is $[(u, f) \in L_2(E_j) \oplus L_2(E_{j+1}); d_j u = f$, where equality in $d_j u = f$ means as distributions]. Let $\tilde{H}^j = \operatorname{Ker} \tilde{d}_j \cap \operatorname{Ker} \tilde{d}^*_{j-1}$ be a closed subspace of $L_2(E_j)$. Let P_j be projection on $\tilde{H}^j$, and let ${}_jS_f = \int_G \tau_j(g) f(g)\, dg$ where τ_j is the representation of G on $L_2(E_j)$ and $f \in C_0^\infty(G)$. (The representation τ_j is weakly continuous.)

THEOREM. *$P_j\, {}_jS_f P_j$ is of trace class. The map $f \to \operatorname{tr}({}_jS_f)$ is a distribution.*

PROOF. We need only to show $P_j {}_jS_f P_j$ is of trace class. Once the trace functional is well defined it is easy to check that it is continuous on $C_0^\infty(G)$ by the closed graph theorem.

We introduce the notion of wave front set $WF(u)$ for u in $C^\infty(E_j)^* = \mathscr{D}'_j$ [**20**, §2.5]. An element $\xi \in T^*(M) - 0$ is not in $WF(u)$ if there exists a pseudodifferential operator $P \ni Pu \in C^\infty$ and $\sigma(P)(\xi)$ is injective. Since $WF(u)$ projects onto sing supp(u) under the projection map $T^*(M) \to M$, $u \in C^\infty(M) \Leftrightarrow WF(u)$ is empty. To show $P_j\, {}_jS_f P_j$ is of trace class, it suffices to show that ${}_jS_f P_j$ maps $L_2(E_j)$ continuously into $C^\infty(E_j)$ so ${}_jS_f P_j$ is of trace class. It suffices, then, to show that ${}_jS_f P_j u \in C^\infty(E_j)$ for any $u \in L_2(E_j)$ because ${}_jS_f P_j: L_2(E_j) \to \mathscr{D}'_j$ is continuous. If the image lies in $C^\infty(E_j)$, the closed graph theorem implies the map into $C^\infty(E_j)$ is continuous. Finally, then, we want to show that $u \in L_2(E_j)$ implies that $WF({}_jS_f P_j u)$ is empty.

First observe that $WF({}_jS_f u) \subseteq T_G^*(M) - 0$. For if ξ is not in $T_G^*(M) - 0$ and $\xi \neq 0$, there exists an $X \in \mathfrak{g}$, the Lie algebra of $G \ni \xi(\tilde{X}) \neq 0$, i.e., $\sigma_{\tilde{X}}(\xi)$ is injective. But $\tilde{X}^n({}_jS_f u) = {}_jS_{X^n f} u$ so that ${}_jS_f u$ is already smooth in the $\tilde{X}$ direction. It is easy to see this means $\xi \in WF({}_jS_f u)$.

Next observe that if $u \in \tilde{H}_j$, then $WF(u) \subset C_j$ for $(d_j^* d_j + d_{j-1} d_{j-1}^*)u = 0$ and C_j by definition is the characteristic set for this Laplacian. If g is close to $e \in G$, then $WF(\tau_j(g)u)$ is close to $WF(u)$. Since by transversal ellipticity $WF(u) \cap T_G^*(M) - 0$ is empty, so is $WF(\tau_j(g)u) \cap T_G^*(M) - 0$. Hence, if f has support near e, $WF({}_jS_f P_j v) \cap T_G^*(M) - 0$ is empty.

Our two observations combine to make the wave front set of ${}_jS_f P_j v$ empty when f has support near e. By translations and a partition of unity we conclude that $WF({}_jS_f P_j v)$ is empty. Q.E.D.

The theorem now allows us to define the transversal index of E as the distribution $\operatorname{ind}_T E(f) = \sum (-1)^j \operatorname{tr}({}_jS_f)$.

REMARK. Suppose that the representations ρ_j are in fact unitary, relative to our choice of hermitian metrics and volume element. Then $\tilde{d}_j$ and $\tilde{d}^*_{j-1}$ are G-operators so that G acts on $\tilde{H}^j$, say via μ_j. Then $P_j\, {}_jS_f P_j = \int_G \mu_j(g) f(g)\, dg$. Since these operators are compact operators, the representations μ_j are discrete direct sums of irreducible representations.

BIBLIOGRAPHY

1. M. F. Atiyah, *Vector fields on manifolds*, Arbeitsgemeinschaft für Forschung des Landes Nord-

rhein-Westfalen, Heft 200, Westdeutscher Verlag, Cologne, 1970. MR **41** #7707.

2. ———, *Riemann surfaces and spin structures*, Ann. Sci. École Norm. Sup. (4) **4** (1971), 47–62.

3. ———, *Topology of elliptic operators*, Global Analysis, Proc. Sympos. Pure Math., vol. 16, Amer. Math. Soc., Providence, R.I., 1970, pp. 101–119. MR **41** #9291.

4. M. F. Atiyah and R. Bott, *A Lefschetz fixed point formula for elliptic complexes*. I, Ann. of Math. (2) **86** (1967), 374–407. MR **35** #3701.

5. ———, *A Lefschetz fixed point formula for elliptic complexes*. II. *Applications*, Ann. of Math. (2) **88** (1968), 451–491. MR **38** #731.

6. M. F. Atiyah, R. Bott and A. Shapiro, *Clifford modules*, Topology **3** (1964), suppl. 1, 3–38. MR **29** #5250.

7. M. F. Atiyah and J. L. Dupont, *Vector fields with finite singularities*, Acta Math. **128** (1972), 1–40.

8. M. F. Atiyah and F. Hirzebruch, *Spin manifolds and group actions*, Essays on Topology and Related Topics (Mémoirs dédiés à Georges de Rham), Springer, New York, 1970.

9. M. F. Atiyah and G. B. Segal, *The index of elliptic operators*. II, Ann. of Math. (2) **87** (1968), 531–545. MR **38** #5244.

10. M. F. Atiyah and I. M. Singer, *The index of elliptic operators*. I, Ann. of Math. (2) **87** (1968), 484–530. MR **38** #5243.

11. ———, *The index of elliptic operators*. III, Ann. of Math. (2) **87** (1968), 546–604. MR **38** #5245.

12. ———, *The index of elliptic operators*. IV, Ann. of Math. (2) **93** (1971), 119–138.

13. ———, *The index of elliptic operators*. V, Ann. of Math. (2) **93** (1971), 139–149.

14. ———, *Index theory for skew-adjoint Fredholm operators*, Inst. Hautes Études Sci. Publ. Math. No. 37 (1969), 305–326.

15. P. F. Baum and R. Bott, *On the zeros of meromorphic vector-fields*, Essays on Topology and Related Topics (Mémoires dédiés à Georges de Rham), Springer, New York, 1970, pp. 29–47. MR **41** #6248.

16. P. F. Baum and J. Cheeger, *Infinitesimal isometries and Pontryagin numbers*, Topology **8** (1969), 173–193. MR **38** #6627.

17. P. Gilkey, *The index of an elliptic differential complex and the asymptotic behavior of the eigenvalues* (to appear).

18. F. Hirzebruch, *Neue topologische Methoden in der algebraischen Geometrie*, Ergebnisse der Math. und Grenzgebiete, Heft 9, Springer-Verlag, Berlin, 1956; English transl., Die Grundlehren der math. Wissenschaften, Band 131, Springer-Verlag, New York, 1966. MR **18**, 509; MR **34** #2573.

19. W. C. Hsiang, *A splitting theorem and the Künneth formula in algebraic K-theory*, Algebraic *K*-theory and its Geometric Applications (Conf., Hull, 1969), Springer, Berlin, 1969, pp. 72–77. MR **40** #6560.

20. L. Hörmander, *Fourier integral operators*, Acta Math. **127** (1971), 79–183.

21. C. Kosniowski, *Applications of the holomorphic Lefschetz formula*, Bull. London Math. Soc. **2** (1970), 43–48. MR **41** #6249.

22. A. Lichnerowicz, *Spineurs harmoniques*, C. R. Acad. Sci. Paris **257** (1963), 7–9. MR **27** #6218.

23. G. Lusztig, *Novikov's higher signature and families of elliptic operators*, Thesis, Princeton University, Princeton, N.J. (to appear).

24. J. Milnor, *Spin structures on manifolds*, Enseignement Math. **9** (1936), 198–203.

25. ———, *On simply connected* 4-*manifolds*, Symposium international de topologia algebraica, Universidad Nacional Autonoma de Mexico and UNESCO, Mexico City, 1958, pp. 122–128. MR **21** #2240.

26. S. P. Novikov, *Manifolds with free Abelian fundamental groups and their applications* (*Pontrjagin classes, smoothnesses, multidimensional knots*), Izv. Akad. Nauk SSSR Ser. Mat. **30** (1966), 207–246; English transl., Amer. Math. Soc. Transl. (2) **71** (1968), 1–42. MR **33** #4951.

27. R. S. Palais, *Seminar on the Atiyah-Singer index theorem*, Ann. of Math. Studies, no. 57, Princeton Univ. Press, Princeton, N.J., 1965. MR **33** #6649.

28. V. K. Patodi, *Curvature and the eigenforms of the Laplace operator*, J. Differential Geometry **5** (1971), 233–249.

29. ———, *Analytic proof of the Riemann-Roch-Hirzebruch theorem for Kahler manifolds*, J. Differential Geometry **5** (1971), 251–283.

30. D. B. Ray and I. M. Singer, *Reidemeister torsion and the Laplacian on Riemannian manifolds*, Advances in Math. 7 (1971), 145–210.

31. R. T. Seeley, *Complex powers of an elliptic operator*, Singular Integrals, Proc. Sympos. Pure Math., vol. 10, Amer. Math. Soc., Providence, R.I., 1967, pp. 288–307. MR **38** #6220.

32. I. M. Singer, *Elliptic operators on manifolds*, Pseudo-Differential Operators (C.I.M.E., Stresa, 1968), Edizioni Cremonese, Rome, 1969, pp. 333–375. MR **41** #7317.

33. ———, *Operator theory and periodicity*, Indiana Univ. Math. J. **20** (1971), 949–951.

34. E. Thomas, *Vector fields on manifolds*, Bull. Amer. Math. Soc. **75** (1969), 643–683. MR **39** #3522.

MASSACHUSETTS INSTITUTE OF TECHNOLOGY

ON THE EXISTENCE AND REGULARITY OF SOLUTIONS OF LINEAR PARTIAL DIFFERENTIAL EQUATIONS

F. TREVES

Foreword. This is an attempt to provide the reader with a view on recent developments in the *local* theory of linear partial differential equations. The body of results gathered so far is considerable and a narrow selection has been unavoidable. In this selection I have tried to fulfill three requirements. First, the reader should be given some hint about the origins of the theory and the motivations for its present development. I hope that the simple example studied in §1 from a variety of viewpoints will be of help in this purpose. Secondly, the reader ought to be precisely informed on the results obtained so far (at the time of this writing) and also on the conjectures which orient the research currently carried out. On this see §2. Lastly, the reader might wish to learn something of the methods. To this purpose §§3 and 4 describe the construction of fundamental solutions of an important class of partial differential equations of principal type. It was chosen for inclusion here because of its similarities and differences, both striking, with the methods of geometric optics, currently of widespread use in the theory of linear partial differential equations, and also because I believe that constructions akin to this one will play an important role in future developments of the theory.

0. **The basic definitions.** Let $P(x, D)$ denote a differential operator of order m in an open subset of $\boldsymbol{R}^N$; this tacitly presumes that its coefficients are complex-valued C^∞ functions in Ω. The case $m = 0$ is almost always trivial, therefore we shall assume that $m > 0$. The principal symbol of $P(x, D)$ is denoted by $P_m(x, \xi)$.

DEFINITION 0.1. *The operator $P(x, D)$ is said to be of principal type in Ω if, given any $x \in \Omega$ and any $\xi \in \boldsymbol{R}_N$, $\xi \neq 0$, $d_\xi P_m(x, \xi) \neq 0$.*

Observe that, by Euler's homogeneity relation, $mP_m(x, \xi) = \xi \cdot \mathrm{grad}_\xi P_m(x, \xi)$ and therefore any zero of $d_\xi P_m(x, \cdot)$ would be a zero at least double of $P_m(x, \cdot)$. In

AMS 1970 *subject classifications.* Primary 35A05, 35H05, 35S05; Secondary 35C15, 35C05, 35D05, 35D10, 35F25.

other words, to say that $P(x, D)$ is of principal type is to say that its *real* characteristics are *simple*. All elliptic differential operators, all hyperbolic differential operators are of principal type; the parabolic ones are not. All first-order differential operators whose principal part does not vanish identically at any point are also of principal type.

DEFINITION 0.2. *The operator $P(x, D)$ is said to be locally solvable at the point x_0 of Ω if there is an open neighborhood $U \subset \Omega$ of x_0 such that, given any function $f \in C^\infty(U)$, there is a distribution $u \in \mathscr{D}'(U)$ such that $P(x, D)u = f$ in U.*

The operator $P(x, D)$ is locally solvable in a subset S of Ω if it is locally solvable at every point of S; note that it is then locally solvable in some *open* subset of Ω containing S.

DEFINITION 0.3. *The operator $P(x, D)$ is said to be hypoelliptic in Ω if, given any open subset U of Ω and any distribution u in U, $P(x, D)u \in C^\infty(U)$ implies $u \in C^\infty(U)$ (i.e., $P(x, D)$ preserves the singular supports).*

DEFINITION 0.4. *The operator $P(x, D)$ is said to be analytic-hypoelliptic in Ω if it is hypoelliptic in Ω and if, given any open subset U of Ω and any function $f \in C^\infty(U)$, the fact that $P(x, D)f$ is analytic in U implies that f itself is analytic in U (i.e., $P(x, D)$ preserves the analytic singular supports).*

DEFINITION 0.5. *The operator $P(x, D)$ is said to be subelliptic at $x_0 \in \Omega$ if there is a number $\delta > 0$ and an open neighborhood U of x_0 in Ω such that, for some constant $C > 0$ and all $u \in C_c^\infty(U)$,*

$$\|u\|_{m-1+\delta} \leqq C\|P(x, D)u\|_0. \tag{0.1}$$

We have denoted by $\| \ \|_s$ the norm in the sth Sobolev space on $\boldsymbol{R}^N$. An equivalent definition of subellipticity is that, to every compact subset K of Ω, there is a number $\delta_K > 0$ and a positive constant C_K such that, for all $u \in C_c^\infty(\Omega)$ with support in K,

$$\|u\|_{m-1+\delta_K} \leqq C_K\{\|P(x, D)u\|_0 + \|u\|_0\}. \tag{0.2}$$

Of course, if (0.1) holds with $\delta = 1$, $P(x, D)$ is elliptic at x_0. If $P(x, D)$ is subelliptic at a point, it is so in a full neighborhood of this point.

1. **A simple first-order equation.** The simplest differential operators of *principal type* (Definition 0.1) are the first-order ones whose principal part does not vanish at any point. Among the latter, the simplest nontrivial examples are provided by the operators of the form

$$\boldsymbol{L} = \partial/\partial t + ib(t)\,\partial/\partial x, \tag{1.1}$$

where x and t are real variables and $b(t)$ a *real-valued* C^∞ function in some open interval $-T < t < T$ $(T > 0)$. It turns out that studying the operators (1.1) can be very instructive: it can give many cues on the phenomena which dominate the theory of differential operators of principal type of any order, in any number of variables.

In solving the equation $Lu = f$, where f is, say, a C^∞ function with compact support in the slab $\{(x, t) \in \boldsymbol{R}^2; |t| < T\}$, it is natural to perform a Fourier transform with respect to x. This leads to the equation

$$\hat{u}_t - b(t)\xi\hat{u} = \hat{f}(\xi, t), \tag{1.2}$$

where the "hats" denote Fourier transforms with respect to x. Now, all the solutions of (1.2) are known. Each is the sum of a solution of the homogeneous equation and of a solution of the inhomogeneous equation, of the form

$$\hat{u}(\xi, t) = \int_{t_0}^{t} e^{B(t,t')\xi}\hat{f}(\xi, t')\,dt'. \tag{1.3}$$

We have used the notation

$$B(t, t') = \int_{t'}^{t} b(s)\,ds. \tag{1.4}$$

Of course, we are not primarily interested in solving (1.2): it is $Lu = f$ which we want to solve. We must therefore ask ourselves if we can revert, by the inverse Fourier transformation with respect to x, from a solution $\hat{u}$ of (1.2) to a solution u of $Lu = f$. We are willing to accept distribution solutions of the latter equation, therefore $\hat{u}$ must be *tempered* with respect to ξ. Observe that the limit of integration t_0 in (1.3) can be made to depend on ξ. We may therefore ask whether we could choose $t_0 = t_0(\xi)$ in such a way that $\hat{u}(\xi, t)$ will be of slow growth with respect to ξ. It is clear that this requires

$$B(t, t')\xi \leqq 0, \tag{1.5}$$

for all t, $|t| < T$, and all t' in the interval joining t_0 to t. Since only the sign of $\xi \neq 0$ is relevant, we may as well take $\xi = \pm 1$. Our question can therefore be rephrased as follows: can one choose t_0, $|t_0| \leqq T$, so as to have $B(t, t') \leqq 0$ (resp. $B(t, t') \geqq 0$) for all t, $|t| < T$, and all t' between t_0 and t? A short thinking will easily convince the reader that this is possible if and only if $B(t, t')$ is a *monotone* function of either t or t', in other words, if

(1.6) *$b(t)$ does not change sign in the interval $|t| < T$.*

Suppose for instance $b \geqq 0$ in $]-T, T[$. Then we may take $t_0 = T$ when $\xi > 0$ and $t_0 = -T$ when $\xi < 0$ and if we take the inverse Fourier transform with respect to x of the solution of (1.2) thus obtained, we find

$$\begin{aligned} u(x, t) = {} & \frac{1}{2\pi}\int_{-\infty}^{0}\int_{-T}^{t} e^{ix\xi + B(t,t')\xi}\hat{f}(\xi, t')\,dt'\,d\xi \\ & - \frac{1}{2\pi}\int_{0}^{+\infty}\int_{t}^{T} e^{ix\xi + B(t,t')\xi}\hat{f}(\xi, t')\,dt'\,d\xi. \end{aligned} \tag{1.7}$$

This is indeed a solution of $Lu = f$.

Suppose now that (1.6) does not hold. It may be that we cannot find, then, a solution of the kind (1.4) but are we certain that the equation $Lu = f$ is not solvable? That this is indeed the case is a consequence of the following particular case of a functional-analytic lemma of Hörmander [3, Lemma 6.1.2]. Let Ω be an open rectangle in R^2 of the form $|x| < r, |t| < T$. If to every $f \in C_c^\infty(\Omega)$ there were a distribution u in Ω satisfying $Lu = f$, then given any compact subset K of Ω, there would be constants $C, M, m > 0$ such that, for all $v, f \in C_c^\infty(R^2)$ with support contained in K,

$$\left|\iint fv\,dx\,dt\right| \leqq C \sup_{R^2} \sum_{|\alpha|\leqq m} |D^\alpha f| \sup_{R^2} \sum_{|\beta|\leqq M} |D^\beta(Lv)| \tag{1.8}$$

(observe that the formal transpose of L is ${}^tL = -L$; $\alpha = (\alpha_0, \alpha_1)$, $\beta = (\beta_0, \beta_1)$ are pairs of nonnegative integers).

Suppose now that $B(t, t')$ is not a monotone function of t for every t'; for instance, that, for some t', $t \mapsto B(t, t')$ has a true *minimum* (we assume throughout that t and t' belong to $]-T, T[$). This is equivalent to saying that, for some t_0, t_1, t_2 such that $t_1 < t_0 < t_2$, we will have $B(t, t_0) \geqq 0$ for all $t \in (t_1, t_2)$ and $B(t_1, t_0) > 0$, $B(t_2, t_0) > 0$. Set now $z = x - iB(t, t_0)$ and observe that any analytic function of z, $h(z)$, verifies $Lh = 0$. In particular, this is true of $w = z - i\theta z^2$ and of $e^{i\xi w}$, where ξ and θ are numbers. Let $g_0 \in C_c^\infty(]-T, T[)$ be equal to one in a neighborhood of t_0 and such that $B(t, t_0) \geqq 0$ on the support of g_0 and $B(t, t_0) > 0$ on that of g_0'. Let $g_1 \in C_c^\infty(]-r, r[)$ be equal to one in a neighborhood of the origin and set $g(x, t) = g_0(t)g_1(x)$. We choose, in (1.8), $v = ge^{i\xi w}$ and we choose the number θ to be > 0 and small enough so as to have $\theta B(t, t_0)^2 \leqq \frac{1}{2}B(t, t_0)$ on the support of g. Therefore, in this set,

$$\operatorname{Im} w = -B(t, t_0) - \theta x^2 + \theta B(t, t_0)^2 \leqq -\tfrac{1}{2}B(t, t_0) - \theta x^2.$$

Now, $Lv = (Lg)\, e^{i\xi w}$ and consequently, there is a constant $c > 0$ such that, on the support of Lg (which is contained in the support of grad g), we have, for all $\xi < 0$,

$$\operatorname{Re}(i\xi w) \leqq -c|\xi|.$$

From this we conclude that, for some $C' > 0$ and all $\xi < 0$,

$$\sup_{R^2} \sum_{|\beta|\leqq M} |D^\beta(Lv)| \leqq C'|\xi|^M e^{-c|\xi|}. \tag{1.9}$$

If we take then

$$f(x, t) = F(|\xi|x, |\xi|(t - t_0)), \qquad F \in C_c^\infty(R^2),$$

we obtain

$$\sup_{R^2} \sum_{|\alpha|\leqq m} |D^\alpha f| \leqq C''|\xi|^m, \tag{1.10}$$

$$(1.11) \qquad |\xi|^2 \iint e^{i\xi w} fg \, dx \, dt \to \iint e^{-iy} F(y, s) \, dy \, ds = \hat{\hat{F}}(1, 0)$$

when $\xi \to -\infty$.

We have denoted by $\hat{\hat{F}}$ the Fourier transform of $F(y, s)$ with respect to both variables y, s. The assertion (1.11) is easily proved by making the change of variables $y = |\xi|x$, $s = |\xi|(t - t_0)$ in $\iint fv \, dx \, dt$. We may choose F so as to have $\hat{\hat{F}}(1, 0) \neq 0$; also note that, for $|\xi|$ sufficiently large, the support of f will be contained in any neighborhood of $(0, t_0)$, for instance in the support of g. In view of (1.9), (1.10), (1.11) we reach the conclusion that (1.8) cannot be true, i.e., that $\boldsymbol{L}u = f$ cannot be solvable in Ω.

To summarize, we have shown that *the equation* $\boldsymbol{L}u = f$ *is solvable in* Ω *if and only if* (1.6) *holds*; if this is the case, a solution is obtained by a formula like (1.8).

So much for the solvability of the equation $\boldsymbol{L}u = f$. Let us look now at the regularity of its solutions.

Suppose that $b(t) = 0$ in an open nonempty subinterval of $)-T, T($, J. Then any function or distribution $u(x)$, independent of t, will satisfy the homogeneous equation $\boldsymbol{L}u = 0$ in the region $|x| < r$, $t \in J$: we cannot expect any kind of regularity of the solutions. In other words, if we want $\boldsymbol{L}$ to be hypoelliptic we better have

(1.12) *$b(t)$ does not vanish identically on any open nonempty subinterval of $)-T, T($.*

On the other hand [**12**, Theorem 52.2], the transpose of a hypoelliptic operator is locally solvable and, consequently, if $\boldsymbol{L}$ is hypoelliptic, it must be locally solvable (recalling that ${}^t\boldsymbol{L} = -\boldsymbol{L}$) and therefore condition (1.6) must hold. We may use formula (1.7). But under the hypothesis (1.12), (1.7) can be rewritten in a convenient fashion. Indeed

$$u(x, t) = \frac{1}{2\pi} \int_{-\infty}^{+\infty} \int_{-\infty}^{0} \int_{-T}^{t} e^{i(x-y)\xi + B(t,t')\xi} f(y, t') \, d\xi \, dt' \, dy$$
$$- \frac{1}{2\pi} \int_{-\infty}^{+\infty} \int_{0}^{+\infty} \int_{t}^{T} e^{i(x-y)\xi + B(t,t')\xi} f(y, t') \, d\xi \, dt' \, dy.$$

We have performed first the integration with respect to ξ. This is permitted provided we interpret the result in the correct sense of distribution theory. It can be written

$$(1.13) \qquad u(x, t) = \frac{1}{2\pi i} pv \iint \frac{f(y, t')}{x - y - iB(t, t')} \, dy \, dt'.$$

The kernel $E = (2\pi i)^{-1}(x - y - iB(t, t'))^{-1}$ can be regarded as a continuous function of $t \neq t'$ valued in the space of distributions in the variables x, y. It has a discontinuity at $t = t'$; the jump there is

$$\lim_{\varepsilon\to+0}\frac{1}{2\pi i}\left\{\frac{1}{x-y-i\varepsilon}-\frac{1}{x-y+i\varepsilon}\right\}=\delta(x-y)$$

(δ is the Dirac measure). By the elementary formula for distribution derivatives of such functions we have

$$(\partial/\partial t)E=\{(\partial/\partial t)E\}-\delta(x-y)\delta(t-t'),$$

where $\{(\partial/\partial t)E\}$ is the "classical derivative," that is to say, the function equal to $(\partial/\partial t)E$ when $t\neq t'$. Of course, we have

$$\{(\partial/\partial t)E\}=-ib(t)(\partial/\partial x)E,$$

whence

$$LE=\delta(x-y)\delta(t-t'); \tag{1.14}$$

in other words, E is a *fundamental kernel* of $\boldsymbol{L}$ (it is checked likewise that

$$-(\partial/\partial t'+ib(t')\,\partial/\partial y)E=\delta(x-y)\delta(t-t') \tag{1.15}$$

which means that if we view E as an operator, the operator defined by the kernel E, then $\boldsymbol{L}E=\delta(x-y)\delta(t-t')$: thus E is a *two-sided* fundamental kernel of $\boldsymbol{L}$).

Now, if (1.12) holds, $B(t,t')\neq 0$ as soon as $t\neq t'$. But this implies at once that $E=E(x,y,t,t')$ is a C^∞ function in the complement of the diagonal, i.e., in the region

$$(x-y)^2+(t-t')^2\neq 0.$$

This is well known to imply the hypoellipticity of $\boldsymbol{L}$.

Thus we have shown that *for* $\boldsymbol{L}$ *to be hypoelliptic, it is necessary and sufficient that both conditions* (1.6) *and* (1.12) *be satisfied.*

Let us go one step further and assume that $b(t)$ is an *analytic* function of t in $]-T,T[$. Note then that the conjunction of (1.6) and (1.12) can be stated as follows:

(1.16) *every zero of* $b(t)$ *in* $]-T,T[$ *is of finite even order.*

In this case, the above kernel E is not only a C^∞ function, but also an analytic function in the complement of the diagonal. And this is known to imply that $\boldsymbol{L}$ is *analytic-hypoelliptic* (Definition 0.3). Thus the latter property is equivalent with (1.16).

This brings us to the question of *subellipticity* (Definition 0.4). Suppose that $b(t)$ is a C^∞ function satisfying (1.16). Then the zeros of $b(t)$ are necessarily isolated. Suppose that the origin is one of them and that it is of order $2k$; then, for all practical purposes, $b(t)$ behaves like $\pm t^{2k}$ (as a matter of fact, a change of the variable t about the origin can transform $\boldsymbol{L}$ into $b_0(t)(\partial/\partial t+it^{2k}\,\partial/\partial x)$ with $b_0(0)\neq 0$). Let us therefore assume that $b(t)=t^{2k}$. Then

$$B(t,t')=(t^{2k+1}-t'^{2k+1})/(2k+1).$$

Let us go back to the Fourier transform of $u(x, t)$ given by (1.8). It is

$$
\begin{aligned}
\hat{u}(\xi, t) &= -\int_t^T e^{B(t,t')\xi}\hat{f}(\xi, t')\,dt' \quad \text{if} \quad \xi > 0, \\
&= \int_{-T}^t e^{B(t,t')\xi}\hat{f}(\xi, t')\,dt' \quad \text{if} \quad \xi < 0.
\end{aligned}
\tag{1.17}
$$

Note that since we assume that the support of $f(x, t)$ is contained in a compact neighborhood of the origin, we may as well take $T = +\infty$. If we denote by $H(t)$ the Heaviside function, we see that

$$
\hat{u}(\xi, t) = \int K_\xi(t, t')\hat{f}(\xi, t')\,dt' \tag{1.18}
$$

where

$$
K_\xi(t, t') = -\frac{\xi}{|\xi|}H\left(-\frac{\xi}{|\xi|}(t - t')\right)e^{B(t,t')\xi}. \tag{1.19}
$$

We shall now apply the following elementary lemma:

LEMMA 1.1. *Let $K(t, t')$ be a function in $\boldsymbol{R}^2$, separately integrable with respect to t and t'. Suppose that the two functions*

$$
\int |K(t, t')|\,dt', \qquad \int |K(t, t')|\,dt \tag{1.20}
$$

belong to $L^\infty(\boldsymbol{R}^1)$ and that their L^∞ norm is $\leqq M$ $(M > 0)$. Then the operator K, defined by

$$
(Kf)(t) = \int K(t, t')f(t')\,dt', \tag{1.21}
$$

is a bounded linear operator on $L^2(\boldsymbol{R}^1)$ with norm $\leqq M$.

First we apply this with $K = K_\xi$. By making the change of variables $s = t^{2k+1}/(2k + 1)$, $s' = t'^{2k+1}/(2k + 1)$ in the corresponding integrals (1.20), we see easily that we can take

$$
M = \mathrm{const}(1/|\xi|)^{1/(2k+1)}.
$$

As a matter of fact, let us set $\delta = 1/(2k + 1)$. We derive from Lemma 1.1 that

$$
(1 + \xi^2)^\delta \int |\hat{u}(\xi, t)|^2\,dt \leqq \mathrm{const} \int |\hat{f}(\xi, t)|^2\,dt. \tag{1.22}
$$

If we integrate this with respect to ξ over $\boldsymbol{R}^1$ and apply Plancherel's theorem we see that

(1.23)
$$\|u\|_{(\delta,0)} \leqq \text{const}\ \|f\|_0$$

where the meaning of the norms is obvious. We may in fact obtain more. Not surprisingly we have

$$(\partial/\partial t)K_\xi(t, t') = \delta(t - t') + b(t)\xi K_\xi(t, t').$$

We now apply Lemma 1.1 to $K = b(t)\xi K_\xi = t^{2k}\xi K_\xi$. By re-introducing the variables s, s' as indicated above, we see easily that the conditions in Lemma 1.1 are fulfilled and that we may take M to be a constant independent of ξ. We make $(\partial/\partial t)K_\xi$ act on $\hat{f}(\xi, t)$ and reach the conclusion that

(1.24)
$$\int |\hat{u}_t(\xi, t)|^2\, dt \leqq \text{const} \int |\hat{f}(\xi, t)|^2\, dt,$$

whence, by Plancherel's theorem,

(1.25)
$$\|u_t\|_0 \leqq \text{const}\ \|f\|_0.$$

We may also introduce the Sobolev norm

$$\|u\|_\delta = \left\{\iint (1 + \xi^2 + \tau^2)^\delta |\hat{u}(\xi, \tau)|^2\, d\xi\, d\tau\right\}$$

which has the advantage of not distinguishing between the x and t variables, thus preserving some "invariance." We have clearly

$$\|u\|_\delta \leqq \text{const}(\|u\|_{(\delta,0)} + \|u_t\|_0),$$

whence the *subelliptic estimate*

(1.26)
$$\|u\|_\delta \leqq \text{const}\ \|f\|_0.$$

We recall that $\boldsymbol{L}u = f$. What we have shown is that, given any $f \in L^2(\boldsymbol{R}^2)$ with compact support in a suitable open neighborhood U of the origin, the solution u provided by (1.7) belongs to H^δ. A consequence of the fact, already pointed out, that $\boldsymbol{L}E = \delta(x - y, t - t')$, is that if now u is given, say a C^∞ function with compact support in U, and if we set $f = \boldsymbol{L}u$, we recover u by means of (1.7). Combining this remark with (1.26) we conclude that

(1.27)
$$\|u\|_\delta \leqq \text{const}\ \|\boldsymbol{L}u\|_0 \quad \textit{for all } u \in C_c^\infty(U).$$

It is possible to show that *if, for some* $\delta > 0$, (1.27) *is true, then every zero of* $b(t)$ *in a small interval containing the origin must be of even order not exceeding* $1/\delta - 1$ (it is very easy to see that (1.27) can hold only if $\delta \leqq 1$; of course, a "zero-order" zero means no zero at all). That the zeros of $b(t)$ must be of even order is obvious, as (1.27) implies the solvability of $\boldsymbol{L}$ in U. To show that their order is bounded by $1/\delta - 1$, one argues by contradiction, very much like in the proof of the necessity of (1.6) for solvability.

2. **Conditions and conjectures in the general case: What is known so far.** The problem, now, is to extend the results of the previous section to more general differential operators than (1.1). This problem has two aspects: First of all, how to formulate the conditions which ought to generalize (1.6) and (1.12)? Secondly, how to prove the equivalence of the general conditions with the properties under study, local solvability and hypoellipticity? The answer to the first question comes essentially from historical and heuristic considerations, which I shall now briefly indicate.

Let $P(x, D)$ denote a differential operator of order $m > 0$, with complex C^∞ coefficients in an open subset Ω of $\boldsymbol{R}^N$ and let $P_m(x, \xi)$ be its principal symbol; in fact, write $P_m(x, \xi) = A(x, \xi) + iB(x, \xi)$, where A and B are real. With a real symbol such as $A(x, \xi)$ we may associate a vector field in the *cotangent bundle* $T^*(\Omega)$ over Ω, the *Hamiltonian* of A:

$$H_A = \sum_{j=1}^{N} \frac{\partial A}{\partial \xi_j}\frac{\partial}{\partial x^j} - \frac{\partial A}{\partial x^j}\frac{\partial}{\partial \xi_j}. \tag{2.1}$$

The integral curves of H_A are the *bicharacteristic strips* of A; they are defined by the system of $2n$ ordinary (nonlinear) differential equations

$$\dot{x} = \operatorname{grad}_\xi A(x, \xi), \qquad \dot{\xi} = -\operatorname{grad}_x A(x, \xi). \tag{2.2}$$

Note that, along such a strip, $A(x, \xi) = \text{const}$. We shall refer to those along which $A = 0$ as *null bicharacteristic strips* of A. Observe that the principal symbol of the commutator $[A(x, D), B(x, D)]$, which is a differential operator of order $\leqq 2m - 1$, is nothing else but $(H_A B)(x, \xi) = -(H_B A)(x, \xi)$.

It is worth recalling now the following result of L. Hörmander which, together with Hans Lewy's celebrated example, is at the origin of much of what I am discussing here. *Suppose that, at some point* $(x_0, \xi^0) \in \dot{T}^*(\Omega)$,[1] *we have the following situation*:

$$P_m(x_0, \xi^0) = 0, \qquad (H_A B)(x_0, \xi^0) \neq 0. \tag{2.3}$$

Then the equation $\boldsymbol{P}u = f$ *is not locally solvable at* x_0. Clearly, (2.3) can be reformulated as follows:

(2.4) *the restriction of* $B(x, \xi)$ *to the null bicharacteristic strip of* $A(x, \xi)$ *through* (x_0, ξ^0) *vanishes at this point whereas its first derivative along this curve does not.*

If (2.4) is true, the restriction of $B(x, \xi)$ to the null bicharacteristic strip of $A(x, \xi)$ through (x_0, ξ^0) must change sign at that point. This inevitably reminds one of condition (1.6). Indeed, if we apply the last considerations to the operator (1.1) or, better, to $\boldsymbol{P} = i^{-1}\boldsymbol{L}$, we see that $A = \tau$ and $B = b(t)\xi$ (obvious notation for the

[1] We shall systematically denote by $\dot{T}^*(\Omega)$ the complement in $T^*(\Omega)$ of the zero section over Ω.

"covariables"). The bicharacteristic strips of A are the straight-line segments in $T^*(\Omega) \cong \Omega \times \boldsymbol{R}^2$ parallel to the t-axis, and condition (1.6) can be rephrased as follows:

(2.5) *the restriction of B to any null bicharacteristic strip of A does not change sign.*

Because of the peculiar form of A and B in this case, (2.5) is equivalent to saying that B does not change sign along any bicharacteristic strip of A, be it null or not: indeed, these strips are distinguished only by the value of $\tau = A$ along each of them and B is independent of τ. The formulation (2.5) has two great advantages over (1.6): it lends itself to generalization; it is invariant under changes of coordinates in Ω. It has a serious disadvantage which we shall discuss below.

An analogous reformulation of (1.12) is clearly possible:

(2.6) *the restriction of B to any null bicharacteristic strip of A does not vanish identically on any nonempty open interval of that strip.*

With this we come into contact with another body of problems and results, known under the name of *propagation of singularities*, which has grown out of the early studies of the Cauchy problem for the wave equations and for the more general hyperbolic equations (occurring in optics, electromagnetics, etc.). In the very special case (1.1) we have shown that, if (2.6) holds, the solutions of $\boldsymbol{L}u = f \in C^\infty(\Omega)$ has no singularity whatsoever; we have also seen that, if (2.6) does not hold, then some of these solutions will have singularities. If one inspects our argument, one will also find the "location" of these singularities, in the following sense: if some point (x_0, t_0) belongs to the *singular support* of such a solution u, $b(t)$ must vanish identically in an open interval centered at t_0, J, and the whole segment $\{(x_0, t); t \in J\}$ must be contained in that singular support. This is a very particular case of general phenomena.

Consider first the case where the principal symbol of $\boldsymbol{P}$ is real, i.e., $B \equiv 0$, and let (x_0, ξ^0) be a point in $\dot{T}^*(\Omega)$ where $A = P_m$ vanishes whereas $d_\xi A$ does not. If we go back to the Hamilton-Jacobi equations (2.2) we see that, under these conditions, not only is there a true curve which is an integral curve of H_A through (x_0, ξ^0), i.e., that the bicharacteristic strip of A through this point is not reduced to a single point, but also that the projection of this strip into the base space Ω is also a true curve. The latter is called a *bicharacteristic curve* of A through x_0 in U. The following can then be proved:

(1) *if x_0 belongs to the singular support of a solution of $\boldsymbol{P}u = f \in C^\infty(\Omega)$ so does a whole arc of some bicharacteristic curve of A through x_0;*

(2) *there is a C^∞ function f in some open neighborhood U of x_0 and a solution u of the equation $\boldsymbol{P}u = f$ in U whose singular support is identical to a given bicharacteristic curve of A in U through x_0.*

Another relatively simple situation is the following: do not assume that B

vanishes identically; instead assume that *$d_\xi A$ and $d_\xi B$ are linearly independent at every point of $\dot{T}^*(\Omega)$ where A and B vanish*; assume also that (2.5) holds. Under the previous hypothesis, this is easily seen to be equivalent with what is known as *Hörmander's condition*:

(H) $\quad H_A B = -H_B A$ *vanishes at every point of $\dot{T}^*(\Omega)$ where A and B vanish.*

Let (x_0, ξ^0) be a point of $\dot{T}^*(\Omega)$ where both A and B vanish; suppose in fact that B vanishes identically on an open arc Γ of the bicharacteristic strip of A through (x_0, ξ^0). It can then be shown that this remains true if we replace P_m by zP_m where z is any nonzero complex number. This amounts to replacing A by $\alpha A - \beta B$ and B by $\beta A + \alpha B$ ($\alpha = \operatorname{Re} z$, $\beta = \operatorname{Im} z$): in other words, given any pair of real numbers $(\alpha, \beta) \neq (0, 0)$, $\beta A + \alpha B$ vanish identically on an open arc $\Gamma(\alpha, \beta)$ of the bicharacteristic strip of $\alpha A - \beta B$ through (x_0, ξ^0). As (α, β) vary in $\boldsymbol{R}^2 \backslash \{0\}$ or, which amounts to the same, over the unit circle, these arcs make up a *two-dimensional* C^∞ manifold about (x_0, ξ^0). Moreover, in view of the linear independence of $d_\xi A$ and $d_\xi B$ on this manifold, its projection into the base space Ω is also a smooth two-dimensional manifold containing x_0 which we denote by $\mathcal{M}(x_0, \xi^0)$. The following can then be proved:

(1) *if x_0 belongs to the singular support of a solution of $\boldsymbol{P}u = f \in C^\infty(\Omega)$ so does the surface $\mathcal{M}(x_0, \xi^0)$ for some ξ^0*;

(2) *given $(x_0, \xi^0) \in \dot{T}^*(\Omega)$ where P_m vanishes, there is a C^∞ function f in some open neighborhood U of x_0 and a solution of $\boldsymbol{P}u = f$ in U whose singular support is identical to $\mathcal{M}(x_0, \xi^0) \cap U$.*

For a proof of these results, see [**4**]. They are particular cases of more general facts proved by Professor L. Hörmander; much precision can be added to them thanks to his theory of wave-front sets. On this we refer to his lectures. For our purposes here, they indicate that *the projections on the base space of the null bicharacteristic strips of A on which B vanishes identically are "carriers" of singularities*, that in some cases, they are the minimal carriers of singularities and that, when no such strip exists, we may expect not to have singularities at all.

An aspect of the immediately preceding discussions leads us to discuss an important drawback in the statements (2.5) and (2.6). It is clear that the properties under study here, local solvability and hypoellipticity, are not only invariant under changes of coordinates in Ω, but also under multiplication of the differential operator $\boldsymbol{P}$ by a nonvanishing complex C^∞ function (or more generally, by an elliptic operator). It is not obvious at all that (2.5) and (2.6) possess such an invariance; they ought to, if they are to be of use to us. At any rate, we can state them in a more invariant (and also more precise) form.

DEFINITION 2.1. *We say that Property* (P) *holds at a point $(x_0, \xi^0) \in \dot{T}^*(\Omega)$ if there is an open neighborhood $\mathcal{O}$ of (x_0, ξ^0) in $\dot{T}^*(\Omega)$ such that, for any complex number z satisfying*

(2.7) $$d_\xi \operatorname{Re}(zP_m)(x, \xi) \neq 0 \quad \textit{for every } (x, \xi) \in \mathcal{O},$$

the following is true:

(2.8) *the restriction of* Im zP_m *to any null bicharacteristic strip of* Re zP_m *in* $\mathcal{O}$ *does not change sign.*

DEFINITION 2.2. *We say that Property* (Q) *holds at* (x_0, ξ^0) *if there is an open neighborhood* $\mathcal{O}$ *of* (x_0, ξ^0) *in* $\dot{T}^*(\Omega)$ *such that, given any complex number z satisfying* (2.7), *the following is true*:

(2.9) Im zP_m *does not vanish identically on any open (nonempty) arc of null bicharacteristic strip of* Re zP_m *contained in* $\mathcal{O}$.

Definitions 2.1 and 2.2 are acceptable from the viewpoint of "invariance" but lead to new difficulties, in so far as (2.8) and (2.9) must, in principle, be checked for all complex z, not an easy task. The next result is therefore useful.

THEOREM 2.1. *Let* $\mathcal{U}$ *be an open subset of* $\dot{T}^*(\Omega)$; *suppose that* d_ξ Re P_m *does not vanish at any point of* $\mathcal{U}$. *Let* $Q(x, \xi)$ *be a complex-valued* C^∞ *function in* $\mathcal{U}$ *such that neither Q nor* d_ξ Re QP_m *vanish at any point of* $\mathcal{U}$.

Suppose that the following property holds:

(2.10) *the restriction of* Im P_m *to any null bicharacteristic strip of* Re P_m *contained in* $\mathcal{U}$ *does not change sign.*

Then (2.10) *also holds if we substitute* QP_m *for* P_m.

There is an analogous statement if we replace (2.10) by the condition:

(2.11) Im P_m *does not vanish identically on any nonempty open arc of null bicharacteristic strip of* Re P_m *contained in* $\mathcal{U}$.

A consequence of Theorem 2.1 is that it suffices to check (2.8) in Definition 2.1 for only one z such that (2.7) holds: (2.8) is then automatically true for all other such z's. Similarly for (2.9). Theorem 2.1 has other applications, as we shall see soon.

We may now state, as conjectures, the generalizations of the facts established in §1 about $L = \partial/\partial t + ib(t)\,\partial/\partial x$. In every statement below, *the differential operator* P *is assumed to be of principal type*. If it were not so, these statements would not make (in general) any sense.

CONJECTURE 1. *The equation* $Pu = f$ *is locally solvable in* Ω *if and only if Property* (P) *holds at every point of* $\dot{T}^*(\Omega)$.

CONJECTURE 2. *The operator* P *is hypoelliptic in* Ω *if and only if both Properties* (P) *and* (Q) *hold at every point of* $\dot{T}^*(\Omega)$.

To which we shall add the

CONJECTURE 3. *If the coefficients of* P *are analytic, the operator* P *is analytic-hypoelliptic in* Ω *if and only if both Properties* (P) *and* (Q) *hold at every point of* $\dot{T}^*(\Omega)$.

Note that if the coefficients of P_m are analytic, we may restate as follows the

Conditions (P), and (P) & (Q) at the point $(x_0, \xi^0) \in \dot{T}^*(\Omega)$:

(P)′ *There is an open neighborhood* $\mathcal{O}$ *of* (x_0, ξ^0) *such that, for all complex* z *satisfying* (2.7), *the restriction of* $\operatorname{Im} zP_m$ *to any null bicharacteristic strip of* $\operatorname{Re} zP_m$ *does not have any zero of finite odd order.*

(P)′ & (Q)′ *There is an open neighborhood* $\mathcal{O}$ *of* (x_0, ξ^0) *such that, for all complex* z *satisfying* (2.7) *and for all* $(x_1, \xi^1) \in \mathcal{O}$ *such that* $P_m(x_1, \xi^1) = 0$, *the restriction of* $\operatorname{Im} zP_m$ *to the bicharacteristic strip of* $\operatorname{Re} zP_m$ *through* (x_1, ξ^1) *vanishes of finite even order* $2k_z(x_1, \xi^1)$ *at that point.*

Of course these statements make good sense also when the coefficients of P_m are not analytic but only C^∞. There is an invariance statement for (P)′ analogous to Theorem 2.1, and likewise for (P)′ & (Q)′. As a matter of fact, *the order of the zero of* $\operatorname{Im} zP_m$ *at* (x_1, ξ^1) *along the* (*null*) *bicharacteristic strip of* $\operatorname{Re} zP_m$ *through that point is independent of* z (provided that (2.7) is satisfied!). Heretofore we denote it by $\omega(x_1, \xi^1)$: of course this is an integral-valued function which is homogeneous of degree zero with respect to ξ^1. When ξ^1 ranges over the unit sphere of $\boldsymbol{R}_N$ this function reaches its maximum, which we shall denote by $\omega(x_1)$. It is also clear that x_1 has a neighborhood in which ω is $\leqq \omega(x_1)$, i.e., the function ω is *upper semi-continuous.* Thus it is bounded and reaches its maximum on any relatively compact subset S of Ω. We denote this maximum by $\omega(S)$ and refer to it as the *subellipticity degree of* $\boldsymbol{P}$ *in* S. In analogy with the phenomena encountered in §1 (and under the hypothesis that $\boldsymbol{P}$ be of principal type, with C^∞ coefficients), we can make the additional

CONJECTURE 4. *The differential operator* $\boldsymbol{P}$ *is subelliptic in* Ω *if and only if* (P)′ & (Q)′ *holds at every point of* $\dot{T}^*(\Omega)$.

I shall conclude this section by stating the results which have been obtained so far, concerning our four conjectures. It is convenient to run over our list of conjectures in reverse order, as Conjectures 4 and 3 are completely proved, whereas Conjecture 2 is almost completely and Conjecture 1 is only partly proved.

THEOREM 2.2. *Let* $\boldsymbol{P}$ *be a differential operator of principal type, of order* m, *with* C^∞ *coefficients in* Ω. *Suppose that* (P)′ & (Q)′ *holds at every point of* $\dot{T}^*(\Omega)$. *To every compact subset* K *of* Ω *there is a number* $\delta = \delta(K), 0 \leqq \delta(K) < 1$, *such that, to every real number* s, *there is a constant* $C = C(K, s) > 0$ *such that*

$$\text{(2.12)} \qquad \begin{gathered}\textit{for every function } u \in C_c^\infty(\Omega) \textit{ with support in } K,\\ \|u\|_{s+m-\delta} \leqq C(\|\boldsymbol{P}u\|_s + \|u\|_s).\end{gathered}$$

One may take

$$\text{(2.13)} \qquad \delta = \omega(K)/(\omega(K) + 1),$$

where $\omega(K)$ *is the subellipticity degree of* $\boldsymbol{P}$ *on* K.

Conversely suppose that (2.12) *holds for some* $\delta, 0 \leqq \delta < 1$, *and some* $C > 0$. *Let* Ω' *be the interior of* K. *Then* (P)′ & (Q)′ *holds at every point of* $\dot{T}^*(\Omega')$ *and we* *have moreover*

$$\omega(\Omega') \geqq \delta/(1 - \delta). \tag{2.14}$$

In Theorem 2.2 the subellipticity degrees are *even* integers.
We go now to Conjecture 3.

THEOREM 2.3. *Suppose that* P *is a differential operator of principal type, of order* m, *with analytic coefficients in* Ω. *The following properties are equivalent*:

(2.15) P *is hypoelliptic in* Ω;

(2.16) P *is analytic-hypoelliptic in* Ω;

(2.17) P *is subelliptic in* Ω;

(2.18) (P) & (Q) (*or, equivalently*, (P)′ & (Q)′) *holds at every point of* $\dot{T}^*(\Omega)$.

Next, about Conjecture 2.

THEOREM 2.4. *Let* P *be a differential operator of principal type, of order* m, *with* C^∞ *coefficients in* Ω.

If (P) & (Q) *holds at every point of* $\dot{T}^*(\Omega)$, P *is hypoelliptic in* Ω.

Conversely, if P *is hypoelliptic in* Ω, *Property* (Q) *must hold at every point of* $\dot{T}^*(\Omega)$ *and there is no point* (x_0, ξ^0) *in* $\dot{T}^*(\Omega)$ *such that*

$$P_m(x_0, \xi^0) = 0, \qquad \{d_\xi(\operatorname{Re} P_m)\}(x_0, \xi^0) \neq 0 \tag{2.19}$$

and such that the following is true:

(2.20) *the restriction of* Im P_m *to the bicharacteristic strip of* Re P_m *through* (x_0, ξ^0) *has an isolated zero at this point and changes sign there.*

To complete the proof of the necessity of Property (P) in order that the operator P be hypoelliptic, it remains to show that if this is the case, Im P_m cannot change sign, along any null bicharacteristic strip of Re P_m, at every point of a *perfect* set.

At last we come to Conjecture 1, where the situation is much less satisfactory, both in what regards the sufficiency and the necessity of Property (P) in order that the operator P be locally solvable. We shall describe separately the results about necessity and the ones about sufficiency.

THEOREM 2.5. *Let* P *be a differential operator of principal type, of order* m, *with* C^∞ *coefficients in* Ω.

Suppose that for some point $(x_0, \xi^0) \in \dot{T}^*(\Omega)$ *such that* (2.19) *holds, the following is true*:

(2.21) *the restriction of* Im P_m *to the bicharacteristic strip of* Re P_m *through* (x_0, ξ^0) *has a zero of finite odd order at that point.*

Then P *is not locally solvable at* x_0.

It will be noted that (2.21) is a stronger condition than (2.20). In some cases (such as in that of $\boldsymbol{L} = \partial/\partial t + ib(t)\,\partial/\partial x$) one may relax (2.21) and still have the same conclusion, but it has not yet been done in the general case, not even in the general case of a first-order operator in two independent variables $\partial/\partial t + ib(x, t)\,\partial/\partial x$ (b real, C^∞).

Concerning the sufficiency of (P) now:

THEOREM 2.6. *Let* $\boldsymbol{P}$ *be as in Theorem 2.5 and suppose that at least one of the following conditions is satisfied*:

(a) $n = 2$, *i.e., there are only two independent variables*;

(b) $m = 1$, *i.e., the operator* $\boldsymbol{P}$ *is of order one*;

(c) *the coefficients of the principal part* $P_m(x, D)$ *of* $\boldsymbol{P}$ *are analytic in* Ω.

Then, if Property (P) *holds at every point of* $\dot{T}^*(\Omega)$, *the differential operator* $\boldsymbol{P}$ *is locally solvable in* Ω.

Moreover the following is true, whatever the point x_0 *of* Ω:

(2.22) *to every* $\varepsilon > 0$, *there is an open neighborhood* $U = U(x_0, \varepsilon)$ *of* x_0 *in* Ω *such that, for all* $u \in C_c^\infty(U)$, $\|u\|_{m-1} \leqq \varepsilon\|\boldsymbol{P}u\|_0$;

(2.23) *to every real number* s, *there is an open neighborhood* $U = U_s(x_0)$ *of* x_0 *in* Ω *and a constant* $C = C_s(x_0) > 0$ *such that, for all* $u \in C_c^\infty(U)$, $\|u\|_{s+m-1} \leqq C\|\boldsymbol{P}u\|_s$;

(2.24) *to every real number* s *there is an open neighborhood* $U = U_s(x_0)$ *of* x_0 *in* Ω *such that, for every* $f \in H^s(\boldsymbol{R}^N)$ *there is* $u \in H^{s+m-1}(\boldsymbol{R}^N)$ *satisfying the equation* $\boldsymbol{P}u = f$ *in* U.

The following is a corollary of Theorems 2.5 and 2.6.

THEOREM 2.7. *Let* $\boldsymbol{P}$ *be a differential operator of principal type, of order* m, *with* C^∞ *coefficients in* Ω. *Suppose that the coefficients of the principal part* $P_m(x, D)$ *of* $\boldsymbol{P}$ *are analytic in* Ω.

Then $\boldsymbol{P}$ *is locally solvable in* Ω *if and only if Property* (P) *holds at every point of* $\dot{T}^*(\Omega)$.

By combining Theorems 2.4 and 2.5 we also obtain

THEOREM 2.8. *Assume the same hypotheses as in Theorem 2.7.*

Then $\boldsymbol{P}$ *is hypoelliptic in* Ω *if and only if both Properties* (P) *and* (Q) *hold at every point of* $\dot{T}^*(\Omega)$.

Theorem 2.2 is a particular case of general results of Yu. V. Egorov (see [**1**], [**2**]); a proof of it can be found in [**16**]. The proof of Theorem 2.3 is given in [**15**] and [**17**]. That of Theorem 2.4 can be found in [**18**]. Theorem 2.5 is proved in [**9**] (that of Theorem 2.4 is unpublished but resembles very much the proof of Theorem 2.5). The proof of Theorem 2.6 *case* (a) is given in [**13**], *case* (b) in [**8**] and [**14**], *case* (c) in [**10**] and [**11**]. The "invariance" result stated here as Theorem 2.1 is proved in the Appendix to the article [**10**].

3. **Integral representations of the solutions.** In our study of the operator (1.1) much information was extracted from the integral representation (1.7) of the solutions. Analogous representations in the general case of a *solvable* differential operator of principal type should, similarly, yield useful information. This is confirmed in a number of cases where such representations are known. But it is an open problem what these are in the general case (or even in not so general cases, like that of the solvable operators with analytic leading coefficients, Theorem 2.6, *case* (c)). Roughly speaking, the main cases where such representations have been obtained and have yielded useful information are: (1°) when $P_m(x, \xi)$ is real; (2°) when $P_m(x, \xi)$ is nonreal and $d_\xi(\operatorname{Re} P_m)$ and $d_\xi(\operatorname{Im} P_m)$ are linearly independent wherever $P_m(x, \xi)$ vanishes; (3°) under hypotheses (P) and (Q). For the first two cases we refer to [**4**]. I shall describe the method, as briefly as possible, in the third case.

The leading ideas are the same in all cases and go back to works of P. D. Lax (see [**7**]) and others in relation with the propagation of singularities. The difference is that here we have to deal with complex symbols, partaking little of the advantages of hyperbolicity (under the hypothesis (P) & (Q) the operators will be hypoelliptic!). In one important respect, however, they resemble hyperbolic symbols: their real characteristics are simple. We begin by fully exploiting this fortunate circumstance.

Let $(x_0, \xi^0) \in \dot{T}^*(\Omega)$ be such that $P_m(x_0, \xi^0) = 0$. We know that for some j, $1 \leqq j \leqq N$, $\{(\partial/\partial\xi_j)P_m\}(x_0, \xi^0) \neq 0$ and we have therefore the right to apply the implicit function theorem in the variables ξ (the symbol $P_m(x, \xi)$ can be extended as a holomorphic function of ξ in $\boldsymbol{C}_N$!) and we may write, in a suitable open neighborhood $\mathscr{V}$ of (x_0, ξ^0) in $\Omega \times \boldsymbol{C}_N$,

$$P_m(x, \xi) = Q(x, \xi)(\xi_j - \lambda(x, \xi')) \tag{3.1}$$

where $\xi' = (\xi_1, \ldots, \xi_{j-1}, \xi_{j+1}, \ldots, \xi_N)$ and where Q and λ are C^∞ functions of (x, ξ) in $\mathscr{V}$, holomorphic with respect to ξ. Because of the homogeneity of $P_m(x, \xi)$ with respect to ξ, we may assume that $\mathscr{V}$ is *conic*, i.e., $(x, \xi) \in \mathscr{V} \Rightarrow (x, \rho\xi) \in \mathscr{V}$ for all $\rho > 0$. As a matter of fact, we shall take $\mathscr{V}$ to be a product $V \times \Gamma^{\boldsymbol{C}}$ of an open neighborhood of x_0 in Ω with an open cone $\Gamma^{\boldsymbol{C}}$ in $\boldsymbol{C}_N$ (we shall denote by Γ the *real* part of $\Gamma^{\boldsymbol{C}}$: $\Gamma = \Gamma^{\boldsymbol{C}} \cap \boldsymbol{R}_N$). By the uniqueness part in the implicit functions theorem, we may take Q to be homogeneous of degree $m - 1$ and λ homogeneous of degree one with respect to ξ. Furthermore, we have $Q(x_0, \xi^0) \neq 0$ and we may obviously assume that Q does not vanish anywhere in $\mathscr{V}$, and even that, for some $c > 0$,

$$|Q(x, \xi)| \geqq c|\xi|^{m-1} \quad \textit{for every } (x, \xi) \in \mathscr{V}. \tag{3.2}$$

In different words, Q is an *elliptic* (in fact, a uniformly elliptic) symbol in the conic set $\mathscr{V}$.

Clearly we can cover $\{x_0\} \times (\boldsymbol{R}_N \setminus \{0\}) \subset \Omega \times \boldsymbol{C}_N$ with a finite number of sets $V_k \times \Gamma_k^{\boldsymbol{C}}$ $(k = 0, 1, \ldots, r)$, V_k a neighborhood of x_0, $\Gamma_k^{\boldsymbol{C}}$ an open cone in $\boldsymbol{C}_N$, so that the following is true: first of all, in $V_0 \times \Gamma_0^{\boldsymbol{C}}$, $P_m(x, \xi)$ is uniformly elliptic, i.e., there is $c > 0$ such that

$$(3.3) \qquad |P_m(x, \xi)| \geqq c|\xi|^m \quad \textit{for every } x \in V_0, \xi \in \Gamma_0^{\boldsymbol{C}}.$$

Secondly, for $k > 0$, we have in $V_k \times \Gamma_k^{\boldsymbol{C}}$ a factorization of the kind (3.1). It should however be kept in mind that the index j may vary with k.

Clearly we may replace every neighborhood V_k by the intersection of all the V_k's: this we denote by U. What we have thus is a covering of $\dot{T}^*(U)$ by the sets $U \times \Gamma_k^{\boldsymbol{C}} \subset U \times \boldsymbol{C}_N$ $(k = 0, 1, \ldots, r)$. Another useful remark is that we may shrink the cones $\Gamma_k^{\boldsymbol{C}}$ $(k > 0)$ provided that we increase correspondingly their number, r, and the size of $\Gamma_0^{\boldsymbol{C}}$. By "shrinking" a cone $\Gamma_k^{\boldsymbol{C}}$ we mean decreasing the diameter of its intersection with the unit sphere S_{2N-1}. We may perform this operation a finite number of times.

The next step is to introduce partitions of unity $\{g_k\}_{0 \leqq k \leqq r}$ subordinate to the covering $\{\Gamma_k\}$ of $\boldsymbol{R}_N \setminus \{0\}$. The functions g_k must be C^∞, nonnegative in $\boldsymbol{R}_N \setminus \{0\}$; their sum, equal to one in $\boldsymbol{R}_N \setminus \{0\}$; for each k, the support of g_k must be a (relatively) closed subset of Γ_k.[2]

Suppose then that, for each $k = 0, 1, \ldots, r$, we have succeeded in constructing a "nice" symbol $e_k(x, \xi)$ satisfying

$$(3.4) \qquad P(x, D_x + \xi)e_k(x, \xi) = 1, \qquad x \in U, \xi \in \Gamma_k.$$

Let us define an operator E_k, acting on functions $u \in C_c^\infty$, by the formula

$$(3.5) \qquad E_k u(x) = (2\pi)^{-n} \int e^{ix\cdot\xi} e_k(x, \xi) g_k(\xi) \hat{u}(\xi)\, d\xi.$$

In view of (3.4) we have, for $x \in U$,

$$(3.6) \qquad P(x, D_x)E_k u(x) = (2\pi)^{-n} \int e^{ix\cdot\xi} g_k(\xi) \hat{u}(\xi)\, d\xi.$$

By adding the equations (3.6) when $k = 0, 1, \ldots, r$, and setting $E = E_0 + \cdots + E_r$, we obtain at once

$$(3.7) \qquad P(x, D_x)Eu = u \quad \textit{in } U.$$

We have therefore found a local *fundamental kernel* of $P(x, D)$ (in the neighborhood U of x_0).

Unfortunately it is not often that one can solve exactly the equation (3.4). In

[2] In many situations one likes to take each function g_k to be homogeneous of degree zero with respect to ξ. However this may not be possible for certain purposes, such as studying the structure of the singular supports of the solutions: more sophisticated partitions of unity would have to be used. Suitable bounds on the derivatives of the g_k must then be imposed.

most instances we must content ourselves with an approximate solution. We will have

$$P(x, D_x + \xi)e_k(x, \xi) - 1 = \rho_k(x, \xi), \qquad x \in U, \xi \in \Gamma_k, \tag{3.8}$$

where ρ_k will be such that $\rho_k(x, \xi)g_k(\xi)$ is the symbol of a smoothing operator, R_k. As a consequence, equation (3.4) will have to be replaced by

$$P(x, D)E_k u = g_k(D)u + R_k u \tag{3.9}$$

and equation (3.7) by

$$P(x, D)Eu = u + Ru \quad \text{in } U \ (R = R_0 + \cdots + R_r). \tag{3.10}$$

What we have now is not any more a fundamental kernel but what is called a *parametrix*. Although obviously less satisfactory than fundamental kernels, parametrices can be made to yield much information of interest: they can be used to prove "a priori estimates" and thereby existence and regularity theorems; the study of their symbols can be of help in locating with great precision the singularities of the solutions.

The solution of (3.8) in the elliptic region, i.e., when $k = 0$, is standard.[3] We write

$$P(x, D_x + \xi) = P_m(x, \xi) - \sum_{j=1}^{m} P_{m-j}(x, \xi, D_x);$$

$P_{m-j}(x, \xi, D_x)$ is a differential operator of order $\leqq j$ whose coefficients are C^∞ functions of x in Ω and homogeneous polynomials of degree $m - j$ in ξ. We define recursively a sequence of symbols $e_{0,\nu}(x, \xi)$ by the formulas

$$e_{0,0}(x, \xi) = P_m(x, \xi)^{-1}, \tag{3.11}$$

$$\text{if } \nu > 0, \qquad e_{0,\nu}(x, \xi) = P_m(x, \xi)^{-1} \sum_{j=1}^{m} P_{m-j}(x, \xi, D_x)e_{0,\nu-j}(x, \xi). \tag{3.12}$$

For each ν, $e_{0,\nu}$ is homogeneous of degree $-\nu - m$ with respect to ξ. Then let $\zeta(t)$ be a C^∞ function on the real line, $\zeta(t) = 0$ for $t < \frac{1}{2}$ and $\zeta(t) = 1$ for $t > 1$. One can select a sequence of numbers $\rho_\nu \uparrow +\infty$ such that

$$e_0(x, \xi) = \sum \zeta(\rho_\nu^{-1}|\xi|)e_{0,\nu}(x, \xi)$$

has the desired properties: $e_0(x, \xi)g_0(\xi) \in S^{-m}(U \times \boldsymbol{R}_N)$, i.e., is the symbol of a pseudo-differential operator of order $-m$ in U; also,

$$\{P(x, D_x + \xi)e_0(x, \xi) - 1\}g_0(\xi) \in S^{-\infty}(U \times \boldsymbol{R}_N).$$

[3] We recall it here, only in order to make this section more or less self-contained.

Needless to say, the success of the recursive definitions (3.11), (3.12) is due to the validity of (3.3).

From now on we restrict ourselves to the situations where (3.3) is not valid, that is to say, to the cases $k > 0$, where we have a decomposition of the kind (3.1). For simplicity we shall drop all subscripts k. It is also convenient to rename the coordinates. First of all, we may assume that the index j in (3.1) is equal to N. Then, writing $n = N - 1$, we may reserve the notation x for $(x^1, \dots, x^n)$ and ξ for $(\xi_1, \dots, \xi_n)$ (the latter was denoted by ξ' in (3.1)); we may write t instead of x^N, τ instead of ξ_N. It further simplifies matters if we take x_0 (in the old notation) to be the origin in $\boldsymbol{R}^{n+1}$. The factorization (3.1) now reads

$$P_m(x, t, \xi, \tau) = Q(x, t, \xi, \tau)(\tau - \lambda(x, t, \xi)) \tag{3.13}$$

where $(x, t) \in V$, $(\xi, \tau) \in \Gamma^{\boldsymbol{C}}$. The point earlier denoted by (x_0, ξ^0) is now $(0, 0, \xi^0, \tau^0)$ and we have

$$\tau^0 = \lambda(0, 0, \xi^0). \tag{3.14}$$

We shall let x vary in an open neighborhood V_0 of the origin in $\boldsymbol{R}^n$, t in an open interval $)-T, T($, $T > 0$; both V_0 and T will have to be chosen suitably small (thus $V_0 \times)-T, T(\subset V$).

At this point we wish to use our hypotheses (P) & (Q). We shall also use their invariance under multiplication by elliptic symbols (Theorem 2.1 and following remarks). In view of this, Properties (P) & (Q) are valid for

$$Q(x, t, \xi, \tau)^{-1} P_m(x, t, \xi, \tau) \quad \textit{in } V \times \Gamma.$$

We reach the conclusion that $\tau - \lambda(x, t, \xi)$ satisfies (P) & (Q). Observe that $\operatorname{Im} \lambda$ is independent of τ and that, given any $(x, t) \in V$ and any ξ sufficiently near ξ^0, we may find $\tau = \operatorname{Re} \lambda(x, t, \xi)$ such that $(\xi, \tau) \in \Gamma$. The upshot of this is that τ, in a sense, drops out of the picture. Let us denote by Γ' an open cone in $\boldsymbol{R}_N$, with vertex at the origin, with axis passing through ξ^0—and sufficiently "narrow." The conjunction of (P) & (Q) yields:

(3.15) $\operatorname{Im} \lambda$ *does not change sign and does not vanish identically on any nonempty open interval of any integral curve, contained in* $V \times \Gamma'$, *of the vector field*

$$\partial/\partial t - H_{\operatorname{Re} \lambda}. \tag{3.16}$$

(We have denoted by $H_{\operatorname{Re}\lambda}$ the Hamiltonian of $\operatorname{Re} \lambda$.) Clearly, if V is sufficiently small and Γ' sufficiently narrow, the integral curves of (3.16) form a nice fibration of $V \times \Gamma'$. If we suppose both V and Γ' connected, we see that (3.15) implies the following:

(3.17) $\operatorname{Im} \lambda$ *keeps the same sign throughout* $V \times \Gamma'$.

Of course (3.17) is much weaker than (3.15). But, as we shall see, it suffices for part of our purposes: it is enough to enable us to construct the sought parametrix.

It is worth pointing out that, as (3.15) is a translation of an original property of P_m (namely (P) & (Q)), so is (3.17) the translation of a certain property of P_m, which is not difficult to find out. It is the following one:

(R) *Let $\mathcal{O}$ be an open subset of $\dot{T}^*(\Omega)$, z a complex number such that $d_\xi \operatorname{Re} zP_m$ does not vanish at any point of $\mathcal{O}$. Then $\operatorname{Im} zP_m$ keeps the same sign in each connected component of the variety of zeros of $\operatorname{Re} zP_m$ in $\mathcal{O}$ (this variety is a smooth manifold of codimension one, of course).*

In the case of example (1.1), (R) is but a restatement of (P). Let us take another look at this example, in particular at the integral representation (1.7). In our present notation, we may say that we have subdivided the complement of the origin in the (ξ, τ)-plane into three "cones" (or sectors): the half-plane $\xi > 0$, the half-plane $\xi < 0$ and a third cone, defined by $|\tau| > \gamma|\xi|$, $\gamma > 0$. The latter corresponds to what we have called the elliptic region and will not interest us here. Let us consider the contribution to the integral representation (1.7) coming from the half-plane $\xi < 0$. It can be rewritten (with the notation of (1.7)):

$$E_- f(x, t) = (2\pi)^{-2} \iint\limits_{\xi < 0} e^{i(x\xi + t\tau)} e_-(x, t, \xi, \tau) \hat{f}(\xi, \tau)\, d\xi\, d\tau,$$

where $\hat{f}$ stands now for the total Fourier transform, with respect to both x and t, and where

$$e_-(x, t, \xi, \tau) = \int_{-T}^{t} e^{B(t,t')\xi - i(t-t')\tau}\, dt'. \tag{3.18}$$

We remind the reader that formula (1.7) was devised under the assumption that $b(t)$ is nonnegative. On the other hand the symbol of $\boldsymbol{L}$ (defined by (1.1)) is $i(\tau + ib(t)\xi)$. In our present notation, this amounts to writing $\lambda = -ib(t)\xi$. In other words, we are using the representation (3.18) in the region where $\operatorname{Im} \lambda \geqq 0$. Going back to the general case, where (3.17) is assumed to hold, we shall in fact suppose that

$$\operatorname{Im} \lambda \geqq 0 \textit{ throughout } V \times \Gamma'. \tag{3.19}$$

And in analogy to (3.18) we set

$$e(x, t, \xi, \tau) = \int_{-T}^{t} e^{i\varphi(x,t,t',\xi) - i(t-t')\tau} a(x, t, t', \xi, \tau)\, dt'. \tag{3.20}$$

In the new notation, equation (3.8) reads

$$P(x, t, D_x + \xi, D_t + \tau) e(x, t, \xi, \tau) = 1 + \rho(x, t, \xi, \tau). \tag{3.21}$$

We begin by computing the left-hand side of (3.21).

$$P(x, t, D_x + \xi, D_t + \tau)e(x, t, \xi, \tau)$$

$$(3.22) \qquad = \int_{-T}^{t} e^{-i(t-t')\tau} P(x, t, D_x + \xi, D_t)(e^{i\varphi}a)\,dt' - (-1)^{1/2} \sum_{j=1}^{m} \frac{1}{j!} P^{(j)}(x, t, D_x + \xi, D_t)(D_{t'} + \tau)^{j-1}(e^{i\varphi}a)\Big|_{t=t'},$$

where $P^{(j)}(x, t, \xi, \tau) = (\partial/\partial\tau)^j P(x, t, \xi, \tau)$. We shall try to determine the *phase function* φ and the *amplitude function* a in such a way as to have the integral, on the right-hand side of (3.22), equal to zero and the integrated term equal to one. This will follow if we solve the equations

$$(3.23) \qquad P(x, t, D_x + \xi + \varphi_x, D_{t'} + \varphi_t)a = 0,$$

$$(3.24) \qquad \varphi = 0 \quad \textit{when } t = t',$$

$$(3.25) \qquad \sum_{j=1}^{m} \frac{1}{j!} P^{(j)}(x, t, D_x + \xi + \varphi_x, D_t + \varphi_t)(D_{t'} + \varphi_{t'} + \tau)^{j-1}a\Big|_{t=t'} = (-1)^{1/2}.$$

If these equations could be solved exactly, we would have (3.21) without the "error" ρ. In general they have no exact solutions and this is why there is the error ρ. Nevertheless it pays to proceed formally for a while, just as if we were going to construct exact solutions of (3.23)—(3.24).

We take φ to be homogeneous of degree one with respect to ξ, a to be asymptotically equal to a series $\sum_{\nu=0}^{+\infty} a_\nu$ where, for each ν, a_ν is homogeneous of degree $-(m-1)-\nu$ with respect to (ξ, τ). Observe that we have

$$(3.26) \qquad \begin{aligned} P(x, t, D_x + \xi + \varphi_x, D_t + \varphi_t) = {} & P_m(x, t, \xi + \varphi_x, \varphi_t) \\ & + \{(\partial/\partial\tau)P_m(x, t, \xi + \varphi_x, \varphi_t)\} \cdot D_t \\ & + \{\operatorname{grad}_\xi P_m(x, t, \xi + \varphi_x, \varphi_t)\} \cdot D_x \\ & + \tilde{C}(x, t, t', \xi) + \sum_{k=2}^{m} \tilde{Q}_k(x, t, t', \xi, D_x, D_t), \end{aligned}$$

where $\tilde{C}(x, t, t', \xi)$ is homogeneous of degree $m - 1$ with respect to ξ whereas, for each $k = 2, \dots, m$, $\tilde{Q}_k$ is a differential operator of order $\leq k$ whose coefficients are homogeneous of degree $m - k$ with respect to ξ (of course, all the coefficients are smooth with respect to x, t, t'). In view of (3.26) the term of highest degree with respect to (ξ, τ) on the left-hand side of (3.23) is $P_m(x, t, \xi + \varphi_x, \varphi_t)a_0(x, t, t', \xi, \tau)$, which leads us to require

$$(3.27) \qquad P_m(x, t, \xi + \varphi_x, \varphi_t) = 0.$$

Equation (3.27) is a consequence of the next one:

$$(3.28) \qquad \varphi_t = \lambda(x, t, \xi + \varphi_x).$$

Observe at this point that, because of (3.24), φ_x is close to the origin in $\boldsymbol{C}^n$; therefore, if ξ is close to ξ^0, so is $\xi + \varphi_x$ and $\varphi_t = \lambda(x, t, \xi + \varphi_x)$ must be close to $\lambda(x, t, \xi^0) = \tau^0$. In summary, $(\varphi_x + \xi, \varphi_t)$ may be supposed to belong to the complex cone $\Gamma^{\boldsymbol{C}}$—which is helpful if (3.28) is to make sense and to imply (3.27) via the factorization (3.13).

Incidentally, let us note that we cannot have $\xi^0 = 0$. For this would imply $\tau^0 = 0$. We may as well assume that, for some $\gamma > 0$,

$$|\tau| < \gamma|\xi| \quad \text{in } \Gamma^{\boldsymbol{C}}. \tag{3.29}$$

In particular, this implies that functions which are independent of τ but are homogeneous with respect to ξ can be viewed as homogeneous functions of (ξ, τ), smooth with respect to the latter variable, in Γ.

Thus the first term on the right-hand side of (3.26) vanishes identically. Let us look at the next terms. In view of (3.28), we have

$$\begin{aligned}\{(\partial/\partial\tau)P_m(x, t, \xi + \varphi_x, \varphi_t)\}\cdot D_t + \{\operatorname{grad}_\xi P_m(x, t, \xi + \varphi_x, \varphi_t)\}\cdot D_x \\ = Q(x, t, \xi + \varphi_x, \varphi_t)\{D_t - \lambda_\xi(x, t, \xi + \varphi_x)\cdot D_x\}.\end{aligned} \tag{3.30}$$

This induces us to introduce the following notation:

$$\mathscr{L} = D_t - \sum_{j=1}^{n} \frac{\partial\lambda}{\partial\xi_j}(x, t, \xi + \varphi_x)D_{x^j} + C(x, t, t', \xi), \tag{3.31}$$

where

$$C(x, t, t', \xi) = \tilde{C}(x, t, t', \xi)Q(x, t, \xi + \varphi_x, \varphi_t)^{-1}.$$

Also

$$Q_k(x, t, t', \xi, D_x, D_t) = Q(x, t, \xi + \varphi_x, \varphi_t)^{-1}\tilde{Q}_k(x, t, t', \xi, D_x, D_t). \tag{3.32}$$

Note that the coefficients of $\mathscr{L}$ are homogeneous of degree *zero* while those of Q_k are homogeneous of degree $1 - k$ with respect to ξ $(2 \leqq k \leqq m)$. Taking all this into account and trying to equate the terms with the same homogeneity with respect to ξ in both members of (3.23) leads to the sequence of equations

$$\mathscr{L}a_\nu + \sum_{k=2}^{m} Q_k a_{\nu-k+1} = 0, \tag{3.33$_\nu$}$$

with the tacit agreement that $a_\nu \equiv 0$ if $\nu < 0$. Equation (3.33)$_\nu$ must be viewed as a (first-order linear partial differential) equation in the unknown a_ν depending on the functions $a_{\nu-k+1}$ determined earlier. We must adjoin to it an initial condition derived from (3.25). We observe that the term of highest degree with respect to (ξ, τ) in

$$\sum_{j=1}^{m} \frac{1}{j!} P^{(j)}(x, t, D_x + \xi + \varphi_x, D_t + \varphi_t)(D_{t'} + \varphi_{t'} + \tau)^{j-1}$$

is equal to

$$(\tau + \varphi_{t'})^{-1}\{P_m(x, t, \xi + \varphi_x, \tau + \varphi_t + \varphi_{t'}) - P_m(x, t, \xi + \varphi_x, \varphi_t)\}$$
$$= Q(x, t, \xi + \varphi_x, \tau + \varphi_t + \varphi_{t'}).$$

But when $t = t'$ we have $\varphi_x = 0, \varphi_t + \varphi_{t'} = 0$. Consequently, condition (3.25) can be rewritten as

$$\left\{Q(x, t, \xi, \tau) + \sum_{k=1}^{m-1} \tilde{R}_k(x, t, \xi, \tau, D_x, D_t, D_{t'})\right\}a\bigg|_{t=t'} = (-1)^{1/2},$$

where R_k is a differential operator of order $\leqq k$ whose coefficients are homogeneous of degree $m - 1 - k$ with respect to (ξ, τ). Setting

$$R_k(x, t, \xi, \tau, D_x, D_t, D_{t'}) = Q(x, t, \xi, \tau)^{-1}\tilde{R}_k(x, t, \xi, \tau, D_x, D_t, D_{t'}),$$

(3.25) is seen to be equivalent with

$$(3.34)_0 \qquad a_0|_{t=t'} = (-1)^{1/2}Q(x, t, \xi, \tau)^{-1},$$

$$(3.34)_\nu \qquad a_\nu + \sum_{k=1}^{m-1} R_k(x, t, \xi, \tau, D_x, D_t, D_{t'})a_{\nu-k}\bigg|_{t=t'} = 0 \qquad (\nu > 0).$$

In summary, we have a set of equations which should, in principle, enable us to determine the phase and the amplitude functions: these are equations (3.24) and (3.28) for the phase, equations $(3.33)_\nu$ and $(3.34)_\nu$ ($\nu = 0, 1, \ldots$) for the amplitude. The question now arises whether we can solve this system of equations and if not (which is the general case), how to find approximate solutions in a way that the construction of our local parametrix can still be carried out.

4. **Extending the methods of geometrical optics.** As we have already said, the trouble with the equations defining the phase and amplitude functions in the symbol of the parametrix E is that, in general, they have no solution. Consider for instance the equations defining the phase

$$(4.1) \qquad \varphi_t = \lambda(x, t, \xi + \varphi_x),$$

$$(4.2) \qquad \varphi|_{t=t'} = 0.$$

It is a first-order, in general nonlinear, partial differential equation. In general the Cauchy problem (4.1)–(4.2) has no solution. This is has nothing to do with the nonlinearity of λ: it might very well happen even when λ is linear. It has to do with the fact that λ is not real, and is not analytic. Notice indeed that when λ is either analytic or real, (4.1)–(4.2) has a unique solution (which is either analytic or C^∞ and real). However, if we go back to equation (3.26) and the reason for seeking to satisfy (4.1) and (4.2), we discover that we do not really need an exact solution of (4.1) (although (4.2) must hold). If we are willing to tolerate a remainder ρ in equation (3.21), with the proviso that it be the symbol of an operator which is regularizing of degree M (i.e., transforms compactly supported distributions of order M into C^M functions) for a preassigned but arbitrarily large integer M, then what we need is that $e^{i\varphi}P_m(x, t, \xi + \varphi_x, \varphi_t)a$ be such a symbol.[4]

For this it suffices to make sure that, given any integer $M > 0$, we can find a function φ so as to have, for every $(n + 2)$-tuple (α, l, l') such that $|\alpha| + l + l' \leqq M$,

$$|D_x^\alpha D_t^l D_{t'}^{l'}\{[\varphi_t - \lambda(x, t, \xi + \varphi_x)]e^{i\varphi}\}| \leqq C(\alpha, l, l')|(\xi, \tau)|^{-M}. \tag{4.3}$$

Recalling that φ is homogeneous of degree one with respect to ξ and that (ξ, τ) satisfies (3.29), we see that (4.3) is a consequence of the following inequality, valid for the same $(n + 2)$-tuples:

$$|D_x^\alpha D_t^l D_{t'}^{l'}[\varphi_t - \lambda(x, t, \xi + \varphi_x)]|e^{-\operatorname{Im}\varphi} \leqq C_1|\xi|^{-2M}. \tag{4.4}$$

In turn (4.4) will require

$$\operatorname{Im}\varphi \geqq 0 \quad in\ V \times \Gamma, \tag{4.5}$$

and will then be a consequence of

$$|D_x^\alpha D_t^l D_{t'}^{l'}[\varphi_t - \lambda(x, t, \xi + \varphi_x)]| \leqq C_2(\operatorname{Im}\varphi)^{2M+1}|\xi|^{-2M} \tag{4.6}$$

(observe that the two members in (4.6) have the same homogeneity degree with respect to ξ).

A similar reasoning shows that we do not need an exact solution of $(3.33)_\nu$. It will suffice to have, for all (α, l, l') such that $|\alpha| + l + l' \leqq M$,

$$|D_x^\alpha D_t^l D_{t'}^{l'}(\mathscr{L}a_\nu - b_\nu)|e^{-\operatorname{Im}\varphi} \leqq C_3|(\xi, \tau)|^{-2M}, \tag{4.7}$$

where

$$b_\nu = -\sum_{k=2}^{m} Q_k a_{\nu-k+1}. \tag{4.8}$$

It ought to be noted that, in view of (4.5), condition (4.7) will be satisfied regardless of our choice of the amplitude terms a_ν, as soon as ν is large enough (how large is decided by the choice of the integer M). We may for instance take $a_\nu \equiv 0$ for large ν. This introduces an additional error by way of the "initial" conditions $(3.34)_\nu$; but it is very easy to check that the homogeneity degree with respect to (ξ, τ) of this error, and of its derivatives of order $\leqq M$ with respect to x, t, will not exceed $-M$. For the lower values of the index ν, we shall continue to require that the condition $(3.34)_\nu$ be satisfied. Thus it is only with a finite number of inequalities (4.7), and a corresponding number of Cauchy conditions $(3.34)_\nu$, that we have to contend.

[4] We are not using the word "symbol" in the customary restricted sense.

Observe that the homogeneity degree with respect to (ξ, τ) of the factor of $e^{-\operatorname{Im}\varphi}$ on the left-hand side of (4.7) is $-\nu - m + 1$. Consequently (4.7) will be a consequence of

$$|D_x^\alpha D_t^l D_{t'}^{l'}(\mathscr{L}a_\nu - b_\nu)| \leqq C_4(\operatorname{Im}\varphi)^{2M-\nu-m+1}|\xi|^{-2M} \tag{4.9}$$

(keeping in mind that (3.29) holds).

In solving (4.6) and (4.9) we may restrict ourselves to the case $|\xi| = 1$. In a sense this is equivalent to solving (4.1) and $(3.33)_\nu$ (under the Cauchy conditions (4.2) and $(3.34)_\nu$) with a certain degree of approximation, degree measured by the powers of $\operatorname{Im}\varphi$ on the right-hand sides of (4.6) and (4.9).

In order to solve (4.6) and (4.9) we shall apply a general result on approximate solutions to Cauchy problems (for a proof, see [**19**]; also [**18**, Appendices])

$$u_t = f(x, t, u, u_x), \qquad x \in U, |t| < \delta, \tag{4.10}$$

$$u(x, 0) = u_0(x), \qquad x \in U. \tag{4.11}$$

Here the data f and u_0, as well as the unknown u, are valued in a real Euclidean space $\boldsymbol{R}^\mu$. For instance, in the particular case of (4.1)–(4.2) we would take $\mu = 2$, $f(x, t, u, p)$ would be the two-vector

$$(\operatorname{Re}\lambda(x, t, \xi + p_1 + ip_2), \operatorname{Im}\lambda(x, t, \xi + p_1 + ip_2)).$$

In general we make the assumption that

(4.12) *$f(x, t, u, p)$ is a C^∞ function of (x, t, u, p) (in an appropriate open subset of $\boldsymbol{R}^{n+1} \times \boldsymbol{R}^{\mu(n+1)}$).*

With the Cauchy problem (4.10)–(4.11) we associate another Cauchy problem but this time for a system of *ordinary* differential equations

$$v_t = f(x, t, v, q), \qquad q_t = f_x(x, t, v, q) + f_u(x, t, v, q)q, \tag{4.13}$$

$$v|_{t=0} = u_0(x), \qquad q|_{t=0} = u_{0x}(x). \tag{4.14}$$

Of course f_x, f_u, u_{0x}, q, etc. are tensor-valued functions; we have multiplied them according to the standard rules of index-contraction.

By the fundamental theorem on ordinary differential equations, the problem (4.13)–(4.14) has a unique solution $(v(x, t), q(x, t))$ which is a C^∞ function of (x, t) in a neighborhood of $\bar{U} \times (-\delta, \delta)$ if U and $\delta > 0$ are suitably chosen. Let us set then

$$F(x, t) = \left| \int_0^t |f_p(x, s, v(x, s), q(x, s))| \, ds \right|. \tag{4.15}$$

The result on approximate solutions to Cauchy problems of the kind (4.10)–(4.11), which we want to use, is the following one:

LEMMA 4.1. *Under the preceding hypothesis, if the open set $U \subset \mathbf{R}^n$ is suitably chosen, to every integer $J > 0$ there is a number $\delta > 0$ such that the following is true:*

There is a C^∞ function of (x, t) in $U \times]-\delta, \delta[$, $u(x, t)$, and positive constants $C_J, C_{J,\alpha,l}$ ($\alpha \in \mathbf{N}^n$, $|\alpha| + l \leqq J$) such that, whatever $x \in U$ and $|t| < \delta$,

$$|D_x^\alpha D_t^l\{u_t - f(x, t, u, u_x)\}| \leqq C_{J,\alpha,l} F^{J+1-|\alpha|-l}, \tag{4.16}$$

$$|u - v| \leqq C_J F^2. \tag{4.17}$$

Let us show how this lemma is applied to our Cauchy problems and, first of all, to (4.1)–(4.2). Observe that the vector ξ is *fixed* (arbitrarily, of course, in a certain open cone of $\mathbf{R}_n$). Depending on ξ, we may perform a change of the variables x, t about the origin in $\mathbf{R}^{n+1}$ which brings us to the situation where

$$\operatorname{Re} \lambda(x, t, \xi) \equiv 0, \qquad d(\operatorname{Re} \lambda)(x, t, \xi) \equiv 0 \tag{4.18}$$

(there is also the variable t' which must be taken into account; here, for the sake of simplicity, we shall assume that $t' = 0$). We may assume, on the other hand, that the data and the solutions are complex-valued rather than valued in $\mathbf{R}^2$. The equations (4.13) now become

$$v_t = \lambda(x, t, \xi + q), \qquad q_t = \lambda_x(x, t, \xi + q), \tag{4.19}$$

and the initial values are equal to zero. We have at once

$$v(x, t) = \int_0^t \lambda(x, s, \xi + q)\, ds. \tag{4.20}$$

As for $q(x, t)$, Picard's iteration method in solving ordinary differential equations shows at once that we have, for a suitable constant $C > 0$,

$$|q(x, t)| \leqq C\left|\int_0^t |\lambda_x(x, s, \xi)|\, ds\right|. \tag{4.21}$$

By combining (4.20) and (4.21) we obtain

$$\left|v(x, t) - \int_0^t \lambda(x, s, \xi)\, ds\right| \leqq C'\left|\int_0^t |\lambda_x(x, s, \xi)\, ds \int_0^t |\lambda_\xi(x, s, \xi)\, ds\right|. \tag{4.22}$$

As for the function F, in the present context it is equal to

$$\begin{aligned} F(x, t) &= \left|\int_0^t |\lambda_\xi(x, s, \xi + q(x, s))|\, ds\right| \\ &\leqq \left|\int_0^t |\lambda_\xi(x, s, \xi)|\, ds + C'' \int_0^t |q(x, s)|\, ds\right| \end{aligned}$$

whence, by (4.21),

$$F(x, t) \leqq C^{(3)}\left|\int_0^t |d_{x,\xi}\lambda(x, s, \xi)|\, ds\right|. \tag{4.23}$$

By virtue of (4.18), we may replace λ by $\operatorname{Im}\lambda$ everywhere in (4.22) and (4.23). At this point we use our hypothesis (3.19): $\operatorname{Im}\lambda$ remains nonnegative in $V \times \Gamma'$. From this one derives easily that, in the domain under consideration,

$$|d_{x,\xi}\operatorname{Im}\lambda| \leqq C^{(4)}(\operatorname{Im}\lambda)^{1/2} \tag{4.24}$$

(for a proof of this assertion, see [**8**, Lemma 1.1].) If we avail ourselves of (4.24) in the inequalities (4.22) and (4.23), we obtain

$$\left|v(x, t) - i\int_0^t \operatorname{Im}\lambda(x, s, \xi)\, ds\right| \leqq C^{(5)}t\int_0^t \operatorname{Im}\lambda(x, s, \xi)\, ds, \tag{4.25}$$

$$F(x, t) \leqq C^{(5)}|t|^{1/2}\left|\int_0^t \operatorname{Im}\lambda(x, s, \xi)\, ds\right|. \tag{4.26}$$

At last we combine these facts with the conclusion in Lemma 4.1. We must revert to the original coordinates x, t and take into account the fact that the Cauchy condition is set at $t = t'$, not at $t = 0$ (see (4.2)). Thus we obtain, for $x \in U$ and $t, t' \in)-\delta, \delta($, if U is a sufficiently small open neighborhood of the origin in $\boldsymbol{R}^n$ and δ a sufficiently small number > 0,

$$|\operatorname{Im}\varphi - B| \leqq \tfrac{1}{2}B, \tag{4.27}$$

$$|D_x^\alpha D_t^l\{\varphi_t - \lambda(x, t, \xi + \varphi_x)\}| \leqq C_{J,\alpha}[(t - t')B]^{(J+1-|\alpha|-l)/2}, \qquad |\alpha| + l \leqq J, \tag{4.28}$$

where $B = B(x, t, t', \xi)$ denotes the integral of $\operatorname{Im}\lambda$ on the following arc of curve $\mathscr{C}$: $\mathscr{C}$ is the projection on $V \times \Gamma'$ of the segment of null bicharacteristic strip of $\tau - \operatorname{Re}\lambda$ joining the point (y, t', ξ, τ') to the point (x, t, η, τ). It should be noted that this determines completely τ and τ': we must have

$$\tau' = \operatorname{Re}\lambda(y, t', \xi), \qquad \tau = \operatorname{Re}\lambda(x, t, \eta).$$

It is less obvious that it also determines uniquely y and η: on this we refer to Professor Hörmander's lectures on canonical relations (see also [**18**, §2]).

Clearly (4.27) and (4.28) imply (4.6), at least when $l' = 0$. In order to also cover the case $l' \neq 0$ one needs a straightforward extension of Lemma 4.1 in which a parameter such as t' is introduced (on this subject, see the Appendices of [**18**]).

The solution of (4.9) proceeds along similar lines: it is made possible by the relationship between the function $\lambda(x, t, \xi)$ and the first-order differential operator $\mathscr{L}$ (see (3.31)). In the present case, the function $F(x, t)$ defined by (4.15) is equal to

$$\left|\int_0^t \lambda_\xi(x, s, \xi)q(x, s)\, ds\right|,$$

and q is a solution of the equation

$$q_t = \lambda_x(x, t, \xi + q) + Cq.$$

It is these facts that enable us to apply Lemma 4.1 and derive from it the solution of (4.9). For the details we refer to [**18**, §3].

References

1. Ju. V. Egorov, *On subelliptic pseudodifferential operators*, Dokl. Akad. Nauk SSSR **188** (1969), 20–22 = Soviet Math. Dokl. **10** (1969), 1056–1059. MR **41** # 630.

2. ———, *Nondegenerate subelliptic pseudodifferential operators*, Math. Sb. **82 (124)** (1970), 323–342 = Math. USSR Sb. **11** (1970), 291–310. MR **42** # 3406.

3. L. Hörmander, *Linear partial differential operators*, 3rd rev. ed., Die Grundlehren der math. Wissenschaften, Band 116, Springer-Verlag, New York, 1969. MR **40** # 1687.

4. ———, *On the singularities of solutions of partial differential equations*, Comm. Pure Appl. Math. **23** (1970), 329–358. MR **41** # 7251.

5. ———, *Fourier integral operators*, Acta Math. **127** (1971), 79–183.

6. ———, *On the existence and the regularity of solutions of linear partial differential equations* (to appear).

7. P. D. Lax, *Asymptotic solutions of oscillatory initial value problems*, Duke Math. J. **24** (1957), 627–646. MR **20** # 4096.

8. L. Nirenberg and F. Treves, *Solvability of a first order linear partial differential equation*, Comm. Pure Appl. Math. **16** (1963), 331–351. MR **29** # 348.

9. ———, *On local solvability of linear partial differential equations.* I: *Necessary conditions*, Comm. Pure Appl. Math. **23** (1970), 1–38. MR **41** # 9064a.

10. ———, *On local solvability of linear partial differential equations.* II: *Sufficient conditions*, Comm. Pure Appl. Math. **23** (1970), 459–509. MR **41** # 9064b.

11. ———, *A correction to "On local solvability of linear partial differential equations.* II: *Sufficient conditions"*, Comm. Pure Appl. Math. **24** (1971), 279–288.

12. F. Treves, *Topological vector spaces, distributions and kernels*, Academic Press, New York, 1967. MR **37** # 726.

13. ———, *On the local solvability of linear partial differential equations in two independent variables*, Amer. J. Math. **92** (1970), 174–204. MR **41** # 3974.

14. ———, *Local solvability in L^2 of first order linear PDE's*, Amer. J. Math. **92** (1970), 369–380. MR **41** # 8805.

15. ———, *Hypoelliptic partial differential equations of principal type with analytic coefficients*, Comm. Pure Appl. Math. **23** (1970), 637–651. MR **42** # 6420.

16. ———, *A new method of proof of the subelliptic estimates*, Comm. Pure Appl. Math. **24** (1971), 71–115.

17. ———, *Analytic hypoelliptic partial differential equations of principal type*, Comm. Pure Appl. Math. **24** (1971), 537–570.

18. ———, *Hypoelliptic partial differential equations of principal type. Sufficient conditions and necessary conditions*, Comm. Pure Appl. Math. **24** (1971), 631–670.

19. ———, *Approximate solutions to Cauchy problems*, J. Differential Equations **11** (1972), 349–363.

Rutgers University

PSEUDO-DIFFERENTIAL OPERATORS AND HYPOELLIPTICITY

J. J. KOHN

This is an introductory expository paper. We wish to summarize here some elementary facts about pseudo-differential operators and to apply to this the study of the hypoellipticity of the second order differential operator

$$(1) \qquad Pu = \sum_{j=1}^{k} X_j^2 u + X_0 u + cu$$

where $X_j = \sum_{l=1}^{n} a_j^m \, \partial/\partial x_m$, $j = 0, \ldots, k$, and a_j^m are real-valued C^∞ functions defined in a neighborhood U of the origin in $\boldsymbol{R}^n$.

DEFINITION. P is hypoelliptic if whenever the equation

$$(2) \qquad Pu = f$$

is satisfied with u and f distributions on U then the following condition holds: If V is an open subset of U such that $f|_V \in C^\infty(V)$ then $u|_V \in C^\infty(V)$. In [**4**] Hörmander proved the following.

THEOREM H. *If every vector field on U can be expressed as a linear combination (with C^∞ coefficients) of* $X_0, X_1, \ldots, X_k, \ldots, [X_i, X_j], \ldots, [X_i, [X_j, X_m]], \ldots, [X_{i_1}, [X_{i_2}, \ldots, X_{i_p}], \ldots]$ *then the operator P defined by* (1) *is hypoelliptic.*

Hörmander's original proof gives very precise estimates and is based on the analysis of the one-parameter groups generated by these vector fields. The proof described below is based on much rougher estimates which are derived by use of the calculus of pseudo-differential operators and thus the method of proof can be applied to much more general operators. The basic idea for this proof was discovered independently by E. V. Radkevič and the author, see [**6**] and [**10**]. However E. V. Radkevič first gave the complete proof along these lines and O. A. Oleĭnik and

AMS 1969 *subject classifications*. Primary 3519, 3523, 3524, 3548.

Radkevič have investigated (1) and related equations in great detail in [**9**].

The starting point of the proof of the theorem is the following "energy" estimate.

PROPOSITION. *There exists a constant* $C > 0$ *such that*

$$\sum_{j=1}^{k} \|X_j u\|^2 \leqq C(|(Pu, u)| + \|u\|^2) \tag{3}$$

for all $u \in C_0^\infty(U)$.

PROOF. Note that $X_j^* u = -X_j u + f_j u$ where $f_j = -\sum \partial a_j^m / \partial x_m$. Hence we have

$$(X_j^2 u, u) = -\|X_j u\|^2 + O(\|u\|\,\|X_j u\|)$$

for $j = 1, \ldots, k$; and

$$(X_0 u, u) = -(u, X_0 u) + O(\|u\|^2)$$

so that

$$|(X_0 u, u)| = O(\|u\|^2).$$

Observing that

$$\|u\|\,\|X_j u\| \leqq \text{large const}\|u\|^2 + \text{small const}\|X_j u\|^2$$

and combining this with the above we deduce the desired estimate.

Now recall that if $u \in C_0^\infty(U)$ and $s \in \boldsymbol{R}$ we define the norm $\|u\|_s$ by

$$\|u\|_s^2 = \int |\hat{u}(\xi)|^2 (1 + |\xi|^2)^s \, d\xi, \tag{4}$$

where $\hat{u}$ denotes the Fourier transform of u. Under the hypotheses of the theorem we will establish that there exist $\varepsilon > 0$ and $C > 0$ such that

$$\|u\|_\varepsilon \leqq C(\|Pu\| + \|u\|) \tag{5}$$

for all $u \in C_0^\infty(U)$.

For an arbitrary second order operator the estimate (5) would not yield much information. For example if (2) is the wave equation, (5) holds with $\varepsilon = 1$. However, for the operators of form (1) considered here the estimate (5) implies hypoellipticity. The following lemma shows that for operator P of the form (1) the estimate (5) is localizable.

LOCALIZATION LEMMA. *If P is of form* (1) *and if* (5) *holds then for any* $\zeta, \zeta_1 \in C_0^\infty(U)$ *with* $\zeta_1 = 1$ *on the support of* ζ *there exists* C' *such that*

$$\|\zeta u\|_\varepsilon \leqq C'(\|\zeta_1 Pu\| + \|\zeta_1 u\|) \tag{6}$$

for all $u \in C^\infty(U)$.

PROOF. Replacing u with ζu in (5) we see that it suffices to estimate $\|[P, \zeta]u\|$ by the right-hand side of (6). We have

$$[P, \zeta]u = 2 \sum_{j=1}^{k} [X_j, \zeta]X_j u + \sum_{j=1}^{k} [X_j, [X_j, \zeta]]u + [X_0, \zeta]u. \tag{7}$$

Clearly the L_2-norm of the last two terms can be estimated by $\text{const}\|\zeta_1 u\|$. Now we have, by (3),

$$\|[X_j, \zeta]X_j u\|^2 \leqq \text{const}\|X_j \zeta_1^2 u\|^2 \leqq \text{const}[(P\zeta_1^2 u, \zeta_1^2 u) + \|\zeta_1^2 u\|^2].$$

So it will suffice to estimate $([P, \zeta_1^2]u, \zeta_1^2 u)$ and this can be done easily by replacing ζ by ζ_1^2 in (7) and by the usual integration by parts.

Before proceeding with the proof of Theorem H we will discuss an algebra of pseudo-differential operators. A good expository account is given by L. Nirenberg in [**8**]. Here we wish to call attention to some basic simple properties of these operators which suffice to prove Theorem H. The following mappings of $C_0^\infty(U)$ to $C^\infty(U)$ are examples of pseudo-differential operators

(i) multiplication by $a \in C^\infty(\boldsymbol{R}^n)$,

(ii) $D^\alpha = \partial^{|\alpha|}/\partial x_1^{\alpha_1} \cdots \partial x_n^{\alpha_n}$, $|\alpha| = \sum \alpha_j$,

(iii) for $s \in \boldsymbol{R}$, $\Lambda^s : C_0^\infty(U) \to C^\infty(U)$ defined by $\widehat{\Lambda^s u}(\xi) = (1 + |\xi|^2)^{s/2}\hat{u}(\xi)$.

Let $\mathscr{P}$ be the algebra generated by the above under sums, composition and formal L_2-adjoints. Strictly speaking we cannot compose these operators since $\Lambda^s C_0^\infty \not\subset C_0^\infty$; however this is a technical point which can be easily overcome and which we will ignore here.

DEFINITION. If $P: C_0^\infty(U) \to C^\infty(U)$ is a linear operator we say that P is of *order m* if for each $s \in \boldsymbol{R}$ there exists a constant C_s such that

$$\|Pu\|_s \leqq C_s \|u\|_{s+m} \tag{8}$$

for all $u \in C_0^\infty(U)$.

The basic properties of $\mathscr{P}$ which we use here are summarized in the following theorem.

THEOREM (Ψ). *If P and P' are in $\mathscr{P}$ of orders m and m' respectively, then PP' is of order $m + m'$, $[P, P'] = PP' - P'P$ is of order $m + m' - 1$ and P^* is of order m. The multiplication operator of* (i) *is of order zero, the differentiation D^α is of order $|\alpha|$ and Λ^s is of order s.*

OUTLINE OF PROOF THAT (6) FOR P GIVEN BY (1) IMPLIES HYPOELLIPTICITY. First we show that given $s \in \boldsymbol{R}$ and $N > 0$ there exist constants $C_{s,N}$ such that

$$\|u\|_{s+\varepsilon} \leqq C_{s,N}(\|Pu\|_s + \|u\|_{-N}) \tag{9}$$

for all $u \in C_0^\infty(U)$. We wish to apply (3) to $\Lambda^s u$, again strictly speaking we cannot do this since $\Lambda^s u \notin C_0^\infty(U)$; however, this step is easily justified and we obtain

$$\|u\|_{s+\varepsilon} = \|\Lambda^s u\|_\varepsilon \leqq C(\|P\Lambda^s u\| + \|\Lambda^s u\|).$$

For any δ there exists $C(s, N, \delta)$ such that

$$\|u\|_s \leqq \delta\|u\|_{s+\varepsilon} + C(s, N, \delta)\|u\|_{-N} \tag{10}$$

for all $u \in C_0^\infty(U)$.

Thus it suffices to prove

$$\|[P, \Lambda^s]u\| \leqq \text{const}(\|Pu\|_s + \|u\|_s). \tag{11}$$

From the basic properties of pseudo-differential operators listed in the above theorem it follows that

$$[P, \Lambda^s] = \sum_{j=1}^{k} T_j^s X_j + T^s \tag{12}$$

where T_j^s and T^s are pseudo-differential operators of order s. Thus the inequality (9) is reduced to proving that

$$\sum_{j=1}^{k} \|X_j u\|_s \leqq \text{const}(\|Pu\|_s + \|u\|_s). \tag{13}$$

We have

$$\begin{aligned}\sum \|X_j u\|_s^2 &\leqq \sum \|X_j \Lambda^s u\|^2 + \text{const}\|u\|_s^2 \\ &\leqq |(P\Lambda^s u, \Lambda^s u)| + \text{const}\|u\|_s^2 \\ &\leqq \text{const}\{|([P, \Lambda^s]u, \Lambda^s u)| + \|Pu\|_s^2 + \|u\|_s^2\}.\end{aligned}$$

Applying (12) to the first term on the right, we bound that term by

$$\text{const} \sum \|X_j u\|_s \|u\|_s \leqq \text{small const}\|X_j u\|_s^2 + \text{large const}\|u\|_s^2.$$

This establishes (13) and hence the desired estimate (9).

Now we can localize the estimate (9) by the same method as in the Localization Lemma and we obtain that, for all $u \in C^\infty(U)$,

$$\|\zeta u\|_{s+\varepsilon} \leqq \text{const}(\|\zeta_1 Pu\|_s + \|\zeta_1 u\|_{-N}) \tag{14}$$

where $\zeta, \zeta_1 \in C_0^\infty(U)$ and $\zeta_1 = 1$ on the support of ζ. Suppose that u is a distribution which satisfies (2) and that $f|_V \in C^\infty(V)$ (V an open subset of U); we wish to prove that $u|_V \in C^\infty(V)$. If $x_0 \in V$ we will show that u is C^∞ in some neighborhood of x_0. First, for $\delta > 0$, let

$$\begin{aligned} S_\delta u(x) &= \int u(x + \delta y)\varphi(y)\,dy \\ &= \delta^{-n} \int u(y)\varphi\left(\frac{y-x}{\delta}\right) dy, \end{aligned} \tag{15}$$

where $dy = dy_1 \cdots dy_n$ and $\varphi \in C_0^\infty(\boldsymbol{R}^n)$ with

$$\int \varphi(y)\, dy = 1. \tag{16}$$

Then $S_\delta u \in C^\infty$ for $\delta > 0$ and if u is a distribution then u is in H_s in a neighborhood of x_0 if and only if there exists $\zeta \in C_0^\infty(V)$, $\zeta = 1$ in a neighborhood of x_0 such that $\|S_\delta \zeta u\|_s$ is bounded independently of δ. Further if u is locally in H_s (in a neighborhood of x_0) and if T^1 is a pseudo-differential operator of order 1, then

$$\|[S_\delta \zeta, T^1]u\|_s \leqq \text{const}\|u\|_s,$$

where the constant is independent of δ. By arguments similar to those above we obtain

$$\|S_\delta \zeta u\|_{s+\varepsilon} \leqq \text{const}(\|S_\delta \zeta_1 f\|_s + \|S_\delta \zeta_1 u\|_{-N} + \|\zeta_1 u\|_s) \tag{17}$$

whenever u satisfies (2) and $\zeta u \in H_s$. Thus if u is a distribution then for some N we have $\zeta_1 u \in H_{-N}$ and of course we can choose ζ and ζ_1 so that ζu has small support and $\zeta_1 f \in C_0^\infty(V)$. Now setting $s = -N$ in (17) we conclude that $\zeta u \in H_{-N+\varepsilon}$. Iterating this argument we conclude that there is a neighborhood W of x_0 such that for any $\zeta \in C_0^\infty(W)$, $\zeta u \in H_s$ for all s and hence $u|_W \in C^\infty(W)$. This completes the outline of the proof; of course, many details have been omitted but we have tried to present the main ideas.

PROOF OF THEOREM H. The proof of this theorem is now reduced to proving (5). First we observe that

$$\|u\|_\varepsilon \leqq \text{const} \sum_{j=1}^{u} \|D_j u\|_{\varepsilon - 1}. \tag{18}$$

Now by the assumption of the theorem we can express D_j as

$$D_j = \sum a_j^{i_1 \dots i_p} F_{i_1 \dots i_p} \tag{19}$$

where $0 \leqq i_m \leqq k$ and

$$\begin{aligned} F_{i_1 \dots i_p} &= X_{i_p}, & \text{if } p = 1, \\ &= [X_{i_p}, F_{i_1 \dots i_{p-1}}], & \text{if } p > 1. \end{aligned} \tag{20}$$

Thus it will suffice to bound $\|F_{i_1 \dots i_p} u\|_{\varepsilon - 1}$ by the right side of (5). For simplicity we will write $F^p = [X, F^{p-1}]$ and since F^p is an operator of order one we have

$$\begin{aligned} \|F^p u\|_{\varepsilon-1}^2 &= (F^p u, T^{2\varepsilon - 1} u) \\ &= (XF^{p-1}u, T^{2\varepsilon-1}u) - (F^{p-1}Xu, T^{2\varepsilon-1}u) \end{aligned} \tag{21}$$

where $T^{2\varepsilon-1}$ denotes a pseudo-differential operator of order $2\varepsilon - 1$. Now we will estimate each term on the right-hand side separately.

Case I. $X = X_j$ with $1 \leqq j \leqq k$.

$$\begin{aligned}(XF^{p-1}u, T^{2\varepsilon-1}u) &= -(F^{p-1}u, XT^{2\varepsilon-1}u) + O(\|F^{p-1}u\|_{2\varepsilon-1}\|u\|) \\ &= -(F^{p-1}u, T^{2\varepsilon-1}Xu) + O(\|F^{p-1}u\|_{2\varepsilon-1}\|u\|) \\ &= O(\|F^{p-1}u\|_{2\varepsilon-1}(\|Xu\| + \|u\|)).\end{aligned}$$

Hence, using (3),

$$|(XF^{p-1}u, T^{2\varepsilon-1}u)| \leq \text{const}(\|F^{p-1}u\|_{2\varepsilon-1}^2 + \|Pu\|^2 + \|u\|^2). \tag{22}$$

Similarly, we have

$$(F^{p-1}Xu, T^{2\varepsilon-1}u) = -(Xu, T^{2\varepsilon-1}F^{p-1}u) + O(\|Xu\|\,\|u\|_{2\varepsilon-1}).$$

So we obtain, in case $X = X_j$, with $1 \leqq j \leqq k$,

$$\|F^p u\|_{\varepsilon-1} \leqq \text{const}(\|F^{p-1}u\|_{2\varepsilon-1} + \|Pu\| + \|u\|). \tag{23}$$

Case II. $X = X_0$. First we write $X_0 = -P^* + \sum_{j=1}^k X_j^2 + \sum_{j=1}^k b_j X_j + f$. Substituting in the first term in the right of (21) we have

$$\begin{aligned}(X_0F^{p-1}u, T^{2\varepsilon-1}u) &= -(P^*F^{p-1}u, T^{2\varepsilon-1}u) + \sum_{j=1}^k (X_j^2F^{p-1}u, T^{2\varepsilon-1}u) \\ &\quad + \sum_{j=1}^k (b_jX_jF^{p-1}u, T^{2\varepsilon-1}u) + (fF^{p-1}u, T^{2\varepsilon-1}u).\end{aligned} \tag{24}$$

It is easy to see that the last two terms on the right can be estimated by the right side of (23). Now we have

$$(P^*F^{p-1}u, T^{2\varepsilon-1}u) = (F^{p-1}u, T^{2\varepsilon-1}Pu) + (F^{p-1}u, [P, T^{2\varepsilon-1}]u)$$

and since

$$[P, T^{2\varepsilon-1}]u = \sum_{j=1}^k T_j^{2\varepsilon-1}X_ju + T_{k+1}^{2\varepsilon-1}u \tag{25}$$

we conclude that the first term on the right in (24) is bounded by the right-hand side of (23). Now we analyze the second term on the right in (24).

$$\begin{aligned}(X_j^2F^{p-1}u, T^{2\varepsilon-1}u) &= -(X_jF^{p-1}u, X_jT^{2\varepsilon-1}u) + \cdots \\ &= -(X_jT^{2\varepsilon-1}F^{p-1}u, X_ju) + \cdots \\ &= O(\|X_jT^{2\varepsilon-1}F^{p-1}u\|^2) + \cdots,\end{aligned}$$

where the dots denote terms that can be bounded by the right-hand side of (23). Now, by (3),

$$\sum_{j=1}^{k} \|X_j T^{2\varepsilon-1} F^{p-1} u\|^2 \leqq \text{const}(|(PT^{2\varepsilon-1}F^{p-1}u, T^{2\varepsilon-1}F^{p-1}u)| + \|F^{p-1}u\|_{2\varepsilon-1}^2)$$

and

$$\begin{aligned}(PT^{2\varepsilon-1}F^{p-1}u, T^{2\varepsilon-1}F^{p-1}u) &= (PT^{2\varepsilon}u, T^{2\varepsilon-1}F^{p-1}u)\\ &= ([P, T^{2\varepsilon}]u, T^{2\varepsilon-1}F^{p-1}u) + (Pu, T^{4\varepsilon-1}F^{p-1}u).\end{aligned} \tag{26}$$

Again we have

$$[P, T^{2\varepsilon}] = \sum_{j=1}^{k} T_j^{2\varepsilon} X_j + T_{k+1}^{2\varepsilon}$$

so that (26) is bounded by

$$\text{const}(\|F^{p-1}u\|_{4\varepsilon-1}^2 + \|Pu\|^2 + \|u\|^2). \tag{27}$$

Thus we conclude that in Case II the first term on the right of (21) is bounded by (27). To deal with the second term we write

$$X_0 u = Pu - \sum_{j=1}^{k} X_j^2 u - cu.$$

Thus we have

$$\begin{aligned}(F^{p-1}X_0u, T^{2\varepsilon-1}u) = (F^{p-1}Pu, T^{2\varepsilon-1}u) &- \sum_{j=1}^{k} (F^{p-1}X_j^2u, T^{2\varepsilon-1}u)\\ &- (F^{p-1}cu, T^{2\varepsilon-1}u).\end{aligned} \tag{28}$$

Clearly the first and third terms in the right-hand side are bounded by (27) (in fact, by the right-hand side of (23)). To bound the second term we have

$$(F^{p-1}X_j^2u, T^{2\varepsilon-1}u) = (X_j^2F^{p-1}u, T^{2\varepsilon-1}u) + ([F^{p-1}, X_j^2\, u, T^{2\varepsilon-1}u). \tag{29}$$

Here the first term on the right was bounded above, and for the second term we write

$$\begin{aligned}[F^{p-1}, X_j^2] &= [F^{p-1}, X_j]X_j + X_j[F^{p-1}, X_j]\\ &= -\tilde{F}^p X_j - X_j \tilde{F}^p\end{aligned}$$

where $\tilde{F}^p$ is a commutator of the type treated in Case I. So we see that the last term in (29) is bounded by $\text{const}(\|\tilde{F}^p u\|^2_{2\varepsilon-1} + \sum \|X_j u\|^2 + \|u\|^2)$. Applying (23) with ε replaced by 2ε we obtain

$$\|\tilde{F}^{\neg} u\|_{2\varepsilon-1} \leqq \text{const}(\|F^{p-1}u\|_{4\varepsilon-1} + \|Pu\| + \|u\|).$$

Thus we conclude that (21) is bounded by (27). Putting Cases I and II together we have

(30) $$\|F^p u\|_{\varepsilon-1} \leqq \text{const}(\|F^{p-1}u\|_{4\varepsilon-1} + \|Pu\| + \|u\|).$$

Inductively we then obtain

(31) $$\|F^p u\|_{\varepsilon-1} \leqq \text{const}\left(\sum_{j=0}^{k} \|X_j u\|_{4^{p-1}\varepsilon-1} + \|Pu\| + \|u\|\right).$$

Now we claim that

(32) $$\|X_j u\|_{-1/2} \leqq \text{const}(\|Pu\| + \|u\|).$$

Thus the desired estimate is obtained for any $\varepsilon \leqq 2/4^p$. If $1 \leqq j \leqq k$ then (32) is a consequence of (3). To estimate $\|X_0 u\|_{-1/2}$ we have

$$\begin{aligned}\|X_0 u\|^2_{-1/2} &= (X_0 u, T^0 u)\\ &= (Pu, T^0 u) - \sum_{j=1}^{k} (X_j^2 u, T^0 u) - (cu, T^0 u).\end{aligned}$$

Now

$$\left|\sum_{j=1}^{k} (X_j^2 u, T^0 u)\right| \leqq \text{const}\left(\sum_{j=1}^{k} \|X_j u\|^2 + \|u\|^2\right).$$

So applying (3) we obtain the desired estimate and thus complete the proof of Theorem H.

REMARKS. (1) By a more detailed analysis one can improve the size of ε in the above. The most precise estimates are obtained by Hörmander in [**4**] by a different method.

(2) In this proof we only used the algebra $\mathscr{P}$. It is useful to enlarge the algebra as follows. If $P: C_0^\infty(U) \to C^\infty(U)$ is of order m and if there exists a sequence $\{P_j\}_{j=1}^\infty$ with $P_j \in \mathscr{P}$ such that $P - P_j$ is of order $m - j$ then we say $P \in \tilde{\mathscr{P}}$. $\tilde{\mathscr{P}}$ is then an algebra with the same properties as in Theorem (Ψ). Furthermore every elliptic operator has a local inverse in $\tilde{\mathscr{P}}$. This coupled with the fact that operators in $\tilde{\mathscr{P}}$

are pseudo-local (i.e. if $u|_V \in C^\infty(V)$ then $Pu|_V \in C^\infty(V)$) shows that elliptic operators are hypoelliptic. One important problem is to find an invariant class of pseudo-local operators which contains $\tilde{\mathscr{P}}$ and which contains local inverses of the operators described here. In [**3**] Hörmander introduces larger classes of pseudo-local operators which are very useful and for which the properties discussed above also hold. Recently regularity properties have been studied by means of Fourier integral operators; these give a much more precise description of the phenomenon of hypoellipticity. For an account of this and also for a very beautiful treatment of pseudo-differential operators we refer to Hörmander [**5**].

(3) In Theorem H the assumption that P is real plays a crucial role. The only general result known about hypoellipticity of complex operators was obtained by Trèves (see [**11**]). In his work Trèves assumes that the operator has simple characteristics; in the multiple characteristics case very little is known. We call attention to [**2**] and [**7**] for some results in this direction.

(4) Finally we mention that little is known about the analytic hypoellipticity of the equations discussed here; we refer to [**1**] for a very interesting example and related matters.

References

1. M. S. Baouendi and C. Goulaouic, *Not analytic hypoelliptic for some class of hypoelliptic second order operators*, Bull. Amer. Math. Soc. **78** (1972), 485–486.

2. V. V. Grusin, *On a class of hypoelliptic operators*, Mat. Sb. **83** (**125**) (1970), 456–473 = Math. USSR Sb. **12** (1970), 458–476.

3. L. Hörmander, *Pseudo-differential operators and hypoelliptic equations*, Proc. Sympos. Pure Math., vol. 10, Amer. Math. Soc., Providence, R.I., 1966, pp. 138–183.

4. ———, *Hypoelliptic second order differential equations*, Acta Math. **119** (1967), 147–171. MR **36** # 5526.

5. ———, *Fourier integral operators*. I, Acta Math. **127** (1971), 79–183.

6. J. J. Kohn, *Pseudo-differential operators and non-elliptic problems*, Pseudo-Differential Operators (C.I.M.E., Stresa, 1968), Edizioni Cremonese, Rome, 1969, pp. 157–165. MR **41** #3972.

7. ———, *Complex hypoelliptic equations*, Proc. Conf. on Hypoelliptic Equations, Rome, 1971.

8. L. Nirenberg, *Pseudo-differential operators*, Proc. Sympos. Pure Math., vol. 16, Amer. Math. Soc., Providence, R.I., 1970, pp. 149–167. MR **42** #5108.

9. O. A. Oleinik and E. V. Radkevic, *Second order equations with non-negative characteristic form*, Math. Analysis 1969, Itogi Nauk, Moscow, 1971; English transl., Symposia Math., vol. 7, Academic Press, 1971, pp. 459–468.

10. E. V. Radkevic, *Hypoelliptic operators with multiple characteristics*, Mat. Sb. **79** (**121**) (1969), 193–216 = Math. USSR Sb. **8** (1969), 181–206. MR **41** #5763.

11. F. Trèves, *Hypoelliptic partial differential equations of principal type with analytic coefficients*, Comm. Pure Appl. Math. **23** (1970), 637–651.

PRINCETON UNIVERSITY

NODAL AND CRITICAL SETS FOR EIGENFUNCTIONS OF ELLIPTIC OPERATORS

J. H. ALBERT

The general philosophy behind the theorems discussed here is to try to describe solutions of elliptic equations by means of geometric properties, such as the layout of the zero set and the nature of the critical points. This idea is not new; the well-known Sturm-Liouville theory is a study of the zeros of eigenfunctions of second-order elliptic ordinary differential equations and our approach is motivated by it.

Geometric definitions. M will be a C^∞ manifold of dim ν, compact, connected and without boundary. $C^\infty(M)$ denotes the real-valued C^∞ functions on M. Suppose $u \in C^\infty(M)$. The *zero* or *nodal* set of u is $\{x \in M : u(x) = 0\}$. $a \in M$ is a *critical point* if $du(a) = 0$, $du = (\partial u/\partial x_1, \ldots, \partial u/\partial x_\nu)$. A critical point $a \in M$ is *nondegenerate* if the Hessian matrix $((\partial^2 u(a)/\partial x_i \partial x_j))$ is nonsingular.

The nicest geometric properties u can have are the following: (1) all the critical points are nondegenerate (i.e., u is a Morse function); (2) the zeros are all regular points (i.e., there are no nodal critical points). (2) implies that the nodal set is a $(\nu - 1)$-dimensional submanifold.

DEFINITION. We will call u *generic* if properties (1) and (2) hold or $u \equiv 0$.

REMARK. The set of generic functions is open and dense in the C^s topology on $C^\infty(M)$ if $s \geqq 2$. This is essentially the Morse lemma. We may rephrase it as "most C^∞ functions are generic." [By the C^s topology, we mean the following: For U open $\subset \boldsymbol{R}^\nu$, define $|f|_s = \sup_{x \in U} \sum_{|\alpha| \leqq s} |D^\alpha u(x)|$. On a compact manifold, one can define a global s-norm by using a partition of unity.]

Eigenfunctions. Let P be a second-order, selfadjoint, C^∞ elliptic differential operator on M (in the future we will just say elliptic operator and mean this). Let $Pu = \lambda u$, i.e., u is an eigenfunction of P with eigenvalue λ. Since requiring u to satisfy a differential equation imposes an additional constraint, it is plausible to

---

AMS 1970 *subject classifications*. Primary 58G99, 35B05, 35J15; Secondary 57D70, 35P05, 35B20.

ask whether or not eigenfunctions of an elliptic operator are generic. If dim $M = 1$, they are. (The uniqueness theorem for ordinary differential equations says that no consecutive derivatives of a nonzero solution of a second-order elliptic equation can vanish at a point.) However, for dim $M = 2$, life is different. Let us look at the Laplacian on S^2. Its eigenfunctions are well-known; they are the spherical harmonics. The eigenvalues take the form $-n(n-1)$, for $n = 1, 2, \ldots$, and the nth eigenvalue has multiplicity $2n - 1$ (multiplicity = dimension of the eigenspace). On pp. 76–77, we have drawn some diagrams to indicate what can go wrong with the eigenfunctions.

In the diagrams of the eigenfunctions of the Laplacian on S^2, the nodal critical points are isolated and the nodal set exhibits specific patterns, namely rays emanating from the critical point. In dimension 2, this is always true.

DEFINITION. Let $u \in C^\infty(M)$ and let $a \in M$ be a zero of order k (i.e., $D^\alpha u(a) = 0$ for all $\alpha : |\alpha| < k$ but there exists $\alpha : |\alpha| = k$ and $D^\alpha u(a) \neq 0$). We say that a is a *weakly degenerate zero* if there exists $c > 0$ such that $|du(x)| \geqq c|x - a|^{k-1}$ for x in a neighborhood of a.

This implies a is an isolated critical point.

THEOREM 0. *If* dim $M = 2$ *and* $Pu = \lambda u$, *all the nodal critical points of* u *are weakly degenerate.*

In the proof you show that $u(z) = \operatorname{Re}(z - a)^k$ + higher order terms in an appropriate coordinate system (using complex notation $z = x_1 + x_2(-1)^{1/2}$). Apply $Pu = \lambda u$ to the Taylor series of u and equate coefficients; the kth order part u_k satisfies $\Delta u_k = 0$, $\Delta = \partial^2/\partial x_1^2 + \partial^2/\partial x_2^2$.

By a theorem due to Kuo, the zeros of u are the image of the zeros of $u_k(z) = \operatorname{Re}(z - a)^k$ under a local homeomorphism about a, and these are just k lines passing through the critical point a.

In general, it seems to be very difficult to prove statements about all elliptic operators, as the variety of different types of nodal and critical sets for Δ on S^2 indicates. An alternate approach is to try to prove theorems for "most" operators.

The main theorem we want to discuss states that most operators have generic eigenfunctions and simple eigenvalues (i.e., multiplicity = 1). We will study operators of the form $P = L + \rho$, where L is a fixed elliptic operator and $\rho \in C^\infty(M)$ is allowed to vary. ρ is called the *forcing function*; one thinks of it as representing the zeroth order part of the operator P. The statement "for most operators P" is made by saying "for most $\rho \in C^\infty(M)$." We express this rigorously as follows. Let

$$\Omega = \{\rho \in C^\infty(M) \colon \text{all eigenfunctions of } L + \rho \text{ have only weakly degenerate zeros}\},$$

$$A = \{\rho \in C^\infty(M) \colon L + \rho \text{ has only simple eigenvalues}\},$$

$$B = \{\rho \in A \colon \text{all eigenfunctions of } L + \rho \text{ are generic}\}.$$

THEOREM 1. *A is a countable intersection of sets which are open and dense in $C^\infty(M)$ in the C^s topology (any s).*

THEOREM 2. *B is a countable intersection of sets which are open and dense in Ω in the C^s topology, provided $s \geqq [\dim M/2] + 1$.*

REMARKS. (1) A is dense in $C^\infty(M)$; B is dense in Ω.

(2) The interesting case is when $\dim M = 2$, where we know $\Omega = C^\infty(M)$ by Theorem 0. Although Theorem 2 as stated is valid in any number of dimensions, nothing is known about Ω except in dimension 2; Ω may even be empty. In fact, it is probable that the notion of weakly degenerate zeros is the wrong notion to look at except in dimension 2.

(3) We separated the result into two theorems because Theorem 1 is true independently of $\dim M$; it is completely free of the restriction of Theorem 2.

(4) The point of requiring simple eigenvalues is to be able to translate "for all" statements into "there exist" statements. For if one nonzero function is generic, then so is any scalar multiple of it.

One achieves the countable intersection as follows: Since $L + \rho$ is selfadjoint, its eigenvalues are real. They can be arranged in a sequence $\lambda_1 \geqq \lambda_2 \geqq \cdots \geqq \lambda_n \geqq \cdots \to -\infty$ counting multiplicity. For Theorem 1, let

$$A_n = \{\rho \in C^\infty(M)\text{: the first } n \text{ eigenvalues of } L + \rho \text{ are simple}\},$$
$$A_0 = C^\infty(M).$$

Then $A = \bigcap_{n=0}^{\infty} A_n$ and we prove: (1) A_n is open in $C^\infty(M)$; (2) A_n is dense in A_{n-1}. (2) implies A_n dense in $C^\infty(M)$. For Theorem 2, let

$$B_n = \{\rho \in A_n\text{: the eigenfunctions in the first } n \text{ eigenspaces are generic}\},$$
$$B_0 = \Omega.$$

Then $B = \bigcap_{n=0}^{\infty} B_n$ and we prove: (3) B_n is open in $C^\infty(M)$ in the C^s topology for $s \geqq [\dim M/2] + 1$; (4) B_n is dense in $A_n \cap B_{n-1} \cap \Omega$. (4) implies B_n is dense in Ω. (1) to (4) prove Theorems 1 and 2.

To indicate some of the techniques involved and to show why (2) and (4) are phrased as they are, we will sketch the proof of (4). Suppose $\rho \in A_n \cap B_{n-1}$. This means that the first n eigenvalues of $L + \rho$ are simple and the first $n - 1$ eigenfunctions are generic. If you perturb ρ by a small enough amount, the same is true of the perturbed operator by (1) and (3). So what must be shown is that you can perturb so that the eigenfunctions in the nth eigenspace become generic. The perturbation is done in two stages: first we show how to eliminate the nodal critical points and second we show how to make the remaining critical points nondegenerate.

Let $\rho \in A_n \cap B_{n-1} \cap \Omega$, $\lambda = \lambda_n$ and $(L + \rho)u = \lambda u$; for simplicity assume that $a \in M$ is the only nodal critical point of u. We will perform a linear perturbation

$\rho(\tau) = \rho - \tau\sigma$ where $\tau \in \boldsymbol{R}$ and $\sigma \in C^\infty(M)$. For such a perturbation (1) the nth eigenvalue $\lambda(\tau)$ of $L + \rho(\tau)$ is an analytic function of τ for small τ and (2) there exists an analytic function $u(\tau)$ for small τ (analytic in the sense of normed space valued functions) such that $u(0) = u$ and $(L + \rho(\tau))u(\tau) = \lambda(\tau)u(\tau)$. The point of doing this perturbation is that σ and the unperturbed λ, u completely control the linear terms of $\lambda(\tau)$ and $u(\tau)$. Write

$$\lambda(\tau) = \lambda + \alpha\tau + \text{higher order terms},$$

$$u(\tau) = u + v\tau + \text{higher order terms}.$$

Then

$$(L + \rho - \sigma\tau - \lambda - \alpha\tau - \cdots)(u + v\tau + \cdots) = 0$$

and equating the coefficients of τ to zero gives

$$(L + \rho - \lambda)v = (\sigma + \alpha)u.$$

There exists a pseudo-differential operator Q such that

(1) $Q(L + \rho - \lambda) = I - \pi_N$ where π_N is the projection on $N = \ker(L + \rho - \lambda)$,

(2) $Q = 0$ on N,

(3) $Qf(x) = \int_M K(x, y)f(y)\,dy$, where for fixed x, $y \mapsto K(x, y)$ is in $L_p(M)$, where $p = 2$ if $v = 2$ and $p = (v - 1)/(v - 2)$ if $v > 2$ ($v = \dim M$).

Using these facts, we have

$$v = Q[\sigma u] + \eta u, \quad \text{for some } \eta \in \boldsymbol{R},$$

i.e.,

$$v(x) = \int_M K(x, y)\sigma(y)u(y)\,dy + \eta u(x).$$

PROPOSITION. *There exists $\sigma \in C^\infty(M)$ such that $v(a) \neq 0$.*

PROOF. If not, we have

$$0 = v(a) = \int_M K(a, y)\sigma(y)u(y)\,dy, \quad \text{for all } \sigma \in C^\infty(M),$$

since $u(a) = 0$. Hence $K(a, y)u(y) = 0$ in $L_p(M)$, i.e., $=0$ almost everywhere. Since $u^{-1}(0)$ has measure zero [this follows from the fact that $\rho \in \Omega$], $K(a, y) = 0$ a.e., hence

$$\int_M K(a, y)f(y)\,dy = 0, \quad \text{for all } f \in C^\infty(M).$$

Thus if $g \in N^\perp \cap C^\infty(M)$, $g(a) = 0$, for Q maps onto $N^\perp$. Also any element of N is a multiple of u and must vanish at a. This gives $g(a) = 0$ for all $g \in C^\infty(M)$, which is absurd.

Now choose such a σ and write $u(\tau) = u + \tau v(\tau)$, so $v(0) = v$. Since $v(a) \neq 0$, there exists a constant $J > 0$ with $|v(\tau)(x)| \geqq J$ uniformly for τ in a neighborhood of 0 and x in a neighborhood of a. Making τ small, by continuity we can insure that no nodal critical points occur outside an arbitrarily small neighborhood of a. Suppose a is a zero of order k. Then for x in a neighborhood of a,

$$|u(x)| \leqq c_1|x - a|^k,$$
$$|du(x)| \geqq c_2|x - a|^{k-1},$$

since $\rho \in \Omega$ implies that a is a weakly degenerate zero. (The c_i's will be positive constants throughout.) We will show that for τ small enough and $\neq 0$, there are no nodal critical points in this neighborhood of a. If x is a critical point,

$$0 = [du(\tau)](x) = du(x) + \tau[dv(\tau)](x),$$
$$|x - a|^{k-1} \leqq |du(x)|/c_2 = |\tau|\,|[dv(\tau)](x)|/c_2 \leqq c_3|\tau|,$$

and

$$\begin{aligned} |u(\tau)(x)| &= |u(x) + \tau v(\tau)(x)| \\ &\geqq |\tau|\,|v(\tau)(x)| - |u(x)| \\ &\geqq J|\tau| - c_1|x - a|^k \\ &\geqq J|\tau| - c_4|\tau|^{k/(k-1)} \\ &\geqq J|\tau|/2 \quad \text{provided } |\tau| \leqq (J/2c_4)^{k-1}, \\ &\neq 0 \quad \text{if } |\tau| \neq 0. \end{aligned}$$

This completes the proof that nodal critical points can be perturbed away.

The argument for perturbing so that the nonnodal critical points become nondegenerate is fairly straightforward. Suppose $(L + \rho)u = \lambda u$, with λ simple and u having no nodal critical points. Let v be a Morse function close to u and write $u' = \phi u + (1 - \phi)v$, where $\phi \equiv 1$ in a neighborhood of $u^{-1}(0)$ and supp ϕ contains no critical points of u. Define

$$\begin{aligned} \rho' &= (\lambda u' - Lu')/u' \quad && \text{on } M - u^{-1}(0), \\ &= \rho && \text{on } \phi^{-1}(1). \end{aligned}$$

This is well-defined and C^∞ on M and u' is automatically an eigenfunction of $L + \rho'$ with eigenvalue λ. Since v can be chosen arbitrarily close to u by the Morse lemma, ρ' can be chosen close to ρ and we can insure that there are no critical points of u' where $\phi \neq 0$. Thus u' is generic.

Final remarks. (1) Once you have a generic eigenfunction, it makes sense to formulate the Sturm oscillation and comparison theorems. Consider the following

In all diagrams except the one at the left, only a planar projection of the front hemisphere is drawn.

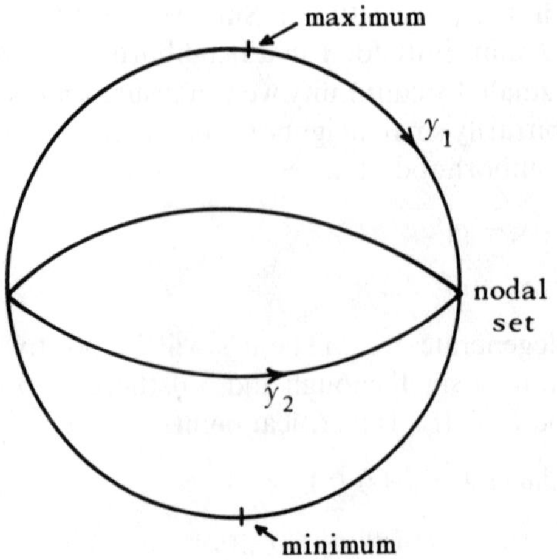

$\cos y_1$; eigenvalue -2
generic

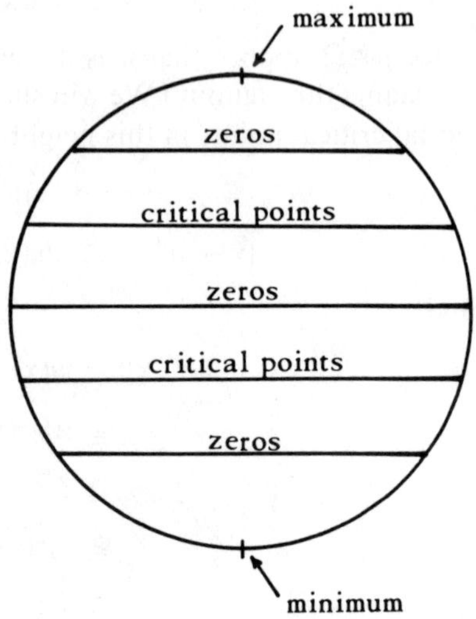

$5 \cos^3 y_1 - 3 \cos y_1$
eigenvalue -12
degenerate, non-isolated critical points
nodal set regular

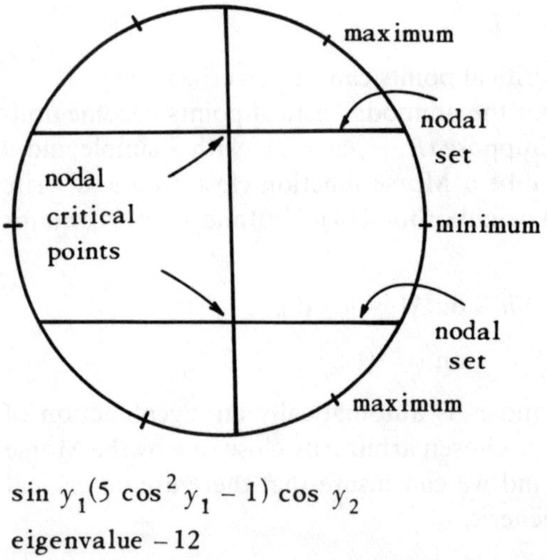

$\sin y_1 (5 \cos^2 y_1 - 1) \cos y_2$
eigenvalue -12
all critical points are non-degenerate
nodal set is not regular

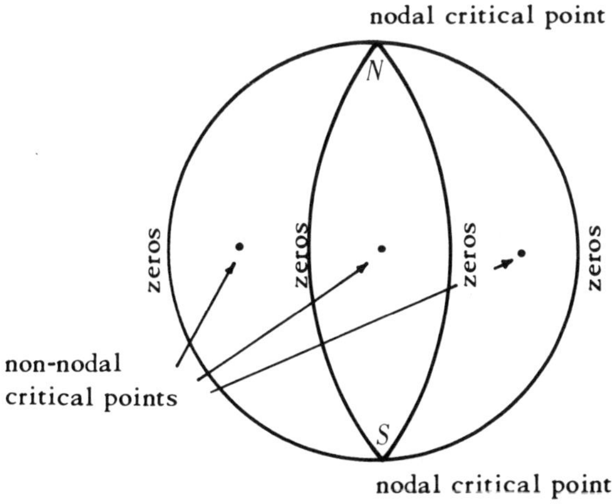

$\sin^2 y_1 \cos 3y_2$
eigenvalue -12
non-nodal critical points are non-degenerate
zeros of order 3 (degenerate critical points) at north and south poles

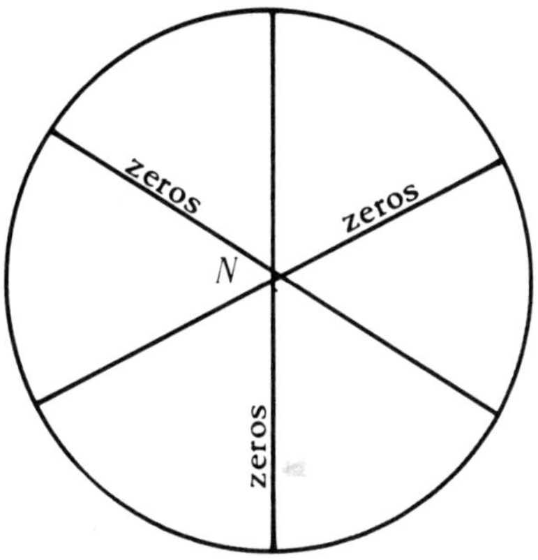

view of above example from above the north pole
nodal set has three lines passing through N

question: Is the number of connected components of the nodal set of an eigenfunction related to the eigenvalue? In dimension 1, the nth eigenfunction has exactly n zeros. In higher dimensions, upper bounds on the number of components are known, but that is all.

(2) Some other natural questions are:

(a) What is the right theorem for dim $M > 2$?

(b) Is the set of operators whose first n eigenvalues are simple and whose first n eigenfunctions are generic open under perturbations by first order operators?

(c) Can these conditions (e.g., weakly degenerate zeros, operators with generic eigenfunctions, etc.) be expressed in the language of transversality?

References

1. J. H. Albert, *Topology of the nodal and critical point sets for eigenfunctions of elliptic operators*, Thesis, M.I.T., Cambridge, Mass., 1971.

2. R. Courant and D. Hilbert, *Methods of mathematical physics*. Vol. I, Interscience, New York, 1953. MR **16**, 426.

3. E. L. Ince, *Ordinary differential equations*, Dover, New York, 1956.

4. T. Kuo, *On C^0-sufficiency of jets of potential functions*, Topology *8* (1969), 167–171. MR *38* #6614.

5. F. Rellich, *Perturbation theory of eigenvalue problems*, Gordon and Breach, New York, 1969. MR **39** #2014.

6. R. Seeley, *Topics in pseudo-differential operators*, Pseudo-Differential Operators (C.I.M.E., Stresa, 1968), Edizioni Cremonese, Rome, 1969, pp. 167–305. MR **41** #3973.

Tufts University

ANALYTICITY FOR DEGENERATE ELLIPTIC EQUATIONS AND APPLICATIONS

M. S. BAOUENDI AND C. GOULAOUIC

1. We consider some classes of elliptic boundary value problems degenerating at the boundary which are generalizations of Legendre's operator on $[-1,1]$:

$$\frac{d}{dx}(1-x^2)\frac{d}{dx}.$$

Let Ω be a bounded open subset of $\boldsymbol{R}^n$, and φ a real analytic function defined in a neighborhood of $\bar{\Omega}$ such that

$$\begin{aligned} \Omega &= \{x \in \boldsymbol{R}^n;\ \varphi(x) > 0\}, \\ \partial\Omega &= \{x \in \boldsymbol{R}^n;\ \varphi(x) = 0\}, \\ d\varphi(x) &\neq 0 \quad \text{when } x \in \partial\Omega. \end{aligned}$$

Let us consider the following sesquilinear form, defined for $u,v \in \mathscr{C}_0^\infty(\Omega)$:

$$a(u,v) = \sum_{|\alpha| \leqq 1;|\beta| \leqq 1} \int_\Omega a_{\alpha\beta}(x)\varphi(x) D^\alpha u(x) \overline{D^\beta v(x)}\, dx.$$

We assume that the coefficients $a_{\alpha\beta}$ are analytic up to the boundary and there exists a constant $\lambda > 0$ such that for every $u \in \mathscr{C}_0^\infty(\Omega)$ we have

$$\operatorname{Re} a(u,u) \geqq \lambda \sum_{|\alpha| \leqq 1} \|\varphi^{1/2} D^\alpha u\|^2_{L^2(\Omega)}.$$ [1]

For $1 \leqq j \leqq n$ and $1 \leqq k \leqq n$, let $\Lambda_{k,j}$ denote the following first order operator:

$$\Lambda_{k,j} = \frac{\partial\varphi}{\partial x_k} D_j - \frac{\partial\varphi}{\partial x_j} D_k.$$

AMS 1970 *subject classifications*. Primary 35J70; Secondary 35H05, 35J25, 35P20.

[1] In fact, this coercive condition is not necessary; we assume it for simplicity. It may be replaced by suitable a priori estimates.

Finally we consider the following second order degenerate elliptic operators:

$$A_1 = A_1(x,D) = \sum_{|\alpha| \leqq 1; |\beta| \leqq 1} D^\beta a_{\alpha\beta} \varphi D^\alpha, \tag{2}$$

$$A_2 = A_2(x,D) = A_1 + \sum_{k,j} \Lambda_{k,j}^* \Lambda_{k,j}. \tag{3}$$

We have

THEOREM 1. *The operators* A_i $(i = 1,2)$ *realize isomorphisms from* $\mathscr{C}^\infty(\bar{\Omega})$ *into itself and from* $\mathscr{A}(\bar{\Omega})$ *into itself.*

We have denoted by $\mathscr{C}^\infty(\bar{\Omega})$ (resp. $\mathscr{A}(\bar{\Omega})$) the space of functions defined in Ω which can be extended in $\mathscr{C}^\infty$ (resp. analytic) functions in some neighborhood of $\bar{\Omega}$.

In fact, these results of regularity are local. The analytic case may be considered as a generalization of [7]. For details and proofs, we refer to [**1**] and [**2**].

2.[2] Now we are interested in characterizing the analytic functions (and eventually Gevrey's classes) on real analytic compact manifolds (namely $\bar{\Omega}$) using successive powers of a differential operator. This characterization is possible using A_2 (but not A_1 (see §4)). We recall that, in the case of an analytic manifold without boundary, such a result is obtained by using an elliptic operator, for example the Laplace-Beltrami operator (see [**5**], [**6**]).

We have to consider some subspaces of $\mathscr{C}^\infty(\bar{\Omega})$ related to Gevrey's classes.

DEFINITION. *Let s be a real number* $\geqq 1$. *A function u is said to be in* $\mathscr{A}_s(\bar{\Omega})$ *if and only if*

(i) $u \in \mathscr{C}^\infty(\bar{\Omega})$.

(ii) $u \in G_s(\Omega)$ *(Gevrey's class of order s).*

(iii) *For every point in* $\partial\Omega$, *there exists a neighborhood V with local coordinates* $x = (x_1, \ldots, x_{n-1})$ *(tangential) and y (normal), and a constant* $M > 0$ *such that, for every* $\alpha \in N^{n-1}$ *and* $k \in N$,

$$\|(D_y y D_y)^k D_x^\alpha u\|_{L^2(V)} \leqq M^{|\alpha|+k+1}((2k + |\alpha|)!)^s.$$

We can prove easily, using Hardy's inequality,

$$G_s(\bar{\Omega}) \subset \mathscr{A}_s(\bar{\Omega}) \subset G_{2s-1}(\bar{\Omega}).$$

Therefore, for $s = 1$ we obtain $\mathscr{A}_1(\bar{\Omega}) = \mathscr{A}(\bar{\Omega})$ (for $s > 1$, $\mathscr{A}_s(\bar{\Omega}) \neq G_s(\bar{\Omega})$).

THEOREM 2. *Let s be a real number* $\geqq 1$ *and* $u \in \mathscr{C}^\infty(\bar{\Omega})$. *The following conditions are equivalent*:

(i) $u \in \mathscr{A}_s(\bar{\Omega})$.

(ii) *There exists* $L > 0$ *such that, for every integer* $k \geqq 0$,

[2]The details and the proofs of §§2 and 3 are in [**2**].

$$\|A_2^k u\|_{L^2(\Omega)} \leqq L^{k+1}((2k)!)^s$$

(*where the operator* A_2 *is defined by* (3)).

3.[2] In this section we will give some applications of Theorem 2. We assume that the form a defined by (1) is selfadjoint. Therefore, the spectrum of A_2 consists of a positive increasing sequence of eigenvalues λ_j. We denote by w_j an orthonormal basis of eigenfunctions (which are analytic).

We can prove the following asymptotic result.

THEOREM 3. *There exists a constant K such that* $\lambda_j \sim K^{j^{2/n}}$ *as* $j \to +\infty$.

(Note that we obtain the same asymptotic growth as in the elliptic case. We can show that the eigenvalues of A_1 do not have the same property.)

As a consequence of Theorems 2 and 3, we obtain the following results.

COROLLARY 1. *The mapping J, which associates to each u in* $L^2(\Omega)$ *the sequence* (u_j) *of its Fourier's coefficients on the basis* (w_j), *is an isomorphism from* $\mathscr{C}^\infty(\Omega)$ *on the space* $\mathscr{S}$ *of fastly decreasing sequences, and for every* $s \geqq 1$, *from* $\mathscr{A}_s(\overline{\Omega})$ *on* $\operatorname{indlim}_{a>1} l^2_{a^{j1/sn}}$.[3]

COROLLARY 2. *For* $s > 1$, *the space* $\mathscr{A}_s(\overline{\Omega})$ *is an interpolation space between* $\mathscr{A}(\overline{\Omega})$ *and* $\mathscr{C}^\infty(\overline{\Omega})$.

Let us note that $G_s(\overline{\Omega})$ for $s > 1$ is not an interpolation space between $\mathscr{A}(\overline{\Omega})$ and $\mathscr{C}^\infty(\overline{\Omega})$.

4. It is possible to show that Theorem 2 becomes false if we replace A_2 by A_1. Using a similar argument, we will prove, in this section, a nonanalytic result. We limit ourselves to a local problem as simple as possible.

We consider an operator

$$A = yP + Q \tag{4}$$

where P is a second order elliptic (nondegenerate) operator and Q is a first order operator; we assume the coefficients of P and Q are analytic in some neighborhood $\mathcal{O}$ of the origin in $\boldsymbol{R}^n = \{(x,y); x \in \boldsymbol{R}^{n-1}$ and $y \in \boldsymbol{R}\}$. We suppose $[P,Q] = PQ - QP = 0$.

We obtain the following result:

PROPOSITION 1. *Let V be a neighborhood of the origin in* $\boldsymbol{R}^{n-1} \times \overline{\boldsymbol{R}}_+ = \{(x,y);$ $x \in \boldsymbol{R}^{n-1}$ *and* $y \in [0,\infty[\}$ *which is relatively compact in* $\mathcal{O}$. *There exists a function* $u \in G_2(\overline{V})$, *whose restriction to any neighborhood of the origin is nonanalytic, such that*

there exists a constant $C > 0$ *with*

---

[3]If $P(j)$ is a sequence of nonnegative real numbers we denote $l^2_{P(j)} = \{(f_j) \in \boldsymbol{C}^N : \sum_{j=0}^{\infty} |f_j|^2 P(j) < +\infty\}$.

(5) $$\|D^\alpha A^k u\|_{L^2(V)} \leqq C^{|\alpha|+k+1}(2k)!(2\alpha)!$$

for each $k \in \boldsymbol{N}$ *and* $\alpha \in \boldsymbol{N}^n$.

PROOF. We note $\Gamma = \overline{\mathcal{O}} \cap \{(x, y) \in \boldsymbol{R}^{n-1} \times \boldsymbol{R}; y = 0\}$. Let g be in $G_2(\Gamma)$ and nonanalytic in any neighborhood of the origin in $\boldsymbol{R}^{n-1}$. We construct a function u in some neighborhood of $\overline{V}$ in $\boldsymbol{R}^{n-1} \times \overline{\boldsymbol{R}}_+$ such that

$$Pu = 0,$$
$$u|_{\Gamma \cap \overline{V}} = g|_{\Gamma \cap \overline{V}},$$

by solving a Dirichlet problem in some neighborhood of $\overline{V}$ in $\boldsymbol{R}^{n-1} \times \overline{\boldsymbol{R}}_+$. Then $u \in G_2(\overline{V})$ and is nonanalytic in any neighborhood of the origin (see [**6**]). So, it is easy to prove (5).

We give, now, a nonanalyticity result.

PROPOSITION 2. *There exists a neighborhood* W *of the origin in* $\boldsymbol{R}^{n-1} \times \overline{\boldsymbol{R}}_+ \times \boldsymbol{R}$ *and a function* $w \in G_2(\overline{W})$ *whose restriction to any neighborhood of the origin is not analytic, such that*

$$Bw = 0 \quad \text{in } W.$$

PROOF. Let us consider the series

(6) $$w(x,y,t) = \sum_{m=0}^{\infty} t^{2m} \frac{A^m u(x,y)}{(2m)!},^4$$

where u is given by Proposition 1. By using (5) it is easily seen that the function w is defined in $W = V \times [-\delta, +\delta]$ where δ is some suitable strictly positive number, and satisfies $Bw = 0$ in W and there exists $M > 0$ such that

$$\|D^\alpha_{x,y} D^k_t w\|_{L^2(W)} \leqq M^{|\alpha|+k+1} k!(2\alpha)!$$

for each $k \in \boldsymbol{N}$ and $\alpha \in \boldsymbol{N}^n$.

Furthermore we have

$$w(x,y,0) = u(x,y),$$

then w is nonanalytic in any neighborhood of the origin.

EXAMPLES. Let us consider, for example, the following simple case (with $n = 2$):

$$P = D_x^2 + 4D_y^2,$$
$$Q = -2miD_y \quad \text{with } m \text{ an integer} \geqq 1,$$

then

(7) $$B = y(D_x^2 + 4D_y^2) - 2miD_y + D_t^2.$$

[4]Such a series is also used in [**6**].

We use the change of variables

$$y = z_1^2 + \dots + z_m^2. \tag{8}$$

We denote by

$$\tilde{w}(x,z_1,\dots,z_m,t) = w(x,z_1^2 + \cdots + z_m^2,t)$$

where w is given by Proposition 2.

The function $\tilde{w}$ is in the Gevrey class of order 2 in some neighborhood of the origin in $\boldsymbol{R}^{m+2}$ and nonanalytic.

By the change of variables (8), the operator B defined by (7) becomes

$$H = (z_1^2 + \cdots + z_m^2)D_x^2 + D_{z_1}^2 + \cdots + D_{z_m}^2 + D_t^2$$

which can be written also in the form

$$H = \sum_{j=1}^{m} (z_j D_x)^2 + \sum_{j=1}^{m} D_{z_j}^2 + D_t^2. \tag{9}$$

In some neighborhood of the origin in $\boldsymbol{R}^{m+2}$ we have

$$H\tilde{w} = 0.$$

Therefore the following result is proved.

THEOREM 4. *Let m be an integer $\geqq 1$. The following operator*

$$H = \sum_{j=1}^{m} (z_j D_x)^2 + \sum_{j=1}^{m} D_{z_j}^2 + D_t^2$$

is not analytic-hypoelliptic in $\boldsymbol{R}^{m+2}$. More precisely, one can find a function $\tilde{w}$ defined in some neighborhood of the origin, belonging to the Gevrey class of order 2, nonanalytic and such that $H\tilde{w} = 0$.

The operator H is obviously of the form $\sum X_j^2$ and satisfies the Hörmander condition (see [**4**]).

If, in the example (9), we take $m = 1$, it turns out that the operator $z^2D_x^2 + D_z^2 + D_t^2$ is not analytic-hypoelliptic in $\boldsymbol{R}^3$; but one can prove that the operator

$$z^2D_x^2 + D_z^2 \tag{10}$$

is analytic-hypoelliptic in $\boldsymbol{R}^2$. Let us point out that M. Derridj and C. Zuily have also announced recently the analytic-hypoellipticity for some classes of operators which can be considered as generalizations of (10).

BIBLIOGRAPHY

1. M. S. Baouendi and C. Goulaouic, *Étude de la régularité Gevrey pour une classe d'opérateurs elliptiques dégénérés*, Ann. Sci. École Norm. Sup. **4** (1971), fasc. 1, 31–46.

2. ———, *Régularité analytique et itérés d'opérateurs elliptiques dégénérés; applications*, J. Functional Analysis **9** (1972), 208–248.

3. ———, *Nonanalytic-hypoellipticity for some classes of degenerate elliptic operators*, Bull. Amer. Math. Soc. **78** (1972), 483–486.

4. L. Hörmander, *Hypoelliptic second order differential equations*, Acta Math. **119** (1967), 147–171. MR **36** #5526.

5. T. Kotake and N. S. Narasimhan, *Fractional powers of a linear elliptic operator*, Bull. Soc. Math. France **90** (1962), 449–471.

6. J. L. Lions and E. Magenes, *Problemes aux limites non homogenes*, Tome 3, Dunod, Paris, 1970.

7. C. B. Morrey, Jr. and L. Nirenberg, *On the analyticity of the solutions of linear elliptic systems of partial differential equations*, Comm. Pure Appl. Math. **10** (1957), 271–290. MR **19**, 654.

PURDUE UNIVERSITY

UNIVERSITY OF PARIS II, ORSAY, FRANCE

PROLONGEMENT ET EXISTENCE DES SOLUTIONS DES SYSTÈMES HYPERBOLIQUES NON STRICTS À COEFFICIENTS ANALYTIQUES

JEAN-MICHEL BONY ET PIERRE SCHAPIRA

Ce texte est la première rédaction d'un article plus détaillé (et sans doute plus complet) à paraître prochainement. Il ne doit donc pas être considéré comme définitif.

Introduction. Nous étudions sous quelles conditions les solutions analytiques ou les solutions hyperfonctions u d'un système d'équations aux dérivées partielles $P_i u = v_i$ se prolongent à travers la frontière d'un ouvert Ω, de classe C^1 ou convexe. Nous introduisons pour cela une notion d'"hyperbolicité", ne portant que sur la partie principale des opérateurs, et vérifiée par exemple par les opérateurs hyperboliques dont les caractéristiques sont de multiplicité constante, ou par les systèmes holomorphes (dans les directions non caractéristiques).

Les méthodes utilisées sont "géométriques" et permettent d'obtenir aussi des théorèmes d'existence (et d'unicité). Par exemple si S est une hypersurface analytique réelle (resp. complexe) de normale "hyperbolique" (resp. non-caractéristique) pour un opérateur P on montre que S admet un système fondamental de voisinages dans lesquels on peut résoudre le problème de Cauchy pour les fonctions analytiques (resp. holomorphes) et même pour les hyperfonctions (ce qui a un sens grâce à la théorie du faisceau C de M. Sato [**11**]).

Le prolongement des solutions d'une équation à coefficients constants a été étudié par C.-O. Kiselman [**8**] par une méthode entièrement différente.

L'étude des opérateurs hyperboliques de type principaux, dans le cadre des hyperfonctions a été faite récemment par T. Kawaï [**6**], [**7**].

Une partie des résultats exposés ici ont été annoncés dans [**2**], [**3**], [**4**].

I. **Solutions holomorphes.** Dans ce paragraphe $P_i(z, \partial/\partial z)$ désignera une famille finie d'opérateurs différentiels à coefficients holomorphes dans un ouvert U de $\boldsymbol{C}^n$.

AMS 1970 *subject classifications*. Primary 35L99.

Key words and phrases. Hyperbolique, holomorphe, analytique, hyperfonction, problème de Cauchy.

On identifiera $\boldsymbol{C}^n$ pour le produit hermitien $\langle z, \zeta \rangle = \sum_i z_i \bar{\zeta}_i$ à l'espace euclidien $\boldsymbol{R}^{2n}$ muni du produit scalaire $\operatorname{Re}\langle z, \zeta \rangle$. Le mot hyperplan signifiera, sauf mention du contraire, hyperplan réel de $\boldsymbol{R}^{2n}$.

Un hyperplan d'équation $\operatorname{Re}\langle z - z_0, \zeta \rangle = 0$ sera caractéristique en z_0 par rapport à la famille (P_i) si on a pour tout i, $p_i(z_0, \zeta) = 0$ où p_i désigne le symbole principal de P_i, défini sur $U \times \boldsymbol{C}^n$. On dira aussi que c'est le vecteur ζ qui est caractéristique.

(a) *Théorèmes de prolongement.* Le théorème suivant est fondamental dans cette étude. Il a été démontré par M. Zerner.

THÉORÈME 1.1 [**13**]. *Soit Ω un ouvert de U dont la frontière $\partial\Omega$ est de classe C^1 en un point z_0. Supposons que la normale N à Ω en z_0 soit non-caractéristique. Alors si f est holomorphe dans Ω et $P_i f$ est holomorphe dans un voisinage de z_0, f se prolonge en fonction holomorphe au voisinage de z_0.*

DÉMONSTRATION. On peut évidemment supposer la famille réduite à un seul opérateur P.

Soit H_ε l'hyperplan d'équation $\operatorname{Re}\langle z - z_0, N \rangle = -\varepsilon$. L'intersection de H_ε et de Ω contient une boule (dans H_ε) B_a, centrée au point $z_0 + \varepsilon N$ et dont le rayon a est infiniment grand par rapport à ε lorsque ε tend vers 0.

D'après le théorème de Cauchy-Kowalevski "précisé" [**5**] il existe un nombre positif δ indépendant de a et de ε tel que si f est holomorphe au voisinage de $\bar{B}_a$, f se prolonge holomorphiquement dans le cône ouvert (tronqué) de base B_a et de hauteur δa. Ce cône sera un voisinage de z_0 pour ε assez petit.

Nous allons "globaliser" ce théorème en adaptant une méthode dûe à L. Hörmander [**5**, Théorème 5.3.3]. Pour cela si A est une partie de la sphère S^m de $\boldsymbol{R}^m$ (ici $m = 2n$), et L une partie de $\boldsymbol{R}^m$ on appellera "domaine d'influence" de L relativement à A, l'intérieur de l'intersection des demi-espaces de normale appartenant à A, contenant L.

THÉORÈME 1.2. *Soit ω et Ω deux convexes de U, ω étant localement compact, Ω ouvert, avec $\omega \subset \Omega$. Considérons les hyperplans dont la normale N est limite de directions caractéristiques en (au moins) un point de Ω, et supposons que tout hyperplan de ce type qui coupe Ω coupe ω. Alors si f est une fonction holomorphe au voisinage de ω et si $P_i f$ est holomorphe dans Ω, f se prolonge en fonction holomorphe dans Ω.*

DÉMONSTRATION. Soit A l'adhérence dans S^{2n-1} des directions caractéristiques dans Ω. Soit z_0 appartenant à ω, $\tilde{\omega}$ un voisinage ouvert de ω dans lequel f est holomorphe, z_1 un point de Ω et $\eta > 0$ tel que la boule fermée $\bar{B}(z_1, \eta)$ de centre z_1 de rayon η soit contenue dans Ω. Il existe un compact K de ω tel que tout hyperplan de normale appartenant à A qui coupe $B(z_1, \eta)$ coupe K.

Soit ω' un voisinage convexe de K ouvert et relativement compact dans ω, soit $0 < \varepsilon < \eta$ tel que ω'_ε (ensemble des points à la distance inférieure à ε de ω') soit contenu dans $\tilde{\omega}$.

Soit $0 \leqq t \leqq 1$, $z_t = z_0 + t(z_1 - z_0)$, K_t l'enveloppe convexe de ω'_ε et de $B(z_t, \varepsilon)$. La frontière de K_t est de classe C^1 et ses normales en un point hors de $\tilde{\omega}$ n'appartiennent pas à A. Il résulte alors du Théorème 1.1, que le plus petit t_0 tel que f ne se prolonge pas à K_t pour $t > t_0$, est 1.

Pour tout $z \in \Omega$ on a donc obtenu un prolongement holomorphe de f dans un ouvert étoilé en z_0 contenant z_1, ce qui définit un prolongement de f à Ω.

COROLLAIRE. *Soit Ω un ouvert convexe de U avec $\overline{\Omega} \subset U$. Soit z_0 appartenant à $\partial\Omega$ et supposons qu'aucun hyperplan d'appui à $\overline{\Omega}$ en z_0 ne soit caractéristique pour le système P_i. Il existe alors un voisinage V de z_0 dans U, tel que si f est holomorphe dans Ω, et $P_i f$ est holomorphe dans $\Omega \cup V$, f est holomorphe dans $\Omega \cup V$.*

DÉMONSTRATION. Soit Γ le cône convexe fermé des normales à $\overline{\Omega}$ en z_0. On peut trouver un voisinage borné W de z_0 et un cône convexe $\tilde{\Gamma}$ voisinage de $\Gamma - \{0\}$ tel que pour tout z dans $\overline{W}$ et ζ dans $\overline{\tilde{\Gamma}}$ on ait $p_i(z, \zeta) \neq 0$ pour un i.

Soit C le cône de sommet z_0, polaire de $\tilde{\Gamma}$, N un vecteur de $\Gamma - \{0\}$, et H_ε l'hyperplan d'équation $\mathrm{Re}\langle z - z_0, N\rangle = -\varepsilon$. L'intersection de C et de H_ε est une base compacte de C et est contenue dans $\Omega \cap W$ pour ε assez petit.

Soit A l'ensemble (fermé) des $\zeta \in S^{2n-1}$ tel que $p_i(z, \zeta) = 0$ $\forall i$ pour un z appartenant à $\overline{W}$. Si $\zeta \in A$, l'hyperplan $\mathrm{Re}\langle z - z_0, \zeta\rangle = 0$ rencontre $C \cap H_\varepsilon$ pour tout ε, donc rencontre $\Omega \cap W \cap H_\varepsilon$ suivant une partie compacte indépendante de ζ, ce qui montre que le domaine d'influence relativement à A de $\Omega \cap W \cap H_\varepsilon$ est un voisinage de z_0 pour ε assez petit.

(b) *Théorèmes d'existence.*

THÉORÈME 1.3.[1] *Soit $P(z, \partial/\partial z)$ un opérateur différentiel d'ordre m à coefficients holomorphes dans U, S une hypersurface complexe de U non-caractéristique pour P. Il existe un système fondamental de voisinages V de S dans lesquels le problème de Cauchy $Pf = g$, $\gamma(f) = (h)$ a une solution f et une seule, holomorphe dans V, pour g holomorphe dans V et (h) m-uple de fonctions holomorphes dans S.*

DÉMONSTRATION. D'après l'unicité au problème de Cauchy on est ramené à démontrer le théorème pour un ouvert convexe relativement compact d'un hyperplan complexe, et c'est alors une conséquence triviale du théorème de Cauchy-Kowalevski et du Théorème 1.2.

THÉORÈME 1.4. *Soit $P(z, \partial/\partial z)$ un opérateur différentiel sur U. Soit Ω un ouvert convexe de $\mathbf{C}^n$ avec $\overline{\Omega} \subset U$, et soit z_0 appartenant à $\partial\Omega$. Supposons qu'il existe une normale N à $\overline{\Omega}$ en z_0, un voisinage X convexe fermé $\mathbf{C}$-équilibré de 0 dans $\mathbf{C}^n$ ($\zeta \in X \Rightarrow \rho\zeta \in X, \rho \in \mathbf{C}, |\rho| \leqq 1$) tel que si ζ appartient à X, $N + \zeta$ soit non caractéristique pour P et tel que toutes les normales à $\overline{\Omega}$ en z_0 appartiennent à $N + X$. Alors z_0 admet*

[1] Ce théorème vient d'être démontré indépendemment par C. Wagschal qui utilise la méthode des series majorantes.

un système fondamental de voisinages convexes V tels que si g est holomorphe dans $\Omega \cap V$ il existe f holomorphe dans $\Omega \cap V$ solution de $Pf = g$.

L'hypothese se réduit à: N est non caractéristique, si N est l'unique normale à $\overline{\Omega}$ en z_0.

DÉMONSTRATION. Soit W un voisinage borné de z_0 tel que en tout point de $\overline{W}$ les vecteurs de $N + X$ soient non caractéristiques. Soit $\tilde{H}_\varepsilon$ l'hyperplan complexe d'équation $\langle z - z_0, N\rangle = -\varepsilon$, C le cône de sommet z_0 polaire de $N + X$ et soit enfin A l'ensemble des directions caractéristiques en un point de $\overline{W}$.

Si $\zeta \in A$ l'hyperplan d'équation $\operatorname{Re}\langle z - z_0, \zeta\rangle = 0$ rencontre $C \cap \tilde{H}_\varepsilon$ car sinon cela signifierait que $\zeta \pm iN$ appartient au cône engendré par $N + X$ ce qui contredit le fait que X est équilibré et $\zeta \notin N + X$.

Comme $C \cap \tilde{H}_\varepsilon$ est une partie convexe compacte de $\Omega \cap W$ pour ε assez petit, le domaine d'influence de $\Omega \cap W \cap \tilde{H}_\varepsilon$ relativement à A sera un voisinage de z_0 pour ε assez petit.

Il suffit alors de prendre pour V l'enveloppe convexe d'un voisinage de z_0 et d'un voisinage de $\Omega \cap W \cap \tilde{H}_\varepsilon$ suffisamment petits pour être dans ce domaine d'influence.

On résoud ensuite le problème de Cauchy $Pf = g, \gamma(f) = 0$, au voisinage de $\Omega \cap W \cap \tilde{H}_\varepsilon$ et le Théorème 1.2 permet de prolonger f à $\Omega \cap V$.

II. **Solutions analytiques.** Dans ce paragraphe $P_i(x, \partial/\partial x)$ désignera une famille finie d'opérateurs différentiels à coefficients analytiques dans un ouvert U de $\boldsymbol{R}^n$ et se prolongent en fonctions holomorphes dans un voisinage $\tilde{U}$ de U dans $\boldsymbol{C}^n$.

Nous poserons $z = x + iy, \zeta = \xi + i\eta$ et, identifiant $\boldsymbol{C}^n$ à $\boldsymbol{C}^{n-1} \times \boldsymbol{C}$, on désignera par $\zeta' = \xi' + i\eta'$ (resp. $\zeta_n = \xi_n + i\eta_n$) la première (resp. la deuxième) projection de ζ.

(a) *Hypothèses d'hyperbolicité.* Nous serons amenés à faire les hypothèses d'"hyperbolicité" suivantes.

$H_1(x_0, N)$. Le vecteur N de $\boldsymbol{R}^n - \{0\}$ est non caractéristique en x_0 et les racines τ communes aux polynômes $p_i(x, \xi + \tau N)$ sont réelles pour x réel et ξ réel.

$H_2(x_0, N)$. Le vecteur N de $\boldsymbol{R}^n - \{0\}$ est non caractéristique en x_0 et si l'on suppose les coordonnées choisies telles que $N = (0, \dots, 0, 1)$, il existe deux constantes $\alpha > 0$ et $C > 0$ telles que les racines communes aux polynômes $p_i(z, \zeta' + \tau N) = 0$ vérifient

$$|\operatorname{Im} \tau| \leqq C[|\operatorname{Im} z| + |\operatorname{Im} \zeta'|]$$

pour $|\operatorname{Re} \zeta'| = 1$, $|\operatorname{Im} \zeta'| < \alpha$, $|z - x_0| < \alpha$.

REMARQUE 2.1. Sous l'hypothèse $H_2(x_0, N)$ il existe α' et C' tels que pour $|z - x_0| < \alpha'$ et pour ζ appartenant à $\boldsymbol{C}^n$ on ait

$$|\operatorname{Im} \tau| \leqq C'[|\xi'|\,|y| + |\eta|]$$

si $p_i(z, \zeta + \tau N) = 0$ quelque soit i.

En effet cette inégalité se déduit par homogénéité de la précédente si $|\eta'| \leqq \alpha|\xi'|$. D'autre part N étant non caractéristique pour l'un des P_i au voisinage de x_0, on a une majoration $|\tau + \zeta_n| \leqq C''|\zeta'|$ et donc $|\mathrm{Im}\, \tau| \leqq C''[|\zeta'| + |\eta_n|]$ ce qui entraîne $|\mathrm{Im}\, \tau| \leqq C'''|\eta|$ pour $|\eta'| \geqq \alpha|\xi'|$.

REMARQUE 2.2. L'hypothèse $H_2(x_0, N)$ entraîne $H_2(x, N')$ pour N' voisin de N et x voisin de x_0.

(b) *Discussion de l'hypothèse H_2.* Les hypothèses faites ne portent que sur les parties principales des opérateurs. Signalons deux situations très différentes où l'hypothèse H_2 est vérifiée.

THÉORÈME 2.1. *Soit P un opérateur différentiel à coefficients analytiques dont la partie principale est hyperbolique dans la direction N au voisinage de x_0. Si les racines en τ de $p(x, \xi + \tau N)$ ont une multiplicité constante pour x voisin de x_0 et $\xi \in \boldsymbol{R}^n - \{0\}$, l'opérateur P vérifie l'hypothèse $H_2(x_0, N)$. Il en est de même si la partie principale de P est un produit de tels opérateurs.*

D'après [**10**], cette hypothèse est équivalente à: p est un produit de symboles d'opérateurs strictement hyperboliques (voir [**9**]).

En effet le polynôme $p(x, \xi + \tau N)$ se décompose sous la forme $p(N)\Pi(\tau - \tau_i(x, \xi))^{\alpha_i}$ au voisinage de chaque point (x_0, ξ_0), les τ_i étant des fonctions analytiques à valeurs réelles. Les fonctions τ_i se prolongent en fonctions holomorphes de (z, ζ) réelles pour (z, ζ) réel et donc vérifiant $|\mathrm{Im}\, \tau_i| \leqq C[|\mathrm{Im}\, z| + |\mathrm{Im}\, \zeta|]$. D'autre part on a toujours $p(z, \zeta + \tau N) = p(N)\Pi(\tau - \tau_i(z, \zeta))^{\alpha_i}$ d'où $H_2(x_0, N)$.

La proposition peut être généralisée à un système d'opérateurs (P_i) dont les parties principales sont à coefficients réels. Dans ce cas l'hypothèse $H_1(x_0, N)$ entraîne $H_2(x_0, N)$ si de plus pour chaque racine commune τ à $p_i(x_0, \xi_0 + \tau N) = 0$, la racine τ a une multiplicité constante pour l'un au moins des $p_i(x, \xi + \tau N)$ au voisinage de (x_0, ξ_0).

THÉORÈME 2.2. *Supposons que le vecteur N soit non caractéristique et que le système d'équations $p_i(x_0, \xi + \tau N) = 0$ n'ait aucune racine τ complexe commune pour ξ non proportionnel à N. Alors (P_i) vérifie $H_2(x_0, N)$.*

Choisissons les coordonnées de manière que l'on ait $N = (0, \ldots, 0, 1)$. D'après la continuité des racines de chaque p_i, le système $p_i(z, \zeta' + \tau N) = 0$ n'a pas de racines communes pour $|z - x_0| \leqq \alpha$, $|\xi'| = 1$ et $|\eta'| \leqq \alpha$ ce qui entraîne trivialement $H_2(x_0, N)$.

EXEMPLES. Les systèmes satisfaisant aux hypothèses de la Proposition 2 sont sur-déterminés elliptiques. Donnons deux exemples:

(1) Dans $\boldsymbol{C}^m$ identifié à $\boldsymbol{R}^{2m}$ le système $\{\partial/\partial\bar{z}_1, \ldots, \partial/\partial\bar{z}_m, P(z, \partial/\partial z)\}$ pour $p(z_0, N) \neq 0$.

(2) Dans $\boldsymbol{R}^n$, le système (P, Q), où le vecteur $N = (0, \ldots, 0, 1)$ est non caractéristique par rapport à P et où la partie principale de Q se réduit à un opérateur en $\partial/\partial x_1, \ldots, \partial/\partial x_{n-1}$, elliptique en ces variables.

(c) *Théorème de prolongement* (1).

THÉORÈME 2.3. *Soit N le vecteur $(0, \ldots, 0, 1)$ et supposons que la famille (P_i) vérifie l'hypothèse $H_2(x, N)$ en tout point x de la boule fermée $\bar{B}(0, r)$ de centre 0 de rayon r. Il existe alors une constante $\delta > 0$ telle que pour tout $a < r$, toute fonction f analytique au voisinage de $\bar{B}(0, a) \cap \{x_n = 0\}$ ayant la propriété que $P_i f$ se prolonge analytiquement dans le cône tronqué de sommet δaN et de base $B(0, a) \cap \{x_n = 0\}$, se prolonge elle-même analytiquement dans ce cône.*

La constante δ peut être choisie ne dépendant que de r et de la constante C' de la Remarque 2.1.

DÉMONSTRATION. Soit $\varepsilon > 0$. Désignons par A_ε l'ensemble des directions caractéristiques au système en un point $z = x + iy$ avec $|x| \leqq r$, $|y| \leqq \varepsilon$. D'après le Théorème 1.2 il suffit de vérifier que si $\zeta \in A_\varepsilon$, l'hyperplan de normale ζ passant par le point δaN rencontre l'ouvert $|x'| < a$, $|y| < \varepsilon$ de $\{x_n = 0\}$.

Il faut donc vérifier que le système d'inéquations $\mathrm{Re}\langle z - \delta aN, \zeta\rangle = 0$, $x_n = 0$, $|x'| < a$, $|y| < \varepsilon$ a une solution sachant que

$$|\xi_n| \leqq C'[\varepsilon|\eta| + |\xi'|]$$

d'après la Remarque 2.1 appliquée au vecteur $i\zeta$.

On peut par exemple supposer $\xi_n \geqq 0$ et d'après la convexité des inéquations il suffit de trouver x' et y avec $|x'| < a$, $|y| < \varepsilon$ et $\langle x', \xi'\rangle - \langle y, \eta\rangle \geqq \delta a\xi_n$.

Or on peut trouver x' tel que $\langle x', \xi'\rangle = a/2|\xi'|$ et y tel que $\langle y, \eta\rangle = -\varepsilon|\eta|/2$, d'où l'inégalité

$$\tfrac{1}{2}[a|\xi'| + \varepsilon|\eta|] \geqq \delta a\xi_n$$

qui sera vérifiée pourvu que l'on ait $\delta \leqq 1/2rC'$.

(d) *Théorèmes de prolongement* (2).

THÉORÈME 2.4. *Soit Ω un ouvert de U dont la frontière est de classe C^1. Soit x_0 appartenant à $\partial\Omega$ et N la normale à $\bar{\Omega}$ en ce point. Supposons que l'hypothèse $H_2(x_0, N)$ soit vérifiée. Il existe alors un voisinage V de x_0 tel que si f est analytique dans Ω et $P_i f$ dans $\Omega \cup V$, f se prolonge analytiquement à $\Omega \cup V$.*

THÉORÈME 2.5. *Soit ω et Ω deux convexes de U, ω étant localement compact, Ω ouvert, avec $\omega \subset \Omega$. Considérons les hyperplans dont la normale N est limite de direction N' ne vérifiant pas $H_2(x, N')$ en un point (au moins) x de Ω, et supposons que tout hyperplan de ce type qui coupe Ω coupe ω. Alors si f est une fonction analytique au voisinage de ω et si $P_i f$ est analytique dans Ω, f se prolonge en fonction analytique dans Ω.*

COROLLAIRE. *Soit Ω un ouvert convexe avec $\bar{\Omega} \subset U$. Soit x_0 appartenant à $\partial\Omega$ et supposons que toutes les normales N en x_0 à $\bar{\Omega}$ vérifient $H_2(x_0, N)$. Il existe alors un voisinage V de x_0 tel que si f est analytique dans Ω et $P_i f$ est analytique dans $\Omega \cup V$, f se prolonge analytiquement dans $\Omega \cup V$.*

Les démonstrations sont les mêmes que dans le cas holomorphe.

REMARQUE 2.3. Les Théorèmes 2.4 et 2.5 contiennent les Théorèmes 1.1 et 1.2.

(e) *Théorèmes d'existence.*

THÉORÈME 2.6. *Soit $P(x, \partial/\partial x)$ un opérateur différentiel d'ordre m à coefficients analytiques dans U, S une hypersurface analytique de U. On suppose qu'en chaque point x_0 de S, la normale N à S vérifie l'hypothèse $H_2(x_0, N)$ relativement à P (cf. le Théorème 2.1). Alors S admet un système fondamental de voisinages V tels que si g est analytique dans V et (h) est un m-uple de fonctions analytiques dans S, le problème de Cauchy $Pf = g$, $\gamma(f) = (h)$ admet une solution et une seule f, analytique dans V.*

La démonstration est la même que dans le cas holomorphe.

THÉORÈME 2.7. *Soit $P(x, \partial/\partial x)$ un opérateur différentiel à coefficients analytiques dans U. Soit Ω un ouvert convexe de U avec $\overline{\Omega} \subset U$. Soit x_0 appartenant à $\partial\Omega$ et supposons que toutes les normales N à $\overline{\Omega}$ en x_0 vérifient l'hypothèse $H_2(x_0, N)$ pour P. Alors x_0 admet un système fondamental de voisinages convexes V tels que si g est analytique dans $\Omega \cap V$ il existe f analytique dans $\Omega \cap V$, solution de $Pf = g$.*

Ce théorème est "meilleur" que le Théorème 1.4 et sa démonstration (dont le principe est le même) un peu plus simple, grâce au fait que l'on peut résoudre le problème de Cauchy au voisinage d'une hypersurface réelle, donc de codimension réelle 1.

III. **Solutions hyperfonctions.** Nous nous plaçons dans le cadre du §II. Bien qu'elle n'apparaisse pas explicitement ici, la théorie du faisceau C de M. Sato est sous-jacente à cette étude [**11**].

(a) *Théorèmes d'unicité (Holmgren).*

THÉORÈME 3.1 [**12**]. *Soit Ω un ouvert de U dont la frontière $\partial\Omega$ est de classe C^1. Supposons que la normale N à Ω en un point x_0 soit non caractéristique pour le système (P_i). Soit u une hyperfonction sur U solution du système $P_i u = 0$, et supposons u nulle dans Ω. Alors u est nulle au voisinage de x_0.*

THÉORÈME 3.2. *Soit ω et Ω deux convexes de U, ω étant localement compact, Ω ouvert avec $\omega \subset \Omega$. Considérons les hyperplans dont la normale N est limite de directions caractéristiques en un point au moins de Ω, et supposons que tout hyperplan de ce type qui coupe Ω coupe ω. Alors si u est une hyperfonction sur Ω solution du système $P_i u = 0$, et si u est nulle au voisinage de ω, u est nulle dans Ω.*

On pourrait évidemment énoncer un corollaire analogue à celui du Théorème 2.5.

On pourrait aussi affaiblir l'hypothèse que les normales sont non caractéristiques [**1**].

(b) *Un nouveau théorème de prolongement des fonctions holomorphes.* Désignons par $B'(0, a)$ l'intersection avec l'hyperplan $x_n = 0$ de la boule $B(0, a)$ de centre 0 de rayon a. Soit $N = (0, \ldots, 0, 1)$ et $K(a, \delta)$ le cône ouvert tronqué de base $B'(0, a)$ de sommet $\delta a N$.

Si Γ est un cône de $\boldsymbol{R}^n$ on désignera par Γ_ε l'intersection de ce cône et de la boule de rayon ε.

THÉORÈME 3.3. *Supposons que l'hypothèse $H_2(x, N)$, avec $N = (0, \ldots, 0, 1)$, soit vérifiée en tout point x de $\bar{B}(0, r)$. Soient Γ et Γ' deux cônes ouverts convexes de $\boldsymbol{R}^n$ tels que l'on ait $\bar{\Gamma}' - \{0\} \subset \Gamma$. Il existe alors une constante δ telle que, si f est holomorphe au voisinage de $\bar{B}'(0, a) \times i\Gamma_\varepsilon$, et si $(P_i f)$ est holomorphe dans $K(a, \delta) \times i\Gamma_\varepsilon$ (avec $a < r$), la fonction f se prolonge holomorphiquement dans $K(a, \delta) \times i\Gamma'_{\varepsilon/2}$.*

DÉMONSTRATION. Soit A_ε l'ensemble des directions caractéristiques en au moins un point de $\bar{B}(0, r) \times i\{y < \varepsilon\}$. Il résulte des hypothèses que l'on peut trouver δ_0 tel que tout hyperplan de normale appartenant à A_ε et passant par le point $\delta_0 a N$ rencontre $B'(0, a) \times i\{|y| < \varepsilon\}$ (voir la démonstration du Théorème 2.3). On en déduit qu'il existe $\alpha > 0$ tel que, pour tout vecteur γ de norme l'appartenant à Γ', tout hyperplan de normale appartenant à A_ε et passant par le point $\alpha\delta_0 a N + i(\varepsilon/2)\gamma$ coupe $B'(0, a) \times i\Gamma_\varepsilon$. Si on désigne par $M_{\gamma,\varepsilon}$ l'intérieur de l'enveloppe convexe de $B'(0, a) \times i\Gamma_\varepsilon$ et de $\alpha\delta_0 a N + i(\varepsilon/2)\gamma$, la fonction f sera holomorphe dans $M_{\gamma,\varepsilon}$ s'il en est ainsi des $P_i f$ (cf. Théorème 1.2). La réunion des $M_{\gamma,\varepsilon'}$ pour γ dans Γ' et $\varepsilon' < \varepsilon$ contient l'ouvert voulu, en posant $\delta = \alpha\delta_0$.

(c) *Problème de Cauchy.* Si u est une hyperfonction sur un ouvert Ω, et si N est un vecteur de S^{n-1}, nous dirons avec M. Sato [**11**] que u est analytique dans la direction N s'il existe des parties convexes propres fermées I_α de S^{n-1} avec $\pm N \notin I_\alpha$, et des fonctions holomorphes f_α dans $\Omega \times i\Gamma_\alpha$ (où Γ_α est le cône polaire de I_α) au voisinage de Ω, tels que u soit somme des valeurs aux bord des f_α. Si G est un cône ouvert contenant N tel que G et $-G$ soient disjoint des I_α, nous dirons que f est G-analytique.

Si H est un hyperplan de normale N, on peut alors définir la restriction $\gamma_0(u)$ de u à $H \cap \Omega$ comme étant la somme des valeurs au bord des restrictions des f_α à $\tilde{H}$, complexifié de H. Cette restriction ne dépend pas des f_α choisis.

Si N est noncaractéristique pour un opérateur P sur Ω, et si Pu est analytique dans la direction N, l'hyperfonction u est analytique dans la direction N (théorème de Sato [**11**]).

THÉORÈME 3.4. *Soit $P(x, \partial/\partial x)$ un opérateur différentiel d'ordre m sur U. Soit N le vecteur $(0, \ldots, 0, 1)$, soit G un voisinage conique de N, et supposons l'hypothèse $H_2(x, N)$ vérifiée en tout point x de $\bar{B}(0, r)$. Il existe alors une constante $\delta > 0$, telle que pour tout $a < r$, toute hyperfonction v G-analytique définie dans la réunion d'un voisinage de $\bar{B}'(0, a)$ et de $K(a, \delta)$, tout m-uple (w) d'hyperfonctions définies au voisinage de $\bar{B}'(0, a)$ dans $\boldsymbol{R}^{n-1}$, le problème de Cauchy $Pu = v, \gamma(u) = (w)$ admet une et une seule solution hyperfonction, définie dans la réunion d'un voisinage de $\bar{B}'(0, a)$ et de $K(a, \delta)$.*

DÉMONSTRATION. On peut trouver un recouvrement (I_α) du complémentaire d'un voisinage de $\pm N$ dans S^{n-1} par des parties convexes propres fermées telles que toute hyperfonction G-analytique soit somme de valeurs au bord de fonctions g_α holomorphes dans $(V \cup K(a, \delta)) \times i\Gamma_{\alpha,\varepsilon}$ (où Γ_α est le cône polaire de I_α, et où V est un voisinage de $\bar{B}'(0, a)$) et que tout m-uple (w) soit comme de valeurs au bord de m-uples (h_α) de fonctions holomorphes dans $V' \times i\Gamma'_{\alpha,\varepsilon} = (V \times i\Gamma_{\alpha,\varepsilon}) \cap \{z_n = 0\}$.

Soit f_α la solution holomorphe du problème de Cauchy $Pf_\alpha = g_\alpha$, $\gamma(f_\alpha) = (h_\alpha)$. D'après la forme précisée du théorème de Cauchy-Kowalevski, f_α est définie dans un voisinage de $\bar{B}'(0, a) \times i\Gamma^1_{\alpha,\varepsilon}$, où Γ^1_α est un cône ouvert dont la trace sur $\mathbf{R}^{n-1}$ est Γ'_α. On peut alors appliquer le Théorème 3.3 et prolonger f dans $K(a, \delta) \times i\Gamma^2_{\alpha,\varepsilon/2}$. On prend alors comme solution u la somme des valeurs au bord des f_α.

(REMARQUE. Le nombre δ trouvé ne dépend que de P et de G. En particulier, pour le problème de Cauchy homogène $Pu = 0$, $\gamma(u) = (w)$, le domaine d'existence $K(a, \delta)$ ne dépend pas des données de Cauchy (w)).

L'unicité de u peut se démontrer de la même manière. Supposons en effct que $Pf_\alpha = \sum_{\beta \neq \alpha} g_{\alpha,\beta}$; $\gamma(f_\alpha) = \sum_{\beta \neq \alpha} (h_{\alpha,\beta})$, où $g_{\alpha,\beta} = -g_{\beta,\alpha}$, $h_{\alpha,\beta} = -h_{\beta,\alpha}$. On peut résoudre au voisinage de $\bar{B}'(0, a) \times i\Gamma'_{\alpha,\varepsilon}$ les équations $Pf_{\alpha,\beta} = g_{\alpha,\beta}$, $\gamma(f_{\alpha,\beta}) = (h_{\alpha,\beta})$. Les valeurs au bord des fonctions $f_{\alpha,\beta}$ sont bien définies d'après le Théorème 3.3 et ces fonctions vérifieront d'après l'unicité du problème de Cauchy

$$f_{\alpha,\beta} = -f_{\beta,\alpha}; \qquad \sum_{\beta \neq \alpha} f_{\alpha,\beta} = f_\alpha.$$

La somme des valeurs au bord des f_α sera donc nulle.

En fait, l'unicité de u peut se démontrer à l'aide et sous les seules hypothèses du Théorème 3.1.

(d) *Un théorème d'existence.*

THÉORÈME 3.5. *Soit $P(x, \partial/\partial x)$ un opérateur différentiel à coefficients analytiques qui vérifie l'hypothèse $H_2(x_0, N)$ en un point x_0 de U pour un vecteur N. Il existe alors un voisinage Ω de x_0 tel que pour tout ouvert ω de Ω et toute hyperfonction v sur ω, il existe une hyperfonction u sur ω solution de $Pu = v$.*

DÉMONSTRATION. D'après le Théorème 3.4, il suffit de montrer que l'on peut résoudre l'équation $Pu = v$ quand v est somme de valeurs au bord de fonctions g^+ et g^- respectivement holomorphes dans des ouverts $\omega \times i\Gamma^+$ et $\omega \times i\Gamma^-$, où $\Gamma^\pm$ est le polaire d'un voisinage que l'on peut prendre arbitrairement petit de $\pm N$. Le théorème résulte alors du Théorème 1.4.

REMARQUE 3.1. Dans le cas où les caractéristiques de P sont simples, les théorèmes 3.4 et 3.5 ont été démontrés par T. Kawaï. La méthode de Kawaï consiste à construire une solution élémentaire en résolvant des problèmes de Cauchy

$$P(z, \partial/\partial z)E_j(z, \tilde{z}, \eta) = 0,$$
$$\left(\frac{\partial}{\partial z_n}\right)^k E_j(z, \tilde{z}, \eta)\Big|_{z_n = \tilde{z}_n = 0} = \delta_{j,k} \frac{1}{\langle z' - \tilde{z}', \eta\rangle^{n-1}},$$

ceci grâce à un théorème de Hamada.

(e) *Théorèmes de prolongement.*

THÉORÈME 3.6. *Soit $P(x, \partial/\partial x)$ un opérateur différentiel à coefficients analytiques sur U. Soit Ω un ouvert de U dont la frontière $\partial\Omega$ est de classe C^1 en un point x_0. Supposons que la normale N en x_0 vérifie l'hypothèse $H_2(x_0, N)$ pour P. Alors si u est une hyperfonction sur Ω solution de l'équation $Pu = 0$, u se prolonge de manière unique au voisinage de x_0 en solution hyperfonction de l'équation $Pu = 0$.*

DÉMONSTRATION. Ce théorème résulte du Théorème 3.5 de la même manière que le Théorème 1.1 résulte du théorème de Cauchy-Kowalevski précisé puisque si $Pu = 0$, u est analytique dans la direction N.

THÉORÈME 3.7. *Soit ω et Ω deux convexes de U, ω étant localement compact, Ω ouvert avec $\omega \subset \Omega$. Considérons les hyperplans dont la normale N est limite de directions N' ne vérifiant pas l'hypothèse $H_2(x, N')$ pour P, en un point (au moins) de Ω et supposons que tout hyperplan de ce type qui coupe Ω coupe ω. Alors si u est une hyperfonction au voisinage de ω solution de $Pu = 0$, u se prolonge (de manière unique) en solution hyperfonction sur Ω de l'équation $Pu = 0$.*

DÉMONSTRATION. Soit A l'adhérence des directions N ne vérifiant pas $H_2(x, N)$ pour un x de Ω. Soit $z \in \Omega$. Il résulte du Théorème 3.6 que si ω_0 est un ouvert relativement compact de ω, u se prolonge au voisinage de l'enveloppe convexe de $\overline{\omega}_0$ et de z (la démonstration est la même que pour le Théorème 1.2).

Soit alors ω_t $(0 \leqq t < 1)$ une famille d'ouverts convexes relativement compacts dans ω tels que $\bigcup_{t<t_0} \omega_t = \omega_{t_0}$, $(\omega_t)_{t>t_0}$ est un système fondamental de voisinages de $\overline{\omega}_{t_0}$ et $\bigcup_{t<1} \omega_t = \Omega$.

Soit L_t l'intérieur de l'enveloppe convexe de ω_t et de z.

Soit t_0 le plus petit t tel que u ne se prolonge pas à L_t pour $t > t_0$. Il résulte du Théorème 3.6 que $t_0 = 1$ car les normales à L_t hors de ω et de z_1 n'appartiennent pas à A.

Pour tout z de Ω, u se prolonge donc en solution hyperfonction de l'équation $Pu = 0$ dans l'intérieur de l'enveloppe convexe de z et de ω. Les différents prolongements se recollent quand z parcourt Ω d'après le Théorème 3.2 et définissent le prolongement cherché.

Nous laissons au lecteur le soin d'énoncer l'analogue du corollaire du Théorème 1.2.

AJOUTÉ À LA CORRECTION DES ÉPREUVES. (1) Dans le cas d'un seul opérateur, les hypotheses $H_1(x, N)$ et $H_2(x, N)$ sont en fait équivalentes comme nous l'a fait remarquer M. Kashiwara.

(2) Les conclusions du Théorème 1.4 restent valides sous la seule hypothese: aucune normale à $\overline{\Omega}$ en Z_0 n'est caractèristique.

(3) La constante δ du Théorème 3.4 est indépendante de G.

Bibliographie

1. J.-M. Bony, *Une extension de théorème de Holmgren sur l'unicité du problème de Cauchy*, C. R. Acad. Sci. Paris Sér. A-B **268** (1969), A1103–A1106. MR **39** #3143.

2. J.-M. Bony et P. Schapira, *Sur le prolongement des solutions holomorphes d'équations aux dérivées partielles définies dans des ouverts convexes*, C. R. Acad. Sci. Paris Sér. A-B **273** (1971), A360.

3. ———, *Existence et prolongement des solutions analytiques des systèmes hyperboliques non stricts*, C. R. Acad. Sci. Paris **274** (1972), 86–89.

4. ———, *Problème de Cauchy, existence et prolongement pour les hyperfonctions solutions d'équations hyperboliques non strictes*, C. R. Acad. Sci. Paris **274** (1972), 188–191.

5. L. Hörmander, *Linear partial differential operators*, Die Grundlehren der math. Wissenschaften, Band 116, Academic Press, New York; Springer-Verlag, Berlin, 1963. MR **28** #4221.

6. T. Kawai, *Construction of elementary solutions for I-hyperbolic operators and solutions with small singularities*, Proc. Japan Acad. **46** (1970), 912–915.

7. ———, *On the global existence of real analytic solutions of linear differential equations*. I, II, Proc. Japan Acad. **47** (1971), 537–540, 643–647.

8. C.-O. Kiselman, *Prolongement des solutions d'une équation aux dérivées partielles à coefficients constants*, Bull. Soc. Math. France **97** (1969), 329–356. MR **42** #2161.

9. J. Leray et Y. Ohya, *Systèmes linéaires hyperboliques non stricts*, Deuxième Colloq. Anal. Fonct. (Liège, 1964), Centre Belge Recherches Math., Librairie Universitaire, Louvain, 1964. MR **32** #7956.

10. S. Matsuura, *On non strict hyperbolicity*, Proc. Internat. Conf. on Functional Analysis and Related Topics (Tokyo, 1969), Univ. of Tokyo Press, Tokyo, 1970, pp. 171-176. MR **42** #2174.

11. M. Sato, *Regularity of hyperfunction solutions of partial differential equations*, Proc. Internat. Congress Math., Nice, 1970.

12. P. Schapira, *Théorème d'unicité de Holmgren et opérateurs hyperboliques dans l'espace des hyperfonctions*, An. Acad. Brasil. Sci. **43** (1971), 38–44.

13. M. Zerner, *Domaine d'holomorphie des fonctions vérifiant une équation aux dérivées partielles*, C. R. Acad. Sci. Paris Sér. A-B **272** (1971), A1646–A1648.

Université de Paris VI

Université de Paris VII

GROWTH PROPERTIES OF SOLUTIONS OF CERTAIN "CANONICAL" HYPERBOLIC EQUATIONS WITH SUBHARMONIC INITIAL DATA

ROBERT CARROLL[1] AND HOWARD SILVER

1. For certain Lie groups G and compact subgroups K we show that the Darboux equation (cf. [**5**]), or the Euler-Poisson-Darboux equation (cf. [**8**]), on G/K can be embedded in a "canonical" infinite sequence of hyperbolic equations analogous to an EPD sequence. Canonical operations in the representation theory of G lead to canonical recurrence relations which yield formulas connecting solutions with different parameter values. This gives group theoretic meaning to some of the operations of EPD theory developed by Weinstein and others and yields new classes of equations where positive resolvents exist and a parallel theory can be constructed (we omit general references to EPD theory since such a list would be quite long). In particular we obtain some new growth and convexity theorems for suitable subharmonic initial values similar to those of Weinstein (cf. [**1**], [**4**], [**8**], [**9**]). In this article we indicate briefly the group theoretic situation for an example of the EPD theory where $G = \boldsymbol{R}^2 \times_\gamma SO(2)$ (natural semidirect product) with $K = SO(2)$ and then we develop briefly the parallel theory for $G = SL(2, \boldsymbol{R})$ and $K = SO(2)$. Full details and some extensions will appear elsewhere. For background material on groups and special functions we refer to [**5**], [**6**], [**7**] to which our debt is evident.

2. Let $G = \boldsymbol{R}^2 \times_\gamma SO(2)$ and $K = SO(2)$ where $(x, \alpha)(y, \beta) = (x + \gamma(\alpha)y, \alpha + \beta)$ with $\gamma(\alpha)y = (y_1 \cos\alpha - y_2 \sin\alpha, y_1 \sin\alpha + y_2 \cos\alpha)$. Elements of G can be represented by matrices

$$(2.1) \qquad g = \begin{pmatrix} \cos\alpha & -\sin\alpha & x_1 \\ \sin\alpha & \cos\alpha & x_2 \\ 0 & 0 & 1 \end{pmatrix}$$

AMS 1970 *subject classifications*. Primary 35Q05, 33A75; Secondary 35L15, 43A85.
[1] Research supported in part by a grant from the University of Illinois Research Council.

and as generators of the Lie algebra $\tilde{g}$ of G one can take

$$(2.2)\qquad a_1 = \begin{pmatrix} 0 & 0 & 1 \\ 0 & 0 & 0 \\ 0 & 0 & 0 \end{pmatrix}; \quad a_2 = \begin{pmatrix} 0 & 0 & 0 \\ 0 & 0 & 1 \\ 0 & 0 & 0 \end{pmatrix}; \quad a_3 = \begin{pmatrix} 0 & -1 & 0 \\ 1 & 0 & 0 \\ 0 & 0 & 0 \end{pmatrix},$$

with multiplication table

$$(2.3)\qquad [a_1, a_2] = 0; \quad [a_2, a_3] = a_1; \quad [a_3, a_1] = a_2$$

(thus $\tilde{g}$ is solvable). Writing $M = G/K$ we consider the quasi regular representation of G on $L^2(M)$ defined by $L(g)f(x) = f(g^{-1}x) = f(\gamma(-\alpha)(x - y))$ for $g = (y, \alpha)$. Then L induces a representation of $\tilde{g}$ on any subspace of C^∞ functions $V \subset L^2(M)$ by the rule

$$(2.4)\qquad L(a_i)f(x) = \frac{d}{dt} f(\exp(-a_i t)x)\Big|_{t=0}.$$

Writing $A_i = L(a_i)$ this yields $A_1 = -\partial/\partial x_1$, $A_2 = -\partial/\partial x_2$, and $A_3 = x_2\partial/\partial x_1 - x_1\partial/\partial x_2$. One defines $H_+ = A_1 + iA_2$, $H_- = A_1 - iA_2$, and $H_3 = iA_3$ so that, in geodesic polar coordinates (θ, t) in M $(ds^2 = dt^2 + t^2 d\theta^2)$,

$$(2.5)\qquad \begin{aligned} &H_+ = -e^{i\theta}\left[\frac{\partial}{\partial t} + \frac{i}{t}\frac{\partial}{\partial \theta}\right]; \qquad H_- = -e^{-i\theta}\left[\frac{\partial}{\partial t} - \frac{i}{t}\frac{\partial}{\partial \theta}\right]; \\ &\Delta = H_+ H_- = \frac{\partial^2}{\partial t^2} + \frac{1}{t}\frac{\partial}{\partial t} + \frac{1}{t^2}\frac{\partial^2}{\partial \theta^2}. \end{aligned}$$

Now if $p, q \in M$ with $\pi(g) = p$ ($\pi: G \to G/K$ is the natural map) then the orbit of $g \cdot q$ under the isotropy subgroup gKg^{-1} of G at p is gKq and one defines the mean value of a function f over this orbit by

$$(2.6)\qquad F(p, q) = [M^q(f)](p) = \int_K f(gkq)\, dk,$$

where dk is the normalized invariant measure on K (cf. [**5**]). In the present situation if $p = x$ and $q = y = (\theta, t)$ this reduces to the mean value of f over the circle with center x and radius t which can also be written as $M(x, t, f) = \mu_x(t) * f$ where

$$\langle \mu_y(t), \varphi \rangle = (1/2\pi) \int_0^{2\pi} \varphi(t \cos \theta, t \sin \theta)\, d\theta$$

(cf. [**1**] for notation). The Darboux equation on M is $\Delta_t M(x, t, f) = \Delta_x M(x, t, f)$ which becomes

(2.7)
$$M_{tt}(x, t, f) + \frac{1}{t} M_t(x, t, f) = \Delta_x M(x, t, f);$$
$$M(x, 0, f) = f; \qquad M_t(x, 0, f) = 0.$$

Further, one knows that $\Delta_x M(x, t, f) = M(x, t, \Delta f)$ which is trivial here since $\Delta_x \delta * \mu_x(t) * f = \mu_x(t) * \Delta f$, but the relation is also true in more general situations (cf. [**4**], [**5**], [**9**]).

To decompose L we consider irreducible representations of $\tilde{g}$ on function spaces W over M induced by irreducible unitary representations of G; with certain provisos (cf. [**6**], [**7**]) these are characterized by the choice of basis vectors $f_m \in W$ such that

(2.8)
$$H_3 f_m = m f_m, \qquad H_+ f_m = i\nu f_{m+1}, \qquad H_- f_m = i\nu f_{m-1},$$

where $m \in Z$ and ν is real. Thus $\Delta f_m = H_+ H_- f_m = -\nu^2 f_m$ and we can take $f_m = \exp(im\theta) u^m(t)$ where $u^m(t) = (-i)^m J_m(\nu t)$ satisfies

(2.9)
$$u_{tt}^m + u_t^m/t - m^2 u^m/t^2 = -\nu^2 u^m.$$

For $m = 0$ we think of (2.9) as the partial Fourier transform of (2.7) in x $(x \to \xi)$ where $\nu = 2\pi|\xi|$, $f = \delta$, and thus $u^0(t) = \mathscr{F}\mu_x(t)$. We shall embed the $m = 0$ equation in what will be called a canonical resolvent sequence $(\nu = 2\pi|\xi|)$

(2.10)
$$\hat{R}_{tt}^m + (2m + 1)\hat{R}_t^m/t + \nu^2 \hat{R}^m = 0,$$

where $\hat{R}^m(\nu, t) = i^m \Gamma(m + 1) 2^m (\nu t)^{-m} u^m(t)$ is chosen so that the resolvent initial values $\hat{R}^m(\nu, 0) = 1$ and $\hat{R}_t^m(\nu, 0) = 0$ hold. This is an EPD sequence (following a partial Fourier transform) and for simplicity we only consider $m \geqq 0$; both $\hat{R}^m$ and $R^m = \mathscr{F}^{-1}\hat{R}^m$ will be called resolvents.

THEOREM 2.1. *The recurrence relations* $H_+ f_m = i\nu f_{m+1}$ *and* $H_- f_m = i\nu f_{m-1}$ *are equivalent to*

(2.11)
$$\hat{R}_t^m = \frac{-t\nu^2}{2(m + 1)} \hat{R}^{m+1}; \qquad \hat{R}_t^m + \frac{2m}{t} \hat{R}^m = \frac{2m}{t} \hat{R}^{m-1}.$$

The first equation in (2.11) is the same as one of Weinstein's recursion relations (with initial data also specified) and the second equation is similarly related to a recursion formula used by Payne (which also follows from a formula of Weinstein). Both equations were used systematically together in [**1**] in studying questions of existence, uniqueness, and growth.

THEOREM 2.2. *The second equation in* (2.11) *yields directly*

(2.12)
$$\hat{R}^m(\nu, t) = 2m \int_0^1 \xi(1 - \xi^2)^{m-1} \hat{R}^0(\nu, \xi t)\, d\xi.$$

This and related formulas also derivable along the way from the second equation in (2.11) are versions of the Sonine integral formula. They can be used in EPD theory to show that $\mathscr{F}^{-1}\hat{R}^m$ is a positive resolvent (since $\mathscr{F}^{-1}\hat{R}^0(v, \xi t) = \mu_x(\xi t) \geqq 0$). Thus (2.11)–(2.12), which lead to various results in EPD theory, are really consequences of the group framework and one is led to exploit their counterparts in other such situations. We will not expand upon the use of (2.11)–(2.12) here but refer to [**1**], [**8**] for details. Variations on this technique appear in §§3 and 4.

3. Let now $G = SL(2, \boldsymbol{R})$ and $K = SO(2)$ with Lie algebras $\tilde{g}$ and $\tilde{k}$ respectively. Let $\tilde{g} = \tilde{k} + \tilde{p}$ be a Cartan decomposition of the semisimple $\tilde{g}$ so that in particular the Killing form $B(\xi, \eta) = \operatorname{tr}(\operatorname{ad}\xi \operatorname{ad}\eta)$ is positive definite on $\tilde{p}$ and $G = PK$ is the polar decomposition of G ($P = \exp(\tilde{p})$). Let

$$(3.1) \qquad X = \frac{1}{2}\begin{pmatrix} 0 & 1 \\ 1 & 0 \end{pmatrix}, \qquad Y = \frac{1}{2}\begin{pmatrix} 1 & 0 \\ 0 & -1 \end{pmatrix}, \qquad Z = \frac{1}{2}\begin{pmatrix} 0 & 1 \\ -1 & 0 \end{pmatrix}$$

be a basis of $\tilde{g}$ where X, Y form an orthonormal basis of the vector space $\tilde{p}$ under the scalar product $(\xi, \eta) = \frac{1}{2}B(\xi, \eta) = 2\operatorname{tr}(\xi\eta)$ and Z generates $\tilde{k}$. We note that, in $sl(2, \boldsymbol{C})$, $X_+ = -X + iY$, $X_- = -X - iY$, and $H = -2iZ$ form a canonical triple with $[X_+, H] = -2X_+$, $[X_-, H] = +2X_-$, and $[X_+, X_-] = H$ relative to the root space decomposition $\tilde{g} = \boldsymbol{C}H + \boldsymbol{C}X_+ + \boldsymbol{C}X_-$, where $\boldsymbol{C}H$ is a Cartan subalgebra. If $QU(2)$ is the group of quasi-unitary unimodular matrices of second order (cf. [7]) there is an isomorphism $Q: SL(2, \boldsymbol{R}) \to QU(2)$ sending

$$g = \begin{pmatrix} \alpha & \beta \\ \gamma & \delta \end{pmatrix} \to \hat{g} = \begin{pmatrix} a & b \\ \bar{b} & \bar{a} \end{pmatrix},$$

where $a = \frac{1}{2}[\alpha + \delta + i(\beta - \gamma)]$ and $b = \frac{1}{2}[\beta + \gamma + i(\alpha - \delta)]$ with $\det g = \det \hat{g} = 1$. There is a natural parametrization of $QU(2)$ in terms of generalized Euler angles (φ, τ, ψ) so that any $\hat{g} \in QU(2)$ can be written

$$(3.2) \qquad \hat{g} = \begin{pmatrix} \cosh\frac{\tau}{2}\, e^{i/2(\varphi+\psi)} & \sinh\frac{\tau}{2}\, e^{i/2(\varphi-\psi)} \\ \sinh\frac{\tau}{2}\, e^{-i/2(\varphi-\psi)} & \cosh\frac{\tau}{2}\, e^{-i/2(\varphi+\psi)} \end{pmatrix}.$$

Writing $w_1(t) = \exp(tX)$, $w_2(t) = \exp(tY)$, and $w_3(t) = \exp(tZ)$ it follows that $Qw_1(t) = (0, t, 0)$, $Qw_2(t) = (\pi/2, t, -\pi/2)$, and $Qw_3(t) = (t, 0, 0) = (0, 0, t)$.

Now the Killing form induces a Riemannian structure on $M = G/K$ in a standard manner under left translation where $g_\omega(\xi, \eta) = \frac{1}{2}B(\xi, \eta)$ for $\omega = \pi(e)$ and $\xi, \eta \in \tilde{p}$ ($g_q(\cdot, \cdot)$ denotes the metric tensor at $q \in M$) while $ds^2 = dt^2 + \sinh^2 t\, d\theta^2$ in geodesic polar coordinates (cf. [**5**]). If $p = \exp(\alpha X + \beta Y) \in P \subset G$, one has $Q(p) = (\theta, t, -\theta)$ in Euler angles where $t^2 = \alpha^2 + \beta^2$ and $\tan\theta = \beta/\alpha$; on the other hand we can take (α, β) as coordinates on M since $\pi \circ \exp: \tilde{p} \to G/K$ is a diffeomorphism (cf. [**5**]) and in the Riemannian structure indicated it follows that (θ, t) of this form represents geodesic polar coordinates. Thus if $g = pk$ the geodesic polar coordinates of $\pi(g) = pK$ can be read off directly from the Euler angles of $Q(p)$. This makes certain calculations relatively easy in view of the composition formulas for Euler angles indicated in [**7**]. In particular, upon considering the quasiregular representation of G on $L^2(M)$ as in §2, we find for the induced representation of $\tilde{g}$ on a dense subspace V that

$$A_1 = L(X) = \coth t \sin\theta\, \partial/\partial\theta - \cos\theta\, \partial/\partial t,$$

$$A_2 = L(Y) = -\coth t \cos\theta\, \partial/\partial\theta - \sin\theta\, \partial/\partial t \quad \text{and} \quad A_3 = L(Z) = -\partial/\partial\theta.$$

We set $H_+ = -A_1 + iA_2$, $H_- = -A_1 - iA_2$, and $H_3 = -iA_3$ so that $H_3 = i\partial/\partial\theta$ with

$$H_+ = e^{-i\theta}[-i \coth t\, \partial/\partial\theta + \partial/\partial t];$$

(3.3) $$H_- = e^{i\theta}[i \coth t\, \partial/\partial\theta + \partial/\partial t];$$

$$H_+H_- + H_3H_3 - H_3 = \Delta = \partial^2/\partial t^2 + \coth t\, \partial/\partial t + \operatorname{csch}^2 t\, \partial^2/\partial\theta^2.$$

Again we consider irreducible representations of $\tilde{g}$ on W induced by suitable unitary irreducible representations of G and with certain provisos these are specified in [**6**]. Taking the case $A^{0,\rho}$ of [**6**] for example, we want to realize this in W with basis vectors $f_m \in W$ satisfying

(3.4) $$H_3 f_m = m f_m; \qquad H_+ f_m = (m - l) f_{m+1}; \qquad H_- f_m = -(m + l) f_{m-1},$$

where $m \in Z$ and $l = -\frac{1}{2} + i\rho$ with $\rho > 0$. We can take $f_m = \exp(-im\theta) u^m(t)$ where the requirement $\Delta f_m = l(l + 1) f_m$ resulting from (3.4) implies that

(3.5) $$u^m_{tt} + \coth t\, u^m_t - m^2 \operatorname{csch}^2 t\, u^m = l(l + 1) u^m.$$

The recursion relations in (3.4) will be satisfied by taking $u^m(t) = (-1)^m P^l_{0,m}(\cosh t)$ (see [**7**] for notation). Following the prescription of §2 we want to embed the $m = 0$ equation in a canonical resolvent sequence by multiplying $u^m(t)$ by a factor so that the resulting function $\hat{R}^m(l, t)$ satisfies the resolvant initial values $\hat{R}^m(l, 0) = 1$ and $\hat{R}^m_t(l, 0) = 0$. This is achieved by writing ($z = \cosh t$)

(3.6)
$$\begin{aligned} \hat{R}^m(l, t) &= \frac{\Gamma(m + 1)\Gamma(l + 1)2^m}{\Gamma(l + m + 1)(z^2 - 1)^{m/2}} P^l_{0,m}(z) \\ &= \frac{\Gamma(l + 1)\Gamma(m + 1)\Gamma(l - m + 1)}{2^{l-m}(z + 1)^{m-l}} \sum_{s=m}^{\infty} c_s \left(\frac{z - 1}{z + 1}\right)^{s-m} \end{aligned}$$

where the coefficients c_s are given by

$$c_s = 1/\Gamma(s - m + 1)\Gamma(l + m - s + 1)\Gamma(s + 1)\Gamma(l - s + 1).$$

Again keeping $m \geqq 0$ we have easily

THEOREM 3.1. *The canonical resolvent sequence associated with* $\hat{R}^0_{tt} + \coth t\, \hat{R}^0_t = l(l+1)\hat{R}^0$ *is*

$$\hat{R}^m_{tt} + (2m+1)\coth t\, \hat{R}^m_t + m(m+1)\hat{R}^m = l(l+1)\hat{R}^m. \tag{3.7}$$

The recursion formulas in (3.4) now induce "canonical" recursion relations among the $\hat{R}^m$ which we collect as

THEOREM 3.2. *The* $\hat{R}^m$ *satisfy the canonical recursion relations*

$$\begin{aligned} \hat{R}^m_t &= \frac{\sinh t}{2(m+1)}[l(l+1) - m(m+1)]\hat{R}^{m+1}; \\ \frac{1}{2m}\hat{R}^m_t &+ \coth t\, \hat{R}^m = \operatorname{csch} t\, \hat{R}^{m-1}. \end{aligned} \tag{3.8}$$

Analogous to Theorem 2.2 the second formula in (3.8) yields various "Sonine" formulas beginning with a first integral

$$\sinh^{2m} t\, \hat{R}^m(l,t) = 2m \int_0^t \sinh^{2m-1}\eta\, \hat{R}^{m-1}(l,\eta)\, d\eta. \tag{3.9}$$

THEOREM 3.3. *The second formula in* (3.8) *yields directly*

$$\sinh^{2m} t\, \hat{R}^m(l,t) = m2^m \int_0^t (\cosh t - \cosh\eta)^{m-1} \sinh\eta\, \hat{R}^0(l,\eta)\, d\eta. \tag{3.10}$$

4. The Darboux equation on M can be written as

$$\begin{aligned} M_{tt}(q,t,f) + \coth t\, M_t(q,t,f) &= \Delta_q M(q,t,f); \\ M(q,0,f) = f; \qquad M_t(q,0,f) &= 0; \end{aligned} \tag{4.1}$$

and one knows that $\Delta_q M(q,t,f) = M(q,t,\Delta f)$. Here $M(q,t,f)$ denotes the mean value of a function $f \in C^2$ over a geodesic sphere with center q and radius t (cf. [**4**], [**9**] and (2.6)). Similarly one can define a mean value operator $\mu(t) \in E'(M)$ as in §2 by the rule $\langle \mu(t), \varphi \rangle = M(\omega, t, \varphi)$ where $\omega = \pi(e)$ and $\varphi \in E(M)$ (E and E' denote standard distribution spaces on M). Now a Fourier transform theory on $QU(2)$ is available in various forms (see e.g., [**3**], [**7**]) and can be exploited to interpret the resolvent equations of §3 in the same way as were the corresponding equations in §2. For simplicity however, we will use another method here to obtain the growth and convexity theorems and will only make a few brief remarks about the Fourier transform. Let us observe, first in this direction, that if f is a function on M and $pK = \pi(g)$ where $g = pk$, then one can write in an obvious notation $f(\theta, t) = f(\pi(g)) = (f \circ \pi)(g)$

$= \tilde{f}(g) = (\tilde{f} \circ Q^{-1})(Q(g)) = F(Q(g)) = F(Q(p)Q(k)) = F((\theta, t, -\theta)(0, 0, \alpha)) = F(\theta, t, \psi)$; thus f can be considered as a function on $\hat{G} = QU(2)$ which is independent of ψ. The Fourier transform $\mathscr{F}$ of [**3**] is defined for any $T \in D'(\hat{G})$, and thus if $f \in D(M)$ for example, $\mathscr{F}f$ will be defined as an infinite collection of functions of a complex variable l, having special properties for $l = -\frac{1}{2} \pm i\rho$ as in §3 (we note here that $P_{m,n}^{-1/2-i\rho}(\cosh t) = (-1)^{m-n}P_{m,n}^{-1/2+i\rho}(\cosh t)$); the lack of ψ dependence effectively integrates out the ψ variable in $\mathscr{F}f$ and this eliminates one parameter in $\mathscr{F}f$ as well as all contributions from the discrete series. Similarly $\mathscr{F}\mu(t)$ could be defined and since in addition $\mu(t)$ effectively integrates out the θ variable it follows that $\mathscr{F}\mu(t)$ would consist of only one term which is $P_{0,0}^{l}(\cosh t) = \hat{R}^0(l, t)$. Since $\mathscr{F}\Delta T = l(l+1)\mathscr{F}T$ (cf. [**3**]), the differential equation for $\hat{R}^0$ would be the partial Fourier transform of the Darboux equation for $\mu(t)$ and similarly $\hat{R}^m$ would be the Fourier transform of a positive resolvent $R^m(t)$ (cf. Theorem 3.3). Then one would "compose" the various resolvant formulas with a function $f \in C^2(M)$ for example to obtain solutions of

$$w_{tt}^m + (2m+1)\coth t\, w_t^m + m(m+1)w^m = \Delta w^m; \tag{4.2}$$
$$w^m(q, 0, f) = f(q); \qquad w_t^m(q, 0, f) = 0.$$

Then one would deduce facts about w^m from Theorems (3.2) and (3.3).

The above comments furnish some interpretation and give a comparison with §2. What we do now, however, does not require the Fourier transform at all. First, given $f \in C^2(M)$ and $q \in M$, we define $w^m(q, t, f)$ in the form suggested by (3.10), namely

$$\sinh^{2m} t\, w^m(q, t, f) = m2^m \int_0^t (\cosh t - \cosh \eta)^{m-1} \sinh \eta\, M(q, \eta, f)\, d\eta. \tag{4.3}$$

Then obviously the relation analogous to (3.9) for w^m will hold and

$$w_t^m/2m + \coth t\, w^m = \operatorname{csch} t\, w^{m-1}. \tag{4.4}$$

On the other hand $M(q, t, f)$ satisfies (4.1) with $M(q, t, \Delta f) = \Delta_q M(q, t, f)$, and from these facts one can prove the analogue of the first equation in (3.8) (the second being taken care of by (4.4)).

LEMMA 4.1. *If w^m is defined by* (4.3) *then* (4.4) *holds and*

$$w_t^m = \frac{\sinh t}{2(m+1)}[\Delta - m(m+1)]w^{m+1}. \tag{4.5}$$

Consequently w^m satisfies (4.2) *and evidently $\Delta w^m(q, t, f) = w^m(q, t, \Delta f)$.*

Now from (4.3) we see that if $f \geqq 0$, then $w^m(q, t, f) \geqq 0$ while from (4.2) one has (cf. [**4**], [**9**])

$$\sinh^{2m+1} t\, w_t^m(q,t,f) = \int_0^t \sinh^{2m+1}\xi\, w^m(q,\xi,\Delta_m f)\, d\xi, \tag{4.6}$$

where $\Delta_m = \Delta - m(m+1)$. Consequently

THEOREM 4.2. *If $f \in C^2(M)$ and $\Delta f \geqq m(m+1)f$ then $w^m(q,t,f)$ is monotone nondecreasing in t.*

This theorem could also be proved using (4.5) directly. Now we take any function $\rho(t)$ such that $d\rho/dt = \operatorname{csch}^{2m+1} t$ so that $d/d\rho = \sinh^{2m+1} t\, d/dt$. Then (4.2) can be written (cf. [**4**], [**9**])

$$\frac{d^2}{d\rho^2} w^m(q,t,f) = \sinh^{4m+2} t\, w^m(q,t,\Delta_m f). \tag{4.7}$$

THEOREM 4.3. *If $f \in C^2(M)$ and $\Delta f \geqq m(m+1)f$ then $w^m(q,t,f)$ is a convex function of $\rho(t)$.*

Theorems 4.2 and 4.3 for $m = 0$ follow from results in [**4**], [**9**]. General existence-uniqueness theorems for singular equations of the type (4.2) can be found in [**2**].

REFERENCES

1. R. W. Carroll, *Some singular Cauchy problems*, Ann. Mat. Pura Appl. (4) **56** (1961), 1–31. MR **27** #1706.

2. ———, *On the singular Cauchy problem*, J. Math. Mech. **12** (1963), 69–102. MR **26** #5284.

3. L. Ehrenpreis and F. I. Mautner, *Some properties of the Fourier transform on semi-simple Lie groups*. I, II, III, Ann. of Math. (2) **61** (1955), 406–439; Trans. Amer. Math. Soc. **84** (1957), 1–55; Trans. Amer. Math. Soc. **90** (1959), 431–484. MR **16**, 1017; MR **18**, 745; MR **21** #1541.

4. B. A. Fusaro, *Spherical means in harmonic spaces*, J. Math. Mech. **18** (1968/69), 603–606. MR **39** #657.

5. S. Helgason, *Differential geometry and symmetric spaces*, Pure and Appl. Math., vol. 12, Academic Press, New York, 1962. MR **26** #2986.

6. W. Miller, Jr., *Lie theory and special functions*, Math. in Sci. and Engineering, vol. 43, Academic Press, New York, 1968. MR **41** #8736.

7. N. Ja. Vilenkin, *Special functions and theory of group representations*, "Nauka", Moscow, 1965; English transl., Transl. Math. Monographs, vol. 22, Amer. Math. Soc., Providence, R.I., 1968. MR **35** #420.

8. A. Weinstein, *On a Cauchy problem with subharmonic initial values*, Ann. Mat. Pura Appl. (4) **43** (1957), 325–340. MR **19**, 656.

9. ———, *Spherical means in spaces of constant curvature*, Ann. Mat. Pura Appl. (4) **60** (1962), 87–91. MR **27** #463.

UNIVERSITY OF ILLINOIS AT URBANA-CHAMPAIGN

TANGENTIAL CAUCHY-RIEMANN COMPLEXES ON SPHERES

G. B. FOLLAND

The *tangential Cauchy-Riemann complex*, or *$\bar{\partial}_b$ complex*, is a complex of differential operators living on the boundary of a complex manifold which arises as follows. Let M be a complex manifold of (complex) dimension n with smooth boundary bM, embedded in a slightly larger open manifold M'; we assume bM is defined by the equation $R = 0$ where R is a C^∞ real-valued function on a neighborhood of bM with $R < 0$ inside M, $R > 0$ outside M, and $dR \neq 0$ on bM. Let A^{ij} be the vector bundle of differential forms of type (i, j) on M'; then we have the Dolbeault complex

$$0 \to \Gamma(A^{i0}) \xrightarrow{\bar{\partial}} \Gamma(A^{i1}) \xrightarrow{\bar{\partial}} \cdots \xrightarrow{\bar{\partial}} \Gamma(A^{in}) \to 0, \tag{1}$$

where Γ denotes the space of smooth sections. Assuming given a hermitian metric on M', we define B^{ij} to be the orthogonal complement of $(A^{i(j-1)}|\mathrm{b}M) \wedge \bar{\partial}R$ in $A^{ij}|\mathrm{b}M$. Since $\bar{\partial}(\theta \wedge \bar{\partial}R) = \bar{\partial}\theta \wedge \bar{\partial}R$, there is induced by the usual quotient process a complex

$$0 \to \Gamma(B^{i0}) \xrightarrow{\bar{\partial}_b} \Gamma(B^{i1}) \xrightarrow{\bar{\partial}_b} \cdots \xrightarrow{\bar{\partial}_b} \Gamma(B^{i(n-1)}) \to 0, \tag{2}$$

the $\bar{\partial}_b$ complex. It is easily seen that this complex is independent of the choice of R, up to isomorphism.

The operator $\bar{\partial}_b$ is defined explicitly as follows. If $\theta \in \Gamma(B^{ij})$, let θ' be an extension of θ to M'. Then $\bar{\partial}_b\theta$ is the orthogonal projection of $\bar{\partial}\theta'|\mathrm{b}M$ onto $\Gamma(B^{i(j+1)})$, and this is independent of the choice of θ'.

The tangential Cauchy-Riemann operators were studied by H. Lewy [**6**] in the case $n = 2$ in connection with the problem of finding a holomorphic function in a region of $\boldsymbol{C}^2$ with given boundary values. This work was later extended by Kohn and Rossi [**5**], who formalized the notion of "$\bar{\partial}_b$ complex." Meanwhile, however, Lewy had been led by his work to the discovery of a smooth differential equation

AMS 1970 *subject classifications*. Primary 35H05, 35N15; Secondary 33A40, 33A45, 43A75.

which is not locally solvable [7], which has had vast repercussions in the theory of partial differential equations. In fact, Lewy's example is the $\bar{\partial}_b$ operator for a certain strongly pseudoconvex domain in $\boldsymbol{C}^2$, and it can easily be seen from Hörmander's criterion that the $\bar{\partial}_b$ operator for any strongly pseudoconvex 2-manifold is not locally solvable.

If M is compact, the $\bar{\partial}_b$ complex is intimately connected with the *$\bar{\partial}$-Neumann problem* on M which has important applications to the theory of several complex variables. In fact, the $\bar{\partial}_b$ complex is the boundary complex associated to the $\bar{\partial}$-Neumann problem in accordance with Spencer's general theory of Neumann problems for overdetermined elliptic systems (cf. Sweeney [**8**], also Kohn and Rossi [**5**]). We shall have more to say about this later.

The $\bar{\partial}_b$ complex is not elliptic; one can easily verify that the cotangent vectors $i(\bar{\partial} - \partial)R$ are characteristic. However, under suitable pseudoconvexity conditions, it has been shown by Kohn [3] that the Laplacian $\Box_b = \bar{\partial}_b \vartheta_b + \vartheta_b \bar{\partial}_b$ (where ϑ_b is the formal adjoint of $\bar{\partial}_b$) satisfies the "$\frac{1}{2}$-estimate" $\|u\|^2_{1/2} \leqq c((\Box_b u, u) + \|u\|^2)$. Operators satisfying such "subelliptic" estimates have many of the qualitative properties of elliptic operators, such as regularity of weak solutions and compactness of the Green's operator (cf. Kohn and Nirenberg [**4**]), and they have recently attracted much attention from Hörmander, Egorov, and others.

It is our purpose here to investigate in detail the case $M' = \boldsymbol{C}^n$, $M = B_n = \{z : |z| \leqq 1\}$, $\mathrm{b}M = S_n = \{z : |z| = 1\}$, $R = r - 1$ where r is the distance from the origin. In this situation the $\bar{\partial}_b$ complex has an added significance: it is the prototype example of the "transversally elliptic" operators currently being studied by Atiyah and Singer. An operator on the manifold X is said to be *transversally elliptic* with respect to a group action on X if it commutes with the action and the cotangent vectors orthogonal to the orbits of the group are noncharacteristic. In our case, the circle group S^1 acts (as a subset of $\boldsymbol{C}$) by scalar multiplication on S_n, and the characteristics are precisely the cotangent vectors to the orbits of the action.

Our program will be to exploit the symmetry of the $\bar{\partial}_b$ complex with respect to the unitary group $U(n)$. The sphere is a homogeneous space, $S_n = U(n)/U(n-1)$, and all the vector bundles B^{ij} are homogeneous bundles since $U(n)$ commutes with $\bar{\partial}$ and preserves the radial function r. For the same reason, $U(n)$ commutes with $\bar{\partial}_b$. We will therefore obtain very precise information about $\bar{\partial}_b$ by decomposing the spaces of sections under the group action. In this paper we merely state the main results; details and proofs will appear in [**1**].

We restrict our attention to forms of purely antiholomorphic type, i.e., $i = 0$, as the holomorphic part carries no additional information. Let $\mathscr{B}^j$ be the Hilbert space completion of $\Gamma(B^{0j})$; we then write the $\bar{\partial}_b$ complex as

$$0 \to \mathscr{B}^0 \xrightarrow{\bar{\partial}_b} \mathscr{B}^1 \xrightarrow{\bar{\partial}_b} \cdots \xrightarrow{\bar{\partial}_b} \mathscr{B}^{n-1} \to 0 \tag{3}$$

where now $\bar{\partial}_b : \mathscr{B}^j \to \mathscr{B}^{j+1}$ is considered as a closed unbounded operator.

The first task is to decompose the spaces $\mathscr{B}^j$ with respect to the $U(n)$ action. The abstract decomposition, that is, the statement of which irreducible representations occur in $\mathscr{B}^j$ and with what multiplicity, follows immediately from two classical theorems, the Frobenius Reciprocity Theorem and the Branching Theorem. To identify the irreducible subspaces explicitly requires a particular construction of the representations of $U(n)$, namely the theory of Young diagrams. The result is as follows:

THEOREM. *$\mathscr{B}^0 = \bigoplus_{p \geqq 0, q \geqq 0} \Phi_{pq0}$, $\mathscr{B}^j = \bigoplus_{p \geqq 0, q \geqq 1} [\Phi_{pqj} \oplus \Psi_{pqj}]$ for $1 \leqq j \leqq n-2$, and $\mathscr{B}^{n-1} = \bigoplus_{p \geqq -1, q \geqq 1} \Psi_{pq(n-1)}$ where each Φ_{pqj} and Ψ_{pqj} is irreducible. A typical element of Φ_{pqj} is*

$$\phi_{pqj} = \bar{z}_1^{q-1} z_n^p \sum_{i=1}^{j+1} \bar{z}_i \zeta_1 \wedge \cdots \hat{\zeta}_i \cdots \wedge \zeta_{j+1},$$

and a typical element of Ψ_{pqj} is

$$\psi_{pqj} = \bar{z}_1^{q-1} z_n^p \zeta_1 \wedge \cdots \wedge \zeta_j,$$

where $\zeta_i = \bar{\partial}_b \bar{z}_i = d\bar{z}_i - \bar{z}_i \sum_{j=1}^n z_j d\bar{z}_j$. $\Phi_{pqj} \cong \Psi_{pq(j+1)}$, and otherwise these subspaces are all inequivalent.

Note that for $j = 0$, $\phi_{pq0} = \bar{z}_1^q z_n^p$ is a harmonic polynomial. Thus we have obtained a refinement of the usual decomposition of functions on the sphere by spherical harmonics: we have bigraded the spherical harmonics according to their holomorphic and antiholomorphic degree. Likewise, the definitions of ϕ_{pqj} and ψ_{pqj} for $j > 0$ show that the decomposition of forms of higher degree follows the pattern of spherical harmonics.

It is now easy to compute the action of $\bar{\partial}_b$ on each irreducible subspace. Since $\bar{\partial}_b$ commutes with $U(n)$, it suffices to compute the action on a single vector in each subspace. In fact, $\bar{\partial}_b \phi_{pqj} = (q + j)\psi_{pq(j+1)}$ and $\bar{\partial}_b \psi_{pqj} = 0$. Thus we see that the $\bar{\partial}_b$ complex is exact except in degrees 0 and $n - 1$, where the cohomology is $\bigoplus_{p \geqq 0} \Phi_{p00}$ and $\bigoplus_{q \geqq 1} \psi_{(-1)q(n-1)}$ respectively.

By Schur's lemma, $\bar{\partial}_b : \Phi_{pqj} \to \Psi_{pq(j+1)}$ is a constant multiple of a unitary map. We may take this constant to be real and positive; when so determined, it will be denoted by γ_{pqj} and called the *eigenvalue* of $\bar{\partial}_b$ on Φ_{pqj}. (The "eigenvalue" of $\bar{\partial}_b$ on the Ψ_{pqj}'s is 0.) By the formula for $\bar{\partial}_b$, $\gamma_{pqj} = (q + j)\|\psi_{pq(j+1)}\|/\|\phi_{pqj}\|$. The norms on the right-hand side can be determined by elementary calculations, and the result is that $\gamma_{pqj} = (2(q + j)(p + n - 1 - j))^{1/2}$.

$\bar{\partial}_b$ is thus a weighted shift operator on $\bigoplus_j \mathscr{B}^j$ with weights γ_{pqj}, so its adjoint ϑ_b is also a weighted shift operator with the same weights but shifting in the other direction. There follows:

THEOREM. *The eigenvalue (in the usual sense) of $\square_b = \bar{\partial}_b \vartheta_b + \vartheta_b \bar{\partial}_b$ on Φ_{p00} and $\Psi_{(-1)q(n-1)}$ is zero, and on Φ_{pqj} and $\Psi_{pq(j+1)}$ $(p \geqq 0, q \geqq 1)$ is $2(q + j)(p + n - 1 - j)$. The ranges of $\bar{\partial}_b$, ϑ_b, and $\square_b$ are closed since these operators are bounded away from* *zero on the orthogonal complements of their nullspaces, and the Green's operator G_b (defined by $G_b = 0$ on the nullspace of $\Box_b$ and $G_b = \Box_b^{-1}$ on the orthogonal complement) is compact, being the norm limit of operators of finite rank.*

(*Note.* The closed range property is strictly global. The Lewy example shows that the range of $\bar\partial_b$ on a small open set is generally not closed.)

The spaces Φ_{pqj} and Ψ_{pqj} are also eigenspaces of the ordinary Laplacian Δ, and it is easily seen that its eigenvalue is of the order of magnitude of $(p + q)^2$. Since the Sobolev norms on S_n may be defined by $\|u\|_s = \|(\Delta + I)^{s/2}u\|$, we obtain immediately all the global L^2-regularity results for $\bar\partial_b$ by comparing powers of $p + q + 1$ with powers of γ_{pqj}, which is of the order of magnitude of $(pq)^{1/2}$. In particular, since $(pq)^{1/2} \sim (p + q + 1)^{1/2}$ as $q \to \infty$ when p is fixed or as $p \to \infty$ when q is fixed, we see that the $\frac{1}{2}$-estimate is the best possible. On the other hand, since $(pq)^{1/2} \sim p + q + 1$ as $p, q \to \infty$ when $p - q$ is fixed, there are infinite-dimensional spaces of forms on which $\Box_b$ is as strong as Δ. Moreover, since global real analyticity on S_n is equivalent to the exponential decrease of the spherical harmonic coefficients, we see that the solution u of $\bar\partial_b u = v$ which is orthogonal to the nullspace must be analytic whenever v is; likewise for ϑ_b and $\Box_b$.

At this point we mention another geometrical interpretation of $\bar\partial_b$. The quotient of S_n by the action of S^1 mentioned earlier is the complex manifold $\boldsymbol{C}P^{n-1}$. Since the "bad" direction for $\bar\partial_b$ is the direction which is annihilated by the projection $S_n \to \boldsymbol{C}P^{n-1}$, we expect that there should be an intimate connection between $\bar\partial_b$ on S_n and $\bar\partial$ on $\boldsymbol{C}P^{n-1}$.

The fibration $S^1 \to S_n \to \boldsymbol{C}P^{n-1}$ exhibits S_n as a principal bundle over $\boldsymbol{C}P^{n-1}$ with structure group S^1. By a well-known geometrical construction, to each unitary representation of S^1 there is associated a holomorphic vector bundle on $\boldsymbol{C}P^{n-1}$. In particular, the irreducible representations of S^1 are the one-dimensional representations $\{\rho_m : m \in \boldsymbol{Z}\}$ where $\rho_m(e^{i\theta})$ = multiplication by $e^{im\theta}$, and we denote the line bundle associated to ρ_m by η^m. (η^{-1} is the hyperplane section bundle.) On each η^m we have the Dolbeault complex

$$0 \to \Gamma(\eta^m) \xrightarrow{\bar\partial} \Gamma(\eta^m \otimes \lambda^1) \xrightarrow{\bar\partial} \cdots \xrightarrow{\bar\partial} \Gamma(\eta^m \otimes \lambda^{n-1}) \to 0 \tag{4}$$

where λ^j is the bundle of $(0, j)$-forms on $\boldsymbol{C}P^{n-1}$.

Now there is a one-to-one correspondence between sections of η^m and functions on the principal bundle S_n satisfying an appropriate equivariance condition, which extends to a correspondence between sections of $\eta^m \otimes \lambda^j$ and equivariant sections of B^{0j}. But this equivariance condition just says that the forms transform via the representation ρ_m under the action of S^1. If for each m we define $\mathscr{B}^0(m) = \bigoplus_{q-p=m}\Phi_{pq0}$, $\mathscr{B}^j(m) = [\bigoplus_{q-p+j=m}\Phi_{pqj}] \oplus [\bigoplus_{q-p+j-1=m}\Psi_{pqj}]$ for $1 \leq j \leq n - 2$, and $\mathscr{B}^{n-1}(m) = \bigoplus_{q-p+n-2=m}\Psi_{pqj}$, it is not hard to show that $\mathscr{B}^j(m)$ is the subspace of $\mathscr{B}^j$ which transforms via ρ_m under the S^1 action. We are then led to the following theorem:

THEOREM. *$\bar\partial_b(\mathscr{B}^j(m)) \subset \mathscr{B}^{j+1}(m)$. The diagram*

$$\begin{array}{ccccccccc} 0 \to & \mathscr{B}^0(m) & \xrightarrow{\bar\partial_b} & \mathscr{B}^1(m) & \xrightarrow{\bar\partial_b} \cdots \xrightarrow{\bar\partial_b} & \mathscr{B}^{n-1}(m) & \to 0 \\ & \uparrow & & \uparrow & & \uparrow & \\ 0 \to & \Gamma(\eta^m) & \xrightarrow{\bar\partial} & \Gamma(\eta^m \otimes \lambda^1) & \xrightarrow{\bar\partial} \cdots \xrightarrow{\bar\partial} & \Gamma(\eta^m \otimes \lambda^{n-1}) & \to 0 \end{array}$$

commutes, where the vertical arrows are the natural injections given by the correspondence of sections with equivariant forms.

Thus the $\bar\partial_b$ complex on S_n is isomorphic to the direct sum of the Dolbeault complexes of the bundles η^m on CP^{n-1}, and information about one is equivalent to information about the other. In particular, we can read off from our calculations all the eigenvalues for the Dolbeault complexes.

We now state an analogue of the Sobolev theorem for norms related to $\Box_b$. Recall that the ordinary Sobolev theorem says that on a manifold of real dimension N, $H_s \subset C^r$ if $s > r + N/2$.

We could define a norm on functions by $|\,\|u\|\,| = \|(\Box_b + I)u\|$, but this is not very satisfactory because $\Box_b$ has a large harmonic space for which this norm provides no information. To remedy the lopsidedness while preserving the essential regularity properties of $\Box_b$, we introduce the conjugate operator $\bar\Box_b$ and define $|\,\|u\|\,|_{s,0} = \|(\Box_b + \bar\Box_b + I)^{s/2}u\|$. Moreover, since $\Box_b + \bar\Box_b$ is weaker in the direction tangent to the circle orbits than in the others, we can obtain more precise results by measuring differentiation in this direction directly. Thus let X be the vector field which generates the circle action; it is easy to see that X has eigenvalue $i(p-q)$ on Φ_{pq0}. Letting $|X|$ be the operator whose eigenvalue on Φ_{pq0} is $|p-q|$, then, we define $|\,\|u\|\,|_{s,\sigma} = \|(\Box_b + \bar\Box_b + I)^{s/2}(|X| + I)^{\sigma}u\|$, and $B_{s,\sigma}$ = the completion of C^∞ with respect to $|\,\|\;\|\,|_{s,\sigma}$. The basic facts about these spaces are as follows:

(A) For $s, \sigma \geqq 0$ we have $B_{s+2\sigma,0} \subset B_{s,\sigma} \subset B_{s,0}$ and $H_{s+\sigma} \subset B_{s,\sigma} \subset H_{\min(s,s/2+\sigma)}$. These inclusions are sharp and are continuous but not compact.

(B) $B_{s,\sigma}$ is naturally dual to $B_{-s,-\sigma}$ so the inclusion relations for $s, \sigma < 0$ are obtained by dualizing those in (A).

(C) (Rellich lemma.) $|\,\|\;\|\,|_{s,\sigma}$ is compact with respect to $|\,\|\;\|\,|_{s',\sigma'}$ if and only if $s > s'$ and $\sigma \geqq \sigma'$.

There is a direct proof of the ordinary Sobolev theorem on S_n by using the theory of spherical harmonics. By modifying this proof, we obtain

THEOREM. *$B_{s,\sigma} \subset C^0$ if $s > n - 1$ and $\sigma > \frac{1}{2}$.*

COROLLARY. *$B_{s,0} \subset C^0$ if $s > n$.*

(Recall that the real dimension of S_n is $2n - 1$.)

This theorem has the following geometrical interpretation: $\Box_b$ is in some sense the pullback of the Laplacian on CP^{n-1}, and $-X^2$ restricted to an orbit is just the Laplacian on S^1. The ordinary Sobolev theorem says that $H_s(CP^{n-1}) \subset C^0(CP^{n-1})$ if $s > n - 1$ and $H_\sigma(S^1) \subset C^0(S^1)$ if $\sigma > \frac{1}{2}$. These two phenomena are combined by the fibration $S^1 \to S_n \to CP^{n-1}$ to yield our result.

It seems highly likely that the analogous result should hold for C^r, $r > 0$, namely:

Conjecture. $B_{s,\sigma} \subset C^r$ if $s > r + n - 1$ and $\sigma > \frac{1}{2}(r + 1)$, and $B_{s,0} \subset C^r$ if $s > 2r + n - 1$.

However, technical difficulties have so far precluded our finding a satisfactory proof.

We now turn our attention to the $\bar{\partial}$-Neumann problem on the unit ball B_n. Let $\mathscr{A}^j$ denote the Hilbert space of square-integrable $(0, j)$-forms on B_n, so that $\mathscr{A}^j$ is the completion of $\Gamma(A^{0j})$. A form $u \in \Gamma(A^{0j})$ is said to satisfy the *$\bar{\partial}$-Neumann conditions* if $u|S_n \in \mathscr{B}^j$ and $\bar{\partial}u|S_n \in \mathscr{B}^{j+1}$, which is equivalent to the conditions $(u, \bar{\partial}v) = (\vartheta u, v)$ for all $v \in \Gamma(A^{0j})$ and $(\bar{\partial}u, \bar{\partial}v) = (\vartheta\bar{\partial}u, v)$ for all $v \in \Gamma(A^{0(j+1)})$, where ϑ is the formal adjoint of $\bar{\partial}$. The restriction of $\bar{\partial}\vartheta + \vartheta\bar{\partial}$ to forms satisfying the $\bar{\partial}$-Neumann conditions is a positive hermitian operator; we denote its Friedrichs extension by $\square$ (and use the symbol $\square$ only for this purpose). We will solve the following strong form of the $\bar{\partial}$-Neumann problem on B_n: determine the spectral decomposition of $\mathscr{A}^j$ under $\square$, that is, find the eigenvectors and eigenvalues for $\square$.

First we define extensions of the forms ϕ_{pqj} and ψ_{pqj} to B_n, which we still denote by ϕ_{pqj} and ψ_{pqj}, namely

$$\phi_{pqj} = \bar{z}_1^{q-1} z_n^p \sum_{i=1}^{j+1} (-1)^{i-1} \bar{z}_i d\bar{z}_1 \wedge \cdots \widehat{d\bar{z}_i} \cdots \wedge d\bar{z}_{j+1}$$

and

$$\psi_{pqj} = r\bar{z}_1^{q-1} z_n^p \zeta_1 \wedge \cdots \wedge \zeta_j \quad \text{where } \zeta_i = d\bar{z}_i - (2\bar{z}_i/r)\bar{\partial}r.$$

It is easy to check that

$$\psi_{pqj} = r\bar{z}_1^{q-1} z_n^p d\bar{z}_1 \wedge \cdots \wedge d\bar{z}_j + (-1)^j 2\phi_{pq(j-1)} \wedge \bar{\partial}r \quad \text{for } p \geqq 0,$$

and

$$\psi_{(-1)q(n-1)} = (\bar{z}_1^{q-1}/r) \sum_{i=1}^{n} (-1)^{i+n} \bar{z}_i d\bar{z}_1 \wedge \cdots \widehat{d\bar{z}_i} \cdots \wedge d\bar{z}_n.$$

Except at the origin, every $(0, j)$-form θ can be expressed as $\theta = \theta_1 + \theta_2 \wedge \bar{\partial}r$ where θ_1 and θ_2 are pointwise orthogonal to forms divisible by $\bar{\partial}r$. θ_1 and θ_2 can then be expanded in terms of the ϕ's and ψ's with coefficients depending on r. Requiring that θ be an eigenvector for $\square$ then yields an ordinary differential equation for these coefficients which is a variant of the Bessel equation, and the $\bar{\partial}$-Neumann conditions reduce in each case to a single boundary condition for this equation. The theory of Bessel functions then guarantees the existence of a complete orthogonal expansion in eigenfunctions. The results are as follows:

THEOREM.

$$\mathscr{A}^0 = \left\{ \bigoplus_{p \geqq 0} \Phi_{p000} \right\} \oplus \left\{ \bigoplus_{p \geqq 0, q \geqq 0, m \geqq 1} \Phi_{pqm0} \right\},$$

$$\mathscr{A}^1 = \left\{ \bigoplus_{p \geqq 0, q \geqq 1, m \geqq 1} [\Phi_{pqm1} \oplus \vartheta(\Psi_{pqm2})] \right\} \oplus \left\{ \bigoplus_{p \geqq 0, q \geqq 0, m \geqq 1} \bar{\partial}(\Phi_{pqm0}) \right\},$$

$$\mathscr{A}^j = \bigoplus_{p \geqq 0, q \geqq 1, m \geqq 1} [\Phi_{pqmj} \oplus \Psi_{pqmj} \oplus \bar{\partial}(\Phi_{pqm(j-1)}) \oplus \vartheta(\Psi_{pqm(j+1)})] \qquad \text{for } 2 \leqq j \leqq n-2,$$

$$\mathscr{A}^{n-1} = \left\{ \bigoplus_{p \geqq 0, q \geqq 1, m \geqq 1} [\Psi_{pqm(n-1)} \oplus \bar{\partial}(\Phi_{pqm(n-2)})] \right\} \oplus \left\{ \bigoplus_{p \geqq -1, q \geqq 1, m \geqq 1} \vartheta(\Psi_{pqmn}) \right\},$$

$$\mathscr{A}^n = \bigoplus_{p \geqq -1, q \geqq 1, m \geqq 1} \Psi_{pqmn}.$$

Each Φ_{pqmj} and Ψ_{pqmj} is an irreducible subspace. The typical element of Φ_{p000} is z_n^p. The typical element of Φ_{pqmj} ($m \geqq 1$) is $r^{1-n-p-q} J_{p+q+n-1}(\lambda_{pqm}^{j1} r)\phi_{pqj}$ where λ_{pqm}^{j1} is the mth positive zero of $\lambda J'_{p+q+n-1}(\lambda) + (q - p + 2j - n + 1)J_{p+q+n-1}(\lambda)$. The typical element of Ψ_{pqmj} is $r^{1-n-p-q} J_{p+q+n-1}(\lambda_{pqm}^{j2} r)\, \psi_{pq(j-1)} \wedge \bar{\partial} r$ where λ_{pqm}^{j2} is the mth positive zero of $J_{p+q+n-1}(\lambda)$. We have $\bar{\partial}(\Psi_{pqmj}) = 0$, $\vartheta(\Phi_{pqmj}) = 0$. The eigenvalue of $\Box$ is 0 on Φ_{p000}, $\frac{1}{2}(\lambda_{pqm}^{j1})^2$ on Φ_{pqmj} and $\bar{\partial}(\Phi_{pqmj})$, and $\frac{1}{2}(\lambda_{pqm}^{j2})^2$ on Ψ_{pqmj} and $\vartheta(\Psi_{pqmj})$.

The next question is how big the eigenvalues are; the answer is given by the following theorem:

THEOREM. (1) *For some constant* $c = c(p, q, j)$, $\lambda_{pqm}^{j1} = c + m\pi + O(1/m)$, *and for some constant* $c = c(p + q)$, $\lambda_{pqm}^{j2} = c + m\pi + O(1/m)$.

(2) *Let* $v = p + q + n - 1$, $H = q - p + 2j - n + 1$. *Then*

$$(v(v + 2))^{1/2} < \lambda_{pq1}^{j2} < (2(v + 1)(v + 2))^{1/2},$$

$$(v(v + 2))^{1/2} < \lambda_{pq1}^{j1} < (2(v + 1)(v + 3))^{1/2} \quad \text{if } H \geqq 0,$$

$$((v + 2)(v + H))^{1/2} < \lambda_{pq1}^{j1} < (2(v + 1)(v + H))^{1/2} \quad \text{if } -v < H < 0, \text{ and}$$

$$((v + 1)(v + 3))^{1/2} < \lambda_{p01}^{01} < (2(v + 2)(v + 4))^{1/2} \quad (\text{for which } H = -v).$$

From these estimates we can easily see that $\Box$, $\bar{\partial}$, and ϑ have closed ranges, and that the Neumann operator N ($= 0$ on the nullspace of $\Box$ and $= \Box^{-1}$ on the orthogonal complement) is compact.

From the general theory of the $\bar{\partial}$-Neumann problem (cf. Kohn [**2**], Kohn and Nirenberg [**4**]) it is known that $\|u\|_{s+1} \leqq \|\Box u\|_s \leqq \|u\|_{s+2}$ for $u \in \mathscr{A}^j \cap \operatorname{Dom} \Box$, $j > 0$. Unfortunately, the search for a direct proof of these estimates by the present methods is beset by a series of technical difficulties: in the first place, the Sobolev spaces on the ball cannot be described in terms of operators as neatly as on the sphere; in the second place, the estimates seem to depend on some very delicate information about the distribution of the eigenvalues of $\Box$. However, it is not too difficult to show that the noncoercive estimate $\|u\|_1 \leqq \|\Box u\|$ is sharp. Specifically:

THEOREM. *For each* j, *the estimate* $\|\Box u\| \leqq c_{qj}\|u\|_1$ *holds on* $\bigoplus_{p \geqq 0} \Phi_{pq1j}$ *where* $q = \text{constant} \neq 0$.

The technique is to compare the $\bar{\partial}$-Neumann problem with the coercive d-Neumann problem, which possesses a similar eigenfunction resolution. The essential point is that for $p \gg q$, the eigenvalues λ_{pq1}^{j1} are much smaller than all the other eigenvalues.

On the other hand, for the spaces Ψ_{pqmj} the $\bar{\partial}$-Neumann conditions coincide with the Dirichlet conditions, which are coercive. Hence on $\bigoplus\Psi_{pqmj}$ the estimate $\|\Box u\| \sim \|u\|_2$ holds.

By a slight modification of our procedure, we can also solve the $\bar{\partial}$-Neumann problem on the annulus $A_\rho = \{z \in \boldsymbol{C}^n : \rho \leqq |z| \leqq 1\}$. Here we obtain boundary conditions at both $r = \rho$ and $r = 1$; since the Bessel equation is nonsingular on $[\rho, 1]$, the ordinary Sturm-Liouville theory guarantees the existence of complete eigenfunction expansions. The spaces Φ_{pqmj} and Ψ_{pqmj} are just as before, except that the Bessel functions $J_{p+q+n-1}$ are replaced by linear combinations of $J_{p+q+n-1}$ and $Y_{p+q+n-1}$ (Bessel function of the second kind) satisfying the appropriate boundary conditions. The only important difference is the appearance of a harmonic space in degree $n - 1$, namely $\bigoplus_{q \geqq 1}\Psi_{(-1)q0(n-1)}$ where the typical element of $\Psi_{(-1)q0(n-1)}$ is $r^{3-2n-2q}\psi_{(-1)q(n-1)}$. This comes as no surprise, considering the behavior of the $\bar{\partial}_b$ complex; it is also indicated by general theory, since the basic estimate of Kohn [**2**] fails to hold on A_ρ for $j = n - 1$.

In conclusion, the author wishes to thank J. J. Kohn, E. M. Stein, and D. C. Spencer for their advice and encouragement while this research was being carried out.

References

1. G. B. Folland, *The tangential Cauchy-Riemann complex on spheres*, Ph.D. Thesis, Princeton University, Princeton, N.J., 1971; Trans. Amer. Math. Soc. **171** (1972), 83–133.

2. J. J. Kohn, *Harmonic integrals on strongly pseudoconvex manifolds*. I, II, Ann. of Math. (2) **78** (1963), 112–148; ibid. (2) **79** (1964), 450–472. MR **27** #2999; MR **34** #8010.

3. ———, *Boundaries of complex manifolds*, Proc. Conference Complex Analysis (Minneapolis, 1964), Springer, Berlin, 1965, pp. 81–94. MR **30** #5334.

4. J. J. Kohn and L. Nirenberg, *Non-coercive boundary value problems*, Comm. Pure Appl. Math. **18** (1965), 443–492. MR **31** #6041.

5. J. J. Kohn and H. Rossi, *On the extension of holomorphic functions from the boundary of a complex manifold*, Ann. of Math. (2) **81** (1965), 451–472. MR **31** #1399.

6. H. Lewy, *On the local character of the solutions of an atypical linear differential equation in three variables and a related theorem for regular functions of two complex variables*, Ann. of Math. (2) **64** (1956), 514–522. MR **18**, 473.

7. ———, *An example of a smooth linear partial differential equation without solution*, Ann. of Math. (2) **66** (1957), 155–158. MR **19**, 551.

8. W. J. Sweeney, *The D-Neumann problem*, Acta Math. **120** (1968), 233–277. MR **37** #2250.

PRINCETON UNIVERSITY

COURANT INSTITUTE OF MATHEMATICAL SCIENCES, NEW YORK UNIVERSITY

SEMIBOUNDED BOUNDARY PROBLEMS FOR ELLIPTIC OPERATORS

GERD GRUBB

1. **Introduction.** Let $\bar{\Omega}$ be a compact n-dimensional C^∞ manifold with boundary Γ (and denote $\bar{\Omega}\backslash\Gamma$ by Ω), let E be a C^∞ k-dimensional vector bundle over $\bar{\Omega}$, and let A be a $2m$-order elliptic differential operator from $C^\infty(\bar{\Omega},E)$ into $C^\infty(\bar{\Omega},E)$. We denote the principal symbol of A by a; it is locally of the form

$$a(x,\xi) = \sum_{|\alpha|=2m} a_\alpha(x)\xi^\alpha,$$

where the $a_\alpha(x)$ are $(k \times k)$-matrices, and $a(x,\xi) \neq 0$ for each $x \in \bar{\Omega}$, each $\xi \in T_x^*(\bar{\Omega})\backslash\{0\}$.

We shall assume that the Sobolev spaces $H^s(\Omega,E), H^s(\Gamma,E|_\Gamma)$ $(s \in \boldsymbol{R}), H_0^s(\Omega,E)$ $(s \geqq 0)$ have been suitably defined; the norms will be denoted $\|\cdot\|_s$, and E or $E|_\Gamma$ will usually be omitted. In particular, $H^0(\Omega) = H_0^0(\Omega) = L^2(\Omega)$, $H^0(\Gamma) = L^2(\Gamma)$; and the spaces $H^{-s}(\Omega)$ and $H_0^s(\Omega)$ $(s \geqq 0)$, resp. $H^{-s}(\Gamma)$ and $H^s(\Gamma)$ $(s \in \boldsymbol{R})$, are dual spaces with respect to extensions of the inner products $(\ ,\ )$ in $L^2(\Omega,E)$, resp. $L^2(\Gamma,E|_\Gamma)$. Such dualities will be denoted $\langle\ ,\ \rangle$. (Our notations mainly follow [**10**].)

We are interested in the problem:

For which *homogeneous boundary conditions* does one have

(1) Re $(Au,u) \geqq 0$ for all u satisfying the boundary condition and with $u \in L^2(\Omega)$, $Au \in L^2(\Omega)$?

The precise statement as to what is meant by a boundary condition will be postponed for a moment. Until then we abbreviate the description in (1) to: $u \in$ *bdry cond.*

One interest of the problem is that it discusses whether the operator $u \mapsto -Au$

AMS 1970 *subject classifications*. Primary 35J40, 35B45; Secondary 35J35, 47B44, 58G15.

with domain consisting of the functions $u \in bdry\ cond$ is *dissipative*, and thus has extensions that are infinitesimal generators of strongly continuous semigroups of contraction operators in $L^2(\Omega)$.

A closely related problem consists of asking for criteria determining whether

$$(1)' \qquad \exists\lambda \in \boldsymbol{R} \text{ such that } \operatorname{Re}(Au, u) \geqq -\lambda\|u\|_0^2, \text{ all } u \in bdry\ cond,$$

i.e., whether (1) holds for $A + \lambda$, some $\lambda \in \boldsymbol{R}$.

(1) and (1)′ are examples of a whole score of semiboundedness properties that may be studied. We shall introduce just two other important properties; one that is considerably stronger than (1′) and one that is considerably weaker:

$$(2) \quad \exists c > 0, \lambda \in \boldsymbol{R} \text{ such that } \operatorname{Re}(Au,u) \geqq c\|u\|_m^2 - \lambda\|u\|_0^2, \text{ all } u \in bdry\ cond;$$

$$(3) \qquad \exists\lambda \in \boldsymbol{R} \text{ such that } \operatorname{Re}(Au,u) \geqq -\lambda\|u\|_m^2, \text{ all } u \in bdry\ cond.$$

The property (2) generalizes Gårding's inequality. (In **[6]**, **[7]** we also discuss the inequality where m is replaced by s lying in $[0,m]$, and call the property s-coerciveness.) As is well known, strong ellipticity of A, i.e.

$$(4) \qquad [2 \operatorname{Re} a(x,\xi) =]\ a(x,\xi) + a^*(x,\xi) > 0 \quad \text{on } T^*(\bar\Omega)\backslash 0,$$

is necessary (and sufficient) for (2) to hold even for the $u \in C_0^\infty(\Omega)$. (Property (1) requires $\operatorname{Re} a(x,\xi) \geqq 0$ on $T^*(\bar\Omega)$.) On the other hand, property (3) is a very weak kind of semiboundedness, which holds for $u \in C_0^\infty(\Omega)$ when A is *any* $2m$-order differential operator with smooth coefficients on $\bar\Omega$.

We shall present a theory that treats (1), (2) and (3) at the same time; therefore we will make the assumption that *A is strongly elliptic*. However, we shall later indicate some results that hold without this assumption.

It is now time to be more precise about what is meant by boundary condition. Actually, there are several choices. One of them is to be very general, to operate with a rather abstract concept. We shall begin with describing this briefly, and later take up some more classical definitions.

2. **General boundary conditions.** Define *the maximal operator for A* as the operator A_1 with domain

$$D(A_1) = \{u \in L^2(\Omega) | Au \in L^2(\Omega)\},$$

and mapping $u \in D(A_1)$ into $A_1u = Au$; it is a closed operator in $L^2(\Omega)$. Define the *minimal operator for A* as the operator A_0 which is the closure of A_{00}, where

$$D(A_{00}) = C_0^\infty(\Omega), \qquad A_{00}u = Au \quad \text{for } u \in D(A_{00}).$$

It is clear that $A_0 \subset A_1$. Moreover, one finds by use of the ellipticity of A_1 that

$$(5) \qquad D(A_0) = H_0^{2m}(\Omega), \qquad D(A_1) \subset H_{\mathrm{loc}}^{2m}(\Omega).$$

The linear operators $\tilde A$ with $A_0 \subset \tilde A \subset A_1$ will be called the *realizations of A*.

Because of (5), each realization can be said to represent, in an abstract sense, a homogeneous boundary value problem for A.

This abstract concept of boundary problem (or boundary condition) was introduced by Vishik [**12**, (1952)], who moreover gave an interpretation of a large class of realizations to more usual looking boundary conditions—of the form of linear relations between certain *generalized boundary values*. (We use a theory that differs somewhat from Vishik's; one advantage is that it treats all closed realizations, not just those with closed range, cf. [**5**].) Next, the generalized boundary values are given a precise sense by the work of Lions and Magenes [**9**, (1961–1962)] (see also [**10**]), which enables us to characterize the closed realizations by closed operators (called L below) operating between subspaces of certain Sobolev spaces over the boundary. Finally, the observation of Calderón [**3**, (1963)] (developed further in Seeley [**11**], Hörmander [**8**, (1966)]), that the classical differential boundary problems can be reduced to problems concerning pseudo-differential operators *in* the boundary, has the application here that the discussion of each realization $\tilde{A}$ may be reduced to the discussion of the relation between the operator L in the boundary and certain fixed pseudo-differential operators in the boundary. (In classical cases, L itself will be a ps.d.o.)

The outcome is stated in Theorem 1 below. Before getting to it, we shall have to introduce some notation, elaborating some of the points just mentioned.

Assume, as we may, that a sufficiently large constant has been added to the strongly elliptic operator A so that, for some $c_m > 0$,

$$\text{Re}\,(Au,u) \geqq c_m \|u\|_m^2, \qquad \forall u \in D(A_0). \tag{6}$$

(Cf. the discussion following (2)–(3).) Let D_t denote a first order differential operator on $C^\infty(\bar{\Omega},E)$, which acts transversally to Γ, and define the "normal derivatives" $\gamma_j, j = 0,1,2,\ldots$, as the following operators from $C^\infty(\bar{\Omega},E)$ to $C^\infty(\Gamma,E|_\Gamma)$:

$$\gamma_0 u = u|_\Gamma,$$
$$\gamma_j u = \gamma_0 D_t^j u, \qquad j = 1,2,\ldots.$$

For each nonnegative integer j and each real $s > j + \frac{1}{2}$, γ_j extends by continuity to a continuous operator from $H^s(\Omega)$ to $H^{s-j-1/2}(\Gamma)$ (in a consistent way for varying s). Lions and Magenes showed how these boundary operators may be extended to $D(A_1)$.

PROPOSITION 1. *For $j < 2m$, γ_j extends by continuity to a continuous operator γ_j from $D(A_1)$ into $H^{-j-1/2}(\Gamma)$.*

Denote by γ the Dirichlet boundary operator

$$\gamma = \{\gamma_0,\gamma_1,\ldots,\gamma_{m-1}\}. \tag{7}$$

For this operator we have, as a special case of the theory of Lions and Magenes for elliptic boundary problems (and using the assumption of (6)):

PROPOSITION 2. *$\{A,\gamma\}$ maps $D(A_1)$ isomorphically onto $L^2(\Omega)\times\prod_{j=0}^{m-1}H^{-j-1/2}(\Gamma)$.*

An immediate consequence is that, for the nullspace of A_1,

$$Z(A_1)=\{u\in L^2(\Omega)|Au=0\},$$

γ defines an *isomorphism*

$$\gamma:Z(A_1)\to\prod_{j=0}^{m-1}H^{-j-1/2}(\Gamma);\tag{8}$$

we shall denote it γ_Z. With ν denoting the boundary operator

$$\nu=\{\gamma_m,\gamma_{m+1},\ldots,\gamma_{2m-1}\},\tag{9}$$

we can therefore define

$$P_{\gamma,\nu}=\nu\circ\gamma_Z^{-1}:\prod_{j=0}^{m-1}H^{-j-1/2}(\Gamma)\to\prod_{j=0}^{m-1}H^{-m-j-1/2}(\Gamma).$$

It can be derived from the theory of Calderón-Seeley-Hörmander that $P_{\gamma,\nu}$ is an elliptic pseudo-differential operator in Γ (cf. [7]).

With Green's formula written in the form

$$(Au,v)-(u,A'v)=\langle\mathscr{A}_{01}\nu u+\mathscr{A}_{00}\gamma u,\gamma v\rangle+\langle\mathscr{A}_{10}\gamma u,\nu v\rangle\tag{10}$$

(where A' denotes the formal adjoint of A, and the $\mathscr{A}_{pq}$ are differential operators mapping $\prod_{j=0}^{m-1}H^{-qm-j-1/2}(\Gamma)$ continuously into $\prod_{j=0}^{m-1}H^{pm-2m+j+1/2}(\Gamma)$, cf. [**11**], [**7**]), we finally introduce the special pseudo-differential boundary operator μ:

$$\mu=\mathscr{A}_{01}(\nu-P_{\gamma,\nu}\gamma).$$

It is evidently continuous from $D(A_1)$ into $\prod_{j=0}^{m-1}H^{-2m+j+1/2}(\Gamma)$; actually one finds that it maps $D(A_1)$ continuously onto $\prod_{j=0}^{m-1}H^{j+1/2}(\Gamma)$.

The result can now be stated.

THEOREM 1. *Let $\tilde{A}$ be a closed realization of A. Let*

$$X=\mathrm{Cl}\,(\gamma D(\tilde{A})),\qquad Y=\mathrm{Cl}\,(\gamma D(\tilde{A}^*)),\tag{11}$$

the closures taken in $\prod_{j=0}^{m-1}H^{-j-1/2}(\Gamma)$. Denote the injection $Y\hookrightarrow\prod_{j=0}^{m-1}H^{-j-1/2}(\Gamma)$ by i_Y; its adjoint i_Y^* maps $\prod_{j=0}^{m-1}H^{j+1/2}(\Gamma)$ onto Y'. Then $\tilde{A}$ defines an operator $L:X\to Y'$ with $D(L)=\gamma D(\tilde{A})$ such that

$$D(\tilde{A})=\{u\in D(A_1)|\gamma u\in D(L),L\gamma u=i_Y^*\mu u\}.\tag{12}$$

Conversely, given any pair of closed subspaces X and Y of $\prod_{j=0}^{m-1}H^{-j-1/2}(\Gamma)$, and any closed, densely defined operator $L=X\to Y'$, then (12) defines a closed realization $\tilde{A}$ of A. Hereby is established a 1-1 correspondence between all closed realizations $\tilde{A}$ of A and all such triples X,Y,L.

In the proof (see [**5**] for details), we first characterize the realizations $\tilde{A}$ in terms of operators going from $Z(A_1)$ to $Z(A'_1)$, using elementary Hilbert space theory. These operators are transformed into operators L over the boundary using the isomorphism (8), and the interpretation of $\tilde{A}$ as representing an actual boundary condition (12) is accomplished by use of an extended Green's formula.

The point of Theorem 1 is that *properties* of $\tilde{A}$ correspond to *properties* of L: regularity, dimension of nullspace, codimension of range, adjoints, and much else, cf. [**5**], [**6**], [**7**]. We shall here concentrate on the properties (1)–(3), for which the correspondence in Theorem 1 also gives a profitable discussion. The results to be described are most straightforward when $A = A'$ [**6**]; we shall state them for this case first, and afterwards indicate the modifications to make when $A \neq A'$ [**7**].

THEOREM 2. *Let A be formally selfadjoint, and let $\tilde{A}$ be a closed realization of A, corresponding to $L: X \to Y'$ by Theorem* 1. *Then*

1° $\operatorname{Re}(Au,u) \geqq 0$ *for all $u \in D(\tilde{A})$, if and only if*

$$X \subset Y, \text{ and } \operatorname{Re}\langle L\varphi,\varphi\rangle \geqq 0 \text{ for all } \varphi \in D(L); \tag{13}$$

2° $\exists c > 0, \lambda \in \boldsymbol{R}$, *such that* $\operatorname{Re}(Au,u) \geqq c\|u\|_m^2 - \lambda\|u\|_0^2$ *for all $u \in D(\tilde{A})$, if and only if*

$$\begin{gathered} X \subset Y, \text{ and } \exists c' > 0, \lambda' \in \boldsymbol{R}, \text{ such that for all } \varphi \in D(L), \\ \operatorname{Re}\langle L\varphi,\varphi\rangle \geqq c'\|\varphi\|^2_{\prod H^{m-j-1/2}(\Gamma)} - \lambda'\|\varphi\|^2_{\prod H^{-j-1/2}(\Gamma)}; \end{gathered} \tag{14}$$

3° $\exists \lambda \in \boldsymbol{R}$, *such that* $\operatorname{Re}(Au,u) \geqq -\lambda\|u\|_m^2$, *for all $u \in D(A)$, if and only if*

$$\begin{gathered} X \subset Y, \text{ and } \exists \lambda' \in \boldsymbol{R}, \text{ such that, for all } \varphi \in D(L), \\ \operatorname{Re}\langle L\varphi,\varphi\rangle \geqq -\lambda'\|\varphi\|^2_{\prod H^{m-j-1/2}(\Gamma)}. \end{gathered} \tag{15}$$

There are various relations between the values of c, c', λ and λ', easily deducible from the proof, that we shall not go into here. Part 2° and 3° use also other results of Lions and Magenes than Proposition 2. $D(\tilde{A}) \subset H^m(\Omega)$ is required in 2° but not in 3°.

Modifications in Theorem 2 for the case $A \neq A'$. In 1° and 2° replace L by $L + Q$, where Q is a fixed nonpositive pseudo-differential operator in Γ, derived from A and A'. The validity of 1°–3° is now shown in general only with $D(\tilde{A})$ replaced by $D_1 = D(\tilde{A}) \cap \{u \in L^2(\Omega) \mid A'u \in H^{-m}(\Omega)\}$ (and, accordingly, $D(L)$ replaced by γD_1, the inclusion $X \subset Y$ replaced by the inclusion $\gamma D_1 \subset Y$)—this can of course be disregarded, when suitable density theorems are available.

The theorem looks very natural; it shows e.g. (cf. 1°) that nonnegativity of $\tilde{A}$ corresponds to nonnegativity of L, for which one must have $X \subset Y$ in order for the duality $\langle L\varphi,\varphi\rangle$ to make sense.

However, when we look at more classical types of boundary conditions, we find that the condition $X \subset Y$ has a (perhaps) surprising effect. For one thing, it is highly unstable with respect to otherwise reasonable perturbations. To demonstrate this, let us look at the realizations defined by boundary conditions of the form

$$\sum_{k=0}^{2m-1} B_{jk}\gamma_k u = 0, \qquad j = 0, \ldots, m-1,$$

where the B_{jk} are ps.d.o.'s in Γ. Write the boundary condition as

$$B_0\gamma u + B_1 \nu u = 0$$

(recall (7),(9)), B_0 and B_1 denoting $(m \times m)$-matrices of ps.d.o.'s on $C^\infty(\Gamma, E|_\Gamma)$. We find here that $X = \mathrm{Cl}(\gamma D(\tilde{A}))$ is essentially of the form ($R(S)$ denoting the range of S):

$$X = \text{closure}\,\{\varphi | \exists\psi : B_0\varphi + B_1\psi = 0\} = \mathrm{Cl}\,(B_0^{-1}R(B_1)),$$

and, by use of Green's formula (10), that essentially, $Y = \mathrm{Cl}\,(\gamma D(\tilde{A}^*))$ has the form

$$Y = \mathrm{Cl}\,(R((\mathscr{A}_{10}^*)^{-1}B_1^*)).$$

We see that both X and Y will in general be very sensitive to any perturbations (X depends on the boundary condition, Y moreover on A); this is all the more true for the inclusion $X \subset Y$, which must hold *exactly* in order to apply the conclusions of Theorem 2.

3. **Normal pseudo-differential boundary conditions.** The theory takes a particularly convenient shape in the case where $\tilde{A}$ represents a *normal* pseudo-differential boundary condition; the rest of the article will be devoted to showing what Theorem 2 implies for this case. (Detailed proofs are given in [**7**].)

Introduce the notations M, M_0 and M_1 for the respective index sets

$$M = \{0,1,\ldots,2m-1\}, \quad M_0 = \{0,\ldots,m-1\}, \quad M_1 = \{m,\ldots,2m-1\}.$$

Let J be a subset of M with m elements, and let $K = M \backslash J$. Consider boundary conditions of the form

$$\gamma_j u - \sum_{k \in K, k<j} F_{jk}\gamma_k u = 0, \qquad j \in J, \tag{16}$$

where the F_{jk} denote ps.d.o.'s on $C^\infty(\Gamma, E|_\Gamma)$ of order $j-k$. (When $\dim E = 1$, all normal homogeneous boundary conditions can be put in this form, whereas when $\dim E > 1$, certain kinds of normal conditions are not included in (16); these require a modification of the following theory.)

We write (16) in short as

$$\gamma_J u = F_{JK}\gamma_K u, \tag{17}$$

where $\gamma_J = \{\gamma_j\}_{j\in J}$, $\gamma_K = \{\gamma_k\}_{k \in K}$, and F_{JK} is the matrix $(F_{jk})_{j\in J, k\in K}$; in accordance with (16) we here set

$$F_{jk} = 0 \quad \text{for } k > j. \tag{18}$$

To divide (17) into the part purely concerning Dirichlet data and the part with higher order normal derivatives, we introduce the index sets

$$J_0 = J \cap M_0, \quad J_1 = J \cap M_1, \quad K_0 = K \cap M_0, \quad K_1 = K \cap M_1;$$

then (17) may be written (with $F_{J_0K_0} = (F_{jk})_{j\in J_0, k\in K_0}$, etc., empty index sets giving null operators):

$$\begin{aligned} \gamma_{J_0}u &= F_{J_0K_0}\gamma_{K_0}u, \\ \gamma_{J_1}u &= F_{J_1K_0}\gamma_{K_0}u + F_{J_1K_1}\gamma_{K_1}u; \end{aligned} \tag{19}$$

note that $F_{J_0K_1}\gamma_{K_1}u$ disappears because of (18).

We can now describe exactly what $X \subset Y$ means.

THEOREM 3. *Let $\tilde{A}$ be the realization of A with domain*

$$D(\tilde{A}) = \{u \in D(A_1) | (19) \text{ holds}\}. \tag{20}$$

Then $X \subset Y$ implies

$$\begin{aligned} &\text{(i)} \quad K = \{2m - 1 - j | j \in J\}, \\ &\text{(ii)} \quad F_{J_1K_1} = \mathscr{F}(F_{J_0K_0}); \end{aligned} \tag{21}$$

the function $\mathscr{F}$ is explained below (23). *Moreover,* (21) *implies* (*and thus is equivalent with*)

$$X = Y. \tag{22}$$

Explanation of $\mathscr{F}$. With the $(m \times m)$-matrix $\mathscr{A}_{01}$ in Green's formula (10) indexed as $(\mathscr{A}_{jk})_{j\in M_0, k\in M_1}$, we can form the minors $\mathscr{A}_{J_0K_1}$, $\mathscr{A}_{J_0J_1}$, $\mathscr{A}_{K_0K_1}$ and $\mathscr{A}_{K_0J_1}$; then $\mathscr{F}$ denotes the function

$$\mathscr{F}(F_{J_0K_0}) = -(\mathscr{A}_{K_0J_1} + F^*_{J_0K_0}\mathscr{A}_{J_0J_1})^{-1}(\mathscr{A}_{K_0K_1} + F^*_{J_0K_0}\mathscr{A}_{J_0K_1}), \tag{23}$$

which is well defined when (21)(i) holds. One sees that (21)(ii) involves not only the principal symbols or even the complete symbols of the F_{jk}, but the *full operators*.

SKETCH OF PROOF. $X = \text{Cl}\,(\gamma D(\tilde{A}))$ is seen to be a *graph* of $F_{J_0K_0}$. Because of the normality, the functions in $D(\tilde{A}^*)$ satisfy boundary conditions of a kind similar to (19), and $Y = \text{Cl}\,(\gamma D(\tilde{A}^*))$ is also a graph of an operator G. Using (18), one finds that $X \subset Y$ implies that these two graphs are *identical*, so $F_{J_0K_0} = G$; this can be stated in the form (21)(i),(ii), (23). That the graphs are identical means that $X = Y$.

EXAMPLE. When A is of fourth order ($m = 2$), and we consider the normal boundary conditions with $J = \{1,3\}$, i.e., those of the form

$$\gamma_1 u = F_{10}\gamma_0 u, \qquad \gamma_3 u = F_{30}\gamma_0 u + F_{32}\gamma_2 u,$$

we find that (21)(i) is satisfied, and that (21)(ii) takes the form

$$F_{32} = -F_{10}^{*} - \mathscr{B}.$$

(Here, $\mathscr{B}$ is a fixed first order differential operator in Γ, composed of coefficients in Green's formula (10). It is 0, e.g. when $A = \Delta^2$, $\bar{\Omega} = \bar{R}^n_+$.)

Theorem 3 is relevant whether $A = A'$ or $A \neq A'$, for also $\gamma D_1 \subset Y$ (see Modifications of Theorem 2) is equivalent with (21).

The full implication of Theorem 2 is as follows:

THEOREM 4. *Let $A = A'$, and let $\tilde{A}$ be the realization of A with domain* (20). *There is a pseudo-differential operator $\mathscr{L}_1$ in Γ, continuous from $\prod_{k \in K_0} H^{m-k-1/2}(\Gamma)$ into $\prod_{j \in J_1} H^{-m+j+1/2}(\Gamma)$, such that*

1° Re $(Au,u) \geqq 0$ *for all $u \in D(\tilde{A})$ if and only if*

(24) $$(21) \text{ holds, and } \langle \mathscr{L}_1\varphi,\varphi\rangle \geqq 0, \forall \varphi \in \gamma_{K_0} D(\tilde{A});$$

2° $\exists c > 0, \lambda \in \boldsymbol{R}$ *such that* $\operatorname{Re}(Au,u) \geqq c\|u\|_m^2 - \lambda\|u\|_0^2$ *for all $u \in D(\tilde{A})$, if and only if*

(25) $$(21) \text{ holds, and } \operatorname{Re}\sigma^0(\mathscr{L}_1) > 0 \text{ on } T^*(\Gamma)\setminus 0;$$

3° $\exists \lambda \in \boldsymbol{R}$ *such that* $\operatorname{Re}(Au,u) \geqq -\lambda\|u\|_m^2$ *for all $u \in D(\tilde{A})$, if and only if*

(26) $$(21) \text{ holds};$$

this is also equivalent with

(27) $$\exists c > 0 \text{ such that } |(Au,v)| \leqq c\|u\|_m\|v\|_m, \forall u,v \in D(\tilde{A}).$$

REMARKS. The space $\gamma_{K_0} D(\tilde{A})$ in (24) satisfies

$$\prod_{k \in K_0} H^{2m-k-1/2}(\Gamma) \subset \gamma_{K_0} D(\tilde{A}) \subset \prod_{k \in K_0} H^{-k-1/2}(\Gamma),$$

and when (21) holds and $\varphi \in \gamma_{K_0} D(\tilde{A})$, one has in fact $\mathscr{L}_1\varphi \in \prod_{k \in K_0} H^{k+1/2}(\Gamma)$, so the duality $\langle \mathscr{L}_1\varphi,\varphi\rangle$ is well defined.

When (19) runs through all pseudo-differential boundary conditions satisfying (21), $\mathscr{L}_1$ runs through all ps.d.o.'s continuous from $\prod_{k \in K_0} H^{m-k-1/2}(\Gamma)$ to $\prod_{k \in K_0} H^{-m+k+1/2}(\Gamma)$. $\mathscr{L}_1$ is closely related to the operator L in Theorems 1 and 2. For the precise description of $\mathscr{L}_1$, we refer to [**7**].

The inequality $\langle \mathscr{L}_1\varphi,\varphi\rangle \geqq 0$ for general ps.d.o.'s has not so far been completely characterized. There are some deep results by Hörmander, Lax, Nirenberg, Melin and others ("sharp Gårding inequality," subelliptic estimates) that contribute to the question.

Modifications for $A \neq A'$. When $A \neq A'$, $\mathscr{L}_1$ must in the estimates be replaced by

(28) $$\mathscr{K}_1 = \mathscr{L}_1 + Q_1.$$

where Q_1 is a nonpositive ps.d.o. derived from O (cf. Modifications in Theorem 2) and $F_{J_0K_0}$. In 1°, the statement is then valid at least on a slightly smoother domain. As for 2° and 3°, we have completely satisfactory results, which we state for the sake of completeness.

THEOREM 5. *Let A be strongly elliptic of order $2m$, and let $\tilde{A}$ be the realization of A defined by* (20). *Then there exists $\lambda \in \boldsymbol{R}$ such that*

$$\text{(29)} \qquad \operatorname{Re}(Au,u) \geqq -\lambda\|u\|_m^2, \qquad \forall u \in D(\tilde{A})$$

if and only if (21) *holds, and in that case one also has*

$$\text{(30)} \qquad |(Au,v)| \leqq c\|u\|_m\|v\|_m, \qquad \forall u,v \in D(\tilde{A}),$$

for some c.

Furthermore, there exists a ps.d.o. $\mathscr{K}_1$ in Γ (cf. (28)) *such that*

$$\text{(31)} \qquad \operatorname{Re}(Au,u) \geqq c\|u\|_m^2 - \lambda\|u\|_0^2, \qquad \forall u \in D(\tilde{A}),$$

is valid for some $c > 0$, $\lambda \in \boldsymbol{R}$, if and only if

$$\text{(32)} \qquad \text{(21) } \textit{holds, and } \operatorname{Re}\sigma^0(\mathscr{K}_1) > 0 \textit{ on } T^*(\Gamma)\backslash 0.$$

REMARKS. In (32), $\mathscr{K}_1$ is continuous from $\prod_{k\in K_0} H^{m-k-1/2}(\Gamma)$ to $\prod_{k\in K_0} H^{-m+k+1/2}(\Gamma)$; the principal symbol $\sigma^0(\mathscr{K}_1)$ is defined accordingly, and it only depends on the principal symbols of the F_{jk} and of A on Γ. The boundary problem is *elliptic* if and only if $\mathscr{L}_1$ is elliptic; thus, since $Q_1 \leqq 0$, (32) implies ellipticity of the boundary problem.

The equivalence of (21) with (29) and (30) may in fact be extended, by a different method, to the case where A is just a $2m$-order differential operator for which Γ is nowhere characteristic.

In conclusion, we shall set our results in relation to earlier theories.

Previous works on (2) have always been concerned with conditions on principal symbols. In the fundamental paper [**1**], Agmon studied integro-differential sesquilinear forms

$$\text{(33)} \qquad a(u,v) = \int_\Omega \sum_{|\alpha|,|\beta| \leqq m} a_{\alpha\beta}(x) D^\alpha u \overline{D^\beta v}\, dx,$$

defined on smooth domains $\Omega \subset \boldsymbol{R}^n$, and gave a necessary and sufficient condition, on principal symbols, for the existence of $c > 0$, $\lambda \in \boldsymbol{R}$, with which

$$\text{(34)} \qquad \operatorname{Re} a(u,u) \geqq c\|u\|_m^2 - \lambda\|u\|_0^2$$

for all $u \in H^m(\Omega)$ satisfying a number of differential boundary conditions of orders $\leqq m - 1$. This solves our problem (2) in the case where $\tilde{A}$ is of such a kind that a sesquilinear form $a(u,v)$ may be found satisfying

$$\text{(35)} \qquad (Au,v) = a(u,v), \qquad \forall u,v \in D(\tilde{A}).$$

The question of *when* that is the case was not discussed.

Some time afterwards, Agmon gave another principal symbol criterion that applies to a priori *selfadjoint* $\tilde{A}$, but he remarks that these may also be fitted into the above set-up. (That condition, the "complementing condition with respect to $a(x,\xi) - \lambda$, all λ on a ray," has later been extended to nonselfadjoint $\tilde{A}$ to characterize infinitesimal generators of holomorphic (but not necessarily contraction!) semigroups, see [**2**].)

Much later, Shimakura and Fujiwara (see e.g. [**4**]) studied those normal boundary conditions, for which (2) holds *stably* under lower order perturbations of A. For these, J must be of the form

$$J = \{0,1,\ldots,m-p-1,m,m+1,\ldots,m+p-1\} \tag{36}$$

(for some integer $p \in [0,m]$). When (36) holds, the property (2) is then characterized by a condition on principal symbols. (The author contributed to this case in [**6**].)

When comparing these results with Theorem 5, we find that in all the above cases, (21) is a priori satisfied. E.g., when (36) holds, (21)(i) is clearly valid; and (21)(ii) holds in a trivial way, since $F_{J_0K_0}$ and $F_{J_1K_1}$ (and $\mathscr{A}_{K_0K_1}$) are zero, cf. (18). As for Agmon's old result, it is evident that when (35) holds, $\tilde{A}$ satisfies (30), so (21) is valid by the first part of Theorem 5, and there only remains the condition on principal symbols.

We have (in [7]) further investigated the connection with Agmon's work, and found that, when $\tilde{A}$ is a realization determined by a normal *differential* boundary condition (19), then (21) is in fact *necessary and sufficient for the existence* of an integro-differential sesquilinear form (33), with which (35) holds on $D(\tilde{A}) \cap H^m(\Omega)$.

REFERENCES

1. S. Agmon, *The coerciveness problem for integro-differential forms*, J. Analyse Math. **6** (1958), 183–223. MR **24** #A2748.

2. ———, *On the eigenfunctions and on the eigenvalues of general elliptic boundary value problems*, Comm. Pure Appl. Math. **15** (1962), 119–147. MR **26** #5288.

3. A. P. Calderón, *Boundary value problems for elliptic equations*, Outlines Joint Sympos. Partial Differential Equations (Novosibirsk, 1963), Acad. Sci. USSR Siberian Branch, Moscow, 1963, pp. 303–304. MR **34** #3107.

4. D. Fujiwara and N. Shimakura, *Sur les problèmes aux limites elliptiques stablement variationnels*, J. Math. Pures Appl. **49** (1970), 1–28.

5. G. Grubb, *A characterization of the non-local boundary value problems associated with an elliptic operator*, Ann. Scuola Norm. Sup. Pisa (3) **22** (1968), 425–513. MR **39** #626.

6. ———, *Les problèmes aux limites généraux d'un opérateur elliptique, provenant de la théorie variationnelle*, Bull. Sci. Math. **94** (1970), 113–157.

7. ———, *On coerciveness and semiboundedness of general boundary problems*, Israel J. Math. **10** (1971), 32–95.

8. L. Hörmander, *Pseudo-differential operators and non-elliptic boundary problems*, Ann. of Math. (2) **83** (1966), 129–209. MR **38** #1387.

9. J. L. Lions and E. Magenes, *Problèmes aux limites non homogènes*. II, Ann. Inst. Fourier (Grenoble) **11** (1961), 137–178. MR **26** #4047.

———, *Problemi ai limiti non omogenei*. V, Ann. Scuola Norm. Sup. Pisa (3) **16** (1962), 1–44. MR **26** #4049.

10. ———, *Problèmes aux limites non homogènes et applications*. Vol. 1, Travaux et Recherches Mathématiques, no. 17, Dunod, Paris, 1968. MR **40** #512.

11. R. T. Seeley, *Singular integrals and boundary value problems*, Amer. J. Math. **88** (1966), 781–809. MR **35** #810.

12. M. I. Višik, *On general boundary problems for elliptic differential equations*, Trudy Moskov. Mat. Obšč. **1** (1952), 187–246; English transl., Amer. Math. Soc. Transl. (2) **24** (1963), 107–172. MR **14**, 473.

UNIVERSITETSPARKEN 5, 2100 COPENHAGEN, DENMARK

COMPLEXES OF DIFFERENTIAL OPERATORS

VICTOR GUILLEMIN

Let X be a manifold, E^0, E^1, E^2, etc. vector bundles over X and

$$(*) \qquad 0 \longrightarrow E^0 \xrightarrow{D} E^1 \xrightarrow{D} \ldots \xrightarrow{D} E^N \to 0$$

a complex of differential operators. For simplicity we will assume the D's are first order. In this paper I will be concerned with the following two questions:

(a) When is $(*)$ exact on the local (or sheaf) level?

(b) When is the (global) cohomology finite dimensional?

Concerning (a) I will begin by examining a rather naive conjecture: If $(*)$ is formally exact at a point, $x_0 \in X$, is it exact over a neighborhood U of x_0? By formally exact, I mean we replace the coefficients of the D's by their Taylor series expansions at x_0, and think of them as operating on formal power series. One result is

THEOREM 1. *If $(*)$ is a complex of constant coefficient operators on $\mathbf{R}^n$ then formal exactness implies C^∞ exactness on every convex open set of $\mathbf{R}^n$.*

This is a well-known theorem of Malgrange-Ehrenpreis. See Malgrange [**4**].

If $(*)$ is not a constant coefficient, formal exactness is not a terribly useful notion, even for ordinary partial differential equations.

EXAMPLE *The equation $x^2\, df/dx - f = g$.* Given a formal power series, g, f can be solved for recursively. In fact the solution is unique. However, if $g = -x^2$, then $f = \sum_1^\infty n!x^{n+1}$.

There is, however, a useful notion which is stronger than formal exactness:

CRITERION A. Given a formal power series, g, such that $Dg = 0$ and such that the leading term of g is of degree k, there exists a power series, f, with leading term of degree $k + 1$ such that $Df = g$.

AMS 1970 *subject classifications*. Primary 35N10.

About this criterion we can say

THEOREM 2. *If* A *is satisfied at* x_0, *then it is satisfied at all points in a neighborhood of* x_0.

THEOREM 3. *If the coefficients of the D's are real analytic and* A *holds at* x_0 *then given a real analytic g defined on a neighborhood U of* x_0, *there exists real analytic f defined on a neighborhood* $V \subset U$ *of* x_0 *such that* $Df = g$ *on V*.

REMARK. A complex which satisfies A everywhere is called a *Spencer* complex.

We now consider the C^∞ problem. We recall how to define the characteristic variety of the complex $(*)$ at x. To each $\xi \in T_x^* \otimes \mathbf{C}$ one attaches the symbol complex

$$(**) \qquad 0 \longrightarrow E_x^0 \xrightarrow{\sigma(D)(\xi)} E_x^1 \xrightarrow{\sigma(D)(\xi)} \cdots \longrightarrow E_x^N \longrightarrow 0$$

ξ is *characteristic* if this fails to be exact. The set of characteristics is an algebraic subvariety, called the characteristic variety, of $T_x^* \otimes \mathbf{C}$. If there are no real characteristics except 0, $(*)$ is called *elliptic*. Two fairly elementary results about elliptic complexes are the following:

THEOREM 4. *If* $(*)$ *is elliptic, X compact and* $\partial X = \varnothing$ *then the cohomology of* $(*)$ *is finite dimensional.*

THEOREM 5. *If* $(*)$ *is elliptic, the coefficients of the operator D are real analytic, and the Criterion* A *is satisfied at* x_0 *then the conclusions of Theorem* 3 *hold for f and g*, C^∞.

Spencer conjectured several years ago that one can drop the analyticity assumptions, and Theorem 5 will still be true. As of now this conjecture is wide open.

Efforts have been made to extend Theorem 4 in two directions: either retain the topological assumptions but drop the ellipticity, or keep the ellipticity but allow $\partial X \neq \varnothing$. These two problems are somewhat related. If $(*)$ is elliptic then its boundary data are describable by a complex on ∂X which is generally not elliptic. I will devote the rest of this paper to describing some results on the first problem.

One says that $(*)$ is subelliptic (of order $\frac{1}{2}$) in its ith position if

$$(\mathrm{B}) \qquad \|Df\| + \|D^t f\| + \|f\| \geqq C\|f\|_{1/2}$$

for all $f \in C^\infty(E^i)$. If X is compact without boundary, (B) implies that the cohomology of $(*)$ is finite dimensional.

The estimate (B) can be localized. We will say that $(*)$ is *subelliptic at a point* (x_0, ξ_0) in the cotangent bundle if there exists a pseudo-differential operator $P(x, D)$ of degree $\leqq 1$ such that the symbol of $P(x, D)$ is zero on a conic neighborhood of (x_0, ξ_0) and such that

$$\|Df\| + \|D^t f\| + \|P(x, D)f\| \geqq C\|f\|_{1/2}.$$

One can show by a simple partition of unity argument that $(*)$ is subelliptic if and only if it is subelliptic at every real characteristic (x_0, ξ_0).

If the characteristic (x_0, ξ_0) is *simple* (see [**1**] or [**2**]), there exist functions $f_1(x, \xi), \ldots, f_q(x, \xi)$ defined on a neighborhood U of (x_0, ξ_0) in the complex cotangent bundle with the following properties:

(i) The f's are smooth functions of x and holomorphic functions of ξ.

(ii) $\operatorname{grad}_\xi f_1, \ldots, \operatorname{grad}_\xi f_q$ are linearly independent.

(iii) $(x, \xi) \in U$ is characteristic $\Leftrightarrow f_1(x, \xi) = \ldots = f_q(x, \xi) = 0$.

We will call the $q \times q$ Hermitian matrix $(-1)^{-1/2}\{f_i, \bar{f}_j\}$ the *Levi* form at (x_0, ξ_0).

THEOREM 6. *$(*)$ is subelliptic in its ith position at (x_0, ξ_0) if and only if the Levi form has at least $i + 1$ positive eigenvalues or $q - i + 1$ negative eigenvalues.*

The proof depends on some rather deep results of Hörmander [**3**].

BIBLIOGRAPHY

1. V. Guillemin, *On subelliptic estimates for complexes*, Proc. Internat. Congress Math., Nice, 1970.

2. V. Guillemin and S. Sternberg, *Subelliptic estimates for complexes*, Proc. Nat. Acad. Sci. U.S.A. **67** (1970), 271–274.

3. L. Hörmander, *Pseudo-differential operators and non-elliptic boundary problems*, Ann. of Math. (2) **83** (1966), 129–209. MR **38** #1387.

4. B. Malgrange, *Sur les systèmes différentiels à coefficients constants*, Les Équations aux Dérivées Partielles (Paris, 1962), Éditions du Centre National de la Recherche Scientifique, Paris, 1963, pp. 113–122. MR **31** #486.

MASSACHUSETTS INSTITUTE OF TECHNOLOGY

REMOVABLE SINGULARITIES AND STRUCTURE THEOREMS FOR POSITIVE CURRENTS

REESE HARVEY[1]

Suppose A is a closed nowhere dense subset of an open set $\Omega \subset \boldsymbol{C}^n$ and V is a complex subvariety of $\Omega - A$ of pure dimension d (i.e. each irreducible component of V is of dimension d). One would like conditions on the set A which insure that $\bar{V}$ is a subvariety of Ω. Shiffman [**13**] has shown that if the exceptional set A is of $(2d - 1)$-dimensional Hausdorff measure zero then $\bar{V}$ is a pure d-dimensional subvariety of Ω. In particular, if A is a complex subvariety of dimension $d - 1$ or less, then the $(2d - 1)$-dimensional Hausdorff measure of A is zero and Shiffman's theorem is applicable. This special case is due to Remmert-Stein [**12**]. As is well known, it can be used to give a short proof of the fact that every subvariety of complex projective space is algebraic (Chow's theorem).

The hypothesis in Shiffman's theorem cannot be weakened by allowing the $(2d - 1)$-Hausdorff measure of the exceptional set A to be locally finite. For example, suppose f is holomorphic on the open unit disk Δ in $\boldsymbol{C}$ and has boundary values $\varphi \in C^\infty(\partial\Delta)$ but φ is not real analytic at any point of $\partial\Delta$. Let A denote the graph of φ above $\partial\Delta$, V the graph of f above Δ, and $\Omega = \boldsymbol{C}^2$ (here $d = 1$). Then $\Lambda_{2d-1}(A)$ is finite but one can show that $\bar{V}$ is not a subvariety of $\boldsymbol{C}^2$ since φ is not real analytic on $\partial\Delta$.

Now we will change this problem of extending subvarieties to a removable singularity problem for the partial differential operator $\bar{\partial}$ and obtain a proof of Shiffman's theorem. The basic outline consists of three steps. First, replace the complex subvariety V of $\Omega - A$ by a current $[V]$ on $\Omega - A$ (this is standard). Second, extend $[V]$ to a $\bar{\partial}$-closed current U on Ω (we use a removable singularity theorem for $\bar{\partial}$ here). Third, show that the extended current U corresponds to a subvariety of Ω. This third step (which is perhaps the most interesting) is a con-

AMS 1970 *subject classifications*. Primary 32D15, 32D20, 32C25, 28A75.

[1] Partially supported by NSF grant GP-19011.

sequence of a structure theorem for positive currents.

The remainder of the article will be concerned with elaborating on the above outline. Our main purpose is to discuss a removable singularity theorem for $\bar\partial$ and some structure theorems; Shiffman's theorem provides us with a vehicle for this discussion. The proofs of these theorems will appear elsewhere.

We must start with some definitions and a more careful look at step 1. Let $\mathscr{D}'_{k,k}(\Omega)$ denote the space of currents of degree (k, k) (dimension $n - k, n - k$) on Ω, i.e. all currents of the form $u = \sum_{|I|=|J|=k} u_{IJ}\, dz^I \wedge d\bar z^J$ where each $u_{IJ} \in \mathscr{D}'(\Omega)$ is a distribution on Ω (I denotes $(i_1, \ldots, i_k)$ and $dz^I = dz_{i_1} \wedge \cdots \wedge dz_{i_k}$). If $u \in \mathscr{D}'_{k,k}(\Omega)$ then, for each test form $\psi \in \mathscr{D}_{2n-2k}(\Omega)$ (compactly supported C^∞ forms of total type $(2n - 2k)$), $u(\psi)$ is by definition $(\psi \wedge u)(1)$. The usual differential operators $d, \partial, \bar\partial$ extend to currents.

One possible definition of the important concept of a positive current goes as follows.

DEFINITION. A current u on Ω of type (k, k) is said to be positive if for all $\varphi \in C_0^\infty(\Omega)$, $\varphi \geqq 0$, and, for all choices of complex linear coordinates $z = (z_1, \ldots, z_n)$,

$$\varphi u\left(\frac{i}{2} dz_1 \wedge d\bar z_1 \wedge \cdots \wedge \frac{i}{2} dz_{n-k} \wedge d\bar z_{n-k}\right) \geqq 0.$$

This same definition may be expressed in more geometric language by saying for each positive localization φu of u and for each complex linear projection $\pi : \mathbf{C}^n \to \mathbf{C}^{n-k}$, $\pi_*(\varphi u)$ (the push forward of φu) is a positive measure on $\mathbf{C}^{n-k}$. One can show that a current of type (1, 1), say $u = \sum h_{ij}(i/2) dz_i \wedge d\bar z_j$, is positive if and only if the matrix (h_{ij}) of distributions is positive in the sense that $\sum h_{ij} w_i \bar w_j$ is a positive measure for each fixed vector $w \in \mathbf{C}^n$.

It is clear that if $V = \mathbf{C}^d \times \{0\} \subset \mathbf{C}^n$ then $u(\psi) = \int_V \psi$ defines a positive current of type $(n - d, n - d)$. Since the notion of a positive current is a biholomorphic invariant and a local concept, this proves that $[V](\psi) = \int_V \psi$ defines a positive (k, k) current for any $d = (n - k)$-dimensional complex submanifold. It is well known that if V is a pure d-dimensional subvariety of Ω open $\subset \mathbf{C}^n$ then the volume of the manifold part of V (denoted Reg V) is locally finite. As a consequence one has that $[V](\psi) = \int_{\mathrm{Reg}\, V} \psi$ for all $\psi \in \mathscr{D}_{2d}(\Omega)$ defines a current $[V]$ on Ω for each pure d-dimensional subvariety V of Ω. Also, it is true that $[V]$ is (1) of type $(n - d, n - d)$, (2) positive, and (3) $\bar\partial$-closed. (There are other examples of currents satisfying conditions (1)–(3). For instance, $\omega = (i/2) \sum dz_j \wedge d\bar z_j$ is of type (1, 1) positive and $\bar\partial$-closed on $\mathbf{C}^n$.) This completes step 1 in the proof of Shiffman's theorem outlined above. That is, the pure d-dimensional subvariety V of $\Omega - A$ has been replaced by a current on $\Omega - A$ which is of type $(n - d, n - d)$, positive and $\bar\partial$-closed. Now, using the assumption that the exceptional set A is small, step 2 is taken care of by the following removable singularity theorem for $\bar\partial$ (see Harvey [**5**]).

THEOREM. *Suppose A is a closed subset of Ω with $(2d - 1)$-dimensional Hausdorff* *measure zero. If u is a positive, $\bar{\partial}$-closed current on $\Omega - A$ of type $(n-d, n-d)$* (dim d, d) *then there exists a positive, $\bar{\partial}$-closed current U on Ω of type $(n-d, n-d)$ with $U|_{\Omega-A} = u$. Moreover U is unique with measure coefficients U_{IJ} determined by $U_{IJ}(B) = u_{IJ}(B \cap (\Omega - A))$ for all Borel sets $B \subset \Omega$.*

A holomorphic line bundle is said to be semipositive if its refined Chern class in $\{\omega \in \mathcal{D}'_{1,1} : d\omega = 0\}/\partial\bar{\partial}\mathcal{D}'$ has a representative ω which is positive. The following corollary generalizes a result of Shiffman [**14**], [**15**].

COROLLARY. *If A is a closed subset of Ω open $\subset C^n$ and the $(2n-3)$-dimensional Hausdorff measure of A vanishes then each semipositive holomorphic line bundle on $\Omega - A$ has a unique extension to Ω which is semipositive.*

Now back to the proof of Shiffman's theorem outlined above. Let U denote the extension of the current $[V]$ (from $\Omega - A$ to Ω) given by the above theorem. In order to complete step 3 we must show that there exists a pure d-dimensional subvariety W of Ω with $U = [W]$. The facts that U is of type $(n-d, n-d)$, positive, and $\bar{\partial}$-closed are not enough to ensure that $U = [W]$ for some subvariety W. Currents of the type $[W]$ are, in addition, concentrated in a very special way. More precisely, (4) the $2d$-dimensional density of $[W]$ is a positive integer at each point $z \in W$ (see Thie [**16**] for a proof and cf. (1)–(3) above). This density denoted by $\Theta_{2d}(V, z)$ is by definition $\lim_{r\to 0} \mathrm{vol}_{2d}(W \cap B(z,r))/c_d r^{2d}$ where $c_d r^{2d}$ is the volume of a ball in R^{2d} of radius r and $\mathrm{vol}_{2d}(W \cap B(z,r))$ is the volume of the manifold Reg W in the ball $B(z,r)$ about z of radius r. This function of r can be shown to be decreasing as $r \to 0^+$ (Federer [**4**] or Lelong [**11**]). If U is a $\bar{\partial}$-closed positive current of type $(n-d, n-d)$, it is natural to define

$$\Theta_{2d}(U, z) = \lim_{r\to 0^+} (U \wedge \omega^d)(B(z,r))/c_d r^{2d}$$

(where $\omega = (i/2)\sum dz_j \wedge d\bar{z}_j$) since $(U \wedge \omega^d)(B(z,r)) = \mathrm{vol}_{2d}(W \wedge B(z,r))$ if $U = [W]$. Again this function of r is decreasing as $r \to 0^+$. This fact has as an elementary corollary (see Corollary 4 in [**6**]) that $\Theta_{2d}(U, z)$ is an upper semicontinuous function of z. This fact is useful as follows. Since V is a subvariety of $\Omega - A$ and $U = [A]$ on $\Omega - A$, $\Theta_{2d}(U, z)$ is a positive integer on V. Now by the upper semicontinuity of $\Theta_{2d}(U, z)$ this implies that $\Theta_{2d}(U, z) \geqq 1$ on $\bar{V}$. Finally we need the fact that the support of U is $\bar{V}$ (the closure of the support of $[V]$) which follows from the above removable singularity theorem for $\bar{\partial}$. Since $\Theta_{2d}(U, z) \geqq 1$ on $\bar{V} = \operatorname{supp} U$, the following structure theorem provides a pure d-dimensional subvariety W of Ω with $U = [W]$ on Ω and completes the third step in the proof of Shiffman's extension theorem.

THEOREM. *If U is a positive $\bar{\partial}$-closed current of type $(n-d, n-d)$ and $\Theta_{2d}(U, z)$ is locally bounded away from zero on the support of U then there exists a subvariety W with irreducible components W_j and constants $a_j > 0$ such that $U = \sum a_j[W_j]$.*

Actually one only needs $\Theta_{2d}(U, z)$ locally bounded away from zero almost everywhere on supp U with respect to $2d$-dimensional Hausdorff measure.

This theorem was proved in the special case where $\Theta_{2d}(U, z)$ is a positive integer (Hausdorff $2d$-dimensional measure almost everywhere on supp U) in King [**9**] using techniques of Federer [**4**]. The theorem as stated above is proved in [**6**] using the following deep structure theorem of Bombieri [**1**], [**2**], which says that if u is a positive $\bar{\partial}$-closed $(1, 1)$ current on Ω, then for each $c > 0$, $E_c = \{z : \Theta(u, z) \geqq c\}$ is locally contained in a complex hypersurface. Note that in this theorem one does not require that $E_c = \operatorname{supp} u$.

One might conjecture the following. If u is a positive $\bar{\partial}$-closed (k, k) current on Ω then, for each $c > 0$, $E_c = \{z : \Theta_{2d}(u, z) \geqq c\}$ is a subvariety of Ω of dimension $\leqq d$ (where $d = n - k$). This conjecture is not known in the hypersurface case, i.e. $d = n - 1$ (it does not follow from Bombieri's theorem). If $u = [V]$ for some subvariety V then $E_c = \{z \in \Omega : \Theta(u, z) \geqq c\}$ is a subvariety of V since $\Theta([V], z)$ is just the multiplicity of V at z (Draper [**3**]). Additional information concerning this conjecture is that $E_c = \{z \in \Omega : \Theta_{2d}(u, z) \geqq c\}$ has locally finite $2d$-dimensional Hausdorff measure (Federer [**4**]).

King [**9**] proved that if u is a positive, $\bar{\partial}$-closed, integral current of type (k, k) then $u = \sum a_j[V_j]$ where V is a subvariety of Ω with irreducible components V_j and the a_j are positive integers. This result also follows from the structure theorem in [**6**] stated above. A third proof is given in Harvey-Shiffman [**8**] where in addition a stronger theorem is obtained by dropping the hypothesis that u be positive and then only concluding that $u = \sum a_j[V_j] - \sum b_j[W_j]$ where the a_j and b_j are positive integers. This result has been used by Lawson-Simons [**10**]. It also implies the Bishop extension theorem (see [**8**]).

In conclusion we examine the relationship of the above removable singularity theorem (for positive $\bar{\partial}$-closed currents) and extension results for plurisubharmonic functions. A distribution φ is plurisubharmonic if $i\partial\bar{\partial}\varphi$ is a positive current.

THEOREM. *Suppose A is a closed subset of Ω open $\subset \mathbf{C}^n$ and φ is plurisubharmonic on $\Omega - A$.*

(a) *If $\varphi \in \operatorname{Lip}_\delta(\Omega)$ and $\Lambda_{2n-\delta}(A) = 0$ then φ is plurisubharmonic on Ω.*

(b) $(p < \infty)$ *If $\varphi \in L^p_{\text{loc}}(\Omega)$ and $\Lambda_{2n-2p'}(A)$ is locally finite then φ is plurisubharmonic on Ω.*

(c) *If φ is locally bounded above in Ω and $\Lambda_{2n-2}(A)$ is locally finite then φ is plurisubharmonic on Ω.*

(d) *If $\Lambda_{2n-2}(A) = 0$ then φ has a plurisubharmonic extension to Ω.*

SKETCH OF PROOF. To prove (b) (cf. the proof of Theorem 4.1(a) in [**7**]) pick $\psi \in C_0^\infty(\Omega)$, $\psi \geqq 0$, and complex linear coordinates $z = (z_1, \ldots, z_n)$. Picking φ_ε from Lemma 3.2 of [**7**], we have

$$i\partial\bar{\partial}\varphi(\psi\alpha) = i\partial\bar{\partial}\varphi((1 - \varphi_\varepsilon)\psi\alpha) + i\partial\bar{\partial}\varphi(\varphi_\varepsilon\psi\alpha)$$

where $\alpha = (i/2)dz_1 \wedge d\bar{z}_1 \wedge \cdots \wedge (i/2)dz_{n-1} \wedge d\bar{z}_{n-1}$. Now $i\partial\bar{\partial}\varphi(\varphi_\varepsilon\psi\alpha) = \int \varphi i\partial\bar{\partial}(\varphi_\varepsilon\psi\alpha)$ and the right-hand side approaches zero as $\varepsilon \to 0$ (see (4.1) in [7]). Therefore $i\partial\bar{\partial}\varphi(\psi\alpha) = \lim_{\varepsilon\to 0} i\partial\bar{\partial}\varphi((1 - \varphi_\varepsilon)\psi\alpha) \geqq 0$ since $(1 - \varphi_\varepsilon)\psi \in C_0^\infty(\Omega - A)$ and $(1 - \varphi_\varepsilon)\psi \geqq 0$. This proves that $i\partial\bar{\partial}\varphi$ is positive. Similarly, one can prove part (a) using Lemma 3.1 of [7]. Part (c) follows from the analogous fact for subharmonic functions on $\boldsymbol{R}^{2n}$ and the fact that φ is plurisubharmonic on Ω if and only if φ is subharmonic in the underlying $2n$ real-variables for each choice of complex linear coordinates in $\boldsymbol{C}^n$. Finally part (d) is shown by proving that φ is locally bounded above on Ω and using part (c) (see Shiffman [15]).

Note that part (d) with $\Lambda_{2n-3}(A) = 0$ instead of $\Lambda_{2n-2}(A) = 0$ is a consequence of the removable singularity theorem for positive $\bar{\partial}$-closed (1, 1) currents. Also, since one cannot always solve $i\partial\bar{\partial}\varphi = \omega$ (with ω assumed d-closed) on $\Omega - A$ for $\Lambda_{2n-3}(A) = 0$ (take $A = \{0\}$ in $\Omega = \boldsymbol{C}^2$), the removable singularity theorem for positive $\bar{\partial}$-closed (1, 1) currents does not follow from part (d) above.

Bibliography

1. E. Bombieri, *Algebraic values of meromorphic maps*, Invent. Math. **10** (1970), 267–287.

2. ———, *Addendum to my paper algebraic values of meromorphic maps*, Invent. Math. **11** (1970), 163–166.

3. R. Draper, *Intersection theory in analytic geometry*, Math. Ann. **180** (1969), 175–204. MR **40** #403.

4. H. Federer, *Geometric measure theory*, Die Grundlehren der math. Wissenschaften, Band 153, Springer-Verlag, Berlin and New York, 1969. MR **41** #1976.

5. R. Harvey, *Removable singularities for positive currents*, Amer. J. Math. (to appear).

6. R. Harvey and J. King, *On the structure of positive currents*, Invent. Math. **15** (1972), 47-52.

7. R. Harvey and J. Polking, *Removable singularities of solutions of linear partial differential equations*, Acta Math. **125** (1970), 39–56.

8. R. Harvey and B. Schiffman, *A characterization of holomorphic chains* (to appear).

9. J. King, *The currents defined by analytic varieties*, Acta Math. **127** (1971), 185–220.

10. B. Lawson and S. Simons, *On stable currents and their applications to global problems in real and complex geometry*, Ann. of Math. (to appear).

11. P. Lelong, *Fonctions plurisousharmoniques et formes différentielles positives*, Gordon and Breach, New York, 1968. MR **39** #4436.

12. R. Remmert and K. Stein, *Über die wesentlichen Singularitäten analytischer Mengen*, Math. Ann. **126** (1953), 263–306. MR **15**, 615.

13. B. Shiffman, *On the removal of singularities of analytic sets*, Michigan Math. J. **15** (1968), 111–120. MR **37** #464.

14. ———, *Extension of positive holomorphic line bundles*, Bull. Amer. Math. Soc. **76** (1971), 1091–1093.

15. ———, *Extension of positive line bundles and meromorphic maps*, Invent. Math. (to appear).

16. P. Thie, *The Lelong number of a point of a complex analytic set*, Math. Ann. **172** (1967), 269–312. MR **35** #5661.

Rice University

THE CAUCHY PROBLEM FOR $\bar{\partial}$

C. DENSON HILL[1]

1. **Introduction.** Consider a complex analytic manifold X with $\dim_C X = n$. Let U be an open connected set of X, and let S be a closed (in U) C^∞-differentiable submanifold of U of real dimension $2n - 1$. Under the assumption that S has two sides in U, it follows [**4**] that a real-valued function $\rho \in C^\infty(U)$ exists such that

$$S = \{x \in U \mid \rho(x) = 0\}, \qquad d\rho \neq 0 \quad \text{on } S.$$

Set

$$U^+ = \{x \in U \mid \rho(x) \geqq 0\},$$
$$U^- = \{x \in U \mid \rho(x) \leqq 0\},$$

and consider one of the two sides, say U^-: The Cauchy problem for $\bar{\partial}$ in U^-, at level (r,s), is the initial-value problem

$$(1_{r,s}) \qquad \begin{aligned} \bar{\partial} u &= f \quad \text{in } U^-, \\ u|_S &= u_0. \end{aligned}$$

Here the data consist in prescribing $f \in C^\infty_{(r,s+1)}(U^-)$ (a form of type $(r,s+1)$ in U^-) and $u_0 \in C^\infty_{(r,s)}(S)$ (i.e., u_0 is the restriction to S, in the sense of restricting the coefficients—not the pull-back—of a smooth form of type (r,s) defined in a neighborhood of S in U). A solution $u \in C^\infty_{(r,s)}(U^-)$ is sought satisfying $(1_{r,s})$, where the notation $u|_S$ refers to the restriction of the coefficients of u to S. Thus we prescribe smooth data and ask for smooth solutions, up to the boundary S.

Clearly some compatibility conditions are needed on the data; namely,

AMS 1970 *subject classifications*. Primary 35A24; Secondary 32A22.

[1] Research supported by the Office of Scientific Research of the United States Air Force under Contract AF F 44620-72-C-0031. The author is an Alfred P. Sloan Fellow.

$$\bar{\partial} f = 0 \quad \text{in } U^-,$$
$$\bar{\partial}_S u_0 = f \quad \text{on } S,$$

where $\bar{\partial}_S$ is the tangential Cauchy-Riemann operator (defined below) to S.

Note that if $s < (n-1)/2$ there are more equations than unknowns in $(1_{r,s})$, if $s > (n-1)/2$ there are fewer equations than unknowns; the system is exactly determined if and only if n is odd and $s = (n-1)/2$.

In what follows I shall be concerned with the well-posedness (in various senses) of $(1_{r,s})$, how it is related to several other problems of interest, and how it is influenced by the Levi convexity of S. These remarks are amplifications of some recent joint work with A. Andreotti [3], [**4**], [**5**], [**6**]; the reader is referred to those papers for detailed proofs of all statements made here.

2. **Some questions related to the Cauchy problem.** Let $\mathscr{I}_{(r,s)}(U)$ denote the differential ideal

$$\mathscr{I}_{(r,s)}(U) = \{\varphi \in C^\infty_{(r,s)}(U) | \varphi = \rho\alpha + \bar{\partial}\rho \wedge \beta \text{ for some } \alpha \in C^\infty_{(r,s)}(U), \beta \in C^\infty_{(r,s-1)}(U)\}.$$

Since $\bar{\partial}\mathscr{I}_{(r,s)}(U) \subset \mathscr{I}_{(r,s+1)}(U)$ it follows that the complex

$$(*) \qquad \mathscr{I}_{(r,0)}(U) \xrightarrow{\bar{\partial}} \mathscr{I}_{(r,1)}(U) \xrightarrow{\bar{\partial}} \mathscr{I}_{(r,2)}(U) \to \cdots$$

is a subcomplex of the Dolbeault complex

$$(**) \qquad C^\infty_{(r,0)}(U) \xrightarrow{\bar{\partial}} C^\infty_{(r,1)}(U) \xrightarrow{\bar{\partial}} C^\infty_{(r,2)}(U) \to \cdots;$$

so that there is a quotient complex

$$(***) \qquad Q_{(r,0)}(U) \xrightarrow{\bar{\partial}_S} Q_{(r,1)}(U) \xrightarrow{\bar{\partial}_S} Q_{(r,2)}(U) \to \cdots.$$

This gives a convenient and natural definition of the tangential Cauchy-Riemann operator $\bar{\partial}_S$ (see [**4**]). In concrete terms: the compatibility condition "$\bar{\partial}_S u_0 = f$ on S" means that $\bar{\partial}\tilde{u}_0 - \tilde{f} \in \mathscr{I}_{(r,s+1)}(U)$ for any C^∞ extensions $\tilde{u}_0$ and $\tilde{f}$ of u_0 and $f|_S$, respectively.

The above definitions can also be made with U replaced everywhere by either U^+ or U^-. Let $H^{r,s}(U, \mathscr{I})$ (or $H^{r,s}(U^+, \mathscr{I})$ or $H^{r,s}(U^-, \mathscr{I})$) denote the cohomology of the complex (*), let $H^{r,s}(U)$ (or $H^{r,s}(U^+)$ or $H^{r,s}(U^-)$) denote the cohomology of the complex (**), and let $H^{r,s}(S)$ denote the cohomology of the complex (***) with either U, U^+ or U^- (they are all trivially isomorphic). Then $H^{r,s}(U,\mathscr{I}) = H^{r,s}(U^+,\mathscr{I}) \oplus H^{r,s}(U^-,\mathscr{I})$ and $H^{r,s}(U)$ is the usual (Dolbeault) cohomology of U. But $H^{r,s}(U^+)$ and $H^{r,s}(U^-)$ are *not* standard types of cohomology, because the differential forms involved are required to be C^∞ up to the boundary S, but not beyond it. Finally, $H^{r,s}(S)$ is the boundary cohomology on S.

Observe that the Cauchy problem $(1_{r,s})$ does not, generally speaking, have a unique solution. The difference between two solutions represents a class in $H^{r,s}(U^-,\mathscr{I})$. At best, one could speak of equivalence classes of solutions.

Now a number of interesting questions present themselves:

Problem I. Find necessary and sufficient conditions for the existence of solutions to the Cauchy problem $(1_{r,s})$.

Problem II. Prove vanishing theorems, or finiteness theorems, for the spaces $H^{r,s}(U^+)$ and $H^{r,s}(U^-)$. When are they infinite dimensional?

Problem III. What can be said about the boundary cohomology $H^{r,s}(S)$? E.g., when is there a "Poincaré lemma" for the operator $\bar{\partial}_S$?

Problem IV. When is it true that the natural restriction

$$H^{r,s}(U^-) \xrightarrow{\text{restr}} H^{r,s}(S)$$

is an isomorphism? Is surjective? Is injective?

REMARK. Problem IV can be viewed as a generalized formulation of the homogeneous ($f \equiv 0$) Cauchy problem $(1_{r,s})$ in terms of cohomology classes. We call this problem the H. Lewy problem. From this point of view, an isomorphism in IV means that the H. Lewy problem is well-posed, at level (r,s) from the side U^-.

The point of the following theorem [**4**] is to reduce the problems above to corresponding questions about vanishing of the cohomology groups $H^{r,s}(U^\pm)$ or $H^{r,s}(U^\pm, \mathscr{I})$:

THEOREM 0. *Assume U is Stein. Then*

I. *A necessary and sufficient condition for existence in $(1_{r,s})$ is that $H^{r,s+1}(U^-, \mathscr{I}) = 0$.*

II. $(s = 0)$ $H^{r,1}(U^-, \mathscr{I}) \cong H^{r,0}(U^+)/H^{r,0}(U)$.

$(s > 0)$ $H^{r,s+1}(U^-, \mathscr{I}) \cong H^{r,s}(U^+)$.

III. $(s = 0)$ *There is a short exact sequence*

$$0 \to H^{r,0}(U) \xrightarrow[\text{restr}]{} H^{r,0}(U^+) \oplus H^{r,0}(U^-) \xrightarrow[\text{jump}]{} H^{r,0}(S) \to 0.$$

$(s > 0)$ *There is an isomorphism*

$$H^{r,s}(U^+) \oplus H^{r,s}(U^-) \xrightarrow[\text{jump}]{\sim} H^{r,s}(S).$$

IV. *A necessary and sufficient condition that the H. Lewy problem be well-posed is*

$(s = 0)$ *Either* $H^{r,1}(U^-, \mathscr{I}) = 0$ *or* $H^{r,0}(U^+)/H^{r,0}(U) = 0$.

$(s > 0)$ *Either* $H^{r,s+1}(U^-, \mathscr{I}) = 0$ *or* $H^{r,s}(U^+) = 0$.

Note that $H^{r,0}(U^+)/H^{r,0}(U) = 0$ whenever $U^- \subset$ the envelope of holomorphy of $\tilde{U}^+$. Thus the Cauchy problem $(1_{r,0})$, or the H. Lewy problem at level $s = 0$ (the extension of functions u_0 satisfying the tangential Cauchy-Riemann equations on S to holomorphic functions u in U^-) are essentially reduced to a question about envelopes of holomorphy. Using this remark, one can construct examples of $S \subset \boldsymbol{C}^n$ such that: any smooth u_0 on S with $\bar{\partial}_S u_0 = 0$ has a holomorphic extension u to U^-, where $U^- \not\subset \boldsymbol{C}^n$; i.e., U^- is a Riemann domain spread over $\boldsymbol{C}^n$.

In any case we see, from III, that each smooth u_0 with $\bar{\partial}_S u_0 = 0$ can be expressed (uniquely modulo global holomorphic functions in U) as the jump across S of two holomorphic functions, one in U^+ and one in U^-, that are smooth up to S. This is the additive Riemann-Hilbert problem for functions.

Finally, *independent of the "shape"* of S (in the sense of Levi convexity) it follows that the boundary cohomology is precisely the jump in the cohomologies $H^{r,s}(U^+)$ and $H^{r,s}(U^-)$ across S. In particular, $H^{r,s}(S) = 0 \Leftrightarrow$ both $H^{r,s}(U^+) = 0$ and $H^{r,s}(U^-) = 0$.

3. **The local situation.** The *Levi form* of S can be defined as follows: For a point $z_0 \in S$, consider

$$\mathscr{L}_{z_0}(\rho) = \sum_{j,k=1}^{n} \frac{\partial^2 \rho}{\partial z_j \partial \bar{z}_k}(z_0) w_j \bar{w}_k,$$

$$\sum_{j=1}^{n} \frac{\partial \rho}{\partial z_j}(z_0) w_j = 0.$$

(Actually this is the Levi form at z_0 restricted to the holomorphic tangent space to S at z_0.) Its signature (total of $n - 1$ positive, negative, or zero eigenvalues) is a biholomorphic invariant, and is independent of the particular function ρ used to define S.

Let z_0 be a fixed point on S. We want to investigate the local situation in a small neighborhood of z_0; hence we might as well assume that $X = \boldsymbol{C}^n$. The following theorems, which are proved in [**5**], show how the relevant cohomology groups in Theorem 0 are influenced by the signature of the Levi form $\mathscr{L}_{z_0}$ at z_0:

(A) *Let U be an open set in $\boldsymbol{C}^n$ and let $h: U \to \boldsymbol{R}$ be a C^∞ function with $dh \neq 0$ on $S = \{x \in U | h(x) = 0\}$. Let $z_0 \in S$ be a point at which the Levi form $\mathscr{L}_{z_0}(h)$ has at least $p - 1$ positive eigenvalues.*

THEOREM 1. *With assumptions* (A), *then there exists a neighborhood W of z_0 in U such that for any domain of holomorphy ω contained in W, setting $D = \{x \in \omega | h(x) \leqq 0\}$, we have*

$$H^{r,s}(D) = 0 \quad \textit{for } s > n - p.$$

THEOREM 2. *With assumptions* (A), *then there exists a fundamental sequence of open neighborhoods $\{\omega\}$ of z_0 which are domains of holomorphy, and such that, for $D = \{x \in \omega | h(x) \leqq 0\}$, we have*

$$H^{r,s}(D, \mathscr{I}) = 0 \quad \textit{for } s < p.$$

Combining Theorems 1 and 2 with Theorem 0, we obtain, locally for a point z_0 on the original manifold S,

THEOREM 3. *Assume that the Levi form $\mathscr{L}_{z_0}(\rho)$ at z_0 has p positive and q negative eigenvalues, $p + q \leqq n - 1$. Then there exists a fundamental sequence of neighbor-*

hoods $\{\omega_\nu\}_{\nu\in\mathbf{N}}$ *of* z_0 *such that each* ω_ν *is a domain of holomorphy and moreover, setting*

$$\omega_\nu^- = \omega_\nu \cap U^-, \qquad \omega_\nu^+ = \omega_\nu \cap U^+,$$

we have

$$\begin{cases} H^{r,s}(\omega_\nu^+) = 0 \quad \textit{for} \begin{cases} s > n-q-1, \\ \textit{or} \\ 0 < s < p; \end{cases} \\ \nu \textit{ even}, \end{cases} \qquad \begin{cases} H^{r,s}(\omega_\nu^-) = 0 \quad \textit{for} \begin{cases} s > n-p-1, \\ \textit{or} \\ 0 < s < q. \end{cases} \\ \nu \textit{ odd}, \end{cases}$$

COROLLARY. *When the Levi form at* z_0 *is nondegenerate (i.e., when* $q = n - p - 1$*), then*

$$H^{r,s}(\omega_\nu^+) = 0 \quad \textit{for } \nu \neq 0, p\ (\nu \textit{ even}),$$
$$H^{r,s}(\omega_\nu^-) = 0 \quad \textit{for } \nu \neq 0, q\ (\nu \textit{ odd}).$$

The proofs of Theorems 1 and 2 use techniques from Andreotti and Vesentini [**2**], techniques from Hörmander [**7**], and require the regularity theorem of Kohn and Nirenberg [**8**].

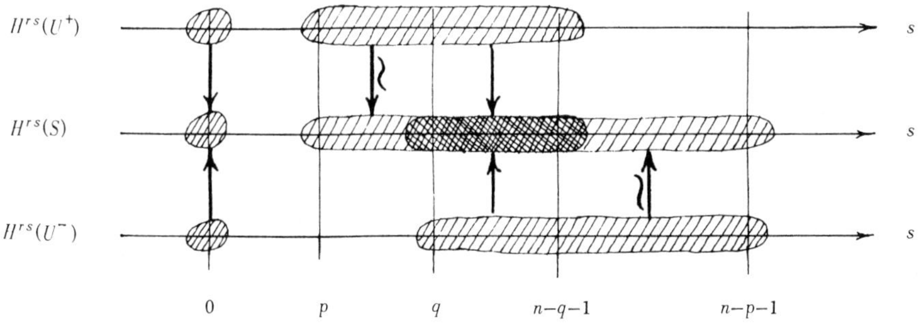

Figure 1. Degenerate Case

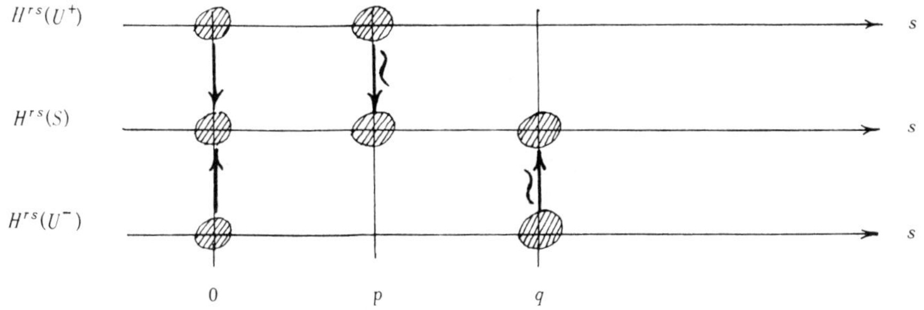

Figure 2. Nondegenerate Case

Under the assumptions of Theorem 3, the local situation with respect to the Problems I–IV mentioned in Theorem 0 is summarized in Fig. 1, where the shaded regions denote the ranges in which there may be some nontrivial cohomology. Fig. 2 shows the much simpler nondegenerate case described in the Corollary. Note that for $s > 0$ the obstruction to the H. Lewy problem on one side of S is the cohomology on the other side. Thus in the overlap region the Cauchy problem may not be well-posed. In the nondegenerate case, there is a "Poincaré lemma" for $\bar{\partial}_S$ except in dimensions p and q. When n is odd and $p = q = (n-1)/2$ (the exactly determined case in $(1_{r,s})$!) then, even in the nondegenerate case, the Cauchy problem is not well-posed on either side of S. In that case the natural problem to consider is the additive Riemann-Hilbert problem.

Finally, at least in the nondegenerate case, the groups that remain are all infinite dimensional.

THEOREM 4. *Let the Levi form $\mathcal{L}_{z_0}(\rho)$ be nondegenerate and have p positive and $q = n - 1 - p$ negative eigenvalues. Then there exists a neighborhood W of z_0 in U such that for any domain of holomorphy $\omega \in z_0$ with $\omega \subset W$, the groups*

$$H^{r,p}(\omega^+), \quad H^{r,q}(\omega^-), \quad H^{r,p}(S \cap \omega), \quad H^{r,q}(S \cap \omega)$$

are all infinite dimensional over $\boldsymbol{C}$.

Thus, even locally, there is some infinite-dimensional boundary cohomology.

4. **An example.** In $\boldsymbol{C}^2$ consider the hypersurface Σ defined by

$$\Sigma = \{(z,\zeta) \in \boldsymbol{C}^2 | \rho(x,y,\xi,\eta) = 0\}, \qquad \rho(x,y,\xi,\eta) = \eta - (x + y)^2,$$

where $z = x + iy$ and $\zeta = \xi + i\eta$. Let U be an open neighborhood of a point $z_0 \in \Sigma$, and let $U^\pm$ and $S = U \cap \Sigma$ be defined as before. In this case the complex (***) is isomorphic to

$$C^\infty(S) \xrightarrow{\bar{\partial}_S} C^\infty(S) \wedge d\bar{z} \xrightarrow{\bar{\partial}_S} 0,$$

where $C^\infty(S)$ denotes C^∞ functions on S. The first $\bar{\partial}_S$ above is given by $\bar{\partial}_S u = (Lu) \wedge d\bar{z}$ where the linear map $L: C^\infty(S) \to C^\infty(S)$, when written in the coordinates (x,y,ξ) on S, is

$$L = \frac{1}{2}\left(\frac{\partial}{\partial x} + i\frac{\partial}{\partial y}\right) - i(x + iy)\frac{\partial}{\partial \xi}.$$

This is the famous example of H. Lewy [**11**] for which the inhomogeneous equation $Lu = f$ is not locally solvable. Moreover, we have the interpretations

$$H^{0,0}(S) \cong \ker \{L: C^\infty(S) \to C^\infty(S)\},$$

$$H^{0,1}(S) \cong \frac{C^\infty(S)}{\operatorname{im}\{L: C^\infty(S) \to C^\infty(S)\}}.$$

It is easily seen that the Levi form $\mathscr{L}_{z_0}(\rho)$ has exactly one negative eigenvalue. Hence it follows from Theorems 0–4 above that, for appropriate U, we have the isomorphisms

$$H^{0,0}(U^+) \xrightarrow[\text{restr}]{\sim} H^{0,0}(S),$$

$$H^{0,1}(U^-) \xrightarrow[\text{restr}]{\sim} H^{0,1}(S),$$

and that these groups are infinite dimensional. Of course $\dim_C H^{0,1}(S) = \infty$ is compatible with the nonsolvability of $Lu = f$. Thus one could take the point of view that Lewy's equation fails to be solvable because there is some nontrivial local boundary cohomology $H^{0,1}(S)$ which S "inherits" from U^-.

5. **The global situation.** We return now to the general situation described in the Introduction. The local results can be pieced together [**5**] using customary arguments via the bump lemma to yield global results.

First consider the case where U^- is compact, so that $U^- - \mathring{U}^- = S = \partial U^-$. Assume that at each point $z_0 \in S$ the Levi form $\mathscr{L}_{z_0}$ has at least p positive and at least q negative eigenvalues ($p + q \leqq n - 1$). Then we have global finiteness theorems:

THEOREM 5. *Under the above assumptions,*

(a) $$\dim_C H^{r,s}(U^-) < \infty \quad \textit{for} \begin{cases} s < q, \\ \textit{or} \\ s > n - p - 1; \end{cases}$$

(b) $$\dim_C H^{r,s}(U^-, \mathscr{I}) < \infty \quad \textit{for} \begin{cases} s < p + 1, \\ \textit{or} \\ s > n - q. \end{cases}$$

With an additional assumption, one obtains global vanishing theorems: Assume that ρ can be chosen such that

$$\sum_{j,k=1}^{n} \frac{\partial^2 \rho}{\partial z_j \partial \bar{z}_k}(z_0) w_j \bar{w}_k$$

has at least $p + 1$ positive eigenvalues at every point $z_0 \in \mathring{U}^-$ (i.e., the assumption is that U^- is $(n - p - 1)$-complete).

THEOREM 6. *With the above assumptions,*

(a) $$H^{r,s}(U^-) = 0 \quad \textit{for } s > n - p - 1,$$

(b) $$H^{r,s}(U^-, \mathscr{I}) = 0 \quad \textit{for } s < p + 1.$$

REMARK. In the nondegenerate case ($q = n - p - 1$) it follows that $H^{r,s}(U^-)$ is finite dimensional for $s \neq q$; moreover it can be shown that

$$\dim_{\mathbf{C}} H^{r,q}(U^-) = \infty.$$

Next consider the case where X is compact and take $U = X$, with $U^{\pm}$, S defined as before. We shall say that the Riemann-Hilbert problem is *almost always solvable* in dimension q, if the natural map

$$H^{r,q}(U^+) \oplus H^{r,q}(U^-) \xrightarrow[\text{jump}]{} H^{r,q}(S)$$

has finite-dimensional kernels and cokernels. We shall say that the Cauchy problem is almost *always solvable* in dimension q, from the side U^-, if the natural map

$$H^{r,q}(U^-) \xrightarrow[\text{restr}]{} H^{r,q}(S)$$

has finite-dimensional kernels and cokernels. Assume now that at every point $z_0 \in S$ the Levi form $\mathscr{L}_{z_0}(\rho)$ is nondegenerate and has p positive and $q = n - p - 1$ negative eigenvalues. Then we have [**5**]

THEOREM 7. *Under the above assumptions,*

(i) *$H^{r,s}(U^-)$ is finite dimensional except for $s = q$, where it is infinite dimensional.*

(ii) *$H^{r,s}(U^+)$ is finite dimensional except for $s = p$, where it is infinite dimensional.*

(iii) *If $p \neq q$, then the Cauchy problem is of interest in dimension q from the side U^-, and in dimension p from the side U^+; in these dimensions it is almost always solvable.*

(iv) *If n is odd and $p = q = (n-1)/2$, then the Riemann-Hilbert problem is of interest, and it is almost always solvable in dimension $p = q = (n-1)/2$.*

Finally, the local results can also be pieced together so as to obtain certain global results in the general case; i.e., for a global $U^{\pm}$ that is not compact, so that S forms only a part of $\partial U^{\pm}$. Also one can obtain some interesting duality theorems. The results alluded to here will be found in [**6**].

This work has also been announced at the International Congress of Mathematicians, in Nice (1970), by Andreotti.

REFERENCES

1. A. Andreotti and H. Grauert, *Théorèmes de finitude pour la cohomologie des espaces complexes*, Bull. Soc. Math. France **90** (1962), 193–259. MR **27** #343.

2. A. Andreotti and E. Vesentini, *Carleman estimates for the Laplace-Beltrami equation on complex manifolds*, Inst. Hautes Études Sci. Publ. Math. No. 25 (1965), 81–130. MR **30** #5333.

3. A. Andreotti and C. D. Hill, *Complex characteristic coordinates and tangential Cauchy-Riemann equations*, Ann. Scuola Norm. Sup. Pisa **26** (1972), 299–324.

4. ———, *E. E. Levi convexity and the Hans Lewy problem.* I: *Reduction to vanishing theorems*, Ann. Scuola Norm. Sup. Pisa **26** (1972), 325–363.

5. ———, *E. E. Levi convexity and the Hans Lewy problem.* II: *Vanishing theorems*, Ann. Scuola Norm. Sup. Pisa (to appear).

6. ———, (to appear).

8. J. J. Kohn and L. Nirenberg, *Non-coercive boundary value problems*, Comm. Pure Appl. Math. **18** (1965), 443–492. MR **31** #6041.

9. J. J. Kohn and H. Rossi, *On the extension of holomorphic functions from the boundary of a complex manifold*, Ann. of Math. (2) **81** (1965), 451–472. MR **31** #1399.

10. H. Lewy, *On the local character of the solutions of an atypical linear differential equation in three variables and a related theorem for regular functions of two complex variables*. Ann. of Math. (2) **64** (1956), 514–522. MR **18**, 473.

11. ———, *An example of a smooth linear partial differential equation without solution*, Ann. of Math. (2) **66** (1957), 155–158. MR **19**, 551.

STANFORD UNIVERSITY

ON HYPOELLIPTICITY OF SECOND ORDER EQUATIONS

O. A. OLEINIK

The sufficient conditions for the hypoellipticity known for linear differential operators of any order do not give good results for second order equations (see for example [**1**], [**2**], [**3**]). The hypoellipticity of second order equations was considered first by L. Hörmander in [**4**]. Another proof of his theorem was given by E. Radkevič [**5**] (see also [**6**] and [**7**]).

According to L. Schwartz [**8**], the operator P is called hypoelliptic in a domain $\Omega \subset \boldsymbol{R}^m$, if given any open subset Ω_1 of Ω and any distribution u in $\Omega_1, Pu \subset C^\infty(\Omega_1)$ implies $u \subset C^\infty(\Omega_1)$.

The operator P is called global hypoelliptic in a domain Ω, if any distribution $u \subset \Omega . Pu \subset C^\infty(\Omega)$ implies $u \subset C^\infty(\Omega)$. L. Hörmander [**4**] considered the class of second order operators of the form

$$Pu = \sum_{j=1}^{r} X_j^2 u + X_0 u + c(x)u \tag{1}$$

where $X_j = a_j^\zeta(x)\partial/\partial x_\zeta$, $j = 0, 1, \ldots, r, x = (x_1, \ldots, x_m)$, the functions a_j^ζ, c are real and belong to $C^\infty(\Omega)$. (Here and in what follows, summations over repeated indices are understood to extend from 1 to m.)

Let $\mathscr{I} = (\alpha_1, \ldots, \alpha_t)$, where $\alpha_l = 0, 1, \ldots, r$, for $l = 1, \ldots, t$, and t is any positive integer. For any multi-index $\mathscr{I}$ we form the operator

$$X_{\mathscr{I}} = \operatorname{ad} X_{\alpha_1} \operatorname{ad} X_{\alpha_2} \cdots \operatorname{ad} X_{\alpha_{t-1}} X_{\alpha_t}$$

where, as usual, $\operatorname{ad} AB = [A,B] \equiv AB - BA$ for any operators A and B. Denote by $\mathscr{L}(X_0, X_1, \ldots, X_r)$ the set of the operators $X_{\mathscr{I}}$ for all multi-indices $\mathscr{I}$. Let $R(x)$ be the number of the operators $X_{\mathscr{I}} \subset \mathscr{L}(X_0, X_1, \ldots, X_r)$ which are linearly independent at the point $x \subset \Omega$. L. Hörmander proved the following [**4**]:

AMS 1970 *subject classifications*. Primary 35H05, 47F05.

THEOREM 1. *The operator* (1) *is hypoelliptic in the domain* Ω, *if* $R(x) = m$ *for all* $x \subset \Omega \subset \boldsymbol{R}^m$.

It can be easily proved that the condition $R(x) = m$ for all $x \subset \Omega$ is also the necessary condition for the hypoellipticity of operators (1) provided that $R(x) = \text{const}$ for all $x \subset \Omega$. M. Derridj [**9**] proved that if $R(x) > 0$ for all $x \subset \Omega, a_j^l, c$ are real analytic functions in Ω and $R(x^0) < m$ for a point $x^0 \subset \Omega$, then operator (1) is not hypoelliptic in Ω.

Therefore, the condition $R(x) = m$ in Ω is necessary and sufficient for the hypoellipticity of operator (1) with real analytic coefficients, if

$$\sum_{j=0}^{r} \sum_{l=1}^{m} |a_j^l(x)| > 0 \quad \text{in } \Omega. \tag{2}$$

As examples show, it is false if condition (2) is not valid. The operator

$$Pu \equiv |x|^{2\nu}\Delta u - u = 0, \qquad |x|^2 = \sum_{j=1}^{m} x_j^2, \tag{3}$$

is hypoelliptic for any integer $\nu > 1$ in any domain $\Omega \subset \boldsymbol{R}^m$. This is proved in [**1**]. This operator has analytic coefficients in any domain $\Omega \subset \boldsymbol{R}^m$; it can be written in the form (1) and $R(x) = 0$ for $x = 0$ and $R(x) = m$ for $x \neq 0$.

One can prove (see [**7**]) that the operator

$$Pu \equiv \sum_{j=1}^{k} \frac{\partial^2 u}{\partial x_j^2} + \sum_{j=k+1}^{m} \exp\left[-1/|x|^2\right] \frac{\partial}{\partial x_j} \exp\left[-1/|x|^2\right] \frac{\partial u}{\partial x_j} \tag{4}$$

is hypoelliptic in any domain $\Omega \subset \boldsymbol{R}^m$, if $k \neq 0$. It is easy to see that $R(x) = k$ for $x = 0$ and $R(x) = m$ for $x \neq 0$.

The hypoellipticity of the operator (4) follows from the following theorem, proved in [**7**]. Let us denote by $\mathcal{M}$ the manifold, which is the union of a finite number of closed smooth $(m - 1)$-dimensional manifolds such that $\mathcal{M} \subset \Omega$.

Suppose that in a neighborhood of any point $x^0 \subset \mathcal{M}$ the manifold can be given by the equation

$$\Phi(x_1, \ldots, x_m) = 0, \qquad \operatorname{grad} \Phi \neq 0.$$

Let M be a set of points, which belong to $\mathcal{M}$.

THEOREM 2. *Suppose that* $R(x^0) = m$ *at any point* $x^0 \in (\Omega \backslash M)$ *and the inequality*

$$\sum_{j=1}^{r} |a_j^s(x)\Phi_{x_s}| + |P(\Phi)| > 0 \tag{5}$$

is valid at any point x^0 *which belongs to* M. *Then operator* (1) *is global hypoelliptic in* Ω. *In addition, if* $u \subset \mathcal{D}'(\Omega), Pu \subset \mathcal{H}_s^{\text{loc}}(\Omega)$ *with* $s \subset \boldsymbol{R}^1$, *then* $u \subset \mathcal{H}_s^{\text{loc}}(\Omega)$ *and the estimate*

$$\|\varphi u\|_s^2 \leq C_1\{\|\varphi_1 Pu\|_s^2 + \|\varphi_1 u\|_\gamma^2\} \tag{6}$$

is valid, where $\varphi,\varphi_1 \subset C_0^\infty(\Omega),\varphi_1 \equiv 1$ *on the support of* $\varphi;\varphi_1,\varphi \geqq 0$ *and either* $\operatorname{supp}\varphi \cap M = \varnothing$ *or* $\varphi = 1$ *on* M*; the constant* $\gamma < s$*, the constant* C_1 *depends on* $\gamma,s,\varphi,\varphi_1$.

If $M = \varnothing$*, then operator* (1) *is hypoelliptic in* Ω *and for any* $u \subset \mathscr{D}'(\Omega)$ *with* $Pu \subset \mathscr{H}_s^{\mathrm{loc}}(\Omega)$ *we have*

$$\|\varphi u\|^2_{s+\varepsilon(K)} \leqq C_2\{\|\varphi_1 Pu\|_s^2 + \|\varphi_1 u\|_\gamma^2\} \tag{7}$$

where K *is compact in* Ω, C_2 *is a positive constant, depending on* $K, s, \gamma, \varphi_1, \varphi$; $\gamma = \text{const} < s + \varepsilon(K)$, $s \subset \boldsymbol{R}^1$, $\varepsilon(K) = \text{const} > 0$, *the functions* φ_1 *and* φ *belong to* $C_0^\infty(K)$ *and* $\varphi_1 \equiv 1$ *on the support of* φ.

Here, as usual, we denote by $\mathscr{H}_s$ the set of distributions $u \subset \mathscr{D}'(\boldsymbol{R}^m)$ such that $\hat{u}(\xi)$ is a function and

$$\|u\|_s^2 \equiv (2\pi)^{-m}\int (1 + |\xi|^2)^s|\hat{u}(\xi)|^2\, d\xi < \infty.$$

Also we say $u \subset \mathscr{H}_s^{\mathrm{loc}}(\Omega)$ if $\varphi u \subset \mathscr{H}_s$ for any function $\varphi \subset C_0^\infty(\Omega)$. A particular case of the Theorem 2 was announced in [**10**].

Let us consider the general linear second order operator with nonnegative characteristic form; that is

$$Lu \equiv a^{lj}(x)u_{x_l x_j} + b^j(x)u_{x_j} + c(x)u \tag{8}$$

where $a^{lj},b^j,c \subset C^\infty(\Omega),a^{lj} = a^{jl}$, and, for any $\xi \subset \boldsymbol{R}^m$ and for any $x \subset \Omega$,

$$a^{lj}(x)\xi_l\xi_j \geqq 0. \tag{9}$$

It is proved in [**4**], that the condition (9) is necessary for the hypoellipticity of the operator L in Ω. The examples show that there exist operators (8) with $a^{lj},b^l,c \subset C^\infty(\Omega)$ which cannot be written in the form (1) with coefficients $a_j^l(x)$, belonging to $C^\infty(\Omega)$ (see [7, Chapter II, 6]).

We can write operator (8) in the form

$$Lu = -\mathscr{D}_j Q_j u + iQ_0 u + c(x)u \tag{10}$$

where $\mathscr{D}_j = -i\partial/\partial x_j, j = 1,\dots,m$, $Q_0 = (b^l - a^{lj}_{x_j})\mathscr{D}_l$, $Q_j = a^{lj}(x)\mathscr{D}_l$ for $j = 1,\dots,m$. Now consider the operators Q_0,Q_j $(j = 1,\dots,m)$. We form the operators $Q_{\mathscr{I}}$ for any multi-index $\mathscr{I} = (\alpha_1,\dots,\alpha_t)$, where $\alpha_l = 0, 1,\dots,m$ for $l = 1,\dots,t$ and t is any positive integer, by the formula

$$Q_{\mathscr{I}} = \operatorname{ad} Q_{\alpha_1} \cdots \operatorname{ad} Q_{\alpha_{t-1}} Q_{\alpha_t}.$$

Denote by $\mathscr{L}(Q_0,Q_1,\dots,Q_m)$ the set of $Q_{\mathscr{I}}$ for all $\mathscr{I}$ and define $\bar{R}(x^0)$ to be the number of the linearly independent operators in $\mathscr{L}(Q_0,Q_1,\dots,Q_m)$, considered at the point $x^0 \subset \Omega$.

In [7] the following theorem is proved.

THEOREM 3. *Suppose that* $\bar{R}(x_0) = m$ *at any points* $x^0 \subset (\Omega\backslash M)$ *and*

$$a^{lj}\Phi_{x_l}\Phi_{x_j} + |L\Phi| > 0 \tag{11}$$

at any point $x^0 \subset M$. *Then the operator* (8) *is global hypoelliptic in* Ω. *In addition, for any distribution* $u \subset \mathscr{D}'(\Omega)$ *such that* $Lu \subset \mathscr{H}_s^{\text{loc}}(\Omega)$ *the inequality*

$$\|\varphi u\|_s^2 \leqq C_3\{\|\varphi_1 Lu\|_s^2 + \|\varphi_1 u\|_\gamma^2\}$$

is valid, where $\varphi, \varphi_1 \subset C_0^\infty(\Omega), \varphi_1 \equiv 1$ *on the support of* φ, *and either* $\varphi \equiv 1$ *on* M *or* $\operatorname{supp} \varphi \cap M = \varnothing$, $\gamma = \text{const} < s$, *the constant* C_3 *depends on* $s, \varphi, \varphi_1, \gamma$.

If $M = \varnothing$ *the operator* (8) *is hypoelliptic in* Ω *and for any* $u \subset \mathscr{D}'(\Omega)$ *such that* $Lu \subset \mathscr{H}_s^{\text{loc}}(\Omega)$ *the inequality*

$$\|\varphi u\|_{s+\varepsilon(K)}^2 \leqq C_4\{\|\varphi_1 Lu\|_s^2 + \|\varphi_1 u\|_\gamma^2\}$$

holds where K *is compact in* $\Omega, \varphi, \varphi_1 \subset C_0^\infty(K)$, *the constant* C_4 *depends on* $K, s, \gamma, \varphi, \varphi_1$; *the constant* γ *is less than* $s + \varepsilon(K)$, *and* $\varphi_1 \equiv 1$ *on the support of* φ.

The proofs of the Theorems 2 and 3 are based on the theory of pseudodifferential operators.

For operators (8) with real analytic coefficients a^{lj}, b^j the following theorem gives the necessary and sufficient condition for the hypoellipticity in the domain Ω provided that

$$\sum_{l,j=1}^m |a^{lj}| + \sum_{j=1}^m |b^j| > 0 \quad \text{in } \Omega. \tag{12}$$

This theorem is proved in the paper [**11**].

THEOREM 4. *Suppose that coefficients* $a^{lj}(x), b^l(x)$ *are real analytic functions in* Ω *and the condition* (12) *is valid. Then for the hypoellipticity of operator* (8) *in* Ω *it is necessary and sufficient that* $\bar{R}(x) = m$ *at each point* $x \subset \Omega$.

To prove the Theorem 4 we note that from the Theorem 3 it follows that the condition $\bar{R}(x) = m$ in Ω is sufficient for hypoellipticity of such operators (8) in Ω. The proof of necessity of the condition $\bar{R}(x) = m$ in Ω in the Theorem 4 is based on the following lemma concerning analytic vector fields.

Let $Y_1(x), \ldots, Y_l(x)$ be analytic vector fields, defined in a domain $\Omega \subset \boldsymbol{R}^m$:

$$Y_j(x) = (a_j^1(x), \ldots, a_j^m(x)), \qquad j = 1, \ldots, l,$$

where $a_j^s(x)$ are real analytic functions in Ω for $s = 1, \ldots, m$. Let us denote by Y_j the differential operator

$$Y_j \equiv \sum_{j=1}^m a_j^s(x) \frac{\partial}{\partial x_s}, \qquad x \subset \Omega,$$

corresponding to the vector field $Y_j(x)$. For any multi-index $\mathscr{I} = (\alpha_1, \ldots, \alpha_t)$, where $\alpha_s = 1, \ldots, l$ for $s = 1, \ldots, t$ and t is any positive integer, we form the operator

$$Y_{\mathscr{I}} = \operatorname{ad} Y_{\alpha_1} \cdots \operatorname{ad} Y_{\alpha_{t-1}} Y_{\alpha_t}.$$

Let $\mathscr{L}(Y_1,\ldots,Y_l)$ be the set of all such operators $Y_{\mathscr{I}}$. We define $R'(x)$ to be the number of linearly independent operators in $\mathscr{L}(Y_1,\ldots,Y_l)$, considered at the point x.

LEMMA 1. *Suppose that in a domain $\Omega \subset \boldsymbol{R}^m$ we have analytic vector fields $Y_1(x),\ldots,Y_l(x)$ and suppose that $R'(x^0) = k, 0 < k < m$, at the point $x^0 \subset \Omega$. Then in a neighborhood of the point x^0 there exists an analytic manifold V of dimension k such that $x^0 \subset V$ and for any point $x \subset V$ the vector $Y_{\mathscr{I}}(x)$ is tangential to V at x for any multi-index $\mathscr{I}$.*

This lemma is proved in [**11**] (see also [**12**], [**13**]).

LEMMA 2. *Suppose that $\bar{R}(x^0) = k$ at the point $x^0 \subset \Omega$ for $\mathscr{L}(Q_0,Q_1,\ldots,Q_m)$, $1 \leqq k < m$, and coefficients $a^{lj}(x), b^l(x)$ of the operator* (8) *are real analytic functions. Then there exist a neighborhood G of x^0 and new independent variables such that*

$$y_s = \mathscr{F}_s(x_1,\ldots,x_m), \qquad s = 1,\ldots,m, \qquad \mathscr{D}y/\mathscr{D}x \neq 0, \tag{13}$$

where $\mathscr{F}_s$ are analytic functions in G; moreover the operator (8) *may be written in the form*

$$Lu \equiv \alpha^{lj}(y)u_{y_l y_j} + \beta^j(y)u_{y_j} + cu = L_1 u + L_2 u + cu \tag{14}$$

where $\alpha^{lj}(y), \beta^j(y)$ are real analytic functions, the operator L_1 is a differential operator with respect to $y_1,\ldots,y_k$ and the operator L_2 has coefficients equal to zero on $G \cap E^k$ and the hyperplane $E^k = \{y : y_{k+1} = 0,\ldots,y_m = 0\}$.

The proof of Lemma 2 is based on Lemma 1. Using Lemma 2 one can prove that if the coefficients of the operator (8) are real analytic functions in Ω and $\bar{R}(x_0) = k$ at a point $x^0 \subset \Omega$ with $1 \leqq k < m$, then the operator (8) is not hypoelliptic in Ω.

To prove that we choose the new independent variables (13) in a neighborhood G of x_0 and we write the operator (8) in the form (14). Then the adjoint operator L^* can be written in the form

$$L^*u = L_1^*u + L_2^*u + cu$$

where L_1^* is a differential operator with respect to $y_1,\ldots,y_k$ and

$$L_2^*u = M_1 u + M_2 u + c^*u$$

where

$$M_1 = \sum_{\rho=k+1}^{m} \sum_{l=1}^{k} \alpha^{l\rho}_{y_\rho}(y)\frac{\partial}{\partial y_l},$$

and the coefficients of the operator M_2 are equal to zero for any point y of G which belongs to E^k. This follows from the fact that $\alpha^{l\rho}(y) = 0$ on E^k, if either $l \geqq k+1$ or $\rho \geqq k+1$, and grad $\alpha^{l\rho} = 0$ if $l,\rho \geqq k+1$. Consider on the hyperplane E^k the operator

$$L^*_{E^k}u = L^*_1 u + M_1 u + c^* u + cu.$$

The operator L_{E^k} which is the adjoint operator to $L^*_{E^k}$ has the form

$$L_{E^k}u = L_1 u + M^*_1 u + c^* u + cu,$$

where M^*_1 is a first order operator with respect to $y_1, \ldots, y_k$. Using the Cauchy-Kovalevsky theorem or the theory of the first boundary value problem one can prove that there exists a nontrivial solution $v(y_1, \ldots, y_k)$ of the equation $L_{E^k}v = 0$ in a neighborhood $G_1 \cap E^k$ of the point $y^0 \subset E^k$. Let us set

$$(U_1, U_2)_0 \equiv \int_{R^k} U_1 U_2 \, dy_1 \cdots dy_k.$$

It is evident that for any function $\varphi(y_1, \ldots, y_k) \subset C_0^\infty(G_1 \cap E^k)$ we have that

$$(v, L^*_{E^k}\varphi)_0 = 0.$$

Let us consider in a neighborhood G_1 of the point y^0 the distribution $u(y)$ defined by the formula

$$(u, \varphi) = (v, \varphi(y_1, \ldots, y_k, 0, \ldots, 0))_0,$$

for any $\varphi \subset C_0^\infty(G_1)$. It is easy to see that $u = 0$ for $y \not\subset E^k, u \not\equiv 0$ in $G_1, u \not\subset C^\infty(G_1)$ and u satisfies the equation $Lu = 0$ in G_1, since

$$\begin{aligned}(u, L^*\varphi) &= (v, (L^*_1 + M_1 + (c^* + c))\varphi(y_1, \ldots, y_k, 0, \ldots, 0))_0 \\ &= (v, L^*_{E^k}\varphi(y_1, \ldots, y_k, 0, \ldots, 0))_0 = 0.\end{aligned}$$

Therefore in this case, operator (8) is not hypoelliptic in Ω.

In addition to the Theorem 4 we note that operator (3) with analytic coefficients and $\bar{R}(0) = 0$ is hypoelliptic in any $\Omega \subset \boldsymbol{R}^m$, but the operator

$$Lu \equiv |x|^2 \Delta u - (\gamma m + \gamma(\gamma - 2))u$$

is not hypoelliptic in a neighborhood of the origin if $\gamma > 0$ is not integer, since $u = |x|^\gamma$ is a solution of the equation $Lu = 0$. Therefore the cases of operator (8), with analytic coefficients which do not satisfy the condition (12) require further investigation.

References

1. L. Hörmander, *Hypoelliptic differential operators*, Ann. Inst. Fourier Grenoble **11** (1961), 477–492. MR **23** #A3368.

2. B. Malgrange, *Sur une classe d'opérateurs différentiels hypoelliptiques*, Bull. Soc. Math. France **85** (1957), 283–306. MR **21** #5063.

3. F. Trèves, *Opérateurs différentiels hypoelliptiques*, Ann. Inst. Fourier Grenoble **9** (1959), 1–73. MR **22** #4886.

4. L. Hörmander, *Hypoelliptic second order differential equations*, Acta Math. **119** (1967), 147–171. MR **36** #5526.

5. E. V. Radkevič, *On a theorem of L. Hörmander*, Uspehi Mat. Nauk **24** (1969), no. 2 (146), 233–234. (Russian) MR **39** #7286.

6. J. J. Kohn, *Pseudo-differential operators and non-elliptic problems*, Pseudo-Differential Operators (C.I.M.E., Stresa, 1968), Edizioni Cremonese, Rome, 1969, pp. 157–165. MR **41** #3972.

7. O. A. Oleĭnik and E. V. Radkevič, *Second order equations with non-negative characteristic form*, Itogi Nauki, Moscow, 1971.

8. L. Schwartz, *Théorie des distributions*. Tomes I, II, Actualités Sci. Indust., nos. 1091, 1122, Hermann, Paris, 1950, 1951. MR **12**, 31; MR **12**, 833.

9. M. Derridj, *Sur l'hypoellipticité d'une classe d'opérateurs différentiels du* 2e *ordre*, Seminar Goulaouic-Schwartz de l'École Polytechnique, Paris, 1971.

10. V. S. Fediĭ, *Estimates in* H_s *norms and hypoellipticity*, Dokl. Akad. Nauk SSSR **193** (1970), 301–303 = Soviet Math. Dokl. **11** (1970), 940–942.

11. O. A. Oleĭnik and E. V. Radkevič, *On the local smoothness of weak solutions and the hypoellipticity of differential second order equations*, Uspehi Mat. Nauk **26** (1971), no. 2, 265–281.

12. E. C. Zachmanoglou, *Propagation of zeros and uniqueness in the Cauchy problem for first order partial differential equations*, Arch. Rational Mech. Anal. **38** (1970), 178–188. MR **41** #5769.

13. T. Nagumo, *Linear differential system with singularities and application to transitive Lie algebras*, J. Math. Soc. Japan **18** (1966), 398–404.

Moscow University, Moscow, U.S.S.R.

ON THE EXTERIOR PROBLEM FOR THE REDUCED WAVE EQUATION

RALPH S. PHILLIPS[1]

1. **Introduction.** I would like to present a very simple but useful method for treating the reduced wave equation in an exterior domain which was developed by Peter Lax and myself. In essence the method consists of treating the exterior problem as a perturbation of the free space problem and effecting the perturbation by solving only an interior problem.

Technically this method has several advantages over the traditional potential theory approach introduced in the fifties by Kupradse [**1**], Müller [**4**] and Weyl [**8**] and later simplified by Werner [**7**]. Perhaps the main advantage is that the burden of proof is shifted to the solution of an interior problem for which a well-developed elliptic theory is available. In this respect it is similar to an approach for the exterior Dirichlet problem which was developed by R. Leis [**3**]. As in the Werner set-up one can easily avoid the familiar difficulties related to eigenvalues of the interior problem. Beyond that, however, there is a great deal of flexibility built into the method. For instance it permits one to "compactify" the set of obstacles contained in, say, the unit ball; this will be made more precise in §3. Also it has been applied to the exterior problem with a potential by Andy Majda, a student of mine, and this has resulted in a substantial extension of the work of Shenk and Thoe [**5**]; a statement of Majda's results is given in §4.

2. **The basic idea.** In order to illustrate our method we consider the following problem in an exterior domain G for complex ζ:

$$\begin{aligned} \Delta u + \zeta^2 u &= f \quad \text{in } G \subset R^3, \\ u &= 0 \quad \text{on } \partial G \text{ of class } C^2. \end{aligned} \tag{2.1}$$

AMS 1970 *subject classifications.* Primary 35J05; Secondary 35P25.

[1] This work was supported in part by the National Science Foundation under Grant GP-28122 and by the United States Air Force under Contract F44620-71-C-0037.

and u outgoing in the sense that for large $|x|$, u is a superposition of the fundamental solutions:

$$\gamma(x - y) \equiv \frac{e^{-i\zeta|x-y|}}{4\pi|x - y|}. \tag{2.2}$$

We choose ρ so that the boundary ∂G is contained in the ρ-ball: $\{|x| < \rho\}$ and set

$$G_\rho = G \cap \{|x| < \rho\}.$$

Finally we shall suppose that f lies in $L_2(G_\rho)$.

The solution to the free space outgoing problem

$$\Delta v + \zeta^2 v = g \quad \text{in } R^3 \tag{2.3}$$

can be given in terms of γ as

$$v(x) = R_\zeta^0 g = \int \gamma(x - y)g(y)\,dy \tag{2.4}$$

for g in $L_2(G_\rho)$.

We now alter v so as to satisfy the boundary condition; we do this by subtracting off a conveniently chosen function $\varphi h \equiv Q_\lambda v$, where

$$\begin{aligned} \Delta h + \lambda^2 h &= 0 \quad \text{in } G_\rho, \\ h &= v \quad \text{on } \partial G, \\ h &= 0 \quad \text{on } \{|x| = \rho\}; \end{aligned} \tag{2.5}$$

and $\varphi \in C_0^\infty$ is chosen to be identically one in a neighborhood of ∂G and identically zero in a neighborhood of $\{|x| = \rho\}$. We note that (2.5) is solvable providing λ^2 is not an eigenvalue for $-\Delta$ in G_ρ with Dirichlet boundary conditions and in particular if $\operatorname{Im} \lambda^2 \neq 0$.

Setting

$$u = v - \varphi h \tag{2.6}$$

we see that the original problem has now been reduced to finding a g for which u satisfies (2.1), that is

$$f = \Delta u + \zeta^2 u = \Delta v + \zeta^2 v - \Delta(\varphi h) - \zeta^2 \varphi h = g - T_\zeta g \tag{2.7}$$

where T_ζ (we suppress the dependence on λ) is defined as

$$T_\zeta g = 2\nabla\varphi\nabla h + (\Delta\varphi)h + (\zeta^2 - \lambda^2)\varphi h. \tag{2.8}$$

We shall show presently that[2]

$$\|h\|_2 \leqq C_\zeta \|g\|_0; \tag{2.9}$$

[2] We denote by $\|h\|_k$ the Sobolev norm $\|h\|_k^2 = \sum_{|\alpha| \leqq k} \int_{G_\rho} |D^\alpha h|^2\,dx$.

and it follows from this, by the Rellich compactness theorem, that T_ζ is a compact linear operator on $L_2(G_\rho)$. In order to invert $(I - T_\zeta)$ and thus find g, it therefore suffices to prove that $(I - T_\zeta)$ is one-to-one. We now fill in these two missing steps.

Step 1. *Proof of the relation* (2.9). Using interior elliptic estimates or the fact that v can be expressed as a mildly singular integral as in (2.4), we see that

$$\|\varphi v\|_2 \leqq C_\zeta \|g\|_0. \tag{2.10}$$

Now $w = h - \varphi v$ satisfies the equation

$$\begin{aligned} \Delta w + \lambda^2 w &= -\Delta(\varphi v) - \lambda^2 \varphi v \quad &&\text{in } G_\rho, \\ w &= 0 &&\text{on } \partial G_\rho. \end{aligned} \tag{2.11}$$

It therefore follows by elliptic theory that $\|w\|_2 \leqq C\|\varphi v\|_2$ and combining this with (2.10) we obtain (2.9).

We note for future reference that

$$\|w\|_0 \leqq C'_\lambda \|\Delta(\varphi v) + \lambda^2 \varphi v\|_0 \tag{2.12}$$

where C'_λ is equal to the reciprocal of the distance from λ^2 to the spectrum of $-\Delta$ on G_ρ. Combining this with (2.10) we obtain

$$\|h\|_0 \leqq C(\lambda, \zeta)\|g\|_0. \tag{2.13}$$

We also note that h can be represented by means of the Green's function K for the boundary value problem (2.11) as

$$h = \varphi v - \int_{G_\rho} K(x, y; \lambda, G_\rho)[\Delta(\varphi v) + \lambda^2 \varphi v]\, dy. \tag{2.14}$$

Step 2. $(I - T_\zeta)$ *is one-to-one for* $\operatorname{Im}\zeta \leqq 0$. Suppose on the contrary that there exists a nontrivial g in $L_2(G_\rho)$ such that $g = T_\zeta g$. Combining this with (2.3) and (2.8) we see that

$$\Delta v + \zeta^2 v = g = \Delta(\varphi h) + \zeta^2 \varphi h,$$

and setting $w = v - \varphi h$ we conclude that w is an outgoing solution of

$$\begin{aligned} \Delta w + \zeta^2 w &= 0 \quad \text{in } G, \\ w &= 0 \quad \text{on } \partial G. \end{aligned}$$

It is easy to show for $\operatorname{Im}\zeta < 0$ that this implies $w \equiv 0$ in G; when $\operatorname{Im}\zeta = 0$ this follows from the Rellich uniqueness theorem. In particular $h \equiv v$ in a neighborhood of ∂G and therefore h is a smooth continuation of v restricted to the complement $\tilde{G}$ of G. Thus

$$\begin{aligned} z &= v \quad \text{in } \tilde{G}, \\ &= h \quad \text{in } G_\rho \end{aligned}$$

is a solution of

$$\begin{aligned} \Delta z + pz &= 0 \quad \text{for } \|x\| < \rho, \\ z &= 0 \quad \text{on } \{|x| = \rho\}; \end{aligned} \tag{2.15}$$

here $p = \zeta^2$ in $\tilde{G}$ and λ^2 in G_ρ. Multiplying (2.15) through by $\bar{z}$ and integrating by parts we obtain

$$\int_{|x|<\rho} [|\nabla z|^2 - p|z|^2]\, dx = 0.$$

If we choose λ so that $\operatorname{Im} \lambda^2 > 0$ then it follows from this relation that

$$\int_{G_\rho} |z|^2\, dx = 0$$

so that h and hence g vanishes, contrary to our assumption on g. The same conclusion follows if it can be shown that (2.15) has only the trivial solution. This concludes the existence proof for (2.1) if $\operatorname{Im} \zeta \leqq 0$.

Finally we prove that (2.1) is solvable for all but a discrete set of ζ's in the upper half-plane. Combining (2.5), (2.6) and (2.7) we set

$$u = R_\zeta f \equiv (I - Q_\lambda) R_\zeta^0 (I - T_\zeta)^{-1} f. \tag{2.16}$$

It is easy to verify that each of the operators R_ζ^0 and T_ζ are entire functions of ζ. Since $(I - T_\zeta)^{-1}$ exists for $\operatorname{Im} \zeta \leqq 0$, it follows (see for instance [**6**]) that R_ζ is meromorphic on C.

3. **On the scattering frequencies of the Laplace operator for exterior domains.** The values of ζ, for which

$$\begin{aligned} \Delta u + \zeta^2 u &= 0 \quad \text{in } G, \\ u &= 0 \quad \text{on } \partial G \end{aligned}$$

has a nontrivial outgoing solution, are called scattering frequencies for the obstacle $\tilde{G}$. These scattering frequencies coincide with the poles of R_ζ and therefore lie in the upper half-plane. It has been shown in [**2**] that there is a complex neighborhood of the interval $(-\pi/\rho', \pi/\rho')$ which is free of scattering frequencies for all obstacles contained in the ρ'-ball. We shall now sketch an alternative proof of this fact based on the method introduced in §2.

THEOREM 3.1. *There is a complex neighborhood of the interval $(-\pi/\rho', \pi/\rho')$ which is free of poles of R_ζ for all scattering obstacles contained in the ball $\{|x| < \rho'\}$.*

If this assertion were false then there would exist a sequence of obstacles $\{O_n\}$ contained in the ρ'-ball for which the resolvent R_ζ^n for G_n has a pole at ζ_n and the sequence $\{\zeta_n\}$ converges to some point χ in $(-\pi/\rho', \pi/\rho')$. By slightly displacing the points ζ_n we may assume that ζ_n is not a pole of R^n_ζ, but that $\lim \|R^n_{\zeta_n}\| = \infty$ and at the same time $\zeta_n \to \chi$. It therefore suffices to prove

$$\liminf_{n\to\infty} \|R^n_{\zeta_n}\| < \infty. \tag{3.1}$$

To this end we choose ρ so that $\rho' < \rho < \pi/\chi$ and α, β so that

$$\rho' < \alpha\rho < \beta\rho < \rho.$$

We also choose $\varphi \equiv 1$ for $|x| < \alpha\rho$ and $\varphi \equiv 0$ for $|x| > \beta\rho$. Finally we set $\lambda_n^2 = \zeta_n^2$; this does not cause any difficulties in the development of §2 since the lowest eigenvalues for the interior problems (2.11) and (2.15) are both greater than χ^2 and it has the important advantage of making $T^n_{\zeta_n}$ depend only on values of h in the shell $\{\alpha\rho < |x| < \beta\rho\}$.

LEMMA 3.2. *The transformations* $\{Q^n_{\zeta_n} R^n_{\zeta_n}\}$ *are uniformly bounded in norm.*

PROOF. This follows from the estimate (2.10) which holds uniformly for ζ in any compact subset and from (2.13) which holds uniformly for all of the obstacles $\{O_n\}$ with $\lambda_n = \zeta_n$ since the ζ_n^2's are uniformly bounded away from the eigenvalues of the $(G_n)_\rho$ interior problems.

This reduces the proof of Theorem 3.1 to that of proving

LEMMA 3.3. $\liminf \|(I - T^n_{\zeta_n})^{-1}\| < \infty$.

PROOF. The fact that $T^n_{\zeta_n} g$ vanishes for $|x| < \alpha\rho$ and $|x| > \beta\rho$, suggests that we decompose g in $L_2((G_n)_\rho)$ into two orthogonal parts:

$$g = g' + g'',$$

where

$$g'(x) = g(x), \quad \text{for } \alpha\rho < |x| < \beta\rho,$$
$$= 0, \qquad \text{elsewhere},$$

and

$$g''(x) = g(x), \quad \text{for } |x| < \alpha\rho \text{ and } |x| > \beta\rho,$$
$$= 0, \qquad \text{elsewhere}.$$

Next we consider the two possibilities:

(1) $\|g''\|_0 > \delta\|g'\|_0$, where δ is a positive number to be chosen later. Now $[(I - T^n_{\zeta_n})g](x) = g(x)$ for $|x| < \alpha\rho$ and $|x| > \beta\rho$ so that in this case

$$\|(I - T^n_{\zeta_n})g\|_0 \geqq \|g''\|_0 \geqq (1 + \delta^2)^{1/2}\|g\|_0/\delta.$$

(2) $\|g''\|_0 \leqq \delta\|g'\|_0$. For this case we shall establish the existence of two positive constants C and ε such that, for all n,

(3.2) $$\|T^n_{\zeta_n} g\|_0 \leqq C\|g\|_0$$

and

(3.3) $$\|(I - T^n_{\zeta_n})g'\|_0 \geqq \varepsilon\|g'\|_0.$$

Assuming for the moment that (3.2) and (3.3) are valid, then it follows that

$$\|(I - T^n_{\zeta_n})g\|_0 \geqq \varepsilon\|g'\|_0 - (1 + C)\|g''\|_0 \geqq (\varepsilon - (1 + C)\delta)\|g'\|_0 \geqq \tfrac{1}{2}\varepsilon(1 + \delta^2)^{1/2}\|g\|_0$$

for $(1 + C)\delta < \frac{1}{2}\varepsilon$. Thus in either case $\|(I - T^n_{\zeta_n})^{-1}\|$ will be bounded.

Thus all that remains to be proved are the inequalities (3.2) and (3.3). With respect to (3.2) it is clear that $T^n_{\zeta_n}$ depends only on the values of h and ∇h in the shell $\{\alpha\rho < |x| < \beta\rho\}$ and in view of the uniform boundedness by (2.13) of $\|h\|_0$ for all g with $\|g\|_0 \leqq 1$ and all domains under consideration, it follows by elliptic theory that h and ∇h will be uniformly bounded in this interior shell for all such g. This establishes the inequality (3.2).

The estimate (3.3) is more difficult to prove. Let T'_n denote the restriction of $T^n_{\zeta_n}$ to the space of square integrable functions on the shell $\{\alpha\rho < |x| < \beta\rho\}$, which we denote by L'_2. Suppose that (3.3) were false, that is suppose that

(3.4) $$\lim \|(I - T'_n)^{-1}\| = \infty.$$

Arguing as above, we see that the nonsingular parts $k(x, y; \zeta_n, (G_n)_\rho)$ of the corresponding Green's functions $K(x, y; \zeta_n, (G_n)_\rho)$ will be uniformly bounded and hence form a normal family on the shell $\{\alpha'\rho < |x| < \rho\}$ for $\alpha\rho < |y| < \beta\rho$; here $\rho'/\rho < \alpha' < \alpha < \beta < \beta' < 1$. In fact

$$w = k(x, y) - \psi(x)\gamma(x - y)$$

satisfies (2.11) with φv replaced by $\psi(x)\gamma(x - y)$ where $\psi \in C^\infty$ is defined as

$$\begin{aligned} \psi(x) &= 0, \quad \text{for } \alpha'\rho < |x| < \beta'\rho, \\ &= 1, \quad \text{for } |x| \leqq \rho' \text{ and } |x| \geqq \rho; \end{aligned}$$

and $\psi(x)\gamma(x - y)$ varies continuously in the parameter y when y is restricted to the shell $\{\alpha\rho < |y| < \beta\rho\}$. It therefore follows from (2.12) that the mapping $y \to k(x, y)$ on this shell to $L_2((G_n)_\rho)$ varies continuously in y, uniformly for all n. The normal family assertion now follows by elliptic theory. Consequently there exists a subsequence of the $\{k_n\}$ which together with its first and second derivatives in x converges uniformly for y in the shell $\{\alpha\rho < |y| < \beta\rho\}$ and x in any compact subset of $\{\rho' < |x| < \rho\}$. Renumbering, it follows from this and the relation (2.14) that the sequence $\{T'_n\}$ converges in the norm topology to a compact operator T' on L'_2. We note that the k_n also converge uniformly in x for $\alpha'\rho \leqq |x| \leqq \rho$.

If (3.4) holds, then $(I - T')$ cannot be invertible on L'_2 and since T' is compact there will exist a nontrivial null vector g' of $(I - T')$. Set $v = R^0_\chi g'$ and $Q^n_{\zeta_n} R^0_{\zeta_n} g' = \varphi h_n$. Then

$$T_n'g' = \Delta(\varphi h_n) + \zeta_n^2\varphi h_n \to T'g' = g' = \Delta v + \chi^2 v.$$

Setting $w_n = v - \varphi h_n$, we see that w_n is outgoing and satisfies

$$\Delta w_n + \chi^2 w_n \equiv f_n \to 0 \quad \text{in } L_2 \text{ by (2.13)},$$
$$\operatorname{supp} f_n \subset \{|x| < \rho\},$$
$$w_n = 0 \quad \text{on } \partial G_n.$$

Now w_n and v have the same asymptotic behavior for large $|x|$, that is

$$w_n(r\theta) = v(r\theta) \sim e^{i\chi r} m(\theta)/r \quad \text{as } r \to \infty.$$

Applying Green's theorem on $(G_n)_R$ we obtain

$$\int_{(G_n)_R} (f_n\bar{w}_n - w_n\bar{f}_n)\, dx = \int_{(G_n)_R} [(\Delta w_n)\bar{w}_n - w_n(\overline{\Delta w_n})]\, dx$$
$$= \int_{|x|=R} \left(\frac{\partial w_n}{\partial n}\bar{w}_n - w_n\frac{\overline{\partial w_n}}{\partial n}\right) dS$$
$$\to 2i\chi \int_{|\theta|=1} |m(\theta)|^2\, d\theta \quad \text{as } R \to \infty.$$

Finally since $f_n \to 0$ in L_2 and the w_n are uniformly bounded in the L_2 norm, again by (2.13), we conclude that $m(\theta) \equiv 0$ and hence by the Rellich uniqueness theorem that $v \equiv 0$ for all $|x| > \rho$.

We are now essentially through. The representation (2.14) shows that the φh_n together with their first and second derivatives converge uniformly for $|x| > \alpha'\rho$. Setting $h = \lim h_n$ and $w = \lim w_n$, it follows that

$$\Delta w + \chi^2 w = 0, \quad \text{for } |x| > \alpha'\rho,$$

so that by unique continuation $w \equiv 0$ for $|x| > \alpha'\rho$. This shows that $h \equiv v$ in the shell $\{\alpha'\rho < |x| < \alpha\rho\}$ where $\varphi \equiv 1$. Again we define

$$z = v, \quad \text{for } |x| < \alpha\rho,$$
$$= h, \quad \text{for } |x| > \alpha'\rho.$$

Then

$$\Delta z + \chi^2 z = 0, \quad \text{for } |x| < \rho,$$
$$z = 0, \quad \text{on } |x| = \rho,$$

and since χ^2 is smaller than the fundamental frequency for this interior problem we may conclude that $z \equiv 0$. As a consequence $h \equiv 0$ for $|x| > \alpha'\rho$ and so is $g' = \Delta(\varphi h) + \chi^2\varphi h$. Thus assuming the lemma to be false has led to a contradiction; this completes the proof of Theorem 3.1.

4. **The exterior problem with a potential.** We conclude with a statement of some results obtained by Andy Majda using the above approach. He considers the operator A on $L_2(G)$ defined as

$$\begin{aligned} Au &= \Delta u - qu && \text{in } G \subset R^n, \\ \partial u/\partial n + \kappa u &= 0 \quad (\text{or } u = 0) && \text{on } \partial G, \end{aligned} \tag{4.1}$$

for $\kappa \geqq 0$, where ∂G is of class C^2. He assumes that

$$\begin{aligned} &q \in L^{2+\delta}(G), && \text{for } n \leqq 3 \text{ and } \delta > 0, \\ &q \in L^p(G), && \text{for } n > 3 \text{ where } p > \tfrac{1}{2}n, \\ &q \text{ decays like } |x|^{-((n+1+\eta)/2)}, && \text{for large } x, \text{ with } \eta > 0. \end{aligned}$$

Without any further assumptions on q, Majda shows that the absolutely continuous part of the spectrum is $[0, \infty)$, that the point spectrum is contained in $(-\infty, 0)$ and that there is no singular continuous spectrum. He constructs the scattered wave solutions to the reduced wave equation and with them obtains the spectral representation for A as well as that of the scattering operator. In the process he shows that the wave operators exist and are complete.

These results extend those of Shenk and Thoe [**5**] who made use of the more traditional potential theory approach but were unable to handle unbounded q.

REFERENCES

1. W. D. Kupradse, *Randwertaufgaben der Schwingungstheorie und Integralgleichungen*, Akadamie Verlag, Berlin, 1956.

2. P. D. Lax and R. S. Phillips, *On the scattering frequencies of the Laplace operator for exterior domains*, Comm. Pure Appl. Math. **25** (1972), 85–101.

3. R. Leis, *Zur Dirichletschen Randwertaufgabe des Aussenraumes der Schwingungsgleichung*, Math. Z. **90** (1965), 205–211. MR **32** #752.

4. C. Müller, *Randwertprobleme der Theorie elektromagnetischer Schwingungen*, Math. Z. **56** (1952), 261–270. MR **14**, 1043.

5. N. Shenk and D. Thoe, *Eigenfunction expansions and scattering theory for perturbations of* $-\Delta$, J. Math. Anal. Appl. (to appear).

6. S. Steinberg, *Meromorphic families of compact operators*, Arch. Rational Mech. Anal. **31** (1968/69), 372–379. MR **38** #1562.

7. P. Werner, *Randwertprobleme der mathematischen Akustik*, Arch. Rational Mech. Anal. **10** (1962), 29–66. MR **26** #5276.

8. H. Weyl, *Die natürlichen Randwertaufgaben im Aussenraum für Strahlungsfelder beliebiger Dimension und beliebigen Ranges*, Math. Z. **56** (1952), 105–119. MR **14**, 933.

STANFORD UNIVERSITY

GENERAL THEORY OF HYPERBOLIC MIXED PROBLEMS

JEFFREY RAUCH

1. In the last three years a general theory of mixed initial boundary value problems for linear hyperbolic systems with variable coefficients has been developed. This is in sharp contrast to earlier work which was restricted to special classes of problems, for example, second order equations with Dirichlet or Neumann boundary conditions, symmetric positive problems and problems in one space variable. The general theory is very satisfactory in many respects but has several gaps. The main one is that the class of problems treated is not broad enough to include all the previously understood cases. We will discuss the main results to date (August 1971) and indicate some directions for future work. One goal will be to point out the main source of difficulty of the hyperbolic boundary value problems by comparing them with elliptic boundary value problems where this obstacle is not present.

2. For simplicity (of exposition, not mathematics) we restrict attention to first order systems. The theory of a single higher order equation has been worked out along the same lines [**6**] and is somewhat easier. Only perseverance should be required to treat the appropriate generalizations to higher order systems.

Our operators are of the form

$$Lu \equiv \partial_t u - \sum_{j=1}^{m} A_j(t, x)\partial_j u - B(t, x)u$$

where u is a k-vector and A_j and B are smoothly varying $k \times k$ matrices defined on $(-\infty, \infty) \times \overline{\Omega}$. We suppose that the basic domain Ω is open with compact C^∞ boundary, $\partial\Omega$. In addition, we require that A_j and B are constant for $|t| + |x|$ sufficiently large. This is not an essential restriction because hyperbolic problems have finite speed of propagation so the behavior of the solution at a fixed point (t, x) is not affected by the coefficients for $|t| + |x|$ very large.

AMS 1970 *subject classifications.* Primary 35L50; Secondary 35B30, 35D05, 35D10, 35F10, 35F15.

The mixed problem we will study is

$$
\text{(I)}\qquad
\begin{aligned}
&Lu = F && \text{in } [0, T] \times \bar{\Omega} && \text{(differential equation)},\\
&u|_{t=0} = f && \text{in } \bar{\Omega} && \text{(initial condition)},\\
&u \in N(t, x) && \text{for } x \in \partial\Omega && \text{(homogenous boundary condition)}.
\end{aligned}
$$

Here $N(t, x)$ is a smoothly varying subspace of $\boldsymbol{C}^k$ defined on $(-\infty, \infty) \times \partial\Omega$ and constant outside a compact set. In practice the homogenous boundary conditions are often given in the form

$$l_j(t, x)u = 0 \quad \text{for } x \in \partial\Omega, \qquad j = 0, 1, 2, \ldots, \mu,$$

where the l_j are smoothly varying linearly independent linear functionals on $\boldsymbol{C}^k$. In this case we have $\dim N = k - \mu$. Inhomogenous boundary conditions can also be treated.

Throughout we assume that the following conditions are satisfied:

Strict hyperbolicity. $\sum A_j\xi_j$ has k distinct real eigenvalues for any $\xi = (\xi_1, \ldots, \xi_m) \in \boldsymbol{R}^m\backslash 0$.

Noncharacteristic boundary. $\sum A_j(t, x)n_j(x)$ is nonsingular for $x \in \partial\Omega$, t where $(n_1(x), \ldots, n_m(x))$ is a normal to $\partial\Omega$ at x.

Main Problem. Given the operator L and region Ω satisfying the above conditions, for which boundary spaces, N, is the mixed problem (I) well-posed?

3. To describe a sufficient condition for well-posedness let me first recall the analogous situation for elliptic boundary value problems. Consider a first order elliptic system of differential operators, P, in Ω with homogenous boundary conditions imposed as above. The good boundary conditions in that case are found by the following recipe: For each $x \in \partial\Omega$ freeze P and N at x and throw away the lower order term in the operator. Replace Ω by the half space bounded by the tangent plane to Ω at x and including the inward pointing normal to Ω at x. It is then required that each of the resulting constant coefficient problems in a half space is elliptic.

The ellipticity of the problems is checked as follows. Introduce coordinates $(x_1, \ldots, x_{m-1}, x_m) \equiv (x', x_m)$ so that the half space under consideration is $\{x | x_m \geqq 0\}$. Let the dual variables be (ξ', η). Denote the frozen operator and boundary space by P_0 and N_0 respectively and the half space by Ω_0. We call a function of the form

$$\phi(x) = \sum_j p_j(x_m)\exp(i(x' \cdot \xi' + x_m\eta_j))V_j,$$

with $\xi' \in \boldsymbol{R}^{m-1}\backslash 0$, $\operatorname{Im} \eta_j > 0$, p_j polynomials, $V_j \in \boldsymbol{C}^k$, a bounded exponential function. Such a function is bounded on Ω_0. Suppose there were a bounded exponential function, ϕ, which satisfied $P_0\phi = 0$, $\phi \in N_0$ at $\partial\Omega_0$. Then since P_0 has no lower order terms (i.e., it is homogeneous) the same would be true of $\phi_\lambda \equiv \phi(\lambda x)$. As

$\lambda \to \infty$ the derivatives of ϕ_λ grow like a polynomial in λ while ϕ does not grow at all. Thus there could be no estimate of the form $\|D\phi_\lambda\| \leqq c(\|P_0\phi_\lambda\| + \|\phi\|)$. The norms here are purposely vague. To make the argument rigorous one would settle on norms ahead of time and get a counterexample by suitably cutting off ϕ_λ. Such an estimate is characteristic of elliptic boundary value problems. Thus, for ellipticity we must require that there are no such bounded exponential solutions. Notice by homogeneity it suffices to consider only exponential solutions with $|\xi'| = 1$.

The above condition can be stated geometrically as follows. For $\xi' \in \boldsymbol{R}^{m-1}\backslash 0$ let $\mathscr{E}_-(\xi') \subset \boldsymbol{C}^k$ be the space of boundary values of a bounded exponential function ϕ which satisfies $P_0\phi = 0$. It can be shown (by Cauchy integral representation for a projection on $\mathscr{E}_-$) that $\mathscr{E}_-$ is a smoothly varying linear subspace. The above condition says that $\mathscr{E}_- \cap N_0 = \{0\}$; that is, a bounded exponential which satisfies $P_0\phi = 0$ must not satisfy the boundary condition $\phi \in N_0$ at $\partial\Omega_0$. Notice decreasing the size of N corresponds to imposing more boundary conditions. For an adequate existence theory we must not impose too many boundary conditions; thus we require that N_0 be maximal with respect to the above restriction. In this case we call the frozen boundary problem elliptic. To summarize

Elliptic frozen problem means. For all $\xi' \in \boldsymbol{R}^{m-1}$, $|\xi'| = 1$ *we have* $\mathscr{E}_-(\xi') \oplus N_0 = \boldsymbol{C}^k$ *where* $\oplus$ *denotes direct sum.*

Elliptic theory then asserts that if each frozen problem is elliptic in this sense then the original problem has the usual properties, for example, finite index and Fredholm alternative provided $\overline{\Omega}$ is compact, regularity theorems, etc.

Recently analogous results have been obtained for mixed problems. There is one important difference which we now describe. As in elliptic theory, the idea is to freeze L and N at the values at a fixed point (t, x) with $x \in \partial\Omega$ (that is, on the "wall" of the cylinder $(-\infty, \infty) \times \overline{\Omega}$). The domain Ω is replaced by a half plane and the lower order terms of L are thrown out. The resulting constant coefficient mixed problem is required to be hyperbolic.

This hyperbolicity can be checked by examining exponential solutions. By a *growing exponential function* is meant a function of the form

$$\psi(t, x) = \sum_j p_j(x_m)\exp(i(x' \cdot \xi' + x_m\eta_j + t\tau))V_j$$

with $\xi' \in \boldsymbol{R}^{m-1}$, $\operatorname{Im}\eta_j > 0$, $\operatorname{Im}\tau < 0$, $V_j \in \boldsymbol{C}^k$, and p_j are polynomials. For fixed t such a function is bounded in Ω_0 but the function grows exponentially in time. If $\psi(t, x)$ is a growing exponential function which satisfies $L_0\psi = 0$, $\psi \in N_0$ at $\partial\Omega_0$ then so does $\psi_\lambda \equiv \psi(\lambda t, \lambda x)$. As $\lambda \to \infty$ the values of ψ_λ and its derivatives at $t = 0$ grow like a polynomial in λ but the values of ψ at $t = T$ grow exponentially with λ. Thus there can be no estimate of the form

$$\|\psi_\lambda(T, \cdot)\| \leqq c(\|L_0\psi_\lambda\|_{[0,T]\times\Omega} + \|\psi_\lambda(0, \cdot)\|)$$

(with the standing ambiguity in the norms). Such an inequality is characteristic of well-posed evolution equations. So for the hyperbolicity of L_0, N_0 it is necessary to require that there be no such growing exponential solution (this argument is due to Agmon). By homogeneity it suffices to check this for $|\tau|^2 + |\xi'|^2 = 1$, $\operatorname{Im} \tau < 0$,

For $\operatorname{Im} \tau < 0$, $\xi' \in \boldsymbol{R}^{m-1}$ let $E_-(\tau, \xi')$ be the space of boundary values of growing exponential solutions of $L_0\psi = 0$. E_- is a smoothly varying linear subspace of $\boldsymbol{C}^k$. and the above condition becomes $E_-(\tau, \xi') \cap N_0 = \{0\}$. As in elliptic theory we must require that N_0 be maximal with respect to this property, that is, $E_- \oplus N_0 = \boldsymbol{C}^k$. To summarize

Hyperbolicity of frozen problem means. For all τ, ξ' *with* $\operatorname{Im} \tau < 0$, $\xi' \in \boldsymbol{R}^{m-1}$, $|\tau|^2 + |\xi'|^2 = 1$, *we have* $E_-(\tau, \xi') \oplus N_0 = \boldsymbol{C}^k$.

Hersh [**1**] has shown that this condition is equivalent to the well-posedness of the frozen mixed problem for smooth data.

Notice the following property of elliptic boundary value problems. If P_0, N_0 is elliptic then so is P_0, $\tilde{N}$ for any $\tilde{N}$ sufficiently close to N_0. This follows from the compactness of $\{\xi \mid |\xi'| = 1\}$, for ellipticity implies that the angle between N_0 and $\mathscr{E}_-$ is greater than some positive constant. The same is not true of hyperbolicity. There are hyperbolic problems L_0, N_0 for which there are boundary spaces $\tilde{N}$ arbitrarily close to N_0 which are not hyperbolic. This can happen because $\{(\tau, \xi') \mid |\tau|^2 + |\xi'|^2 = 1, \operatorname{Im} \tau < 0\}$ is not compact so the angle between N_0 and $E_-(\tau, \xi')$ can approach zero as $\operatorname{Im} \tau \to 0$. Such a boundary value problem will be called marginally hyperbolic. If L_0, N_0 is hyperbolic and not marginally hyperbolic, it will be called *stable hyperbolic.* We can now state the main theorem (mostly due to Kreiss).

THEOREM (KREISS [**2**], RALSTON [**3**], RAUCH [**5**]). *If for each* $x \in \partial\Omega$, t *the frozen mixed problem is stable hyperbolic, then for each* $F \in \mathscr{L}_2([0, T] \times \Omega)$, $f \in \mathscr{L}_2(\Omega)$ *there is a unique strong solution of* (I). *In addition, the solution has the following properties*:

(i) $\|u(t)\|_\Omega \leqq c(\|F\|_{[0,t] \times \Omega} + \|f\|_\Omega)$, $0 \leqq t \leqq T$, *where the norms are* $\mathscr{L}_2$ *norms.*

(ii) *There is finite speed of propagation. That is, there is a number* $s > 0$ *such that for any* (t, x) *the values of* u *in a neighborhood of* (t, x) *are not influenced by the values of* F *and* f *at points* $(\tilde{t}, \tilde{x})$ *with* $|x - \tilde{x}| > s(t - \tilde{t})$.

Notice that property (i) shows that $u(0) \in \mathscr{L}_2(\Omega)$ implies that $u(t) \in \mathscr{L}_2(\Omega)$ for all $t > 0$. That is, square integrability is a continuable initial condition.

One should not think that requiring stable hyperbolicity of the frozen problems eliminates the difficulties of hyperbolic boundary value problems. The noncompactness of the parameter space still haunts us. This comes about as follows. For the frozen problem consider solutions of $L_0\psi = 0$ with ψ of the form

$$p(x_m)\exp(i(x' \cdot \xi' + x_m\eta + t\tau))V,$$

with $\operatorname{Im} \tau < 0$, $\xi' \in \boldsymbol{R}^{m-1}\backslash 0$, $V \in \boldsymbol{C}^k$, and p a polynomial. They fall into two classes; those with $\operatorname{Im} \eta < 0$ and those with $\operatorname{Im} \eta > 0$. Let $E_{\pm}(\tau, \xi')$ be the span of the boundary values of those with $\mp \operatorname{Im} \eta > 0$ (this is consistent with previous notation). Then it is simple to show that $E_{+}(\tau, \xi') \oplus E_{-}(\tau, \xi') = \boldsymbol{C}^k$. Similarly for elliptic problems there are subspaces $\mathscr{E}_{\pm}(\xi')$ with $\mathscr{E}_{+} \oplus \mathscr{E}_{-} = \boldsymbol{C}^k$. In the elliptic case the angle between $\mathscr{E}_{+}$ and $\mathscr{E}_{-}$ is uniformly bounded away from zero on $\{\xi \mid |\xi'| = 1\}$ (compactness). This allows a simple derivation of *a priori* estimates. However for the hyperbolic case the angle between E_{+} and E_{-} may approach zero as $\operatorname{Im} \tau \to 0$ in the set $\{(\tau, \xi') | \operatorname{Im} \tau < 0, |\tau|^2 + |\xi'|^2 = 1\}$. This is the main source of difficulty in the theory.

It is worth mentioning that for parabolic boundary value problems the parameter space is not compact (the occurrence of a time variable is the common feature with hyperbolic problems) and the general theory is based on stable parabolicity of the frozen problems. However the last difficulty with the plus and minus spaces does not occur. In this sense the hyperbolic boundary value problems are more difficult than the elliptic and parabolic ones. This explains why they are relatively less developed.

In the hope of improving this situation I would like to describe some open problems and some problems just nearing solution (as of August 1971) to give a feeling for the present state of the general theory.

Problem 1. *Differentiability of solutions.* The problem here is that more is required than smoothness of the data F and f. Compatibility conditions at the "edge" $\{0\} \times \partial\Omega$ of the cylindrical domain must be imposed. This problem has just been solved by Frank Massey III and myself.

Problem 2. *Characteristic boundary.* If the condition of noncharacteristic boundary is replaced by the condition rank $(\sum A_j n_j(x)) =$ constant for x in a neighborhood of $\partial\Omega$, then the appropriate generalization of the main theorem is known and I have been able to give half a proof. The other half should follow soon.

Problem 3. *Multiple characteristics.* The condition of strict hyperbolicity of L should be weakened to some sort of symmetrizable hyperbolicity. Some partial results indicate that this is possible but nothing is known in general.

Problem 4. *Marginally hyperbolic problems.* The symmetric positive problems of Lax-Phillips-Friedrichs give examples of variable coefficient mixed problems which are well-posed in the L_2 sense, but their frozen problems need only be marginally hyperbolic. To characterize what type of marginal behavior is permissible is an important problem.

Problem 5. *The correct finite speed of propagation.* Show that the speed of propagation for these general mixed problems is not greater than the speed of propagation for the pure Cauchy problem. This has just been proven by Shirota [**7**] in the case of constant coefficient problems.

Problem 6. *Asymptotics as $t \to \infty$.* In exterior problems, sufficient conditions for energy decay on bounded sets would be nice.

Problem 7. *Propagation of discontinuities.* This problem is usually related to Problem 6. The difficulty here is that there may be "rays" which are tangential to $\partial\Omega$. These so-called diffracted rays have eluded successful mathematical treatment for many years. In particular the truth of the Generalized Huyghens' Principle (discontinuities are propagated along rays if suitably reflected when they hit $\partial\Omega$ transversally and continued smoothly if they are tangent to $\partial\Omega$) is only known in special cases. Perhaps the insight afforded by the work of Hörmander on wave front sets will be useful here.

Problem 8. At present the proof of the main theorem is too difficult ([**2**], [**5**]). Find a simpler one.

Problem 9. *Well-posedness but not in the* $\mathscr{L}_2$ *sense.* The problem here is to find sufficient conditions for well-posedness in some sense weaker than the $\mathscr{L}_2$ sense. For variable coefficients, some progress has been made on the pure initial value problem (Mizohata-Ohya, Flashka-Strang). Almost nothing is known for mixed problems.

ADDED IN PROOF (FEBRUARY 1972). We sketch the solution to Problem 5. Using a Holmgren type argument one can prove finite speed of propagation using the fact [**2**] that stable hyperbolicity is not destroyed by small perturbations of the spacelike initial surface. The result of Problem 5 can be obtained by a global type Holmgren argument (F. John) using the fact that stable hyperbolicity is not destroyed by deformations of the initial surface provided the deformed surfaces are always noncharacteristic. This last fact is a consequence of Theorem 3 in [**7**].

BIBLIOGRAPHY

1. R. Hersh, *Mixed problems in several variables*, J. Math. Mech. **12** (1963), 317–334. MR **26** # 5304.

2. H. O. Kreiss, *Initial boundary value problems for hyperbolic systems*, Comm. Pure Appl. Math. **13** (1970), 277–298.

3. J. Ralston, *Note on a paper of Kreiss*, Comm. Pure Appl. Math. **24** (1971), 759–762.

4. J. Rauch, *Energy and resolvent inequalities for hyperbolic mixed problems*, J. Differential Equations **11** (1972), 528–540.

5. ———, $\mathscr{L}_2$ *is a continuable initial condition for Kreiss' mixed problems*, Comm. Pure Appl. Math. Comm. Pure Appl. Math. **15** (1972), 265–285.

6. R. Sakamoto, *Mixed problems for hyperbolic equations.* I, II, J. Math. Kyoto Univ. **10** (1970), 349–373, 403–417.

7. T. Shirota, *On the propagation speed of hyperbolic operators with mixed boundary conditions*, J. Fac. Sci. Hokkaido Univ. **22** (1972), 25–31.

UNIVERSITY OF MICHIGAN

ANALYTIC TORSION[1]

D. B. RAY AND I. M. SINGER

0. **Introduction.** In [7], we have introduced an invariant of a closed[2] oriented manifold W, which we call the analytic torsion. Its definition is the analogue, in terms of the de Rham complex on W, of a formula for a classical combinatorial invariant, the Reidemeister-Franz torsion, or R-torsion.

The purpose of these remarks is to describe the analytic torsion in analogy with R-torsion, and to indicate some extensions of the definition to other situations. The results concerning R-torsion and analytic torsion in §§1 and 2 are generally contained in [7], to which we refer the reader for the details of some proofs. In §3, we define an analytic torsion which depends on a choice of cohomology base for W, and prove that it too is an invariant.

The R-torsion is defined for a cell complex K as a function of certain representations of the fundamental group $\pi_1(K)$. It has been shown [**4**], [**8**] to be invariant under subdivision of K. Hence if K is a smooth triangulation of a closed oriented manifold W, the R-torsion of K defines a manifold invariant of W. It turns out that the R-torsion of K can be expressed in terms of the determinants of the combinatorial Laplacians acting on chain groups of K associated with the representation of $\pi_1(K)$. Giving W a Riemannian metric, we can write down an analogous analytic formula simply by replacing the combinatorial Laplacians by the usual analytic Laplacians on the de Rham complex. This formula defines the analytic torsion of the Riemannian manifold W. If W is closed, the analytic torsion is independent of the metric and hence is a manifold invariant of W.

Because of the formal analogy of the definitions, the analytic torsion has a number of properties in common with the R-torsion. We do not know whether

---

AMS 1970 *subject classifications*. Primary 58G05, 35P20.

[1] Supported in part by National Science Foundation grants GP-22928 and GP-22927, respectively.

[2] By a closed manifold we mean a compact manifold without boundary.

the two are equal for all closed oriented manifolds, but we have conjectured that the ratio of the two torsions is the same for all suitable representations of the fundamental group.

The analytic torsion is defined in §2 only for certain representations of the fundamental group, namely those for which a corresponding homology is trivial. In general, the R-torsion is defined as a function both of the representations and of the choice of generators of these homology groups. In §3 we will define a corresponding analytic torsion, and show that it too is independent of the choice of Riemannian metric. This material is implicitly contained in [7]; the definition given here seems more natural, and the proof of invariance is considerably shorter.

The analytic torsion can be defined in the same way for the $\bar{\partial}$-complex of a complex Kähler manifold W. In this case, a choice of generators for the Dolbeault groups of W is involved; one knows that this choice depends only on the cohomology class of the Kähler form on W. In this case, however, we can prove only that the ratio of values of the $\bar{\partial}$-torsion for two representations is a function of this cohomology class.

We now have the problem of identifying this invariant. One way of doing this would be to express the $\bar{\partial}$-torsion, or at least the ratio of its values for two representations, in terms of known functions of the complex structure of W. In this regard we can only say that for the torus of one complex dimension, the formula for the $\bar{\partial}$-torsion coincides with Kronecker's second limit formula of number theory; hence we can express the $\bar{\partial}$-torsion in this case in terms of the classical theta functions, whose arguments can easily be identified with the complex structure and the representation. For Riemann surfaces of genus greater than one, the torsion can be expressed in terms of Selberg's zeta function.

These results concerning $\bar{\partial}$-torsion will be the subject of another paper.

1. ***R*-torsion.** Before presenting the definition of R-torsion, we will compute the torsion of the lens space of three dimensions as an illustration. The lens space is defined by the identification of points under a rotation of the sphere S_3. Writing the sphere as

$$S_3 = \{(z_1, z_2): |z_1|^2 + |z_2|^2 = 1\},$$

where $z_1 = r_1 e^{i\theta_1}$, $z_2 = r_2 e^{i\theta_2}$ are complex coordinates in R^4, we can say that the rotation is given by

$$\gamma: (z_1, z_2) \to (\omega^{j_1} z_1, \omega^{j_2} z_2),$$

where ω is a primitive root of one and j_1, j_2 are relatively prime integers.

Suppose that ω is a pth root of one. Then the region $\{(z_1, z_2): 0 \leqq \theta_2 < 2\pi/p\}$ of S_3 is a fundamental domain for the lens space. To be more precise, define the cells

$$e_3 = \{(z_1, z_2): 0 \leqq \theta_2 \leqq 2\pi/p\},$$
$$e_2 = \{(z_1, z_2): \theta_2 = 0\}.$$

$$e_1 = \{(z_1, z_2): z_2 = 0, 0 \leqq \theta_1 \leqq 2\pi/p\},$$
$$e_0 = \{(z_1, z_2): z_2 = \theta_1 = 0\},$$

in S_3. This exhibits the lens space as a cell complex with one cell in each dimension. Identifying equivalent points under the rotation γ, the boundary operator for the lens space is given by

$$\partial e_3 = \gamma^{l_2} e_2 - e_2 = 0,$$
$$\partial e_2 = \sum_0^{p-1} \gamma^k e_1 = p e_1,$$
$$\partial e_1 = \gamma^{l_1} e_0 - e_0 = 0,$$
$$\partial e_0 = 0,$$

where $l_1 j_1 = 1, l_2 j_2 = 1 \pmod p$.

We see that this cell complex does not distinguish the various lens spaces, so we must add additional structure. This is done by choosing a representation χ of the fundamental group in the complex numbers, and introducing a chain complex over the complex numbers by identifying γ with $\chi(\gamma)$. That is, each chain group C_k has complex dimension one, and has a preferred generator e_k; supposing for simplicity that $\chi(\gamma) = \omega$, the boundary operator on C_k is given by

$$\partial e_3 = (\omega^{l_2} - 1)e_2,$$
$$\partial e_2 = \left(\sum_0^{p-1} \omega^k\right) e_1 = 0,$$
$$\partial e_1 = (\omega^{l_1} - 1)e_0,$$
$$\partial e_0 = 0.$$

The homology of this complex is trivial. What structure it has is contained in the ratios $\partial e_{2k+1}/e_{2k} = \omega^{l_k} - 1$, which exhibit the twist used in patching the cells together. The torsion of the lens space, as defined in [**8**], is

$$(\partial e_1/e_0) \cdot (\partial e_3/e_2) = (\omega^{l_1} - 1)(\omega^{l_2} - 1) (\mathrm{mod} \pm \omega).$$

The identification of values of the torsion under multiplication by $\pm\omega$ is necessary because we have made an arbitrary choice of the fundamental domain in S_3; a different choice would multiply the torsion by such a factor.

For our purposes it is more suitable to define the R-torsion as the real number

$$\tau(\chi) = |\omega^{l_1} - 1| \cdot |\omega^{l_2} - 1|,$$

which is uniquely determined by the rotation γ and the representation $\chi : \gamma \to \omega$.

In §8 of [**4**], the R-torsion is defined as a function of representations $\gamma \to O$ of the fundamental group by orthogonal $n \times n$ matrices. Such a representation determines a chain group over R^n; the boundary operator is the $n \times n$ matrix obtained by replacing γ by O, and the corresponding value of the R-torsion is defined by interpreting $\partial e_{2k+1}/e_{2k}$ as the determinant of this matrix. What we have called the R-torsion above is clearly equivalent to the special case $n = 2$. To keep the exposition simple, we will restrict ourselves to this special case throughout.

In general, the R-torsion of a cell complex K is defined as follows. Let π_1 be the fundamental group of K, and $\tilde{K}$ be the simply connected covering space, on which π_1 acts as deck transformations. Consider K embedded as a fundamental domain in $\tilde{K}$, so that $\tilde{K}$ consists of the cells of K and their translates by π_1.

Let χ be a representation of π_1 in the complex numbers of modulus one. By identifying γe with $\chi(\gamma)e$, $\gamma \in \pi_1$, we map the cells of $\tilde{K}$ into a complex chain group $C(K, \chi)$, in which we can single out the cells e of K as a preferred base. Since the boundary of a cell in K will in general be a linear combination of translates of cells of K, the boundary operator on $C(K, \chi)$ is a complex matrix relative to this base.

Suppose the homology of the resulting chain complex $C(K, \chi)$ is trivial. As in the example of the lens spaces, we want to define the ratio $\partial C_{q+1}(K, \chi)/C_q(K, \chi)$. This ratio can be given a meaning as follows. For each q pick a base $\mathbf{b}_q = (b_q)$ for the image ∂C_{q+1} in C_q, and for each base element b_q, pick an element $\tilde{b}_q$ in C_{q+1} such that $\partial \tilde{b}_q = b_q$. Since the homology is trivial, the elements $\mathbf{b}_q, \tilde{\mathbf{b}}_{q-1} = (b_q, \tilde{b}_{q-1})$ form a base for $C_q(K, \chi)$. Expressing b_q and $\tilde{b}_{q-1}$ as linear combinations of the cells e_q of K with complex coefficients defines the matrix of the change of base to $\mathbf{b}_q, \tilde{\mathbf{b}}_{q-1}$ from the preferred base $\mathbf{e}_q$. Let $[\mathbf{b}_q, \tilde{\mathbf{b}}_{q-1}/\mathbf{e}_q]$ denote the determinant of this matrix. Note that, given the bases $\mathbf{b}_q$ for each q, the determinant is independent of the choice of representatives $\tilde{b}_{q-1}$ satisfying $\partial \tilde{b}_{q-1} = b_{q-1}$. We define the R-torsion to be the positive real number

(1.1) $$\tau(K, \chi) = \prod_{q \text{ even}} |[\mathbf{b}_q, \tilde{\mathbf{b}}_{q-1}/\mathbf{e}_q]| / \prod_{q \text{ odd}} |[\mathbf{b}_q, \tilde{\mathbf{b}}_{q-1}/\mathbf{e}_q]|.$$

It is not hard to see that $\tau(K, \chi)$ is independent of the choice of bases $\mathbf{b}_q$ for the boundaries, and is independent also of the embedding of K in $\tilde{K}$. It is known ([**8**], [**4**, §7]) that $\tau(K, \chi)$ is invariant under subdivision of the complex K. In particular, if K is a smooth triangulation of a compact oriented manifold W, then $\tau(K, \chi) = \tau(W, \chi)$ does not depend on the triangulation.

Suppose that the homology of $C(K, \chi)$ is not trivial, but suppose that each homology group $H_q(K, \chi)$ has a preferred base $\mathbf{h}_q$. For each base element h_q, pick a representative cycle h'_q in C_q. Choosing a base $\mathbf{b}_q$ for ∂C_{q+1} and corresponding elements $\tilde{\mathbf{b}}_q$ in C_{q+1} as above, we obtain a base $\mathbf{b}_q, \tilde{\mathbf{b}}_{q-1}, \mathbf{h}'_q$ for C_q. The determinant $[\mathbf{b}_q, \tilde{\mathbf{b}}_{q-1}, \mathbf{h}'_q/\mathbf{e}_q]$ of the change to this base from the preferred base $\mathbf{e}_q$ of C_q is independent of the choice of representative cycles $\mathbf{h}'_q$. The R-torsion $\tau(K, \chi, (\mathbf{h}_q))$ is defined in this case by using these determinants in place of $[\mathbf{b}_q, \tilde{\mathbf{b}}_{q-1}/\mathbf{e}_q]$ in (1.1).

Note that the R-torsion $\tau(K, \chi, (\mathbf{h}_q))$, as a function of the preferred base $\mathbf{h}_q$, depends only on the volume element which this base determines in $H_q(K, \chi)$, in the sense that for another base $\mathbf{h}'_q$ satisfying $[\mathbf{h}_q/\mathbf{h}'_q] = 1$ for each q, $\tau(K, \chi, (\mathbf{h}_q)) = \tau(K, \chi, (\mathbf{h}'_q))$.

In order to introduce the analytic torsion, we will rewrite the formula (1.1) in terms which carry over to the de Rham complex.

Let $\langle\ ,\ \rangle$ be the inner product on $C(K, \chi)$ determined by the preferred base. Let $\partial^*: C_q \to C_{q+1}$ be the adjoint of the boundary operator, and define the combinatorial Laplacian by

$$\Delta^{(c)} = -(\partial^*\partial + \partial\partial^*).$$

If the homology of $C(K, \chi)$ is trivial, then $\Delta^{(c)} = \Delta_q^{(c)}$ is represented by a strictly negative hermitian matrix in each dimension.

PROPOSITION 1.2. *If the homology of $C(K, \chi)$ is trivial, then*

$$\log \tau(K, \chi) = \tfrac{1}{2} \sum_q (-1)^{q+1} q \log \det(-\Delta_q^{(c)}).$$

PROOF. For each q, let $B_q = \partial C_{q+1}$, $\tilde{B}_{q-1} = \partial^* C_{q-1}$. Since ∂^* is adjoint to ∂ and $\partial^2 = 0$, B_q and $\tilde{B}_{q-1}$ are orthogonal subspaces of C_q in our inner product; since the homology is trivial, C_q is the direct sum of the two.

The combinatorial Laplacian splits correspondingly into the direct sum of two operators: these are just $-\partial\partial^*$ on B_q and $-\partial^*\partial$ on $\tilde{B}_{q-1}$. The boundary operator ∂ is nonsingular from $\tilde{B}_{q-1}$ onto B_{q-1}, the inverse being given by $\partial^*(-\Delta_{q-1}^{(c)})^{-1}$: $B_{q-1} \to \tilde{B}_{q-1}$. The Laplacians on B_{q-1} and $\tilde{B}_{q-1}$ are equivalent under this isomorphism; in particular, if we denote by d_q and $\tilde{d}_{q-1}$ the determinants of $\partial\partial^*$ acting on B_q and of $\partial^*\partial$ on $\tilde{B}_{q-1}$, then $\tilde{d}_{q-1} = d_{q-1}$ and $\det(-\Delta_q^{(c)}) = d_q \cdot d_{q-1}$.

Now choose a base $\mathbf{b}_q$ of $B_q = \partial C_{q+1}$ for each q. Expressing each base element b_q in terms of the q-cells of K defines a rectangular matrix β_q. For each base element b_{q-1} of B_{q-1}, let

$$\tilde{b}_{q-1} = \partial^*(-\Delta_{q-1}^{(c)})^{-1} b_{q-1}$$

so that $\partial\tilde{b}_{q-1} = b_{q-1}$. Expressing each element $\tilde{b}_{q-1}$ in terms of the q-cells defines a matrix $\tilde{\beta}_{q-1}$. If we think of ∂^* and $\Delta^{(c)}$ as denoting both operators and the matrices which represent them, then as matrices,

$$\tilde{\beta}_{q-1} = \beta_{q-1}(-\Delta_{q-1}^{(c)})^{-1} \partial^*.$$

Let α_q be the matrix representing the change from the preferred base of C_q to the base $\mathbf{b}_q, \tilde{\mathbf{b}}_{q-1}$. The rows of α_q consist of those of β_q and $\tilde{\beta}_{q-1}$, and we can write the adjoint α_q^* as the direct sum of β_q^* and $\tilde{\beta}_{q-1}^*$. Since B_q and $\tilde{B}_{q-1}$ are orthogonal we have

$$\begin{aligned}
\alpha_q\alpha_q^* &= \beta_q\beta_q^* \oplus \tilde{\beta}_{q-1}\tilde{\beta}_{q-1}^* \\
&= \beta_q\beta_q^* \oplus \beta_{q-1}(-\Delta_{q-1}^{(c)})^{-1} \partial^*\partial(-\Delta_{q-1}^{(c)})^{-1}\beta_{q-1}^* \\
&= \beta_q\beta_q^* \oplus \beta_{q-1}(-\Delta_{q-1}^{(c)})^{-1}\beta_{q-1}^*;
\end{aligned}$$

$$\begin{aligned}\det(\alpha_q\alpha_q^*) &= |[\mathbf{b}_q, \tilde{\mathbf{b}}_{q-1}/\mathbf{e}_q]|^2 \\ &= \det(\mathring{\beta}_q\mathring{\beta}_q^*)\cdot\det(\mathring{\beta}_{q-1}(-\Delta_{q-1}^{(c)})^{-1}\mathring{\beta}_{q-1}^*) \\ &= \det(\mathring{\beta}_q\mathring{\beta}_q^*)\cdot\det(\mathring{\beta}_{q-1}\mathring{\beta}_{q-1}^*)\cdot d_{q-1}^{-1}.\end{aligned}$$

In the formula for the R-torsion of K, the factors $\det(\mathring{\beta}_q\mathring{\beta}_q^*)$ cancel (see the remark after (1.1)), and

$$\begin{aligned}\log \tau(K, \chi) &= \tfrac{1}{2}\textstyle\sum(-1)^{q+1}\log d_{q-1} \\ &= \tfrac{1}{2}\textstyle\sum q(-1)^{q+1}(\log d_q + \log d_{q-1}) \\ &= \tfrac{1}{2}\textstyle\sum q(-1)^{q+1}\log\det(-\Delta_q^{(c)}).\end{aligned}$$

Finally, we can rewrite the formula of Proposition 1.2 by expressing the determinant of the combinatorial Laplacian $\Delta_q^{(c)}$ in terms of the zeta function

$$\begin{aligned}\zeta_q^{(c)}(s) &= \frac{1}{\Gamma(s)}\int_0^\infty t^{s-1}\,\mathrm{tr}(e^{t\Delta_q^{(c)}})\,dt \\ &= \textstyle\sum(-\lambda)^{-s},\end{aligned}$$

where λ runs through the eigenvalues of $\Delta_q^{(c)}$. Since

$$(d(-\lambda)^{-s}/ds) = -\log(-\lambda)(-\lambda)^{-s},$$

we have

$$\log\det(-\Delta_q^{(c)}) = -(\zeta_q^{(c)})'(0). \tag{1.3}$$

2. **Analytic torsion.** Let W be a compact oriented C^∞ manifold without boundary, of dimension N. Let χ be a character of the fundamental group $\pi_1(W)$; that is, a representation of $\pi_1(W)$ in the complex numbers of modulus one. Let $L(\chi)$ be the complex line bundle associated with χ, and let $\mathscr{D}(W, \chi) = \mathscr{D} = \sum\mathscr{D}^q$ be the space of differential forms on W with values in $L(\chi)$. If $\tilde{W}$ is the simply connected covering manifold of W, with π_1 acting as deck transformations, a form f in $\mathscr{D}(W, \chi)$ may be identified with the projection of a complex valued form $\tilde{f}$ on $\tilde{W}$ which satisfies $\tilde{f}\circ\gamma = \chi(\gamma)\tilde{f}$, $\gamma\in\pi_1$.

Suppose that W has a Riemannian metric. This determines an inner product on D:

$$(f, g) = \int_W f\wedge *\bar{g};$$

$*:\mathscr{D}^q\to\mathscr{D}^{N-q}$ is the duality operator associated with the metric. Relative to this inner product, the exterior differential $d:\mathscr{D}^q\to\mathscr{D}^{q+1}$ has the formal adjoint $\delta = (-1)^{Nq+N+1}*d*$. The Laplacian $\Delta = \sum\Delta_q:\mathscr{D}^q\to\mathscr{D}^q$ is given by

$$\Delta = -(d\delta + \delta d).$$

On $\mathscr{D}$, the Laplacian is the infinitesimal generator of a semigroup of compact operators $e^{t\Delta}$, $t > 0$ [**2**]. Suppose that the homology of W determined by the representation χ, as in §1, is trivial. Then by the de Rham-Hodge theorem [**2**], Δ is strictly negative on $\mathscr{D}$, and $\mathrm{tr}(e^{t\Delta})$ decreases exponentially for large t. In this case we can define the zeta function $\zeta_{q,\chi}$ of Δ_q on $\mathscr{D}^q(W, \chi)$ by

$$\zeta_{q,\chi}(s) = \frac{1}{\Gamma(s)} \int_0^\infty t^{s-1} \, \mathrm{tr}(e^{t\Delta_q}) \, dt, \tag{2.1}$$

for Re s large. It is known [**5**] that $\zeta_{q,\chi}$ extends to a meromorphic function in the s-plane, which is analytic at $s = 0$.

(Strictly speaking, the results referred to in the preceding paragraph apply to the de Rham complex on W. It is easy to see that the proofs apply also to our complex $\mathscr{D}(W, \chi)$, since the two are locally identical.)

DEFINITION 2.2. Let W be a closed oriented Riemannian manifold of dimension N, and let χ be a character of $\pi_1(W)$ such that the Laplacian Δ is strictly negative on the associated complex $\mathscr{D}(W, \chi)$. The analytic torsion T of W is defined for such χ as the positive real root of

$$\log T(W, \chi) = \tfrac{1}{2} \sum_{q=0}^{N} (-1)^q \zeta'_{q,\chi}(0).$$

(The analytic torsion may be defined as a function of representations of $\pi_1(W)$ by orthogonal $n \times n$ matrices; see Definition 1.6 of [**7**]. The definition given here is equivalent to the special case $n = 2$.)

The definition of analytic torsion is formally analogous to the formulas (1.2) and (1.3) for the R-torsion. Because of this, a number of properties of R-torsion can be proved also for the analytic torsion. Beyond this, we have the deeper analytic result that the analytic torsion is independent of the choice of Riemannian metric on W [**7**]. Finally, for some manifolds, in particular for the lens spaces [**6**], we can compute the analytic torsion, and its values agree with those of the R-torsion.

We will present the proof of one property of the analytic torsion which depends only on formal manipulations of the definition. See [**7**, §2] for other results of this type. The reader may construct an analogous proof for R-torsion, using the framework developed in the proof of Proposition 1.2, or refer to [**3**] for a proof.

THEOREM 2.3. *Let W be a closed oriented Riemannian manifold, of even dimension N. Then for each character χ of the fundamental group $\pi_1(W)$ such that the associated Laplacian is strictly negative,*

$$\log T(W, \chi) = 0.$$

PROOF. For each q, define the subspaces

$$\mathscr{D}_1^q = d\mathscr{D}^{q-1}, \qquad \mathscr{D}_2^q = \delta\mathscr{D}^{q+1},$$

of $\mathscr{D}^q = \mathscr{D}^q(W, \chi)$. Since δ is the adjoint of d and $d^2 = 0$, $\mathscr{D}_1^q$ and $\mathscr{D}_2^q$ are orthogonal subspaces; since the Laplacian is strictly negative on $\mathscr{D}^q$, $\mathscr{D}^q$ is the direct sum of the two.

$\mathscr{D}_1^q$ is invariant under the operator $(d\delta)_q$, and $\mathscr{D}_2^q$ under $(\delta d)_q$. This determines a direct sum decomposition of the Laplacian on $\mathscr{D}^q$, and we have

$$\operatorname{tr}(e^{t\Delta_q}) = \operatorname{tr}(e^{-t(d\delta)_q}) + \operatorname{tr}(e^{-t(\delta d)_q}),$$

where we consider the operators on the right restricted to $\mathscr{D}_1^q$ and $\mathscr{D}_2^q$, respectively.

We want to say, first, that as in the proof of Proposition 1.2, the operators $(d\delta)_q$ and $(\delta d)_{q-1}$ are equivalent. To see this, observe that since $\delta(d\delta)_q = (\delta d)_{q-1}\delta$,

$$-\frac{d}{dt}e^{-t(d\delta)_q} = d\delta e^{-t(d\delta)_q} = de^{-t(\delta d)_{q-1}}\delta.$$

Hence, using $\operatorname{tr}(AB) = \operatorname{tr}(BA)$ repeatedly,

$$\begin{aligned}
\operatorname{tr}(e^{-t(d\delta)_q}) &= \int_t^\infty \operatorname{tr}(de^{-u(\delta d)_{q-1}}\delta)\,du \\
&= \int_t^\infty \operatorname{tr}((e^{-u(\delta d)_{q-1}/2}\delta)(de^{-u(\delta d)_{q-1}/2}))\,du \\
&= \int_t^\infty \operatorname{tr}(\delta de^{-u(\delta d)_{q-1}})\,du \\
&= \operatorname{tr}(e^{-t(\delta d)_{q-1}}),
\end{aligned}$$

which is the equivalence we want.

Because of this,

$$\begin{aligned}
\sum_{q=0}^N (-1)^q q \operatorname{tr}(e^{t\Delta_q}) &= \sum_{q=0}^N (-1)^q q(\operatorname{tr}(e^{-t(\delta d)_q}) + \operatorname{tr}(e^{-t(\delta d)_{q-1}})) \\
&= \sum_{q=0}^N (-1)^q q(\operatorname{tr}(e^{-t(d\delta)_q}) + \operatorname{tr}(e^{-t(d\delta)_{q+1}})) \\
&= \sum_{q=0}^{N-1} (-1)^{q+1} \operatorname{tr}(e^{-t(\delta d)_q}) \\
&= \sum_{q=1}^{N} (-1)^{q} \operatorname{tr}(e^{-t(d\delta)_q}).
\end{aligned}$$

But we also have the equivalence $*(d\delta)_q = (\delta d)_{N-q}*$, which implies

$$\begin{aligned}
\sum_{q=1}^N (-1)^q \operatorname{tr}(e^{-t(d\delta)_q}) &= \sum_{q=1}^N (-1)^q \operatorname{tr}(e^{-t(\delta d)_{N-q}}) \\
&= (-1)^N \sum_{q=0}^{N-1} (-1)^q \operatorname{tr}(e^{-t(\delta d)_q}).
\end{aligned}$$

Together with the preceding, this shows that when N is even,

$$\sum_{q=0}^{N} (-1)^q q \zeta_{q,\chi}(s) \equiv 0,$$

so that $\log T(W, \chi) = 0$.

A deeper result concerning the analytic torsion is the following.

THEOREM 2.4. *Let W be a closed oriented manifold, and χ a character of the fundamental group $\pi_1(W)$ such that the associated homology of W is trivial. Then the value $T(W, \chi)$ of the analytic torsion does not depend on the choice of Riemannian metric on W.*

PROOF. Let ρ_0 and ρ_1 be two metrics on W, and for each u in $[0, 1]$ define the metric

$$\rho_u = u\rho_1 + (1 - u)\rho_0$$

on W. Where necessary, we will designate operations associated with the metric ρ_u by the subscript u; the complex $\mathcal{D}(W, \chi)$ is of course independent of u.

We will show that $\log T(W, \chi)$ has a vanishing derivative with respect to the parameter u. The proof depends on two results concerning the heat kernel $e^{t\Delta}$ on W, the first of which is

PROPOSITION 2.5. *If the metric on the closed manifold W depends smoothly on a parameter u, and if Δ_u denotes the Laplacian on $\mathcal{D}(W, \chi)$ corresponding to the value u of the parameter, then* $\operatorname{tr}(e^{t\Delta_u})$ *is a differentiable function of u and*

$$\frac{d}{du} \operatorname{tr}(e^{t\Delta_u}) = t \operatorname{tr}(\dot{\Delta}_u e^{t\Delta_u}),$$

where $\dot{\Delta}_u = (d\Delta_u/du)$.

Proposition 2.5 can be proved by a fairly straightforward computation based on a standard application of Green's formula to the heat kernel. For the details in a more general setting, see §6 of [**7**].

Accepting 2.5, let $\alpha = *^{-1}\dot{*}: \mathcal{D}^q \to \mathcal{D}^q$, where $\dot{*} = (d/du)*$. The operator α acts algebraically on the components of a form in $\mathcal{D}^q$. Since $** = \pm 1$, we have $\dot{*}*^{-1} = -*^{-1}\dot{*} = -\alpha$. Thus we can write

$$\begin{aligned}
\dot{\Delta} &= \pm \frac{d}{du}(*d*d) \pm \frac{d}{du}(d*d*) \\
&= \pm(\dot{*}*^{-1}*d*d + *d**^{-1}\dot{*}d) \\
&\quad \pm(d\dot{*}*^{-1}*d* + d*d**^{-1}\dot{*}) \\
&= \alpha\delta d - \delta\alpha d + d\alpha\delta - d\delta\alpha.
\end{aligned}$$

Using this in the formula of 2.5, manipulations similar to those of the preceding proof show that

$$(2.6)\qquad \begin{aligned} \operatorname{tr}(\dot{\Delta} e^{t\Delta_q}) &= \operatorname{tr}(\alpha\delta d e^{t\Delta_q}) - \operatorname{tr}(\delta\alpha d e^{t\Delta_q}) \\ &\quad + \operatorname{tr}(d\alpha\delta e^{t\Delta_q}) - \operatorname{tr}(d\delta\alpha e^{t\Delta_q}) \\ &= \operatorname{tr}(\alpha\delta d e^{t\Delta_q}) - \operatorname{tr}(\alpha d\delta e^{t\Delta_{q+1}}) \\ &\quad + \operatorname{tr}(\alpha\delta d e^{t\Delta_{q-1}}) - \operatorname{tr}(\alpha d\delta e^{t\Delta_q}). \end{aligned}$$

For instance, the second term transforms as follows:

$$\begin{aligned} \operatorname{tr}(\delta\alpha d e^{t\Delta_q}) &= \operatorname{tr}((e^{t\Delta_q/2})(\delta\alpha d e^{t\Delta_q/2})) \\ &= \operatorname{tr}(\alpha d e^{t\Delta_q}\delta) \\ &= \operatorname{tr}(\alpha d\delta e^{t\Delta_{q+1}}). \end{aligned}$$

From Proposition 2.5,

$$\begin{aligned} &\sum_{q=0}^{N} (-1)^q q \frac{d}{du} \operatorname{tr}(e^{t\Delta_q}) \\ &\quad = t \sum_{q=0}^{N} (-1)^q \operatorname{tr}((q\alpha\delta d + (q-1)\alpha d\delta - (q+1)\alpha\delta d - q\alpha d\delta)e^{t\Delta_q}) \\ &\quad = t \sum_{q=0}^{N} (-1)^q \operatorname{tr}(\alpha\Delta e^{t\Delta_q}) = t \sum_{q=0}^{N} (-1)^q \frac{d}{dt} \operatorname{tr}(\alpha e^{t\Delta_q}). \end{aligned}$$

Hence

$$(2.7)\qquad \begin{aligned} \frac{d}{du} \sum_{q=0}^{N} (-1)^q q \zeta_q(s) &= \sum_{q=0}^{N} (-1)^q \frac{1}{\Gamma(s)} \int_0^\infty t^s \frac{d}{dt} \operatorname{tr}(\alpha e^{t\Delta_q})\, dt \\ &= \sum_{q=0}^{N} (-1)^{q+1} \frac{s}{\Gamma(s)} \int_0^\infty t^{s-1} \operatorname{tr}(\alpha e^{t\Delta_q})\, dt, \end{aligned}$$

where we have integrated by parts; the integrated terms vanish since $e^{t\Delta} = O(t^{-N/2})$ as $t \to 0$, and, under the assumption of trivial homology, $e^{t\Delta} = O(e^{-\lambda t})$ as $t \to \infty$ with $\lambda > 0$.

It remains to show that the right side of (2.7) extends to a meromorphic function of s with vanishing derivative at $s = 0$. When the dimension N of W is odd, this is an immediate consequence of the following asymptotic expansion for the heat kernel on W. Observe that the proof breaks down when N is even; however, we have already seen in Proposition 2.3 that $\log T(W, \chi) = 0$ is independent of the metric in this case.

PROPOSITION 2.8. *Let W be a closed manifold of dimension N, and let Δ be the*

Laplacian on the space $\mathscr{D}(W, \chi)$ for a given representation χ of $\pi_1(W)$. For each $t > 0$, let $P_t(x, y)$, $x, y \in W$, be the kernel of $e^{t\Delta}$. Then

$$P_t(x, x) = \sum_{m=0}^{n} t^{m-N/2}\Phi_m(x) + o(t^{n-N/2}),$$

where each Φ_m is smooth on W and the convergence of the remainder term is uniform on W.

The proof of Proposition 2.8 for functions on W was given in [**5**], as well as its application to the zeta function. Since it depends only on the construction of a local parametrix for the heat kernel, the proof remains valid for the complex $\mathscr{D}(W, \chi)$. The construction is given also in §5 of [**7**].

To apply the result to (2.7), note first that since the algebraic operator α acts locally, the asymptotic expansion holds also for the kernel of $\alpha e^{t\Delta}$. Integrating over the closed manifold W, we obtain

$$\operatorname{tr}(\alpha e^{t\Delta}) = \sum_{m=0}^{n} t^{m-N/2} C_m + o(t^{n-N/2}),$$

with $C_m = \operatorname{tr}(\alpha\Phi_m)$. Hence, choosing $n > N/2$,

$$\begin{aligned} \int_0^\infty t^{s-1} \operatorname{tr}(\alpha e^{t\Delta})\, dt &= \int_0^1 t^{s-1} \operatorname{tr}(\alpha e^{t\Delta})\, dt + \int_1^\infty t^{s-1} \operatorname{tr}(\alpha e^{t\Delta})\, dt \\ &= \sum_{m=0}^{n} C_m \int_0^1 t^{s+m-N/2-1}\, dt + R(s) \qquad (2.9) \\ &= \sum_{m=0}^{n} \frac{C_m}{s + m - N/2} + R(s), \end{aligned}$$

where $R(s)$ is analytic for $s > N/2 - n$. Thus the right side defines the extension to this region of the left side, and is seen to be analytic at $s = 0$. But this implies that the extension

$$\frac{s}{\Gamma(s)} \int_0^\infty t^{s-1} \operatorname{tr}(\alpha e^{t\Delta})\, dt$$

has a zero of order two at $s = 0$, completing the proof of Theorem 2.4.

3. **Analytic torsion as a function of cohomology.** We have seen in §1 that the R-torsion of a complex K can be defined in general as a function of the pair $(\chi, (\mathbf{h}_q))$, where χ is a character of the fundamental group $\pi_1(K)$ and $\mathbf{h}_q$ is a choice of generators for the qth homology group $H_q(K, \chi)$, $q = 0, 1, \ldots, N$.

We can also define the analytic torsion in this way. By pairing, a choice of generators for the homology groups is equivalent to a choice of bases $\mathbf{h}^q$ for the cohomology groups $H^q(W, \chi)$. These preferred bases will remain fixed throughout the discussion. Let $\mathscr{H}^q(W, \chi)$ be the space of harmonic q-forms in $\mathscr{D}(W, \chi)$, and let $\boldsymbol{\eta}^q$ be an orthonormal base for $\mathscr{H}^q$. The de Rham map $A^q: \mathscr{H}^q \to H^q$ sends $\boldsymbol{\eta}^q$ into a base $\mathbf{A}^q(\boldsymbol{\eta}^q)$ of H^q. Let $[\mathbf{A}^q(\boldsymbol{\eta}^q)/\mathbf{h}^q]$ be the determinant of the change to this base from the preferred base.

DEFINITION 3.1. Let W be a closed oriented Riemannian manifold of dimension N, and let χ be a character of $\pi_1(W)$. For each q, let $\mathbf{h}^q$ be a base for the cohomology group $H^q(W, \chi)$.

The analytic torsion in this situation is given by

$$\log T(W, \chi, (\mathbf{h}^q)) = \sum_{q=0}^{N} (-1)^q (\tfrac{1}{2} q \zeta_q'(0) + \log |[\mathbf{A}^q(\boldsymbol{\eta}^q)/\mathbf{h}^q]|),$$

where $\boldsymbol{\eta}^q$ is an orthonormal base for the harmonic forms in $\mathscr{D}^q(W, \chi)$, A^q is the de Rham map, and the zeta function ζ_q is defined by

$$\zeta_q(s) = \frac{1}{\Gamma(s)} \int_0^\infty t^{s-1} \operatorname{tr}(e^{t\Delta_q} - P_q)\, dt$$

for Re s large, $P_q: \mathscr{D}^q \to \mathscr{H}^q$ being the orthogonal projection.

As in the case of R-torsion, the analytic torsion $T(W, \chi, (\mathbf{h}^q))$ as a function of the preferred bases $\mathbf{h}^q$ of $H^q(W, \chi)$ depends only on the volume element, in the sense that $T(W, \chi, (\mathbf{h}^q)) = T(W, \chi, (\mathbf{h}'^q))$ if $|[\mathbf{h}^q/\mathbf{h}'^q]| = 1, q = 0, 1, \ldots, N$.

It is clear that the analytic torsion does not depend on which orthonormal base we pick for $\mathscr{H}^q$. The notation $T(W, \chi, (\mathbf{h}^q))$ suggests that the torsion may depend only on the choice of cohomology base. This is the case.

THEOREM 3.2. *Let W be a closed oriented manifold of dimension N, and let χ be a character of the fundamental group $\pi_1(W)$. For each q, let $\mathbf{h}^q$ be a base for the cohomology groups $H^q(W, \chi)$.*

Then the value $T(W, \chi, (\mathbf{h}^q))$ of the analytic torsion does not depend on the choice of Riemannian metric on W.

PROOF. As in the proof of Theorem 2.4, we will first compute the variation of the zeta function as the metric on W changes smoothly. Note that tr (P_q) is equal to the qth betti number of the complex $C(W, \chi)$, and so is independent of the choice of metric. Hence we can follow the steps of the proof of Theorem 2.4 to obtain

$$\begin{aligned}
\frac{d}{du} \sum_{q=0}^{N} (-1)^q q \zeta_q(s) &= \frac{1}{\Gamma(s)} \sum_{q=0}^{N} (-1)^q q \int_0^\infty t^{s-1} \frac{d}{du} \operatorname{tr}(e^{t\Delta_q})\, dt \\
&= \frac{1}{\Gamma(s)} \sum_{q=0}^{N} (-1)^q \int_0^\infty t^s \frac{d}{dt} \operatorname{tr}(\alpha e^{t\Delta_q})\, dt \\
&= \frac{1}{\Gamma(s)} \sum_{q=0}^{N} (-1)^q \int_0^\infty t^s \frac{d}{dt} \operatorname{tr}(\alpha(e^{t\Delta_q} - P_q))\, dt \\
&= \frac{s}{\Gamma(s)} \sum_{q=0}^{N} (-1)^{q+1} \int_0^\infty t^{s-1} \operatorname{tr}(\alpha(e^{t\Delta_q} - P_q))\, dt.
\end{aligned}$$

If N is odd we can use Proposition 2.8. But the result, instead of (2.9), is now

$$\int_0^\infty t^{s-1}\operatorname{tr}(\alpha(e^{t\Delta}-P))\,dt$$

$$= \int_0^1 t^{s-1}\operatorname{tr}(\alpha(e^{t\Delta}-P))\,dt + \int_1^\infty t^{s-1}\operatorname{tr}(\alpha(e^{t\Delta}-P))\,dt$$

$$= \sum_{m=0}^{n} C_m \int_0^1 t^{s+m-N/2-1}\,dt - \operatorname{tr}(\alpha P)\int_0^1 t^{s-1}\,dt + R(s)$$

$$= \sum_{m=0}^{n} \frac{C_m}{s+m-N/2} - \frac{1}{s}\operatorname{tr}(\alpha P) + R(s),$$

where $R(s)$ is analytic for $s > N/2 - n$. From this it follows that the extension of

$$\frac{s}{\Gamma(s)}\int_0^\infty t^{s-1}\operatorname{tr}(\alpha(e^{t\Delta}-P))\,dt$$

has the derivative $-\operatorname{tr}(\alpha P)$ at $s = 0$, and

$$\frac{d}{du}\sum_{q=0}^{N}(-1)^q q\zeta_q'(0) = \frac{1}{2}\sum_{q=0}^{N}(-1)^q\operatorname{tr}(\alpha P_q) \tag{3.3}$$

for N odd.

We will show that (3.3) holds for N even as well: both sides vanish. Let $\mathscr{D}_1^q = d\mathscr{D}^{q-1}$, $\mathscr{D}_2^q = \delta\mathscr{D}^{q+1}$. The Hodge theorem states that $\mathscr{D}^q$ is the direct sum of $\mathscr{D}_1^q$, $\mathscr{D}_2^q$, and of the space $\mathscr{H}^q$ of harmonic forms in $\mathscr{D}^q$. Correspondingly, we have the decomposition

$$\operatorname{tr}(e^{t\Delta_q} - P_q) = \operatorname{tr}(e^{-t(d\delta)_q}) + \operatorname{tr}(e^{-t(\delta d)_q}),$$

as in the proof of Theorem 2.3. The rest of that proof can be used without change to show that when N is even

$$\sum_{q=0}^{N}(-1)^q q\zeta_q(s) \equiv 0.$$

On the other hand, the duality operator $*$ is an isometry of $\mathscr{D}^q$ onto $\mathscr{D}^{N-q}$. Hence for h in $\mathscr{H}^q$ we have

$$\begin{aligned}(*h, \alpha * h) &= (*h, *^{-1}\dot{*}*h)\\ &= -(*h, \dot{*}*^{-1}*h)\\ &= -(*h, **^{-1}\dot{*}h) = -(h, \alpha h).\end{aligned}$$

This implies

$$\operatorname{tr}(\alpha P_{N-q}) = -\operatorname{tr}(\alpha P_q)$$

and

$$\sum_{q=0}^{N}(-1)^q \operatorname{tr}(\alpha P_q) = -\sum_{q=0}^{N}(-1)^q \operatorname{tr}(\alpha P_{N-q})$$
$$= -(-1)^N \sum_{q=0}^{N}(-1)^q \operatorname{tr}(\alpha P_q)$$
$$= 0, \qquad N \text{ even.}$$

To complete the proof of Theorem 3.2 we will show that, for each q,

$$\frac{d}{du}\log|[\mathbf{A}^q(\boldsymbol{\eta}^q)/\mathbf{h}^q]| = -\tfrac{1}{2}\operatorname{tr}(\alpha P_q). \tag{3.4}$$

For together with (3.3) this will imply

$$\frac{d}{du}\log T(W, \chi, (\mathbf{h}^q)) = 0;$$

hence the analytic torsion is constant as the parameter u of the metric varies.

(3.4) is essentially equivalent to Theorem 7.6 of [**7**], the proof of which depended on a relation due to Kodaira between duality in the combinatorial and de Rham settings. We now observe that not all this machinery is needed for the proof.

Let $\mathbf{h} = (h_j)$ denote the fixed preferred cohomology base. Given a parameter value u, there is, for each j, by the de Rham-Hodge theorem, a unique form $f_j = f_j(u)$ which is harmonic in the corresponding metric, and whose image under the de Rham map is h_j:

$$A(f_j(u)) = h_j.$$

These forms comprise a base for the harmonic forms; let $B = B(u)$ be the matrix of the change from this base to an orthonormal base $\boldsymbol{\eta}(u) = (\eta_j(u))$:

$$\eta_j = \sum B_{jk} f_k.$$

$\boldsymbol{\eta}$ being orthonormal, it follows that $B^*B = C^{-1}$, where $C = C(u)$ has entries $C_{lm} = (f_l, f_m)$.

We have, since the de Rham map is linear,

$$|[\mathbf{A}(\boldsymbol{\eta})/\mathbf{h}]|^2 = |[\mathbf{A}(\boldsymbol{\eta})/\mathbf{A}(\mathbf{f})]|^2 = |[\boldsymbol{\eta}/\mathbf{f}]|^2$$
$$= |\det(B)|^2 = \det(B^*B) = (\det(C))^{-1}.$$

So (3.4) will follow if we prove that for each l, m, $C_{lm}(u) = (f_l(u), f_m(u))$ is a differentiable function of u, and

$$\frac{d}{du}\log\det(C(u)) = \operatorname{tr}(\alpha P). \tag{3.5}$$

Let u, u' be two parameter values. Since $A(f_l(u)) = A(f_l(u')) = h_l$, we have

$$f_l(u) = f_l(u') + dg_l.$$

Therefore, denoting by a subscript the metric in which an inner product is taken,

$$(f_l(u), f_m(u))_u = (f_l(u), f_m(u') + dg_m)_u = (f_l(u), f_m(u'))_u$$

since $f_l(u)$ is harmonic in the metric corresponding to u. Similarly

$$(f_l(u'), f_m(u'))_{u'} = (f_l(u) - dg_l, f_m(u'))_{u'} = (f_l(u), f_m(u'))_{u'}.$$

Subtracting,

$$\begin{aligned}(f_l(u'), f_m(u'))_{u'} - (f_l(u), f_m(u))_u &= (f_l(u), f_m(u'))_{u'} - (f_l(u), f_m(u'))_u \\ &= \int f_l(u) \wedge (*' - *)\overline{f_m(u')},\end{aligned}$$

from which it follows that (f_l, f_m) is differentiable and

$$\frac{d}{du}(f_l, f_m) = (f_l, \alpha f_m). \tag{3.6}$$

Finally, it is easy to verify (3.5):

$$\begin{aligned}\frac{d}{du}\log\det(C) &= \operatorname{tr}\left(C^{-1}\frac{d}{du}C\right) = \operatorname{tr}\left(B^*B\frac{d}{du}C\right) = \operatorname{tr}\left(B\left(\frac{d}{du}C\right)B^*\right) \\ &= \sum B_{jl}(f_l, \alpha f_m)\bar{B}_{jm} = \sum(\eta_j, \alpha\eta_j) = \operatorname{tr}(\alpha P).\end{aligned}$$

REFERENCES

1. P. E. Conner, *The Neumann's problem for differential forms on Riemannian manifolds*, Mem. Amer. Math. Soc. No. 20 (1956). MR **17**, 1197.

2. A. N. Milgram and P. C. Rosenbloom, *Harmonic forms and heat conduction*. I. *Closed Riemannian manifolds*, Proc. Nat. Acad. Sci. U.S.A. **37** (1951), 180–184. MR **13**, 160.

3. J. W. Milnor, *A duality theorem for Reidemeister torsion*, Ann. of Math. (2) **76** (1962), 137–147. MR **25** #4526.

4. ———, *Whitehead torsion*, Bull. Amer. Math. Soc. **72** (1966), 358–426. MR **33** #4922.

5. S. Minakshisundaram and Å. Pleijel, *Some properties of the eigenfunctions of the Laplace-operator on Riemannian manifolds*, Canad. J. Math. **1** (1949), 242–256. MR **11**, 108.

6. D. B. Ray, *Reidemeister torsion and the Laplacian on lens spaces*, Advances in Math. **4** (1970), 109–126. MR **41** #2709.

7. D. B. Ray and I. M. Singer, *R-torsion and the Laplacian on Riemannian manifolds*, Advances in Math. **7** (1971), 145–210.

8. G. de Rham, *Complexes à automorphismes et homéomorphie différentiable*, Ann. Inst. Fourier (Grenoble) **2** (1950), 51–67. MR **13**, 268.

MASSACHUSETTS INSTITUTE OF TECHNOLOGY

AN APPLICATION OF VON NEUMANN ALGEBRAS TO FINITE DIFFERENCE EQUATIONS

DAVID G. SCHAEFFER

The theory of elliptic boundary value problems normally has as its starting point the analysis of boundary value problems on a half space $H = \{x \in \boldsymbol{R}^n : \langle x, N \rangle \geqq 0\}$, say

$$(1) \qquad \begin{aligned} P(D)u &= 0 \quad \text{for } x \in H, \\ P_j(D)u &= f_j \quad \text{for } x \in \partial H, \qquad j = 1, \dots, m. \end{aligned}$$

Here we assume that $P(D)$ is a properly elliptic differential operator (with complex coefficients), homogeneous of order $2m$, and we use the convention $D = -i\partial/\partial x$. A Fourier transform of (1) with respect to the boundary variables formally reduces the problem to a family of boundary value problems for ordinary differential equations on a half line. If the number of boundary conditions we impose is half the order of $P(D)$ and if we require that $P(D)$ is properly elliptic, then it follows by a trivial dimension argument that uniqueness of solution in the homogeneous problem is equivalent to existence of solution in the inhomogeneous problem for general data $\{f_j\}$. (See for example [**5**, Chapter II, §4].) In this paper we obtain an analogous result for finite difference approximations of boundary value problems on a half space. Our proof of the equivalence in the difference case is also based on a dimension argument, but it is the dimension function of a von Neumann algebra of type II, rather than the dimension function of linear algebra, which appears.

A simple example may make the appearance of von Neumann algebras in this problem seem more natural. In the so-called "five point approximation" of Laplace's equation in the plane one looks for a function v defined on the set $\boldsymbol{Z}^2$ of lattice points with the property that

$$v(j_1, j_2) = \tfrac{1}{4}\{v(j_1 + 1, j_2) + v(j_1 - 1, j_2) + v(j_1, j_2 + 1) + v(j_1, j_2 - 1)\}.$$

AMS 1970 *subject classifications*. Primary 46A28; Secondary 39A25.

Define an operator on $l^2(\mathbf{Z}^2)$ by

$$Q(D) = T_{(1,0)} + T_{(-1,0)} + T_{(0,1)} + T_{(0,-1)} - 4I.$$

To solve the equation

$$Q(D)v(j) = 0 \quad \text{for } j \in H \cap \mathbf{Z}^2 \tag{2}$$

using the Fourier transform, let us consider the exponential solutions of (2); that is, solutions of the form

$$v(j) = e^{i\langle \xi, j\rangle} w(\langle j, N\rangle) \tag{3}$$

where $\xi \in \mathbf{T}^2$, the torus, and w is a function on $\Sigma = \{\langle j,N\rangle : j \in H \cap \mathbf{Z}^2\}$, the set of projections (up to the scaling of N) of lattice points in H along the normal direction N. If N is a direction in $\mathbf{R}^2$ with rational slope, then Σ is a discrete subset of $[0,\infty)$, but if N has irrational slope, Σ is a countable dense subset of $[0,\infty)$. In the rational case, if we scale $N = (m,n)$ so that m and n are relatively prime integers, then $\Sigma = \mathbf{Z}^+$, the nonnegative integers, and a function on Σ is merely a sequence. Given a sequence w, formula (3) defines a solution of (2) if and only if w satisfies the difference equation

$$e^{i\xi_1}w(k+m) + e^{-i\xi_1}w(k-m) + e^{i\xi_2}w(k+n) + e^{-i\xi_2}w(k-n) - 4w(k) = 0. \tag{4}$$

For each $\xi \in \mathbf{T}^2$ there are $2\max(|m|,|n|)$ linearly independent solutions of (4) corresponding to the roots (in z) of the Laurent polynomial

$$e^{i\xi_1}z^m + e^{-i\xi_1}z^{-m} + e^{i\xi_2}z^n + e^{-i\xi_2}z^{-n} - 4 = 0.$$

Thus for each $\xi \in \mathbf{T}^2$, equation (2) has $2\max(|m|,|n|)$ linearly independent exponential solutions of the form (3). We claim, however, that the same set of solutions is associated to two points in $\mathbf{T}^2$ whose difference is proportional to N. Suppose $\xi = \xi' + tN$ for some $t \in \mathbf{R}$; for any function w,

$$e^{i\langle \xi, j\rangle} w(\langle j,N\rangle) = e^{i\langle \xi', j\rangle} w'(\langle j,N\rangle)$$

where $w'(k) = e^{itk}w(k)$; hence any exponential solution of (2) associated to ξ may also be associated to ξ', and conversely. Let G be the one parameter subgroup of $\mathbf{T}^2$ generated by N; then for each point of $\mathbf{T}^2/G$ there are $2\max(|m|,|n|)$ linearly independent exponential solutions of (2). In the irrational case, say $N = (\alpha,\beta)$ where α and β are incommensurate, for each root of

$$e^{i\xi_1}e^{i\alpha z} + e^{-i\xi_1}e^{-i\alpha z} + e^{i\xi_2}e^{i\beta z} + e^{i\xi_2}e^{-i\beta z} - 4 = 0$$

the function $v(j) = e^{i\langle \xi, j\rangle}e^{iz\langle j,N\rangle}$ is an exponential solution of (2), and the solutions obtained from different roots are linearly independent. Of course a nontrivial analytic almost periodic function has infinitely many roots, so for each $\xi \in \mathbf{T}^2$ there are infinitely many, linearly independent, exponential solutions of (2). As in the rational case, the same set of solutions is associated to two points in the same coset of $G \subset \boldsymbol{T}^2$, but in the irrational case G is a skew line and many more points are identified in forming the quotient $\boldsymbol{T}^2/G$.

In summary, then, the parameterization of the exponential solutions of (2) depends on the action of G on $\boldsymbol{T}^2$—when G is a skew line there are infinitely many, linearly independent, exponential solutions of (2) with a given tangential phase. It is interesting to note the similarities of this example with the group-measure space construction of von Neumann algebras given in §9, Chapter I of Dixmier [**3**]. Perhaps it is not surprising that in the irrational case, when G acts ergodically on $\boldsymbol{T}^2$, a factor of type II enters the problem.

In §1, after certain preliminary definitions concerning elliptic difference operators, we state our main theorem asserting the equivalence of existence and uniqueness for the difference equation. In §2 we introduce the von Neumann algebra which we use in the proof of this theorem—we refer the reader unfamiliar with the von Neumann theory to [**3**]. Finally in §3 we prove our main theorem.

This paper is a slightly condensed version of [**6**]. The techniques here have a great deal in common with our joint paper [**2**]. It is hard to overstate my indebtedness to Robert Seeley, who first called my attention to the continuous dimension theory of von Neumann algebras.

1. **Statement of the theorem.** We consider an approximation of $P(D)$ by a difference operator

$$Q(D) = \sum_{j \in \boldsymbol{Z}^n} c_j e^{i\langle j, D\rangle} = \sum_{j \in \boldsymbol{Z}^n} c_j T_j \quad \text{(finite sum)}$$

where T_j denotes the multi-integer translation $T_j v(x) = v(x + j)$. In contrast to the introduction, we regard $Q(D)$ as an operator on the Lebesgue space $L^2(\boldsymbol{R}^n)$ rather than $l^2(\boldsymbol{Z}^n)$. We use the Fourier transform with the sign convention

$$\tilde{v}(\xi) = \int_{\boldsymbol{R}^n} dx\, e^{-i\langle x, \xi\rangle} v(x),$$

so that in Fourier transform space $Q(D)$ is multiplication by the symbol

$$Q(\xi) = \sum c_j e^{i\langle j, \xi\rangle},$$

a multiply periodic function on $\boldsymbol{R}^n$. We shall say that $Q(D)$ is *elliptic* if $Q(\xi) \neq 0$ for $\xi \in \boldsymbol{R}^n$. We recall that an elliptic differential operator is called properly elliptic if for all linearly independent vectors ξ, N in $\boldsymbol{R}^n$ the polynomial $\tau \to P(\xi + \tau N)$ has its zeros equally distributed between the upper and lower half planes, with none lying on the real axis. This condition on the roots may be expressed in terms of winding numbers as

$$\lim_{T \to \infty} \{\arg P(\xi + TN) - \arg P(\xi - TN)\} = 0. \tag{1.1}$$

For the almost periodic function $\tau \to Q(\xi + \tau N)$ the limits in (1.1) need not exist, so in defining a properly elliptic difference operator we shall replace (1.1) by a condition involving a mean winding number. We may write

$$\arg Q(\xi) = \langle k,\xi\rangle + \psi(\xi), \qquad k \in \boldsymbol{Z}^n, \tag{1.2}$$

where ψ is periodic on $\boldsymbol{R}^n$ and where, for each coordinate direction $\boldsymbol{e}_i$,

$$k_i = (2\pi)^{-1}\{\arg Q(\xi + 2\pi\boldsymbol{e}_i) - \arg Q(\xi)\};$$

of course this definition of k is independent of ξ. It follows from (1.2) that the limit

$$\lim_{T\to\infty} (2T)^{-1}\{\arg Q(\xi + TN) - \arg Q(\xi - TN)\} \tag{1.3}$$

exists and equals $\langle k,N\rangle$. We shall call $Q(D)$ *properly elliptic* if $Q(D)$ is elliptic and (1.3) vanishes for all $\xi, N \in \boldsymbol{R}^n$. Thus an elliptic difference operator $Q(D)$ is properly elliptic if and only if $\arg Q$ is periodic. If $Q(\xi) = \sum c_j e^{i\langle j,\xi\rangle}$ is properly elliptic and if $\{j \in \boldsymbol{Z}^n : c_j \neq 0\}$ is symmetric with respect to the origin, then the zeros of the almost periodic function $\tau \to Q(\xi + \tau N)$ have the same mean density in the two half planes; see for example [**4**, Chapter VI] for a discussion of the zeros of almost periodic functions.

As our approximation of the main equation in the boundary value problem (1) we want to require that the function $v \in L^2(H)$ satisfy $Q(D)v(x) = 0$ for $x \in H$. However $Q(D)$ is a nonlocal operator, and for x near ∂H the domain of dependence of $Q(D)v(x)$ may include points of $\boldsymbol{R}^n \sim H$. Therefore we allow the possibility of modifying the main equation in a layer near ∂H; we require only that

$$Q(D)v(x) = 0 \quad \text{for } \langle x,N\rangle \geqq a \tag{1.4}$$

where $a \geqq 0$. By choosing a large enough we can arrange that the domain of dependence of $Q(D)v(x)$ is entirely contained in H for all x with $\langle x,N\rangle \geqq a$, but we make no assumption on the size of a in this paper; instead we assume that v is extended to be identically zero on $\boldsymbol{R}^n \sim H$ in cases where an evaluation of v outside H is needed. In the boundary layer $S = \{x : 0 \leqq \langle x,N\rangle < a\}$ we imagine that we are given some difference operator $q(x,D) = \sum \gamma_j(x)T_j$,

$$q(x,D)v(x) = \sum_{j\in\boldsymbol{Z}^n} \gamma_j(x)v(x+j) \quad \text{(finite sum)}$$

and in our approximation of (1) we supplement the main equation (1.4) by the boundary condition

$$q(x,D)v(x) = g(x) \quad \text{for } 0 \leqq \langle x,N\rangle < a,$$

where $g \in L^2(S)$. It is natural to suppose that $q(x,D)$ is translationally invariant along directions parallel to ∂H; thus we assume that the coefficients $\gamma_j(x)$ of this operator are bounded, measurable functions of $\langle x,N\rangle$.

The following theorem which extends a property of the differential equation (1) to its difference approximation

$$(1.5)\qquad \begin{aligned} Q(D)v(x) &= 0 && \text{for } x \in H \sim S, \\ q(x,D)v(x) &= g(x) && \text{for } x \in S, \end{aligned}$$

is the main result of this paper.

THEOREM. *If $Q(D)$ is a properly elliptic difference operator, the following two statements are equivalent.*

(i) *The homogeneous problem* (1.5) *with $g = 0$ has the unique solution zero in* $L^2(H)$.

(ii) *For a dense set of functions $g \in L^2(S)$, there is at least one solution of* (1.5) *in* $L^2(H)$.

It is perhaps disappointing that the uniqueness of a solution implies existence only for a dense set of data, but this is a consequence of the infinite-dimensional character of the difference equation. The simplest example of this behavior occurs if $q(x,D)$ is multiplication by a continuous function $m(\langle x,N\rangle)$ where m has isolated zeros in $[0,a]$.

2. **The von Neumann algebra.** In this section we introduce a von Neumann algebra $\mathscr{A}$ of difference operators on $L^2(\boldsymbol{R}^n)$ with variable coefficients depending only on $\langle x,N\rangle$. Let $\mathscr{A}_0$ be the set of operators A on $L^2(\boldsymbol{R}^n)$ which may be expressed as a finite sum

$$(2.1)\qquad A = \sum_{j \in \boldsymbol{Z}^n} m_j(\langle x,N\rangle)T_j,$$

where T_j is a multi-integer translation and $m_j \in L^\infty(\boldsymbol{R}^1)$ has compact support. We claim that $\mathscr{A}_0$ is a *-algebra—it is clear that $\mathscr{A}_0$ is closed with respect to linear combinations, and it follows from the observation that

$$(2.2)\qquad T_j m(\langle x,N\rangle) = m(\langle x + j,N\rangle)T_j$$

that $\mathscr{A}_0$ is closed with respect to products and adjoints. Let $\mathscr{A}$ be the weak closure of $\mathscr{A}_0$. If $\{\psi_k\}$ is an increasing sequence of nonnegative functions on the line having compact support and converging to 1 uniformly on compact sets, then $\{\psi_k(\langle x,N\rangle)\}$ is an approximate identity in $\mathscr{A}_0$; therefore $\mathscr{A}$ is a von Neumann algebra. Alternatively $\mathscr{A}$ may be characterized as the von Neumann algebra on $L^2(\boldsymbol{R}^n)$ generated by multi-integer translations and multiplication by functions of $\langle x,N\rangle$. It is often helpful to think of elements of $\mathscr{A}$ as weakly convergent infinite sums of the form (2.1), but we shall not use this idea in our analysis.

Consider the linear functional, defined on $\mathscr{A}_0$,

$$(2.3)\qquad \operatorname{tr}\left(\sum m_j T_j\right) = \int_{-\infty}^{\infty} m_0(t)\,dt,$$

where $m_0(\langle x,N\rangle)$ is the coefficient of the identity translation in the sum on the left of (2.3). (Since m_j is bounded and has compact support, there is no convergence problem in (2.3).) We note that

(2.4) $$\operatorname{tr} AB = \operatorname{tr} BA$$

if $A, B \in \mathscr{A}_0$, for by linearity it suffices to check (2.4) on a basis, and using (2.2) we see that

$$\operatorname{tr}(mT_j nT_k) = \operatorname{tr}(m(T_j nT_{-j})T_{j+k}) = \delta_{j,-k} \int_{-\infty}^{\infty} m(t)n(t + \langle j,N \rangle)\, dt$$

while

$$\operatorname{tr}(nT_k mT_j) = \operatorname{tr}(n(T_k mT_{-k})T_{j+k}) = \delta_{j,-k} \int_{-\infty}^{\infty} m(t - \langle j,N \rangle)n(t)\, dt.$$

In a similar vein it is readily seen that if $A = \sum m_j T_j$ is in $\mathscr{A}_0$, then

(2.5) $$\operatorname{tr}(A^*A) = \int_{-\infty}^{\infty} \sum_j |m_j(t)|^2\, dt;$$

in particular $\operatorname{tr} A^*A \geqq 0$ and $\operatorname{tr} A^*A = 0$ if and only if $A = 0$. The remainder of §2 is devoted to the proof of the following proposition which extends tr to $\mathscr{A}$. The reader may omit this proof without loss of continuity on a first reading of the paper.

PROPOSITION 2.1. *The linear functional defined on* $\mathscr{A}_0$ *by* (2.3) *may be extended to a faithful, semifinite, normal trace on* $\mathscr{A}$.

PROOF. We begin by showing that for $A \in \mathscr{A}_0$, $\operatorname{tr} A$ may be expressed as a sum of vector functionals $\operatorname{tr} A = \sum (Af_\nu, f_\nu)$, where $f_\nu \in L^2(\boldsymbol{R}^n)$. Let $\sum \varphi_\nu^2(t) = 1$ be a partition of unity on the line such that each term is supported in an interval of length $\frac{1}{3}$, and let f be a function on the $(n-1)$-dimensional subspace $V = N^\perp$ such that f is supported in the ball of radius $\frac{1}{3}$ in V and satisfies $\int_V |f(x')|^2\, dx' = 1$. We define $f_\nu(x) = \varphi_\nu(\langle x,N \rangle) f(\Lambda x)$, where $\Lambda : \boldsymbol{R}^n \to V$ is the orthogonal projection onto V. Then f_ν is supported in some cylinder

$$\{x \in \boldsymbol{R}^n : b_\nu \leqq \langle x,N \rangle \leqq b_\nu + \tfrac{1}{3}, |\Lambda x| \leqq \tfrac{1}{3}\},$$

and as this set is contained in a ball of radius $((\frac{1}{6})^2 + (\frac{1}{3})^2)^{1/2} = 5^{1/2}/6 < \frac{1}{2}$, it follows that for $j \neq 0$ the supports of $T_j f_\nu$ and f_ν do not intersect; thus $(mT_j f_\nu, f_\nu) = 0$ for $j \neq 0$. Moreover for any $b \geqq 0$ the support of f_ν intersects the strip $\{x : |\langle x,N \rangle| \leqq b\}$ only for finitely many ν; but if $m(t) = 0$ for $|t| \geqq b$ then $m(\langle x,N \rangle)T_j f_\nu = 0$ unless the support of f_ν intersects $\{x : |\langle x,N \rangle| \leqq b + |\langle j,N \rangle|\}$; since any operator $A \in \mathscr{A}_0$ is a finite sum $A = \sum m_j T_j$, where m_j has compact support, we conclude that $Af_\nu = 0$ for all except finitely many ν. Thus if $A = \sum m_j T_j \in \mathscr{A}_0$,

$$\sum_\nu (Af_\nu, f_\nu) = \sum_\nu (m_0 f_\nu, f_\nu)$$

$$= \sum_\nu \int_{-\infty}^{\infty} dt \int_V dx'\, m_0(t)\varphi_\nu^2(t)|f(x')|^2 = \int_{-\infty}^{\infty} m_0(t)\, dt.$$

Hence, for $A \in \mathscr{A}_0$,

$$\text{tr } A = \sum_\nu (Af_\nu, f_\nu). \tag{2.6}$$

We also note for use below that if $A = \sum m_j T_j \in \mathscr{A}_0$ then $\|Af_\nu\|^2 = (Mf_\nu, f_\nu)$ where M denotes multiplication by the function $\sum |m_j(\langle x - j, N\rangle)|^2$; therefore if ψ is a function on $\boldsymbol{R}^n$ with $|\psi| \leqq 1$, we have

$$\|A\psi f_\nu\|^2 \leqq \|Af_\nu\|^2. \tag{2.7}$$

For any nonnegative operator $A \in \mathscr{A}^+$ we define tr A by (2.6). It is clear that tr is homogeneous and additive on $\mathscr{A}^+$, so to prove that tr is a trace we need only show that tr $U^*AU =$ tr A if U is a unitary operator in $\mathscr{A}$. We prove this by an approximation argument for which the following lemma is crucial.

LEMMA 2.2. *For any nonnegative operator $A \in \mathscr{A}^+$ there is a sequence $\{A_k\}$ in $\mathscr{A}_0$ such that $\{A_k^* A_k\}$ converges strongly to A and* lim inf tr $A_k^* A_k \leqq$ tr A.

PROOF. Let $\{B_n\}$ be a sequence of selfadjoint operators in $\mathscr{A}_0$ converging strongly to $A^{1/2}$, and let $\{\psi_k\}$ be the approximate identity in $\mathscr{A}_0$ described above. Then by (2.7), for each k and ν,

$$\lim_{n\to\infty} \|B_n \psi_k f_\nu\|^2 \leqq \lim_{n\to\infty} \|B_n f_\nu\|^2 = \|A^{1/2} f_\nu\|^2. \tag{2.8}$$

Now for each k the product $\psi_k f_\nu$ is nonzero only for finitely many ν, say $\psi_k f_\nu = 0$ for $\nu > J_k$. By (2.8) for each k we may choose an integer $n_k \geqq k$ such that

$$\|B_{n_k} \psi_k f_\nu\|^2 \leqq \|A^{1/2} f_\nu\|^2 + 2^{-\nu}/k \tag{2.9}$$

for $\nu = 1, 2, \ldots, J_k$. Of course (2.9) also holds for $\nu > J_k$, since in this case the left-hand side of the inequality vanishes. If we define $A_k = B_{n_k}\psi_k$, then $A_k^* A_k = \psi_k B_{n_k}^2 \psi_k$ converges strongly to A and

$$\text{tr } A_k^* A_k = \sum_\nu \|B_{n_k}\psi_k f_\nu\|^2 \leqq \sum_\nu \|A^{1/2} f_\nu\|^2 + 1/k = \text{tr } A + 1/k,$$

so that lim inf tr $A_k^* A_k \leqq$ tr A. This completes the proof of the lemma.

We now prove that tr is unitarily invariant on $\mathscr{A}^+$. It suffices to prove that tr $U^*AU \leqq$ tr A for $A \in \mathscr{A}^+$ and $U \in \mathscr{A}$ unitary, since the reverse inequality follows from the fact that U is invertible. By the Kaplansky density theorem there is a sequence $\{B_k\}$ in $\mathscr{A}_0$ which converges strongly to U and satisfies $\|B_k\| \leqq 1$. Let A_k be a sequence with the properties given in the above lemma. Note that

$$\sum_\nu \|A_k B_k f_\nu\|^2 = \text{tr}(B_k^* A_k^* A_k B_k) = \text{tr}(A_k B_k B_k^* A_k^*)$$

$$= \sum_\nu \|B_k^* A_k^* f_\nu\|^2 \leqq \sum_\nu \|A_k^* f_\nu\|^2 = \text{tr } A_k A_k^* = \text{tr } A_k^* A_k.$$

Now by Fatou's lemma

$$\operatorname{tr} U^*AU = \sum_\nu \|A^{1/2}Uf_\nu\|^2 \leqq \liminf_{k\to\infty} \sum_\nu \|A_kB_kf_\nu\|^2$$
$$\leqq \liminf \operatorname{tr} A_k^*A_k \leqq \operatorname{tr} A.$$

Therefore tr is indeed a trace on $\mathscr{A}$. It is obvious from (2.6) that tr is normal; for any $A \in \mathscr{A}^+$, the operator $A^{1/2}\psi_k A^{1/2}$ is majorized by A and has finite trace, so tr is semifinite; thus it remains only to show that tr is faithful.

We claim that for any function $g \in C_c(\boldsymbol{R}^n)$, continuous with compact support, there exists a constant C such that

(2.10) $$\|Ag\|^2 \leqq C \operatorname{tr} A^*A$$

for all $A \in \mathscr{A}_0$. First suppose that g is supported in some ball of radius $\frac{1}{2}$ so that supp g does not intersect $T_j(\operatorname{supp} g)$ for $j \neq 0$. In this case, if $A = \sum m_jT_j$ then

$$\|Ag\|^2 = \int_{-\infty}^{\infty} dt \int_V dx'\, M(t)|g(x' + tN)|^2$$

where $M(t) = \sum_j |m_j(t - \langle j,N\rangle)|^2$. Now $\int_V dx'|g(x' + tN)|^2$ is uniformly bounded in t by some constant C so

$$\|Ag\|^2 \leqq C \int_{-\infty}^{\infty} M(t)\, dt = \operatorname{tr} A^*A.$$

In the case of general $g \in C_c(\boldsymbol{R}^n)$, g may be written as a finite sum $g = \sum_{k=1}^K g_k$ of functions with small support and we have

$$\|Ag\|^2 \leqq K^2 \max_k \|Ag_k\|^2 \leqq C' \operatorname{tr} A^*A$$

as claimed.

We may now prove that tr is faithful. If $A \in \mathscr{A}^+$ choose a sequence $\{A_k\} \subset \mathscr{A}_0$ as in Lemma 2.2. Then, for $g \in C_c(\boldsymbol{R}^n)$,

$$(Ag,g) = \lim \|A_kg\|^2 \leqq C \liminf \operatorname{tr} A_k^*A_k \leqq C \operatorname{tr} A.$$

Thus if $\operatorname{tr} A = 0$, then $Ag = 0$ for any g in the dense set $C_c(\boldsymbol{R}^n)$, so $A = 0$. The proof of Proposition 2.1 is complete.

3. **Proof of the main theorem.** Let $\mathscr{B}$ denote the restriction to $L^2(H)$ of the von Neumann algebra $\mathscr{A}$ defined in the preceding section; that is, $\mathscr{B} = E\mathscr{A}E$ where $E \in \mathscr{A}$ is the projection $E = \chi_{[0,\infty)}(\langle x,N\rangle)$. We shall continue to write T_j for translation by $j \in \boldsymbol{Z}^n$ restricted to $L^2(H)$, although more properly this operator should be written ET_jE. Because of this convention the translations $\{T_j\}$ no longer form a group; indeed it is easily seen that

(3.1) $$\begin{aligned} T_jT_k &= \chi_J(\langle x,N\rangle)T_{j+k}, && \text{if} \langle j,N\rangle < 0, \langle k,N\rangle > 0,\\ &= T_{j+k}, && \text{otherwise,}\end{aligned}$$

where χ_J is the characteristic function of the interval $J = [-\langle j,N\rangle,\infty)$. Of course the trace on $\mathscr{A}$ introduced in §2 restricts to a trace on $\mathscr{B}$, which we use to define a (relative) dimension function on $\mathscr{B}$. Note that if $\mathscr{X}_b = L^2\{x: 0 \leqq \langle x,N\rangle < b\}$ then $\mathscr{X}_b$ is a subspace associated to $\mathscr{B}$ and

$$\dim \mathscr{X}_b = \operatorname{tr} \chi_{[0,b)}(\langle x,N\rangle) = b;$$

in particular, if $b < \infty$, then $\dim \mathscr{X}_b < \infty$. Let $\mathscr{K}$ be the uniformly closed ideal in $\mathscr{B}$ generated by operators whose range has finite dimension. Following Breuer [**1**] we shall say that an operator $A \in \mathscr{B}$ is *Fredholm* (relative to $\mathscr{B}$) if A is invertible mod $\mathscr{K}$. It was shown by Breuer that if A is Fredholm, then $\dim\ker A < \infty$ and $\dim\ker A^* < \infty$, so that one may define a real-valued index on the set of Fredholm operators in $\mathscr{B}$,

$$i(A) = \dim\ker A - \dim\ker A^*. \tag{3.2}$$

Breuer also proved that the index so defined has the following properties:

$$\begin{aligned} &\text{(i)} && i(AB) = i(A) + i(B),\\ &\text{(ii)} && i(A^*) = -i(A),\\ &\text{(iii)} && i(A+K) = i(A) \text{ for } K \in \mathscr{K},\\ &\text{(iv)} && i(A+B) = i(A) \text{ for } \|B\| \text{ sufficiently small.}\end{aligned} \tag{3.3}$$

Suppose $Q(D) = \sum c_j T_j$ is a properly elliptic difference operator on $L^2(H)$. As an approximation of the main equation in the boundary value problem (1) we seek a function $v \in L^2(H)$ such that $Q(D)v(x) = 0$ for $\langle x,N\rangle \geqq a$, or more compactly, such that $\chi Q(D)v = 0$, where χ is the characteristic function of $H \sim S$. Of course $\chi Q(D) \in \mathscr{B}$, and the following lemma is the basis of our application of von Neumann algebras to the difference equation.

LEMMA 3.1. *$\chi Q(D)$ is Fredholm relative to $\mathscr{B}$ and has index zero.*

PROOF. If Φ is a continuous, multiply periodic function on $\boldsymbol{R}^n$, let $\Phi(D)$ be the operator which multiplies the Fourier transform of a function by Φ, restricted to $L^2(H)$; that is, let

$$\Phi(D) = E\mathscr{F}^{-1}\Phi\mathscr{F}E$$

where E is the projection onto $L^2(H)$ and $\mathscr{F}$ is the Fourier transform. We shall regard a periodic function Φ as a function on the torus and write $\Phi \in C(\boldsymbol{T}^n)$. It is clear that $\Phi(D)$ is a bounded operator on $L^2(H)$ with

$$\|\Phi(D)\| \leqq \|\Phi : C(\boldsymbol{T}^n)\|, \tag{3.4}$$

and we claim that $\Phi(D) \in \mathscr{B}$. If $\Phi(\xi) = \sum b_j e^{i\langle j,\xi\rangle}$ is an exponential polynomial, then $\Phi(D) = \sum b_j T_j$, so in this case $\Phi(D) \in \mathscr{B}$; for any $\Phi \in C(\boldsymbol{T}^n)$ there is a sequence of exponential polynomials $\{\Phi_k\}$ converging uniformly to Φ and by (3.4), $\{\Phi_k(D)\}$

converges uniformly to $\Phi(D)$; therefore $\Phi(D) \in \mathscr{B}$ as claimed. Since we have restricted these operators to $L^2(H)$, it is not true in general that $\Phi(D)\Psi(D) = \Phi\Psi(D)$; however these two operators are equal modulo the ideal $\mathscr{K}$, as we now prove. It is evident from (3.1) that

$$T_jT_k - T_{j+k} \in \mathscr{K}.$$

Therefore if Φ and Ψ are exponential polynomials then

$$\Phi(D)\Psi(D) - \Phi\Psi(D) \in \mathscr{K}, \tag{3.5}$$

and the same conclusion for general Φ,Ψ follows by a simple approximation argument.

Since $Q(D)$ is an elliptic difference operator, Q is a nonvanishing function on $\boldsymbol{T}^n$ and $Q^{-1} \in C(\boldsymbol{T}^n)$. By (3.5), $Q^{-1}(D)$ is an inverse of $Q(D)$ in $\mathscr{B}/\mathscr{K}$, so $Q(D)$ is Fredholm relative to $\mathscr{B}$. Now it follows from property (iv) in (3.3) that the index of a Fredholm operator is unchanged by a homotopy within the class of Fredholm operators. But if $Q(D)$ is properly elliptic, $Q(D)$ is homotopic to the identity, for in this case $\log Q$ is periodic and one such homotopy is provided by $\{\psi_t(D) : 0 \leq t \leq 1\}$, where

$$\psi_t(\xi) = \exp\{t \log Q(\xi)\}.$$

Therefore $Q(D)$ has index zero. Of course the projection χ is Fredholm with index zero, being of the form $I + K$ where $K \in \mathscr{K}$, and it follows from (i) of (3.3) that $\chi Q(D)$ is Fredholm with index zero. This completes the proof.

REMARK. In the proof of the following lemma we show that $Q(D)$ is invertible. Of course, Lemma 3.1 is an immediate consequence of this fact. We have preferred to give the independent proof above to emphasize the similarity of this approach to difference operators to the theory of pseudodifferential operators.

LEMMA 3.2. *The range of $\chi Q(D)$ is $L^2(H \sim S)$.*

PROOF. Certainly range $\chi Q(D) \subset$ range $\chi = L^2(H \sim S)$. On the other hand we show below that $Q(D)$ is invertible, so that range $\chi Q(D) =$ range χ.

We use a Wiener-Hopf argument to show that $Q(D)$ is invertible. Let $\mathscr{P}_+$ be the projection on $L^2(\boldsymbol{T}^n)$ defined by

$$\begin{aligned} \mathscr{P}_+[e^{i\langle j,\cdot\rangle}] &= e^{i\langle j,\cdot\rangle} \quad \text{if } \langle j,N\rangle \leq 0, \\ &= 0 \quad \text{if } \langle j,N\rangle > 0, \end{aligned}$$

and let $\mathscr{P}_- = I - \mathscr{P}_+$. We define

$$Q_\pm = \exp\{\mathscr{P}_\pm \log Q\}. \tag{3.6}$$

Since $Q(D)$ is properly elliptic, $\log Q$ is a smooth function on $\boldsymbol{T}^n$, and its Fourier coefficients are rapidly decreasing in j. Therefore (3.6) defines $Q_\pm$ as smooth functions on $\boldsymbol{T}^n$. Of course $Q(\xi) = Q_-(\xi)Q_+(\xi)$; note that the Fourier series of $Q_\pm$ contains only terms $e^{i\langle j,\cdot\rangle}$ with $\pm\langle j,N\rangle \leqq 0$; thus according to (3.1) the product $Q_-(D)Q_+(D)$, taken in this order, contains only terms T_jT_k for which the signs are such that $T_jT_k = T_{j+k}$, so $Q_-(D)Q_+(D) = Q(D)$. Also, because the Fourier coefficients of $Q_\pm$ vanish for $\pm\langle j,N\rangle > 0$, the almost periodic functions $t \to Q_\pm(\xi + tN)$ may be extended to bounded, nonvanishing, analytic functions in the half plane $\{\pm \operatorname{Im} z \leqq 0\}$. Since these analytic extensions are nonvanishing, their reciprocals are analytic, so the Fourier coefficients of $Q_\pm^{-1}$ also vanish for $\pm\langle j,N\rangle > 0$. Therefore, by (2.1),

$$Q_\pm(D)Q_\pm^{-1}(D) = Q_\pm^{-1}(D)Q_\pm(D) = I,$$

as no products T_jT_k occur where $\langle j,N\rangle$ and $\langle k,N\rangle$ have the opposite sign. Hence $Q(D) = Q_-(D)Q_+(D)$ is invertible, and the proof of the lemma is complete.

PROOF OF THE MAIN THEOREM. We recall that if $\mathscr{X}$ is a closed subspace of $L^2(H)$ associated to the von Neumann algebra $\mathscr{B}$ then, for any $A \in \mathscr{B}$,

$$\dim(\mathscr{X} \cap \ker A) + \dim \operatorname{Cl} A(\mathscr{X}) = \dim \mathscr{X}. \tag{3.7}$$

We apply (3.7) with $\mathscr{X} = \ker \chi Q(D)$. Since $\chi Q(D)$ is Fredholm with index zero we have $\dim \mathscr{X} < \infty$, and moreover using Lemma 3.2 we see that

$$\dim \mathscr{X} = \dim[\text{range } \chi Q(D)] = \dim L^2(S) = a.$$

First suppose that the homogeneous problem (1.5) has only the zero solution; in other words, suppose that $\mathscr{X} \cap \ker A = \{0\}$, where $A: L^2(H) \to L^2(S)$ is the operator in $\mathscr{B}$ defined by $q(x,D)$. By (3.7), $\operatorname{Cl} A(\mathscr{X})$ is a subspace of $L^2(S)$ whose dimension equals a, the dimension of $L^2(S)$. Since the trace on $\mathscr{B}$ is faithful, we conclude that $\operatorname{Cl} A(\mathscr{X}) = L^2(S)$, or that $A(\mathscr{X})$ is dense in $L^2(S)$. Thus (1.5) is soluble for a dense set of data in $L^2(S)$.

On the other hand, if (1.5) is soluble for a dense set of data in $L^2(H)$, this means that $\operatorname{Cl} A(\mathscr{X}) = L^2(S)$, and by (3.7) it follows that $\dim(\mathscr{X} \cap \ker A) = 0$. Since the trace is faithful on $\mathscr{B}$, we have $\mathscr{X} \cap \ker A = \{0\}$, or the homogeneous problem (1.5) has only the trivial solution zero. This completes the proof of our main theorem.

REFERENCES

1. M. Breuer, *Fredholm theories in von Neumann algebras*. I, II, Math. Ann. **178** (1968), 243–254; ibid. **180** (1969), 313–325. MR **38** #2611; MR **41** #9002.

2. L. Coburn, R. Douglas, D. Schaeffer and I. Singer, *C*-algebras of operators on a half-space*, Inst. Hautes Études Sci. Publ. Math. (to appear).

3. J. Dixmier, *Les algèbres d'opérateurs dans l'espace hilbertien*, Gauther-Villars, Paris, 1969.

4. B. Ja. Levin, *Distribution of zeros of entire functions*, GITTL, Moscow, 1956; English transl., Transl. Math. Monographs, vol. 5, Amer. Math. Soc., Providence, R.I., 1964. MR **19**, 402; MR **28** #217.

5. J. L. Lions and E. Magenes, *Problèmes aux limites non homogènes et applications*. Vol. 1, Travaux et Recherches Mathématiques, no. 17, Dunod, Paris, 1968. MR **40** #512.

6. D. Schaeffer, *An application of von Neumann algebras to finite difference equations*, Ann. of Math. **95** (1972), 117–129.

MASSACHUSETTS INSTITUTE OF TECHNOLOGY

EVOLUTION EQUATIONS NOT OF CLASSICAL TYPE AND HYPERDIFFERENTIAL OPERATORS

STANLY STEINBERG

In this paper we are interested in equations of the form

$$\frac{\partial u(t)}{\partial t} = Au(t), \qquad u(0) = u_0, \tag{1}$$

where u is to be a Banach space valued function. In particular we will be interested in the two examples [**1**], [**2**], [**3**]

$$A = \frac{\partial^2}{\partial x^2} - \frac{\partial^2}{\partial y^2} - x\frac{\partial}{\partial x} + y\frac{\partial}{\partial y}, \tag{2}$$

$$A = \sum_{n=0}^{\infty} a_n \left(\frac{\partial}{\partial x}\right)^n, \tag{3}$$

where a_n is a sequence of complex numbers.

In (2), A is hyperbolic and, in (3), A is of infinite order and consequently Problem 1 is not going to be of classical type.

To study existence and uniqueness for Problem 1 we need to introduce some functional analysis. After that we will discuss the possibility of an *integral* representation for the solutions of Problem 1. To do this we introduce the notion of a symbolic calculus for hyperdifferential operators.

Functional analysis. We first remark that the results we are going to discuss here have appeared in [**4**], [**5**] and in a substantially more general form in [**6**], [**7**].

DEFINITION. Let I be an interval (open, closed, half open, bounded or unbounded) and, for each $s \in I$, let X_s be a Banach space with norm $\|\cdot\|_s$. If, for $s', s \in I$, $s' < s$, we have $X_s \subset X_{s'}$ and $\|x\|_{s'} \leqq \|x\|_s$, then X_s is said to be a scale of Banach spaces on I. (Sometimes it is also required that X_s be dense in $X_{s'}$.)

AMS 1970 *subject classifications*. Primary 35A05; Secondary 35C15.

EXAMPLE. Let $I = (0, \infty)$, $X_s = \{f(z) | f$ is analytic for $|z| < s$, f is bounded for $|z| \leqq s\}$, and $\|f\|_s = \sup_{|z| \leqq s} |f(z)|$.

DEFINITION. Let X_s be a scale on I and let A be a bounded operator from X_s to $X_{s'}$ for $s' < s$, $s', s \in I$. If

$$\|Ax\|_{s'} \leqq C(s - s')^{-d} \|x\|_s,$$

then A is said to be of type d. We restrict d so that $0 \leqq d \leqq 1$.

THEOREM. *If A is of type* 1 *on X_s, then*

$$\partial u(t)/\partial t = Au(t), \qquad u(0) = u_0 \in X_{s_0}$$

possesses a unique solution $u(t) \in X_s$, $s < s_0$ for $|t| < C(s_0 - s)$ where C is a constant. If A is of type d, $d < 1$, then the solution exists for all time.

PROOF. One computes that

$$\|A^n x\|_{s'} \leqq C^n (s - s')^{-n} n^n \|x\|_s$$

and consequently the series for e^{At} converges. The uniqueness goes as in ordinary differential equations.

EXAMPLE. If $f(z)$ is analytic near $z = 0$ and we set $\|f\|_s = \sum_p |f^{(p)}(0)| s^p (p!)^{-1/2}$, $f^{(p)}(0) = d^p f(0)/dz^p$, and $X_s = \{f | \|f\|_s < \infty\}$, then we can compute

$$\|df/dz\|_{s'} \leqq C(s - s')^{-1/2} \|f\|_s,$$

$$\|zf\|_{s'} \leqq C(s - s')^{-1/2} \|f\|_s,$$

where $0 \leqq a < s' < s \leqq b < \infty$. The obvious generalization of this example to two variables [**1**] permits one to solve Problem 1 when A is given by (2).

These techniques have been used to study the Cauchy-Goursat problem [**8**], and infinite systems of ordinary differential equations [**9**]. It is also possible to use spaces of Gevrey functions instead of analytic functions [**10**]. The connection between this material and semigroup theory and the theory of analytic vectors has been studied in [**11**]. Finally, this theory essentially started in F. Trèves' [**6**] extensive study of the Cauchy-Kovalevska and Holmgren's theorems using generalizations of the spaces introduced in the first example.

Hyperdifferential operators. Hyperdifferential operators are formally similar to pseudo-differential operators; however, there are many technical differences. They are designed to study operators of the form e^{At} where A is given by (2) or say $-d^2/dx^2$ (the backwards heat equation). Such operators are clearly not pseudo-differential or even Fourier integral [**12**]. In general hyperdifferential operators operate on spaces of analytic functions or analytic functionals while pseudo-differential operators operate on Sobolev spaces of functions.

We want hyperdifferential operators to be given by an expression of the form

$$A = \sum_{m=0}^{\infty} \sum_{n=0}^{\infty} a_{mn} x^m \left(\frac{d}{dx}\right)^n.$$

We define the symbol of A by

$$\sigma A(x, \xi) = e^{-\xi x} A e^{-\xi x} = \sum \sum a_{mn} x^m \xi^n$$

which is in general an analytic function in (x, ξ) [**6**], [**13**], [**14**].

For certain f we have the representation

$$Af(x) = (2\pi)^{-1/2} \int_{-\infty}^{\infty} e^{ix\xi} \sigma A(x, i\xi) \hat{f}(\xi)\, d\xi,$$

$$\hat{f}(\xi) = (2\pi)^{-1/2} \int e^{-ix\xi} f(x)\, dx,$$

which we refer to as the integral representation.

In general we can compute the symbol of the compose by

$$\sigma(A \circ B)(x, \xi) = \sum_{\alpha=0}^{\infty} (\alpha!)^{-1} \left(\frac{\partial}{\partial \xi}\right)^{\alpha} \sigma A(x, \xi) \left(\frac{\partial}{\partial x}\right)^{\alpha} \sigma B(x, \xi)$$

although there are many technical difficulties. There is a similar formula for the symbol of the transpose of an operator.

These operators have been used to study first order systems with analytic coefficients [**6**], [**13**], and to study [**2**], [**13**] (in fact, compute explicit) solutions to equations governed by (1) and (2). In [**13**] the symbolic calculus is used to rederive many formulas in Weyl's quantization and normal ordering theories.

References

1. M. Miller, *Dynamics in diagonal coherent state representation*, J. Mathematical Phys. **10** (1969), 1406.

2. S. Steinberg and F. Trèves, *Pseudo-Fokker equations and hyper-differential operators*, J. Differential Equations **8** (1970), 333–366.

3. S. Steinberg, *The Cauchy problem for differential equations of infinite order*, J. Differential Equations **9** (1971), 591–607.

4. I. M. Gel'fand and G. E. Šilov, *Generalized functions*. Vol. III: *Some questions in the theory of differential equations*, Fizmatgiz, Moscow, 1958; English transl., Academic Press, New York, 1967. MR **21** #5142b; MR **36** #506.

5. L. V. Ovsjannikov, *A singular operator in a scale of Banach spaces*, Dokl. Akad. Nauk SSSR **163** (1965), 819–822 = Soviet Math. Dokl. **6** (1965), 1025–1028. MR **32** #8164.

6. F. Trèves, *Ovsjannikov theorem and hyperdifferential operators*, Instituto de Mathematica Pura e Aplicada, Rio de Janeiro, 1968.

7. ———, *Banach filtrations*, Purdue University, Lafayette, Ind. (preprint). (This paper discusses sequences of Banach spaces.)

8. P. Du Chateau, University of Kentucky, Lexington, Ky. (in preparation).

9. S. Steinberg, *Infinite systems of ordinary differential equations with unbounded coefficients and moment problems*, J. Math. Anal. Appl. (to appear).

10. P. Du Chateau, *An abstract non-linear Cauchy-Kovalevska theorem with applications to Cauchy problems with coefficients and data of Gevrey type*, Thesis, Purdue University, Lafayette, Ind., 1970.

11. S. Steinberg, *Local groups and analytic vectors* (in preparation).

12. L. Hörmander, *Fourier integral operators*, Acta Math. **127** (1971), 79–183.

13. F. Trèves, *Hyper-differential operators in complex spaces*, Bull. Soc. Math. France **97** (1969), 193–223. MR **41** #6014.

14. M. Miller and S. Steinberg, *Applications of hyperdifferential operators to quantum mechanics*, Comm. Math. Phys. **25** (1971), 40–60.

PURDUE UNIVERSITY

THE CHANGE IN SOLUTION DUE TO CHANGE IN DOMAIN

GILBERT STRANG AND ALAN E. BERGER[1]

1. Suppose we are given Poisson's equation $-\Delta u = f$ in a convex plane domain Ω, with the Dirichlet condition $u = 0$ at the smooth boundary Γ. Then the change in domain which we have in mind is easy to describe: *Ω is to be replaced by an inscribed polygon Ω_h*. The problem is to understand

(i) how much u_h, the solution to the Dirichlet problem on Ω_h, differs from u;

(ii) how this error $u_h - u$ behaves in the interior of Ω_h.

These questions arise naturally—in fact they cannot be avoided—in studying the convergence of discrete approximations to the Dirichlet problem. We shall take this opportunity to describe the approximation technique which has recently become the favorite; it is known as the *finite element method*. Fortunately, this method is not only a great success in actual computations, but also has enough structure to permit a decent mathematical analysis. The method normally begins by triangulating the region Ω, and in most cases the skin $\Omega - \Omega_h$ is simply thrown away. Hence our interest in questions (i) and (ii).

We shall discuss only second-order Dirichlet problems in the plane. Some of the extensions to more general problems are straightforward—the error estimates extend also to Neumann problems and to domains in R^n—but not all. In fact there is a fourth-order example discovered by Babuska [**5**, p. 177] which illustrates that u_h may actually fail to converge to u as h (the maximum length of the edges of Ω_h) approaches zero. For the Dirichlet problem this cannot happen, and convergence can be established by a classical argument based on Green's functions [**4**]. (We note also that Hadamard and Bergman-Schiffer [**1**] were able, for smooth variation of the boundary Γ, to compute the *derivative* of the error e_h with respect to move-

---

AMS 1970 *subject classifications*. Primary 65N15, 65N30, 35A35, 35A40; Secondary 35J05.

[1] This research has been supported by the National Science Foundation under contract GP-13778 and a predoctoral fellowship.

ment of the boundary, i.e. to solve the infinitesimal problem.) The essential point, however, is to find a quantitative estimate for the error with finite change of Γ, in other words *to compute the rate* h^α *at which* $u - u_h$ *decreases.*

We begin with some rough estimates. Since $-\Delta u = f = -\Delta u_h$ in Ω_h, and u_h vanishes on the boundary Γ_h, the difference e_h satisfies

$$(1) \qquad -\Delta e_h = 0 \quad \text{in } \Omega_h, \qquad e_h = u \text{ on } \Gamma_h.$$

Therefore we have first to estimate the boundary data u on Γ_h, and then to examine the consequences for e_h in the interior.

It is easy to check that any point on Γ_h lies within a distance ch^2 of the outer boundary Γ, c depending only on Γ. Since u vanishes on this outer boundary, the mean-value theorem yields

$$(2) \qquad |e_h(x)| = |u(x)| \leqq ch^2 \sup_\Omega |\text{grad } u|, \qquad x \text{ on } \Gamma_h.$$

Then by the maximum principle, this estimate extends to the whole domain Ω_h:

$$(3) \qquad \sup_{\Omega_h} |e_h| \leqq ch^2 \sup_\Omega |\text{grad } u|.$$

The more interesting and more difficult problem is to estimate the derivatives of the error. Here the natural measure, for applications to variational problems, is the Dirichlet integral:

$$D(u - v) = \iint_{\Omega_h} [(u - v)_x^2 + (u - v)_y^2] \, dx \, dy.$$

Suppose for a moment that u is fixed, and this integral is minimized over all v in $H_0^1(\Omega_h)$, i.e. over all functions which satisfy $v = 0$ on Γ_h and $D(v) < \infty$. Then the Euler equation for the minimizing function coincides exactly with the original equation for u_h. This means that $v = u_h$ is the element of $H_0^1(\Omega_h)$ which is closest to u:

$$(4) \qquad D(u - u_h) = \inf_{v \text{ in } H_0^1(\Omega_h)} D(u - v).$$

Therefore any choice of v will yield an upper bound to the actual error $D(u - u_h)$. This is the technique we shall ultimately use to estimate $D(e_h)$, with a choice of v which is suggested by the finite element method.

First, however, we look at the standard estimates for the solution of (1) in terms of the data:

$$(5) \qquad D(e_h) \leqq C(\Gamma_h)|u|_{1/2,\Gamma_h}^2 \leqq C'(\Gamma_h)|u|_{0,\Gamma_h}|u|_{1,\Gamma_h}.$$

Provided u behaves correctly at the corners of Γ_h, the last two norms can be computed by integrating $|u|^2$ and $|u_s|^2$ (u_s = tangential derivative) around the boundary Γ_h. For u itself, we already have the estimate (2), implying

(6) $$|u|_{0,\Gamma_h} = O(h^2).$$

Therefore we need to estimate u_s on Γ_h. Roughly speaking, since u goes from zero to $O(h^2)$ and back along each chord, and the chord length is $O(h)$, we expect the derivative u_s to be of order h. This is confirmed by computing the angle θ between Γ_h and the tangent to Γ at the point P in the figure:

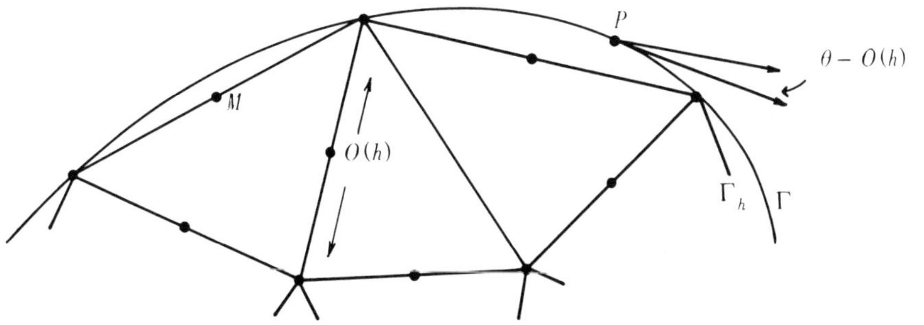

This angle is $O(h)$, and the derivative of u vanishes in the direction of the tangent. Therefore another application of the mean-value theorem, assuming enough smoothness of u, leads to the bound

(7) $$|u|_{1,\Gamma_h} = O(h).$$

These estimates suggest that *the mean square error in the derivatives is of order* h^3:

(8) $$D(u - u^h) = O(h^3).$$

Unfortunately, although this is exactly the result we want, the foregoing arguments hardly constitute a proof. It is still to be established that the constant C' in (5) can be chosen independent of h, and the quantity $|u|_{1,\Gamma_h}$ requires more careful estimation. We believe these steps can be carried out, but we prefer the simpler proof given in the next section.

Before describing that proof, we want to comment on the significance of the estimate (8). Its most striking feature is the following: although the derivatives are of order h around the boundary, as in (7), our estimate means that their average order in the L_2 sense over Ω_h is $h^{3/2}$. In other words, *the derivatives of* $u - u_h$ *must be much smaller in the interior than on the boundary.* Of course this is a reflection of the interior compactness property of elliptic equations, that the solution in the interior is smoother than the boundary data. This property is made explicit by the interior Schauder estimates, and V. Thomée has observed that these estimates lead to a quite different proof of (8). His forthcoming paper will study the behavior of $u - u_h$ in much greater detail and in a variety of norms.

There is still one more viewpoint from which the decay of the derivatives can be investigated. As an illustration we take Laplace's equation on the unit circle: if the boundary data is $\sum c_n e^{in\theta}$, then the solution is $\sum c_n r^{|n|} e^{in\theta}$. It follows that highly oscillatory data, associated with frequencies $|n| \gg 1$, is rapidly damped in the interior. In fact if $n = O(1/h)$, there is a kind of *boundary layer*; each step of length h away from the boundary, from $r = 1$ to $r = 1 - h, 1 - 2h, \ldots$, reduces the solution by more than some constant factor $\alpha > 1$. We noticed earlier that the function u oscillates in exactly this way, between 0 and $O(h^2)$, on each chord of length h. Since u is the boundary data for e_h, the decay of e_h will be very quick—*except for the term giving the mean value of u on the chords*, which is not damped at all. The leading term in the expansion of e_h in the interior appears to be the harmonic function w whose boundary data is

$$w|_\Gamma = \frac{1}{12}\kappa(x)h^2 \frac{\partial u}{\partial n}(x),$$

where h is a uniform chord length, and κ is the curvature of Γ; this is approximately the mean value of u along chords near x. We intend at another time to examine more closely the connection between rapid oscillation of boundary data, and the transition by means of a boundary layer from this data to the very regular solution inside.

2. **The finite element method.** The goal of this method is to approximate the solution to certain elliptic boundary-value problems, and the main ideas can be described in a few words. First the elliptic problem is formulated variationally, in terms of minimizing a quadratic functional $I(v)$ over a class V of admissible functions. The classical Rayleigh-Ritz-Galerkin method is to minimize instead over a finite-dimensional subspace $S_h \subset V$, and *the finite element method is a computationally convenient choice of* S_h. Normally the elements of S_h are piecewise polynomial functions of comparatively low degree. The accuracy of the approximation is improved, not as in the classical (and virtually noncomputable) theory by introducing new and more complex basis functions, but simply by reducing the size of the pieces, or "elements," into which the domain is divided.

We give two examples of the construction of these piecewise polynomials; the second will provide a function v which can be used in (4) to demonstrate that $D(e_h) = O(h^3)$. In both examples we inscribe a polygon Ω_h in Ω, and then subdivide Ω_h into triangles of diameter less than h. (A vertex of one triangle is not permitted to be along an edge of another.) A typical element v of S_h will reduce to a different polynomial within each of the triangles, but it will be continuous across the edges between triangles and vanish on and outside of the polygonal boundary Γ_h. This means that the Dirichlet condition $v = 0$ is satisfied on both Γ and Γ_h, and that the quadratic functional

$$I(v) = \iint (v_x^2 + v_y^2 - 2fv)\,dx\,dy \tag{9}$$

is finite. In other words S_h may be regarded as a subspace both of $V = H_0^1(\Omega)$, which is the admissible space for the original Poisson problem $-\Delta u = f$ in Ω, and of $V_h = H_0^1(\Omega_h)$, the admissible space for the same problem on Ω_h. $I(v)$ is minimized over V by the function u, and over V_h by the function u_h; we shall denote by U_h the minimizing element of S_h. Thus U_h is simultaneously a Rayleigh-Ritz-Galerkin approximation to u and u_h.

Now we construct the spaces S_h. In the simplest example the polynomials are linear: $v = a_1 + a_2 x + a_3 y$ within each triangle. The space S_h then consists of all continuous piecewise linear functions which vanish on Γ_h, and the first question is to find a basis for this finite-dimensional space. Suppose we number the vertices of the triangles by $z_1, \ldots, z_d$, including only those vertices which lie in the interior of Γ_h. Then to each z_j we associate the piecewise linear function φ_j which equals one at z_j and zero at all other vertices (as well as the boundary Γ_h):

$$\varphi_j(z_i) = \delta_{ij}. \tag{10}$$

The function φ_j is uniquely determined within each triangle, since its values at the three vertices are given. Furthermore φ_j is automatically continuous between triangles; each pair of adjacent triangles shares the two vertices which lie on the common edge, and the values of φ_j at these two vertices determine the linear function φ_j along the whole edge. Thus φ_j is in S_h.

These functions $\varphi_1, \ldots, \varphi_d$ form a basis for the space S_h: every v in S_h can be expanded as

$$v(x, y) = \sum_1^d v(z_j)\varphi_j(x, y). \tag{11}$$

We have only to verify that the two sides agree at every vertex, which they do, and then the values at the vertices determine the function within each triangle. This space S_h was proposed by Courant [**3**] in 1943 as being convenient for the minimization of the functional $I(v)$. Substituting (11) into (9), the integrals are as easy as possible to compute, and $I(v)$ reduces to a quadratic function of the scalars $v(z_1), \ldots, v(z_d)$. The minimization of such a quadratic yields a system of d linear equations, with a coefficient matrix which is symmetric, positive-definite, and sparse, i.e. most of the coefficients are zero. Such a system can be solved efficiently even with $d = 10^4$ (not by Cramer's rule!).

The chief drawback to Courant's suggestion is the inaccuracy of piecewise linear approximation. Therefore the development of the method—which was carried out by structural engineers, since everyone else was occupied with the alternative technique of finite differences—has led to polynomials of higher degree. Our second example will be the space of continuous piecewise quadratic functions: $v = a_1 + a_2 x + a_3 y + a_4 x^2 + a_5 xy + a_6 y^2$ within each element. Since there are six coefficients, we now include the midpoints of the edges (see the figure above) together with the vertices in the set of nodes z_j. The values of v at all these nodes uniquely determine a quadratic polynomial in each triangle; the numbering $z_1, \ldots, z_d$ again includes only those nodes which are inside Ω_h, since on the boundary v is always constrained to vanish.

Again we associate with each node z_j the piecewise quadratic function φ_j determined by

$$\varphi_j(z_i) = \delta_{ij}.$$

As before, the continuity of φ_j between triangles is assured: φ_j is a quadratic in one variable along the edge, and is completely determined by the three nodes (two vertices and one midpoint) which lie on the edge and are shared by the two adjacent triangles. Furthermore, the functions $\varphi_1, \ldots, \varphi_d$ form a basis for S_h, the dimension of the space equals the number of interior nodes, and the expansion $v = \sum v(z_j)\varphi_j$ is again valid.

A similar construction is possible for cubics, quartics,... in any number of variables $x_1, \ldots, x_n$. The quadratic example, however, is sufficiently accurate to establish that $D(e_h) = O(h^3)$; this would not be so in the linear case, where the error due to change in domain is submerged in the errors that arise in piecewise linear approximation.

3. **The error in the Dirichlet problem.** We propose to construct a function v_h in $H_0^1(\Omega_h)$ which satisfies $D(u - v_h) = O(h^3)$. It then follows immediately from (4) that $D(e_h) = O(h^3)$.

The function v_h will be continuous and piecewise quadratic, i.e. it will lie in the second of our finite element spaces S_h. It is determined by the following interpolation rule:

$$v_h(z_j) = u(z_j) \quad \text{or} \quad v_h = \sum_1^d u(z_j)\varphi_j(x, y). \tag{12}$$

Thus $v_h = u$ at all the nodes (vertices and midpoints) interior to Γ_h. Of course v_h must vanish along the boundary, to lie in $H_0^1(\Omega_h)$, and therefore v_h agrees with u also at boundary vertices. Only at the boundary midpoints, such as the point M in the figure drawn earlier, is there a discrepancy between v_h and u; the former vanishes and the latter, according to (2), is bounded by

$$|u(M)| \leqq ch^2 \sup_\Omega |\text{grad } u|. \tag{13}$$

It remains to compute $D(u - v_h)$. If v_h interpolated u at *all* vertices and midpoints, including the boundary midpoints M, this would be a standard estimate in approximation theory. In fact if we denote this genuine interpolate by v_I, then

$$D(u - v_I) \leqq C^2h^4\|u\|^2_{H^3(\Omega)}. \tag{14}$$

This estimate is given by Bramble and Zlamal [**2**] and by the first author [**6**]. The constant C depends on the lower bound of the angles which appear in the triangulation; we may assume that all these angles exceed a fixed $\theta_0 > 0$, and in this case C is independent of h. This means that if u is in H^3, i.e. if the original data f is in H^1, then the difference between u and v_I is even of higher order—h^4 rather than h^3 —than we require.

The final step is now to estimate $D(v_h - v_I)$. The functions v_h and v_I are identical on all triangles except those which contain a boundary midpoint M. On such a triangle T the contribution to $D(v_h - v_I)$ is

$$|u(M)|^2 \iint_T (\varphi_x^2 + \varphi_y^2)\, dx\, dy,$$

where φ is the quadratic polynomial which equals one at M and zero at the other nodes. It is easy to show that h grad φ is bounded by a constant, so that the expression above is bounded by $c'h^4 \sup |\mathrm{grad}\, u|^2$. Since the number of such boundary triangles is of order $1/h$, we conclude that

$$D(v_h - v_I) \leqq c''h^3 \sup |\mathrm{grad}\, u|^2. \tag{15}$$

Combined with (14), we finally have the required estimate

$$D(u - v_h) \leqq 2D(u - v_I) + 2D(v_I - v_h) = O(h^3).$$

One further comment: since the Ritz solution U_h is the element of S_h which is closest to u, this argument establishes at the same time the rate of convergence of U_h to u. The whole result can be stated very concisely.

THEOREM I. $D(u - u_h) \leqq D(u - U_h) \leqq D(u - v_h) = O(h^3)$.

REFERENCES

1. S. Bergman and M. Schiffer, *Kernel functions and elliptic differential equations in mathematical physics*, Academic Press, New York, 1953. MR **14**, 876.

2. J. Bramble and M. Zlamal, *Triangular elements in the finite element method*, Math. Comp. **24** (1970), 809–820.

3. R. Courant, *Variational methods for the solution of problems of equilibrium and vibrations*, Bull. Amer. Math. Soc. **49** (1943), 1–23. MR **4**, 200.

4. R. Courant and D. Hilbert, *Methods of mathematical physics*. Vol. II: *Partial differential equations*, Interscience, New York, 1962. MR **25** #4216.

5. J. Nečas, *Les méthodes directes en théorie des équations elliptiques*, Masson, Paris; Academia, Prague, 1967. MR **37** #3118.

6. G. Strang, *Approximation in the finite element method*, Numer. Math. **19** (1972), 81–98.

7. G. Strang and G. Fix, *An analysis of the finite element method*, Prentice-Hall, Englewood Cliffs, N. J., 1973.

MASSACHUSETTS INSTITUTE OF TECHNOLOGY

VARIATIONS OF KORN'S AND SOBOLEV'S INEQUALITIES

MONTY J. STRAUSS

Let Ω be a bounded domain in R^n and $u = (u_1, \ldots, u_n)$ be a real vector-valued function of $x = (x_1, \ldots, x_n)$ defined in Ω. Denote $D_j = \partial/\partial x_j$ for $j = 1, \ldots, n$. It is frequently necessary in the theory of elasticity to estimate the norm of u (in some sense) in terms of norms of linear combinations of the derivatives $D_i u_j$ and several inequalities in this direction are known. Recent work of Duvaut and Lions [**1**] requires a two-dimensional special case of a so far unknown inequality which we will prove in two forms.

THEOREM 1. *For all u in $H_0^{1,p}(\Omega)$,*

$$\|u\|_{L^p(\Omega)} \leqq C \sum_{i,j=1}^{n} \|D_j u_i + D_i u_j\|_{L^1(\Omega)}.$$

THEOREM 2. *Let Ω be a convex subset of R^2 with piecewise C^1 boundary. Then, for all u in $H^{1,2}(\Omega)$ such that $\int_\Omega u_1 = \int_\Omega u_2 = 0$, there exists a function $\alpha(x)$ equal to ± 1 at each point $x \in \Omega$ such that*

$$\|u\|_{L^2(\Omega)} \leqq C[\|D_1 u_1\|_{L^1(\Omega)} + \|D_2 u_2\|_{L^1(\Omega)} + \|D_2 u_1 + \alpha D_1 u_2\|_{L^1(\Omega)}].$$

We use the notation that $p = n/(n-1)$ throughout and $H^{1,p}(\Omega)$ $(H_0^{1,p}(\Omega))$ is the completion of the space of functions in $C^1(\Omega)$ $(C_0^1(\Omega))$ with respect to the norm

$$\|u\|_{1,p}^{\Omega} = \|u\|_{L^p(\Omega)} + \|\text{grad}\, u\|_{L^p(\Omega)}.$$

All constants C are independent of u, but may depend on anything else.

The classical Korn's inequality (see Friedrichs [**4**]) states that

$$\|u\|_{H^{1,2}(\Omega)} \leqq C \sum_{i,j=1}^{n} \|D_i u_j + D_j u_i\|_{L^2(\Omega)}$$

AMS 1970 *subject classifications*. Primary 35F05; Secondary 35B45.

if either $u = 0$ on $\partial\Omega$ or $\int_\Omega D_i u_j - D_j u_i = 0$ for $i, j = 1, \ldots, n$. A Sobolev inequality (see Friedman [3]) says that

$$\|u\|_{L^p(R^n)} \leqq C \sum_{i=1}^{n} \prod_{j=1}^{n-1} \|D_j u_i\|_{L^1(R^n)}^{1/n} \quad \left(\leqq C \sum_{i,j=1}^{n} \|D_j u_i\|_{L^1(R^n)} \right)$$

for any u in $H_0^{1,p}(R^n)$. Finally, de Figueiredo [2] showed that instead of needing $\frac{1}{2}n(n + 1)$ terms on the right side in Korn's inequality, one only has to use $2n - 1$: for he proved that

$$\|u\|_{H_0^{1,2}(\Omega)}^2 \leqq C \int_\Omega \left(\sum_{i=1}^{n} |D_i u_i|^2 + \sum_{k=1}^{n-1} |(\lambda^k \cdot D)(\lambda^k \cdot u)|^2 \right) dx,$$

where the λ^k are n-dimensional vectors with real components such that any minor (of any order) of the matrix

$$\begin{pmatrix} \lambda_1^1 & \cdots & \lambda_n^1 \\ \vdots & & \vdots \\ \lambda_1^{n-1} & \cdots & \lambda_n^{n-1} \end{pmatrix}$$

is nonsingular.

Thus, Korn's inequality bounds $\|D_i u_j\|_{L^2}$ and $\|u\|_{L^2}$ by a small number of L^2 norms involving the $D_i u_j$ and the Sobolev inequality bounds $\|u\|_{L^p}$ by the sum of the $\|D_i u_j\|_{L^1}$. What we do is to bound $\|u\|_{L^p}$ by a small number of L^1 norms involving the $D_i u_j$, and thus produce a hybrid between the two.

We will remark also how the same proof of Theorem 1 with a more judicious collection of terms actually implies

$$\|u\|_{L^p(\Omega)} \leqq C \left(\sum_{i=1}^{n-1} \|D_i u_i\|_{L^1(\Omega)} + \sum_{k=1}^{n-1} \left\| \sum_{i,j=1;\, i \neq k, j \neq k}^{n} D_i u_j \right\|_{L^1(\Omega)} + \left\| \sum_{i,j=1}^{n} D_i u_j \right\|_{L^1(\Omega)} \right),$$

which reduces the $\frac{1}{2}n(n + 1)$ terms of our original inequality to $2n - 1$, analogous to de Figueiredo's reduction of the number of terms involved in Korn's inequality.

PROOF OF THEOREM 1. We may assume without loss of generality that $u \in C_0^1(\Omega)$ since the class of such functions is dense in $H_0^{1,p}(\Omega)$. We may also extend u to be identically zero outside Ω. Then let $x = (x_1, \ldots, x_n) \in R^n$. We have

$$\sum_{i=1}^{n} u_i(x) = \frac{1}{n} \int_{-\infty}^{0} \sum_{i,j=1}^{n} D_i u_j(x + s\delta_n)\, ds = -\frac{1}{n} \int_0^\infty \sum_{i,j=1}^{n} D_i u_j(x + s\delta_n)\, ds,$$

where $\delta_n = (1, 1, \ldots, 1)$ and the integration is performed along the line $\{x_1 + s, x_2 + s, \ldots, x_n + s\}$. Hence,

$$\left| \sum_{i=1}^{n} u_i(x) \right| \leqq \frac{1}{4n} \int_{-\infty}^{\infty} |D_i u_j(x + s\delta_n) + D_j u_i(x + s\delta_n)|\, ds;$$

i.e.

$$\left|\sum_{i=1}^{n} u_i(x)\right|^{1/(n-1)} \leqq C\left(\int_{-\infty}^{\infty} \sum_{i,j=1}^{n} |D_i u_j(x + s\delta_n) + D_j u_i(x + s\delta_n)|\, ds\right)^{1/(n-1)}$$

Next, let $e_k = (0, 0, \ldots, 0, 1, 0, \ldots, 0) =$ the unit vector along the x_k-axis and let $f_k = \delta_n - e_k = (1, 1, \ldots, 1, 0, 1, \ldots, 1)$ for $1 \leqq k \leqq n - 1$. Then

$$\begin{aligned}\sum_{i=1}^{n} u_i(x) &= \frac{1}{n-1}\int_{-\infty}^{0} \sum_{i,j=1;\, i\neq k, j\neq k}^{n} D_i u_j(x + se_k)\, ds + \int_{-\infty}^{0} D_k u_k(x + sf_k)\, ds \\ &= \frac{-1}{n-1}\int_{0}^{\infty} \sum_{i,j=1}^{n} D_i u_j(x + se_k)\, ds - \int_{0}^{\infty} D_k u_k(x + sf_k)\, ds,\end{aligned}$$

where the first integration is along the line $\{x_1 + s, \ldots, x_{k-1} - s, x_{k+1} + s, \ldots, x_n + s, x_k \text{ fixed}\}$; and the second integration is along the line $\{x_k + s$, with x_j fixed for $j \neq k\}$. Thus,

$$\left|\sum_{i=1}^{n} u_i(x)\right|^{1/(n-1)} \leqq C\left[\left(\int_{-\infty}^{\infty} \sum_{i,j=1;\, i,j\neq k}^{n} |D_i u_j(x + se_k) + D_j u_i(x + se_k)|\, ds\right)^{1/(n-1)} + \left(\int_{-\infty}^{\infty} |D_k u_k(x + sf_k)|\, ds\right)^{1/(n-1)}\right],$$

for $k = 1, \ldots, n - 1$.

Since $u \in C_0^1(R^n)$, all these integrals exist, and so we multiply together the n expressions for $|\sum_{i=1}^{n} u_i(x)|^{1/(n-1)}$ and integrate over R^n to get

$$\begin{aligned}&\int_{R^n}\left|\sum_{i=1}^{n} u_i(x)\right|^{n/(n-1)} dx_1 \ldots dx_n \\ &\leqq C\int_{R^n}\left(\int_{-\infty}^{\infty} \sum_{i,j=1}^{n} |D_i u_j(x + s\delta_n) + D_j u_i(x + s\delta_n)|\, ds\right)^{1/(n-1)} \\ &\quad \cdot \prod_{k=1}^{n-1}\left[\left(\int_{-\infty}^{\infty} \sum_{i,j=1, i\neq k, j\neq k}^{n} |D_i u_j(x + se_k) + D_j u_i(x + se_k)|\, ds\right)^{1/(n-1)}\right. \\ &\quad \left. + \left(\int_{-\infty}^{\infty} |D_k u_k(x + sf_k)|\, ds\right)^{1/(n-1)}\right] dx_1 \ldots dx_n \\ &= C\sum_{\sigma}\int_{R^n}\left(\int_{-\infty}^{\infty} \sum_{i,j=1}^{n} |D_i u_j(x + s\delta_n) + D_j u_i(x + s\delta_n)|\, ds\right)^{1/(n-1)} \\ &\quad \cdot \prod_{k=1;\, k\in\sigma}^{n}\left(\int_{-\infty}^{\infty} \sum_{i,j=1;\, i\neq k, j\neq k}^{n} |D_i u_j(x + se_k) + D_j u_i(x + se_k)|\, ds\right)^{1/(n-1)} \\ &\quad \cdot \prod_{k=1;\, k\notin\sigma}^{n}\left(\int_{-\infty}^{\infty} |D_k u_k(x + sf_k)|\, ds\right)^{1/(n-1)} dx_1 \ldots dx_n,\end{aligned}$$

where σ runs over all possible subsets of the integers $1, 2, \ldots, n-1$.

Now we consider separately each term in the summation over σ. It is an easy exercise to show that if we set $\delta_k = e_k$ or f_k, depending upon which appears in that term, the set $\{\delta_1, \ldots, \delta_n\}$ is linearly independent since $\delta_k = (1, 1, \ldots, 1, 0, 1, \ldots, 1)$ or $(0, \ldots, 0, 1, 0, \ldots, 0)$ for $k = 1, \ldots, n-1$ and $\delta_n = (1, 1, \ldots, 1)$. Moreover, if we let ξ_k be a unit vector orthogonal to the $(n-1)$-hyperplane generated by $\{\delta_j, j \neq k\}$, the set $\{\xi_1, \ldots, \xi_n\}$ is also linearly independent.

Hence there is a positive constant γ such that $dx_1 \cdots dx_n = \gamma d\xi_1 \cdots d\xi_n$. Moreover, the integrals

$$\int_{-\infty}^{\infty} \sum_{i,j=1;\, i\neq k, j\neq k}^{n} |D_i u_j(x + se_k) + D_j u_i(x + se_k)|\, ds \quad \text{and} \quad \int_{-\infty}^{\infty} |D_k u_k(x + sf_k)|\, ds$$

are independent of ξ_k for $1 \leqq k \leqq n-1$, and

$$\int_{-\infty}^{\infty} \left| \sum_{i,j=1}^{n} D_i u_j(x + s\delta_n) + D_j u_i(x + s\delta_n) \right| ds \text{ is independent of } \xi_n,$$

as the kth integration is performed in directions orthogonal to all ξ_j for $j \neq k$ $(1 \leqq k \leqq n)$ and therefore they are done with respect to ξ_k; thus the integrals themselves are independent of ξ_k.

Now we use Hölder's inequality repeatedly with exponents $(n-1)$ and $(n-1)/(n-2)$ to get

$$\begin{aligned}\left\| \sum_{i=1}^{n} u_i \right\|_{L^p(R^n)}^{p} &= \int_{R^n} \left| \sum_{i=1}^{n} u_i(x) \right|^{n/(n-1)} dx \\ &\leqq C \sum_{\sigma} \sum_{i,j=1}^{n} \|D_i u_j + D_j u_i\|_{L^1(R^n)}^{1/(n-1)} \\ &\qquad \prod_{k=1;\, k\in\sigma}^{n} \sum_{i,j=1;\, i\neq k, j\neq k}^{n} \|D_i u_j + D_j u_i\|_{L^1(R^n)}^{1/(n-1)} \cdot \prod_{k=1;\, k\notin\sigma}^{n} \|D_k u_k\|_{L^1(R^n)}^{1/(n-1)},\end{aligned}$$

where σ runs over all subsets of $\{1, \ldots, n-1\}$.

Likewise by integrating over lines like $(1, -1, 1, 1, \ldots, -1)$ etc., instead of these, we can get the same bound for any $\|\sum_{i=1}^{n} v_i u_i\|_{L^p(R^n)}^{p}$, where $v_i = \pm 1$. Now let v_i vary by setting $v_i(x) = \operatorname{sgn} u_i(x)$, and then

$$\|u\|_{L^p(R^n)}^{p} = \int_{R^n} \sum_{i=1}^{n} |u_i(x)|^p\, dx \leqq \int_{R^n} \sum_{i=1}^{n} |v_i(x) u_i(x)|^p\, dx$$

has the same bound (up to a constant 2^n).

Finally we use the fact that the geometric mean of nonnegative numbers is no greater than the arithmetic mean, to show that the right-hand side is bounded by

$$\sum_{i,j=1}^{n} \|D_i u_j + D_j u_i\|_{L^1(R^n)} + \sum_{k=1}^{n} \sum_{i,j=1}^{n} \|D_i u_j + D_j u_i\|_{L^1(R^n)} + \sum_{k=1}^{n} \|D_k u_k\|_{L^1(R^n)}. \quad \text{Q.E.D.}$$

It is easy to see that the above proof also shows

$$\left|\sum_{i=1}^{n} u_i(x)\right|^{1/(n-1)} \leqq C \int_{-\infty}^{\infty} \left|\sum_{i,j=1}^{n} D_i u_j(x + s\delta_n)\right| ds$$

and

$$\left|\sum_{i=1}^{n} u_i(x)\right|^{1/(n-1)} \leqq C\left[\left(\int_{-\infty}^{\infty} \left|\sum_{i,j=1;\, i\neq k, j\neq k}^{n} D_i u_j(x + se_k)\right| ds\right)^{1/(n-1)} + \left(\int_{-\infty}^{\infty} |D_k u_k(x + sf_k)|\, ds\right)^{1/(n-1)}\right],$$

for $k = 1, \ldots, n-1$. Finishing the proof with these estimates yields

$$\|u\|_{L^p(R^n)} \leqq C\left\{\left\|\sum_{i,j=1}^{n} D_i u_j\right\|_{L^1(R^n)} + \sum_{k=1}^{n-1}\left[\left\|\sum_{i,j=1;\, i\neq k, j\neq k}^{n} D_i u_j\right\|_{L^1(R^n)} + \|D_k u_k\|_{L^1(R^n)}\right]\right\},$$

which involves only $2n - 1$ norms on the right side, and is thus an improvement similar to de Figueiredo's extension of Korn's inequality.

We turn now to Theorem 2, which we prove only for the case that Ω is a square and $u \in C^1$ piecewise, but can easily be extended to more general regions and functions.

PROOF. We assume without loss of generality that Ω is the square $\{0 \leqq x_1 \leqq \sigma, 0 \leqq x_2 \leqq \sigma\}$. We suppose also, *for the time being*, that $D_1 u_2(x)$ and $D_2 u_1(x)$ have the same sign for all $x \in \Omega$. Let $x = (x_1, x_2)$ and $y = (y_1, y_2)$ be two arbitrary points in Ω. Then

$$u_1(y_1, y_2) + u_2(y_1, y_2) - u_1(x_1, x_2) - u_2(x_1, x_2)$$
$$= \int_{x_1}^{y_1} (D_1 u_1(s, x_2) + D_1 u_2(s, x_2))\, ds + \int_{x_2}^{y_2} (D_2 u_1(y_1, t) + D_2 u_2(y_1, t))\, dt,$$

so that

$$|u_1(y_1, y_2) + u_2(y_1, y_2) - u_1(x_1, x_2) - u_2(x_1, x_2)|$$
$$\leqq \int_{x_1}^{y_1} (|D_1 u_1(s, x_2)| + |D_2 u_2(s, x_2)| + |D_2 u_1(s, x_2) + D_1 u_2(s, x_2)|)\, ds$$
$$+ \int_{x_2}^{y_2} (|D_1 u_1(y_1, t)| + |D_2 u_2(y_1, t)| + |D_2 u_1(y_1, t) + D_1 u_2(y_1, t)|)\, dt,$$

since sgn $D_1u_2D_2u_1$ is nonnegative.

Similarly,

$$
\begin{aligned}
|u_1(y_1, y_2) + u_2(y_1, y_2) - u_1(x_1, x_2) - u_2(x_1, x_2)| \\
\leqq \int_{x_1}^{y_1} (|D_1u_1(s, y_2)| + |D_2u_2(s, y_2)| \\
+ |D_2u_1(s, y_2) + D_1u_2(s, y_2)|)\, ds \\
+ \int_{x_2}^{y_2} (|D_1u_1(x_1, t)| + |D_2u_2(x_1, t)| \\
+ |D_2u_1(x_1, t) + D_1u_2(x_1, t)|)\, dt.
\end{aligned}
$$

Each of these integrals may be extended to go from 0 to σ instead of x_1 to y_1 or x_2 to y_2. We then multiply together the two expressions and integrate with respect to x_1, x_2, y_1, and y_2.

The left side becomes

$$2\sigma^2 \int_\Omega (u_1^2(x_1, x_2) + u_2^2(x_1, x_2) + 2u_1(x_1, x_2)u_2(x_1, x_2))\, dx_1\, dx_2$$

since by hypothesis

$$
\begin{aligned}
\int_\Omega u_i(x_1, x_2)u_j(y_1, y_2)\, dx_1\, dx_2\, dy_1\, dy_2 \\
= \int_\Omega u_i(x_1, x_2)\, dx_1\, dx_2 \int_\Omega u_j(y_1, y_2)\, dy_1\, dy_2 = 0
\end{aligned}
$$

for $i, j = 1, 2$.

The right side is of the form

$$
\begin{aligned}
\left(\int_0^\sigma G(s, x_2)\, ds + \int_0^\sigma H(y_1, t)\, dt\right)\left(\int_0^\sigma G(s, y_2)\, ds + \int_0^\sigma H(x_1, t)\, dt\right) dx_1\, dx_2\, dy_1\, dy_2 \\
= \sigma^2\Big[\int_0^\sigma G(s, x_2)\, ds\, dx_2 \int_0^\sigma G(s, y_2)\, ds\, dy_2 \\
+ \int_0^\sigma G(s, x_2)\, ds\, dx_2 \int_0^\sigma H(x_1, t)\, dt\, dx_1 \\
+ \int_0^\sigma H(y_1, t)\, dt\, dy_1 \int_0^\sigma G(s, y_2)\, ds\, dy_2 \\
+ \int_0^\sigma H(y_1, t)\, dt\, dy_1 \int_0^\sigma H(x_1, t)\, dt\, dx_1\Big]
\end{aligned}
$$

so that the right side is bounded by

$$C\sigma^2[\|D_1u_1\|^2_{L^1(\Omega)} + \|D_2u_2\|^2_{L^1(\Omega)} + \|D_2u_1 + D_1u_2\|^2_{L^1(\Omega)}],$$

where we have again used the fact that the geometric mean of nonnegative numbers is no greater than their arithmetic mean.

Next, we see that

$$\begin{aligned} u_1(y_1, y_2) - u_2(y_1, y_2) - u_1(x_1, x_2) + u_2(x_1, x_2) \\ = \int_{x_1}^{y_1} (D_1u_1(s, x_2) - D_1u_2(s, x_2))\, ds \\ - \int_{y_2}^{x_2} (D_2u_1(y_1, t) - D_2u_2(y_1, t))\, dt \end{aligned}$$

so that

$$\begin{aligned} |u_1(y_1, y_2) - u_2(y_1, y_2) - u_1(x_1, x_2) + u_2(x_1, x_2)| \\ \leqq \int_0^\sigma (|D_1u_1(s, x_2)| + |D_2u_2(s, x_2)| \\ + |D_2u_1(s, x_2) + D_1u_2(s, x_2)|)\, ds \\ + \int_0^\sigma (|D_1u_1(y_1, t)| + |D_2u_2(y_1, t)| \\ + |D_2u_1(y_1, t) + D_1u_2(y_1, t)|)\, dt, \end{aligned}$$

and similarly where we carry out the integrations in reverse order,

Hence

$$\begin{aligned} 2\sigma^2 \int_\Omega (u_1^2(x_1, x_2) + u_2^2(x_1, x_2) - 2u_1(x_1, x_2)u_2(x_1, x_2))\, dx_1\, dx_2 \\ \leqq C\sigma^2[\|D_1u_1\|^2_{L^1(\Omega)} + \|D_2u_2\|^2_{L^1(\Omega)} + \|D_2u_1 + D_1u_2\|^2_{L^1(\Omega)}]. \end{aligned}$$

Adding the two expressions and factoring out the σ^2 yields

$$\|u\|_{L^2(\Omega)} \leqq C[\|D_1u_1\|_{L^1(\Omega)} + \|D_2u_2\|_{L^1(\Omega)} + \|D_2u_1 + D_1u_2\|_{L^1(\Omega)}].$$

Now we allow the signs of $D_1u_2(x)$ and $D_2u_1(x)$ to vary independently of each other. Set $\alpha(x) = \operatorname{sgn} D_1u_2(x)D_2u_1(x)$. Then the above proof shows that

$$\|u\|_{L^2(\Omega)} \leqq C[\|D_1u_1\|_{L^1(\Omega)} + \|D_2u_2\|_{L^1(\Omega)} + \|D_2u_1 + \alpha D_1u_2\|_{L^1(\Omega)}].$$

Q.E.D.

ACKNOWLEDGMENT. I wish to thank L. Nirenberg, whose proof of a special case of Theorem 1 led me to attempt this problem.

BIBLIOGRAPHY

1. C. Duvaut and J. L. Lions, *Sur les inequations en mecanique et en physique*, Dunod, Paris, 1971.

2. D. G. de Figueiredo, *The coerciveness problem for forms over vector valued functions*, Comm. Pure Appl. Math. **16** (1963), 63–94. MR **26** #6578.

3. A. Friedman, *Partial differential equations*, Holt, Rinehart and Winston, New York, 1969.

4. K. O. Friedrichs, *On the boundary-value problems of the theory of elasticity and Korn's inequality*, Ann. of Math. (2) **48** (1947), 441–471. MR **9**, 255.

TEXAS TECH UNIVERSITY

PROBABILITY THEORY AND THE STRONG MAXIMUM PRINCIPLE[1]

DANIEL W. STROOCK AND S. R. S. VARADHAN

1. **Diffusion processes.** We consider the region $[0, T] \times R^d$ and the operator

$$L = \frac{\partial}{\partial s} + \tfrac{1}{2}\sum a^{ij}(s, x)\frac{\partial^2}{\partial x_i \partial x_j} + \sum b^j(s, x)\frac{\partial}{\partial x_j}$$

acting on functions defined there. The coefficients are assumed to be smooth and $\{a^{ij}(s, x)\}$ is a nonnegative definite symmetric matrix. Let $p(s; x, t, dy)$, $0 \leqq s < t \leqq T$, be the fundamental solution corresponding to L in the sense that

$$u(s, x) = \int p(s, x, t, dy)f(y)$$

solves

$$Lu = 0 \quad \text{for } s < t$$

with $u(t, x) = f(x)$.

For each $x \in R^d$ and $0 \leqq s < t \leqq T$, $p(s, x, t, dy)$ is a nonnegative measure with total mass 1 and satisfies

$$p(s, x, t, E) = \int p(s, x, \sigma, dy)p(\sigma, y, t, E)$$

where E is any Borel set and $s < \sigma < t$.

For each s, Ω_s is the space of continuous functions $x(\cdot)$ on $[s, T]$ with values in R^d. We use the fundamental solution to construct a family $\{P_{s,x}\}$ of measures on Ω_s for s in $0 \leqq s \leqq T$ and $x \in R^d$. Let us fix s and x. Then $P_{s,x}$ is defined by the relations

AMS 1970 *subject classifications*. Primary 60J60; Secondary 35K20.

[1] Results obtained at the Courant Institute of Mathematical Sciences, New York University. This research was sponsored by the U.S. Air Force Office of Scientific Research contract AF-49(638)-1719.

$$P_{s,x}[x(\cdot):x(t_1)\in A_1,\ldots,x(t_n)\in A_n]$$
$$= \int_{A_1}\cdots\int_{A_n} p(s,x,t_1,dy_1)\cdots p(t_{n-1},y_{n-1},t_n,dy_n)$$

where $s \leqq t_1 < t_2 < \cdots < t_n \leqq T$. This defines a $P_{s,x}$ uniquely, which is a probability measure on Ω_s. In particular if $a^{ij} = \delta_{ij}$ and $b^j = 0$, p is the fundamental solution of the heat equation and $P_{0,0}$ in that case is the classical Wiener measure. The family $\{P_{s,x}\}$ is closely related to the operator L, considering the fact that it has been concocted out of the fundamental solution for L. For instance if G is a region in $[0, T] \times R^d$ we can define

$$q_G(s,x,t,E) = P_{s,x}[x(t)\in E\colon (\sigma, x(\sigma))\in G \text{ for } s \leqq \sigma \leqq t]. \tag{1}$$

q_G is the candidate for the fundamental solution of L in the region G with zero boundary condition along the walls. In particular, by the ordinary weak maximum principle, if u is a function on G such that

$$u \geqq 0 \quad \text{on } G,$$
$$Lu \geqq 0 \quad \text{on } G,$$

then, for any $0 \leqq s < t \leqq T$ and $x \in G$,

$$u(s,x) \geqq \int_{G_t} u(t,y)q_G(s,x,t,dy), \tag{2}$$

where $G_t = [y\colon(t,y)\in G]$.

2. **The strong maximum principle.** Let $G \subset [0, T] \times R^d$ be a region. Let u be a function on G such that $Lu \geqq 0$ on G and

$$u(s_0, x_0) - \alpha = \sup_{(s,x)\in G} u(s,x),$$

where (s_0, x_0) is some point of G. The strong maximum principle describes a set A_{s_0,x_0}, containing the point (s_0, x_0), on which u is equal to α. Let $q_G(s,x,t,dy)$ be as in §1. For each $t > s$ we consider the measure $q_G(s_0,x_0,t,dy)$ and denote by $B_{s_0,x_0,t}$ its support [i.e., the smallest closed set with no mass outside]. Let us define

$$A_{s_0,x_0} = \overline{\bigcup_{t>s_0}(t, B_{s_0,x_0,t})}.$$

THEOREM 1. *If u satisfies $Lu \geqq 0$ and $u(s_0,x_0) = \alpha = \sup_{(s,x)\in G} u(s,x)$, then $u(s,x) = \alpha$ for all $(s,x) \in A_{s_0,x_0}$.*

PROOF. Replacing u by $v = \alpha - u$ we have

$$0 = v(s_0,x_0) \geqq \int_{G_t} v(t,y)q_G(s_0,x_0,t,dy).$$

[See equation (2).] Since v is nonnegative and continuous, v must vanish identically on the support $(t, B_{s_0,x_0,t})$ and this is true for each $t > s$.

This is not satisfactory because it involves the supports of various $q_G(s, x, t, dy)$ as G runs over all regions. However one can relate all of these to the support of $P_{s,x}$ in the following manner. Since the space Ω_s is separable the measure $P_{s,x}$ has a support which we will denote by $C_{s,x}$.

THEOREM 2. *Let s_0 and x_0 be fixed. Let $G \subset [0, T] \times R^d$ be any region. Then for any $t > s_0$, the set $B_{s_0,x_0,t}$ is the closure of the trace of $f(t)$ as $f(\cdot)$ runs over all the functions in C_{s_0,x_0} such that $(\sigma, f(\sigma)) \in G$ for $s_0 \leqq \sigma \leqq t$.*

PROOF. Obvious from the definition (1) of q_G.

3. **A description of C_{s_0,x_0}.** As we saw earlier the problem of the strong maximum principle involves the description in terms of a and b of the set C_{s_0,x_0} in Ω_{s_0}. Let us define

$$\bar{b}_i = b^i - \tfrac{1}{2} \sum_j a^{ij}_{,j}.$$

[Here and in what follows $a^{ij}_{,j}$ stands for the partial of a^{ij} with respect to x_j.] For each nice function $\phi(t)$ on $[s_0, T]$ with values in R^d we consider the equation

$$f(t) = x_0 + \int_{s_0}^{t} a(\sigma, f(\sigma))\phi(\sigma)\, d\sigma + \int_{s_0}^{t} \bar{b}(\sigma, f(\sigma))\, d\sigma.$$

As ϕ varies over a nice class, f varies over a class which we will denote by D_{s_0,x_0}.

THEOREM 3. $C_{s_0,x_0} = \bar{D}_{s_0,x_0}$.

We shall consider the special case where there is a smooth matrix $\sigma(s, x)$ of order $d \times k$ (k is arbitrary) such that

$$a(s, x) = \sigma(s, x)\sigma^*(s, x).$$

[* denotes the transpose.] This is a restriction on $a(s, x)$ that involves more than smoothness. However by allowing k to be infinite we can recover all smooth coefficients. The case when k is infinite differs from the case of finite k only in a technical sense.

Let us define

$$\tilde{b}^i = b^i - \tfrac{1}{2} \sum_{j=1}^{d} \sum_{l=1}^{k} \sigma^{ij}_{,l}\sigma^{lj}.$$

For each nice function $\psi(t)$ on $[s_0, T]$ with values in R^k we consider the equation

$$f(t) = x_0 + \int_{s_0}^{t} \sigma(\theta, f(\theta))\psi(\theta)\, d\theta + \int_{s_0}^{t} \tilde{b}(\theta, f(\theta))\, d\theta.$$

As ψ varies over a nice class, f varies over a class which we will denote by E_{s_0,x_0}.

THEOREM 4. $C_{s_0,x_0} = \bar{E}_{s_0,x_0}$.

REMARK. Theorems 3 and 4 are really the same. For this we have to show that

$$\bar{D}_{s_0,x_0} = \bar{E}_{s_0,x_0}.$$

To see this we note that σ and a have the same range. Moreover the difference between $\bar{b}$ and $\tilde{b}$ can be computed as follows:

$$\begin{aligned}\bar{b}^i - \tilde{b}^i &= -\tfrac{1}{2}\sum a^{ij}_{,j} + \tfrac{1}{2}\sum\sum \sigma^{ij}_{,l}\sigma^{lj}\\ &= -\tfrac{1}{2}\sum(\sum(\sigma^{il}\sigma^{jl})_{,j}) + \tfrac{1}{2}\sum\sum \sigma^{ij}_{,l}\sigma^{lj}\\ &= -\tfrac{1}{2}\sum\sum \sigma^{il}\sigma^{jl}_{,j} = \sigma\psi_0\end{aligned}$$

where $\psi_0^l = -\frac{1}{2}\sum \sigma^{jl}_{,j}$. This means that any difference between $\bar{b}$ and $\tilde{b}$ can be taken care of by changing ψ. The rest is routine.

4. **Proof of Theorem 4.** In order to get at the support of $P_{s,x}$ we use a probabilistic description of $P_{s,x}$. Let Q be the k-dimensional Brownian motion. That is, Ω' is the space of R^k valued continuous functions $\beta(\cdot)$ on $[0, T]$ such that $\beta(0) = 0$. Q is the Wiener measure on it. We consider the equation

$$f(t) = x_0 + \int_{s_0}^{t} \sigma(\theta, f(\theta))\, d\beta(\theta) + \int_{s_0}^{t} b(\theta, f(\theta))\, d\theta.$$

This equation has to be carefully defined because $\beta \in \Omega'$ is not in general of bounded variation. However Itô has shown that integrals with respect to $d\beta$ can be defined as limits of Riemann sums of special type and the limits can be shown to exist in the sense of convergence in mean square with respect to Q. Once the integrals have been defined, the above equation can be shown to have a unique solution for $t \geqq s_0$. This defines a transformation $T_{s_0,x_0}: \beta(\cdot) \to f(\cdot)$ from $\Omega' \to \Omega_{s_0}$. The connection between this transformation and P_{s_0,x_0} is that

$$P_{s_0,x_0} = QT^{-1}_{s_0,x_0}$$

since the support of Q is all of Ω', if T were continuous, the support of P_{s_0,x_0} will be the closure in Ω_{s_0} of the range of T_{s_0,x_0}. However T_{s_0,x_0} is only defined almost everywhere and is far from continuous. Nevertheless we have the following theorem which serves the purpose.

THEOREM 5. *Let ψ_0 be a nice smooth function in Ω. Then $T_{s_0,x_0}(\psi_0)$ exists in the sense of density and equals f_0 where*

$$f_0(t) = x_0 + \int_{s_0}^{t} \sigma(\theta, f(\theta))\, d\psi_0(\theta) + \int_{s_0}^{t} \tilde{b}(\theta, f(\theta))\, d\theta.$$

COROLLARY. $C_{s_0,x_0} \supset \bar{E}_{s_0,x_0}$.

REMARKS. (1) Theorem 5 proves one half of Theorem 4. For the other half one has to show that

$$P_{s_0,x_0}[C_{s_0,x_0}] = 1.$$

This has to be done independently. Once this is done it will show that A_{s_0,x_0} is the maximal set for which the strong maximum principle holds. It actually provides counterexamples. These counterexamples are not smooth functions. They satisfy the hypothesis $Lu \leqq 0$ only in a weak sense.

(2) The reason that we get $\tilde{b}$ in Theorem 5 rather than b is that β is not of bounded variation and symbolically

$$(d\beta)^2 = I\,dt.$$

This is translated into certain properties of the integrals of $d\beta$ and they are used to prove the theorem.

(3) As a result of our attempt to characterize in what sense the natural counterexamples mentioned in Remark (1) actually satisfy $Lu \leqq 0$, we were led to a study of the first boundary value problem for operators L. The analytically most appealing form of our results is the following: Let $\mathscr{G}$ be a smooth domain in R^d and define Σ_2 and Σ_3 as in [2]. Suppose c is a bounded continuous function on R^d such that $c(x) \geqq \alpha > 0$. Then for each bounded continuous function g on $\mathscr{G}$ and each bounded continuous function h on $\Sigma_2 \cup \Sigma_3$ there is a unique function $f \in L^\infty(\mathscr{G})$ such that

(i) $\int f(L - c)^*\phi\, dx = \int g\phi\, dx$, $\phi \in C_0^\infty(\mathscr{G})$,

(ii) $\operatorname{ess\,lim}_{x \to x_0; x \in \mathscr{G}} f(x) = h(x_0)$, $x_0 \in \Sigma_2 \cup \Sigma_3$.

If there is a T with the property that for each $x \in \mathscr{G}$ there is a bounded measurable $\psi : [0, T] \to R^d$ such that the path

$$\phi(t) = x + \int_0^t a(\phi(s))\psi(s)\, ds + \int_0^t \tilde{b}(\phi(s))\, ds$$

exists from $\overline{\mathscr{G}}$ before time T, then this result remains true even if c is only nonnegative.

5. **References.** For details regarding the transformation T_{s_0,x_0} and its properties see [4]. The strong maximum principle for the elliptic case can be found in [4]. A probabilistic proof appears in [7]. In this case, C_{s_0,x_0} consists of all functions with $f(s_0) = x_0$. Under stringent conditions Bony [**1**] proved a strong maximum principle for the degenerate case. Later on an attempt was made by Hill [**3**] to generalize these results. Redheffer [**6**] has recently proved the strong maximum principle under slightly different conditions by analytical methods. The present approach and in particular Theorem 5 appears in [7]. The transition from the case of $a = \sigma\sigma^*$ with a finite k to the case when k is infinite is made in [**8**]. [**8**] also contains the notion of weak solutions relevant for the construction of counterexamples, along with its relation to the Dirichlet problem.

References

1. J.-M. Bony, *Principe du maximum, inégalité de Harnack et unicité du problème de Cauchy pour les opérateurs elliptiques dégénérés*, Ann. Inst. Fourier (Grenoble) **19** (1969), fasc. 1, 277–304. MR **41** #7486.

2. G. Fichera, *Sulle equazioni differenziali lineari ellittico-paraboliche del secondo ordine*, Atti Accad. Naz. Lincei Mem. Cl. Sci. Fis. Mat. Nat. Sez. I. (8) **5** (1956), 1–30. MR **19**, 658; 1432.

3. C. D. Hill, *A sharp maximum principle for degenerate elliptic-parabolic equations*, Indiana Univ. Math. J. **20** (1970), 213–230.

4. K. Itô, *On stochastic differential equations*, Mem. Amer. Math. Soc. No. 4 (1951). MR **12**, 724.

5. L. Nirenberg, *A strong maximum principle for parabolic equations*, Comm. Pure Appl. Math. **6** (1953), 167–177. MR **14**, 1089; MR **16**, 1336.

6. Ray Redheffer, *The sharp maximum principle for nonlinear inequalities* (to appear).

7. Daniel W. Stroock and S. R. S. Varadhan, *On the support of diffusion processes, with applications to the strong maximum principle*, Proc. Sixth Berkeley Sympos. Probability and Statist. 1970 (to appear).

8. ———, *On degenerate elliptic-parabolic operators of second order and their associated diffusions* (to appear).

COURANT INSTITUTE OF MATHEMATICAL SCIENCES, NEW YORK UNIVERSITY

COERCIVENESS FOR THE NEUMANN PROBLEM

W. J. SWEENEY

Let M be a Riemannian manifold of dimension n, and let

$$(1) \qquad 0 \longrightarrow C^0 \xrightarrow{D} C^1 \xrightarrow{D} C^2 \xrightarrow{D} \cdots \xrightarrow{D} C^n \longrightarrow 0$$

be the Spencer sequence belonging to a formally integrable, elliptic differential operator on M. (For definitions see [**2**].) Let

$$(2) \qquad 0 \longrightarrow C_x^0 \xrightarrow{a(x,\xi)} C_x^1 \xrightarrow{a(x,\xi)} \cdots \xrightarrow{a(x,\xi)} C_x^n \longrightarrow 0$$

denote the symbol sequence corresponding to (1). Our concern here is the coercive estimate

$$(3) \qquad \|u\|_1 \leqq c\{\|D^*u\| + \|Du\| + \|u\|\}$$

for $u \in C^\infty(\Omega, C^k)$ satisfying the boundary condition

$$(4) \qquad a(x,\nu)^*u(x) = 0 \quad \text{for } x \in \partial\Omega,$$

where $\Omega \subset M$ is a compact n-dimensional manifold with boundary smoothly imbedded in M, ν is the interior unit normal cotangent vector field along $\partial\Omega$, $\|\cdot\|$ is the L_2 norm, and $\|\cdot\|_1$ is the Sobolev norm of order 1.

The principal application of this estimate is to establish the solvability of the generalized Neumann problem

$$(5) \qquad \begin{aligned} (DD^* + D^*D)u &= f \quad \text{on } \Omega, \\ a(x,\nu)^*u &= 0 \quad \text{on } \partial\Omega, \\ a(x,\nu)^*Du &= 0 \quad \text{on } \partial\Omega, \end{aligned}$$

which in turn establishes solvability for the equation

AMS 1970 *subject classifications*. Primary 35N15; Secondary 35J55, 58G05.

$$(6) \qquad Dv = f \qquad (Df = 0).$$

More precisely, if the coercive estimate holds, then by classical elliptic theory (5) has a solution $u \in C^\infty(\Omega, C^k)$ for all $f \in C^\infty(\Omega, C^k)$ which are orthogonal to $\boldsymbol{H} = \{u \in C^\infty(\Omega, C^k) | D^*u = 0,\ Du = 0,\ a(x,v)^*u = 0 \text{ on } \partial\Omega\}$, and in case $Df = 0$, then $v = D^*u$ solves (6). Moreover, if the strong estimate

$$(7) \qquad \|u\|_1 \leqq c\{\|D^*u\| + \|Du\|\}$$

holds for all u satisfying (3), then $\boldsymbol{H} = 0$ and (6) can always be solved.

There is a well-known condition, originally given by Ehrenschein, which is necessary and sufficient for coerciveness; namely, (3) holds for all u satisfying (4) if and only if for every $x \in \partial\Omega$ and every $\xi \in T_x^*(\partial\Omega)$ the initial value problem

$$(8) \qquad \begin{aligned} a(x,\xi + vD_t)^*w(t) &= 0, \qquad t \geqq 0, \\ a(x,\xi + vD_t)w(t) &= 0, \qquad t \geqq 0, \\ a(x,v)^*w(0) &= 0 \end{aligned}$$

has no solution $w \in L_2(\boldsymbol{R}_+, C_x^k)$. Here $D_t = -(-1)^{1/2}\partial/\partial t$, $a(x,\xi + vD_t) = a(x,\xi) + a(x,v)D_t$, and $a(x,\xi + vD_t)^* = a(x,\xi)^* + a(x,v)^*D_t$. Our purpose here is to give conditions for coerciveness which are more geometrical in nature; complete proofs of our results will appear in the Journal of Differential Geometry [**3**].

PROPOSITION 1. *If the symbol sequence* (2) *is exact at* C_x^k *for all* $x \in \partial\Omega$ *and all* ξ *of the form* $\xi = \xi' + \lambda v$, *where* $\xi' \in T_x^*(\partial\Omega)$ *and* $\lambda \in \boldsymbol{C}$, *then* (3) *holds for all* $u \in C^\infty(\Omega, C^k)$ *satisfying* (4). *If* (2) *is exact for all* $\xi \in T^*(M) \otimes \boldsymbol{C}$ *then the coercive estimate holds for any* Ω.

The condition in Proposition 1 is not necessary for coerciveness. For example, if $C^k = 0$ for $k > 1$, if $C^0 = C^1$ is the product line bundle over $\boldsymbol{R}^2$, and if $D = \partial/\partial x_1 + i\partial/\partial x_2$, then

$$0 \longrightarrow C^0 \xrightarrow{\ D\ } C^1 \longrightarrow 0$$

is a Spencer sequence. Since $a(x,v)$ is always bijective, $u \in C^\infty(\Omega, C^1)$ satisfies (4) if and only if $u(x) = 0$ for $x \in \partial\Omega$, and for such u the estimate (3) is just Gårding's inequality. On the other hand, the symbol sequence is not exact if $\xi = dx_1 + i\,dx_2$.

To obtain sharper results we shall use

PROPOSITION 2. *The coercive estimate* (3) *holds for all* $u \in C^\infty(\Omega, C^k)$ *satisfying* (4) *if and only if for each* $x \in \partial\Omega$ *and each* $\xi \in T_x^*(\partial\Omega)$ *the sequence*

$$(9) \qquad \mathscr{L}(C_x^{k-1}) \xrightarrow{a(x,\xi + vD_t)} \mathscr{L}(C_x^k) \xrightarrow{a(x,\xi + vD_t)} \mathscr{L}(C_x^{k+1})$$

is exact, where for any vector space V *we write* $\mathscr{L}(V)$ *for the space of* C^∞ *functions* $w: \boldsymbol{R}_+ \to V$ *which are in* L_2. *The coercive estimate holds for all compact* $\Omega \subset M$ *if and only if*

$$(10)\qquad \mathscr{L}(C_x^{k-1}) \xrightarrow{a(x,\xi+\eta D_t)} \mathscr{L}(C_x^k) \xrightarrow{a(x,\xi+\eta D_t)} \mathscr{L}(C_x^{k+1})$$

is exact for all $\xi,\eta \in T^*(M)$ *with* $\xi \wedge \eta \neq 0$.

For $x \in M$ consider the characteristic variety $\mathscr{V}_x = \{\xi \in T_x^*(M) \otimes \mathbf{C} | (2)$ is not exact at $\xi\}$; using the fact that (1) is a Spencer sequence, one can show that $q = \operatorname{codim} \mathscr{V}_x$ is independent of x. Let ${}^{\partial}\mathscr{V}_x$ denote the image of $\mathscr{V}_x$ under the map $T_x^*(M) \otimes \mathbf{C} \to T_x^*(\partial\Omega) \otimes \mathbf{C}$ which is induced by the inclusion $\partial\Omega \subset M$.

THEOREM. *For each* $x \in \Omega$ *assume that* $\mathscr{V}_x$ *contains only Cohen-Macaulay points, and assume* $0 < q < n$. *Then the coercive estimate always holds at* C^k *for* $k \geqq q$, *and it holds at* C^k *for* $k < q$ *if and only if* ${}^{\partial}\mathscr{V}_x \cap T_x^*(\partial\Omega) = 0$ *for all* $x \in \partial\Omega$.

By definition $\xi \in \mathscr{V}_x$ is Cohen-Macaulay if and only if the symbol sequence (2) is exact at C_x^k for $k \geqq q + 1$; and thus the part of the theorem which treats the case $k > q$ follows from Proposition 1. In case $k \leqq q$ the results are more subtle and depend on the canonical form theorem recently obtained by V. W. Guillemin. (See the notes on Spencer's lecture or [**1**].) In particular Guillemin shows that given $x \in \partial\Omega$ and $\xi \in T_x^*(\partial\Omega)$ there exist invertible linear transformations

$$R : C_x^k \to C_x^k \qquad (k = 0,1,\ldots,n)$$

such that $Ra(x,\xi' + \lambda v)R^{-1}$ has a particularly simple form as a function of $\lambda \in \mathbf{C}$. Now in order to check coerciveness for (1) we must compute the homology of (9), and this homology does not change if each operator in (9) is replaced by its R-conjugate $Ra(x,\xi' + D_t v)R^{-1}$. The homology of this sequence can be computed explicitly, and the theorem follows.

A corollary of the theorem is the local exactness of Spencer sequences, in the Cohen-Macaulay case, at the positions $C^q, C^{q+1}, \ldots, C^n$. In fact, at these positions the coercive estimate holds for all Ω, and by using a shrinking argument one can obtain the strong estimate (7) when Ω is a sufficiently small ball. Thus (6) is always solvable when Ω is a small ball and when $f \in C^\infty(\Omega, C^k), k \geqq q$.

The theorem also shows that the coercive estimate rarely holds at the positions $C^k, k < q$. The condition ${}^{\partial}\mathscr{V} \cap T^*(\partial\Omega) = 0$, means that $\operatorname{Im} \xi$ can never be normal to $\partial\Omega$ if $\xi \in \mathscr{V}_x$ and $x \in \partial\Omega$. If Ω is a small disk and if there is a cotangent vector field $\xi : \Omega \to T^*(\Omega) \otimes \mathbf{C}$ which maps into $\mathscr{V}$, then $\operatorname{Im} \xi$ must be normal to $\partial\Omega$ at some $x \in \partial\Omega$, and therefore the coercive estimate must fail on Ω at $C^k, k < q$.

REFERENCES

1. V. W. Guillemin and S. Sternberg, *Subelliptic estimates for complexes*, Proc. Nat. Acad. Sci. U.S.A. **67** (1970), 271–274.

2. D. C. Spencer, *Overdetermined systems of linear partial differential equations*, Bull. Amer. Math. Soc. **75** (1969), 179–239. MR **39** # 3533.

3. W. J. Sweeney, *Coerciveness for the Neumann problem*, J. Differential Geometry **6** (1972), 375–393.

RUTGERS UNIVERSITY

A FREDHOLM THEORY FOR ELLIPTIC PARTIAL DIFFERENTIAL OPERATORS IN R^n

HOMER F. WALKER

The following is primarily a report on material which will soon appear in print elsewhere. Only main results are described here; no proofs are given. For a complete presentation of this work, the reader is referred to [**8**] and [**9**]. Briefly, the theorems in the sequel concern linear first-order partial differential operators

$$Au(x) = \sum_{i=1}^{n} A_i(x)\frac{\partial}{\partial x_i}u(x) + B(x)u(x)$$

with domain $H_1(\boldsymbol{R}^n; \boldsymbol{C}^k)$ in $L_2(\boldsymbol{R}^n; \boldsymbol{C}^k)$ which are *elliptic*, i.e., such that $\det|\sum_{i=1}^n A_i(x)\xi_i| \neq 0$ for all x in $\boldsymbol{R}^n$ and all nonzero ξ in $\boldsymbol{R}^n$, and which have constant coefficients with vanishing zero-order terms outside a bounded subset of $\boldsymbol{R}^n$. It is well-known that if the independent variables are restricted to a compact set or manifold, then an elliptic partial differential operator is *Fredholm*, i.e., its range is closed and the dimension of its null-space and the codimension of its range are both finite. Thus the range of an elliptic operator on a compact set or manifold is determined to be the orthogonal complement of a finite-dimensional subspace of a Hilbert space. Furthermore, the Fredholm *index* of an operator of this type, defined to be the dimension of its null-space minus the codimension of its range, is invariant under small perturbations of the operator and, hence, is of key importance to the perturbation theory of elliptic operators on a compact set or manifold. The results presented in this paper may be summarized by saying that the story is essentially unchanged for elliptic operators of the type considered here, even though the independent variables are allowed to range over all of $\boldsymbol{R}^n$. Specifically, it is asserted that an elliptic operator of the type described above is "practically" Fredholm in

AMS 1970 *subject classifications*. Primary 35A05, 35J45, 47B30; Secondary 35F05, 35P25, 47A40, 47F05.

Key words and phrases. Elliptic operators, elliptic operators on unbounded domains, Fredholm operators.

the following ways:

(1) Such an operator has a finite-dimensional null-space, the dimension of which is an upper-semicontinuous function of the operator in a certain sense.

(2) Such an operator has a finite index which is invariant under small perturbations of the operator, and the range of such an operator can be characterized in terms of the range of an operator with constant coefficients and a finite index-related number of orthogonality conditions.

To describe more precisely the operators under consideration, suppose that there is given a linear elliptic first-order partial differential operator

$$A_0 u(x) = \sum_{i=1}^{n} A_i \frac{\partial}{\partial x_i} u(x)$$

acting on $H_1(\boldsymbol{R}^n; \boldsymbol{C}^k)$ with constant coefficients and no zero-order term. Then for a positive number R, denote by $E(A_0, R)$ the set of linear elliptic first-order partial differential operators

$$Au(x) = \sum_{i=1}^{n} A_i(x) \frac{\partial}{\partial x_i} u(x) + B(x)u(x)$$

acting on $H_1(\boldsymbol{R}^n; \boldsymbol{C}^k)$ with continuously differentiable first-order coefficients $A_i(x)$ and continuous zero-order coefficients $B(x)$ which satisfy the following: The coefficients of A are equal to those of A_0 outside the ball $B_R^n = \{x \in \boldsymbol{R}^n : |x| \leqq R\}$. The operators in $E(A_0, R)$ are those elliptic operators obtained by adding a "perturbing" operator to A_0 whose coefficients vanish outside B_R^n. Accordingly, the operator A_0 will be referred to as the *unperturbed operator* of the set $E(A_0, R)$. Note that if A is an operator in $E(A_0, R)$, then there exist positive constants C_1 and C_2 depending on A for which the standard elliptic estimate

$$(*) \qquad \|u\|_1 \leqq C_1\|u\| + C_2\|Au\|$$

holds for all u in $H_1(\boldsymbol{R}^n; \boldsymbol{C}^k)$ [7]. From this estimate, it follows that $H_1(\boldsymbol{R}^n; \boldsymbol{C}^k)$ is a natural domain for such an operator in the sense that the operator is closed on $H_1(\boldsymbol{R}^n; \boldsymbol{C}^k)$ and its adjoint operator A^* also has domain $H_1(\boldsymbol{R}^n; \boldsymbol{C}^k)$ [**2**]. It is the objective of this paper to describe a Fredholm theory for the operators in $E(A_0, R)$.

Recall that in proving that an elliptic operator on a compact set or manifold is Fredholm, the principal tools used are elliptic estimates of the form of (*) and the Rellich Compactness Theorem [**1**, p. 169]. For an operator in $E(A_0, R)$, the elliptic estimate (*) is valid, but the Rellich Compactness Theorem cannot be applied because the independent variables are allowed to range over all of $\boldsymbol{R}^n$. What seems necessary for the present purposes, then, is to find a subspace of $H_1(\boldsymbol{R}^n; \boldsymbol{C}^k)$ which is sufficiently small that an analog of the Rellich Compactness Theorem holds on the subspace, yet sufficiently large that a useful Fredholm theory for an operator in $E(A_0, R)$ can be constructed using the subspace. Consider the set

$$M(A_0, R) = \{u \in H_1(\boldsymbol{R}^n; \boldsymbol{C}^k) \colon \text{support } A_0 u \subseteq B_R^n\}.$$

Clearly, $M(A_0, R)$ is a closed subspace of $H_1(\boldsymbol{R}^n; \boldsymbol{C}^k)$ and the null-space $N(A)$ of each operator A in $E(A_0, R)$ is contained in $M(A_0, R)$. Furthermore, the set $M(A_0, R)$ is a sufficiently small subspace of $H_1(\boldsymbol{R}^n; \boldsymbol{C}^k)$ for the following Rellich-type result to hold.

LEMMA. *Every subset of $M(A_0, R)$ which is bounded in $H_1(\boldsymbol{R}^n; \boldsymbol{C}^k)$ is relatively compact in $L_2(\boldsymbol{R}^n; \boldsymbol{C}^k)$.*

The lemma is proved using Fourier transforms and the Rellich Compactness Theorem. Having the lemma, one can proceed as in the case of an elliptic operator on a compact set or manifold to obtain the following theorem, proved by Lax and Phillips [**6**] in the case of an odd number of independent variables.

THEOREM. *If A is an operator in $E(A_0, R)$, then the dimension of $N(A)$ is finite.*

One can also use the lemma and the estimate (*) to obtain the following estimate.

LEMMA. *If A is an operator in $E(A_0, R)$, then there exists a positive constant C for which the estimate $\|u\| \leqq C\|Au\|$ holds for all u in $M(A_0, R)$ orthogonal in $L_2(\boldsymbol{R}^n; \boldsymbol{C}^k)$ to $N(A)$.*

By applying standard perturbation theory techniques as found in [**4**] to the estimate (*) and the estimate of the above lemma, one can prove the following theorem.

THEOREM. *Relative to the topology on $E(A_0, R)$ induced by the norm topology on the set of bounded linear operators from $H_1(\boldsymbol{R}^n; \boldsymbol{C}^k)$ to $L_2(\boldsymbol{R}^n; \boldsymbol{C}^k)$, the dimension of $N(A)$ is an upper-semicontinuous function of the operators A in $E(A_0, R)$.*

The two theorems above may be summarized by saying that an operator in $E(A_0, R)$ is *semi-Fredholm*, i.e., has a null-space the dimension of which is finite and an upper-semicontinuous function of the operator. As the reader may have guessed, the full Fredholm theory of such an operator will be derived by considering the restriction of the operator to the space $M(A_0, R)$. Since the image Au of a function u in $M(A_0, R)$ has support in B_R^n for an operator A in $E(A_0, R)$, one may consider the restriction of such an operator to $M(A_0, R)$ to be a bounded operator from $M(A_0, R)$ to $L_2(B_R^n; \boldsymbol{C}^k)$. The fundamental result underlying the theorems which follow is that the restriction of an operator A in $E(A_0, R)$ to $M(A_0, R)$ is a bounded Fredholm operator from $M(A_0, R)$ to $L_2(B_R^n; \boldsymbol{C}^k)$. In the case of at least three independent variables, it is the remarkable fact that the Fredholm index of this restricted operator is equal to the dimension of $N(A)$ minus the dimension of $N(A^*)$. In the case of two independent variables, the index of the restricted operator cannot be so elegantly prescribed. Nevertheless, the index can be bounded.

The precise formulation of the Fredholm theory of operators in $E(A_0, R)$ will now begin. In the case of at least three independent variables, the basic theorem is the following

THEOREM. *If $n \geqq 3$ and if A is an operator in $E(A_0, R)$, then the restriction of A to $M(A_0, R)$ is a bounded Fredholm operator from $M(A_0, R)$ to $L_2(B_R^n; C^k)$, the index of which is equal to the dimension of $N(A)$ minus the dimension of $N(A^*)$.*

Of fundamental importance to the proof of this theorem—and to the distinction of the case $n \geqq 3$ from the case $n = 2$—are the following lemmas, the first of which is found in [7].

LEMMA 1. *If $n \geqq 3$, then the estimate*

$$\int_{|x| \leqq R} |u(x)|^2 \, dx \leqq \frac{R^2}{2(n-2)} \left(\sum_{i=1}^{n} \left\| \frac{\partial}{\partial x_i} u \right\|^2 \right)$$

holds for all u in $H_1(R^n; C^k)$.

LEMMA 2. *If $n \geqq 3$, then $L_2(B_R^n; C^k)$ is contained in the range of A_0.*

The theorem gives a finite index for an operator A in $E(A_0, R)$ when $n \geqq 3$, namely the dimension of $N(A)$ minus the dimension of $N(A^*)$, which is invariant under small perturbations of A in $E(A_0, R)$. It is also satisfying to note that the index of an operator A in $E(A_0, R)$ is equal to the negative of the index of the adjoint operator A^* in $E(A_0^*, R)$. Denote by X_R the characteristic function of B_R^n, i.e.,

$$\begin{aligned} X_R(x) &= 0, \quad \text{if } |x| > R, \\ &= 1, \quad \text{if } |x| \leqq R. \end{aligned}$$

The following theorem is a consequence of the basic theorem above; it provides the promised characterization of the range of an operator in $E(A_0, R)$ in terms of the range of an operator with constant coefficients and a finite number of orthogonality conditions.

THEOREM. *If $n \geqq 3$ and if A is an operator in $E(A_0, R)$, then an element v of $L_2(R^n; C^k)$ is in the range of A if and only if $(1 - X_R)v$ is in the range of A_0 and $\{X_R v - X_R A A_0^{-1}[(1 - X_R)v]\}$ is orthogonal to $N(A^*)$.*

In the case of two independent variables, there is, unfortunately, no analog of the estimate of Lemma 1. It is also not true that $L_2(B_R^2; C^k)$ is contained in the range of A_0. However, the following amended version of Lemma 2 does hold.

LEMMA. *Let v be an element of $L_2(B_R^2; C^k)$. Then the following conditions are equivalent:*

(1) *v is in the range of A_0.*
(2) *The Fourier transform of v vanishes at the origin.*
(3) *v is orthogonal in $L_2(R^2; C^k)$ to every function of the form cX_R, where c is a*

vector in C^k and where X_R denotes the characteristic function of B_R^2.

It is seen from this lemma that the image set $A_0(M(A_0, R))$ has a k-dimensional orthogonal complement in $L_2(B_R^2; C^k)$, a fact which introduces an uncertainty of k in the following theorems describing the index of an operator in $E(A_0, R)$. If A is an operator in $E(A_0, R)$, then the loss of Lemma 1 is circumvented by considering first the operator A_f consisting of the first-order terms of A. One can prove the following lemma.

LEMMA. *If $n = 2$ and if A is an operator in $E(A_0, R)$, then the restriction of the first-order operator A_f to $M(A_0, R)$ is a bounded Fredholm operator from $M(A_0, R)$ to $L_2(B_R^2; C^k)$, the index of which is at most* [dimension $N(A_f)$ − dimension $N(A_f^*)$] *and at least* [dimension $N(A_f)$ − dimension $N(A_f^*)$ − k].

Now adding the zero-order terms of A to A_f has the effect on the restrictions of these operators to $M(A_0, R)$ of adding a compact operator to a bounded Fredholm operator. The theorem below then follows from the basic stability properties of Fredholm operators with respect to compact perturbations [**3**] and from the fact that the orthogonal complement of the image set $A(M(A_0, R))$ in $L_2(B_R^2; C^k)$ has dimension at least equal to the dimension of $N(A^*)$.

THEOREM. *If $n = 2$ and if A is an operator in $E(A_0, R)$, then the restriction of A to $M(A_0, R)$ is a bounded Fredholm operator from $M(A_0, R)$ to $L_2(B_R^2; C^k)$, the index of which is at least* [dimension $N(A_f)$ − dimension $N(A_f^*)$ − k] *and at most*

min{[dimension $N(A)$ − dimension $N(A^*)$], [dimension $N(A_f)$ − dimension $N(A_f^*)$]}.

This theorem again allows the association of a finite perturbation-invariant index with an operator A in $E(A_0, R)$, although the index is no longer equal to [dimension $N(A)$ − dimension $N(A^*)$]. In fact, both A_0 and A_0^* have index $(-k)$. Letting X_R again denote the characteristic function of B_R^2, the range characterization theorem for an operator in $E(A_0, R)$ in the case of two independent variables is the following

THEOREM. *If $n = 2$ and if A is an operator in $E(A_0, R)$, then an element v of $L_2(R^n; C^k)$ is in the range of A if and only if there exists a vector c in C^k such that $[(1 - X_R)v + X_R c]$ is in the range of A_0 and $\{X_R v - X_R A A_0^{-1}[(1 - X_R)v + X_R c]\}$ is in $A(M(A_0, R))$, a closed subspace of $L_2(B_R^2; C^k)$ whose orthogonal complement in $L_2(B_R^2; C^k)$ has dimension equal to the dimension of $N(A)$ minus the index of A.*

This completes the description of the Fredholm theory of the operators in $E(A_0, R)$. In conclusion, it should be mentioned that the semi-Fredholm theory described here has been shown valid for elliptic operators in R^n of arbitrary order whose coefficients approach constant limiting values as $|x|$ grows large at least as fast as a certain power of $1/|x|$ [**10**]. It is anticipated that the full Fredholm theory will be shown to hold for these operators. A desirable subsequent generalization of the Fredholm theory would be an extension of the theorems presented here to include the elliptic operators in some algebra of pseudo-differential operators in $\boldsymbol{R}^n$, e.g., the Kohn-Nirenberg algebra of pseudo-differential operators [**5**]. A difficulty which might be encountered in such an extension is the following: The proofs of the theorems in this paper rely to a certain extent on the fact that partial differential operators are *local* operators; pseudo-differential operators are not local. Nevertheless, an extension of the Fredholm theory to some class of elliptic pseudo-differential operators would allow full use to be made of the perturbation-invariance of the index of a Fredholm operator and, hence, would be valuable indeed.

REFERENCES

1. L. Bers, F. John and M. Schecter, *Partial differential equations*, Proc. Summer Seminar (Boulder, Col., 1957), Lectures in Appl. Math., vol. 3, Interscience, New York, 1964. MR **29** #346.

2. K. O. Friedrichs, *The identity of weak and strong extensions of differential operators*, Trans. Amer. Math. Soc. **55** (1944), 132–151. MR **5**, 188.

3. I. C. Gohberg and M. G. Kreĭn, *The basic propositions on defect numbers, root numbers and indices of linear operators*, Uspehi Mat. Nauk **12** (1957), no. 2 (74), 43–118; English transl., Amer. Math. Soc. Transl. (2) **13** (1960), 185–264. MR **20** #3459; MR **22** #3984.

4. T. Kato, *Perturbation theory for linear operators*, Die Grundlehren der math. Wissenschaften, Band 132, Springer-Verlag, New York, 1966. MR **34** #3324.

5. J. J. Kohn and L. Nirenberg, *An algebra of pseudo-differential operators*, Comm. Pure Appl. Math. **18** (1965), 269–305. MR **31** #636.

6. P. D. Lax and R. S. Phillips, *Lectures on scattering theory*, Summer Institute on Scattering Theory, Flagstaff, Arizona, 1969.

7. ———, *Scattering theory*, Pure and Appl. Math., vol. 26, Academic Press, New York, 1967. MR **36** #530.

8. H. F. Walker, *On the null-spaces of first-order elliptic partial differential operators in* $\boldsymbol{R}^n$, Proc. Amer. Math. Soc. **30** (1971), 278–286.

9. ———, *A Fredholm theory for a class of first-order elliptic partial differential operators in* $\boldsymbol{R}^n$, Trans. Amer. Math. Soc. **165** (1972), 75–86.

10. ———, *On the null-spaces of elliptic partial differential operators in* $\boldsymbol{R}^n$, Trans. Amer. Math. Soc. (to appear).

TEXAS TECH UNIVERSITY

AN INTRODUCTION TO REGULARITY THEORY FOR PARAMETRIC ELLIPTIC VARIATIONAL PROBLEMS[1]

W. K. ALLARD AND F. J. ALMGREN, JR.

1. **Geometric measure theory.** For the past thirty or forty years mathematicians increasingly have been attracted to problems in the calculus of variations in higher dimensions and codimensions. However, prior to 1960 (with one or two notable exceptions) there was relatively little fundamental progress in the calculus of variations in higher dimensions and codimensions and essentially no progress on the so called parametric problems (those which are of an essentially geometric character). Beginning about ten years ago, however, (in particular with the work of De Giorgi, Federer, Fleming, and Reifenberg) new ideas began to be introduced into the subject with surprising success in these higher dimensions and codimensions. Indeed, in these higher dimensions and codimensions, the calculus of variations seems to have passed from a classical period in its development into a modern era. Many of these new methods and ideas are included in the collection of mathematical results known as *geometric measure theory* (see, in particular, the treatise [**FH1**]). This article is intended as a brief introduction to the regularity theory for parametric elliptic variational problems in the context of geometric measure theory. Some of the results discussed are new.

2. **Variational problems in parametric form.** Historically the calculus of variations has been concerned, of course, with extremals of suitable functionals defined over suitable domains. Traditionally, perhaps, the domain has been a space of mappings f from a fixed m-dimensional manifold M into R^{m+n} (or perhaps into an $(m + n)$-dimensional manifold), and the functional in question most commonly would be the integral over M of a suitable integrand depending on f and its derivatives. If the integrand depends only on $x \in M$, $f(x)$, and $Df(x)$ and the integral of this integrand is independent of the parametrization of M (as is the case, for example, with the area

AMS 1970 *subject classifications*. Primary 49F22; Secondary 49-02.

[1] This work was supported in part by National Science Foundation grant GP-29046.

integrand, but is not the case for most "energy" integrands), the variational problem is said to be in *parametric form.* Problems in parametric form are precisely those problems for which the necessary integration can be performed over the image $f(M)$ in R^{m+n}. These problems are also those of the most intrinsic geometric significance since the value of functional depends only on the geometric properties of the set $f(M)$ (counting multiplicities if necessary) and not on the particular mapping f which produced it.

Especially in the past ten years or so, however, parametric problems in the calculus of variations have been viewed increasingly from a different perspective. The domain of the problems has been changed to consist of m-dimensional surfaces S already lying intrinsically in R^{m+n} and the integrand has become a function

$$F: R^{m+n} \times G(m + n, m) \to R^+$$

where $G(m + n, m)$ denotes the Grassmann manifold of m-plane directions in R^{m+n} (which can be regarded as the space of all unoriented m-dimensional planes through the origin in R^{m+n}). In case S is a reasonably nice surface of dimension m in R^{m+n}, one defines the functional $\boldsymbol{F}$ on S as the integral

$$\boldsymbol{F}(S) = \int_{x \in S} F(x, \boldsymbol{S}(x))\, d\mathscr{H}^m x,$$

where $\boldsymbol{S}(x)$ denotes the tangent m-plane direction to S at x and $\mathscr{H}^m$ denotes m-dimensional Hausdorff measure. (Hausdorff m-measure gives a precise meaning to the notion of m-dimensional area in R^{m+n} and is the basic measure used in defining a theory of integration over m-dimensional surfaces in R^{m+n} which may have singularities. The Hausdorff m-measure of a smooth m-dimensional submanifold of R^{m+n} agrees with any other reasonable definition of the m-area of such a manifold.) The *m-dimensional area integrand* is thus the function

$$A: R^{m+n} \times G(m + n, m) \to \{1\}.$$

An integrand of the form $F: G(m + n, m) \to R^+$ (i.e., F is independent of the point in R^{m+n}) is called an *integrand with constant coefficients.* If $F: R^{m+n} \times G(m + n, m) \to R^+$ and $p \in R^n$, one sets

$$F^p: R^{m+n} \times G(m + n, m) \to R^+,$$

$$F^p(q, \pi) = F(p, \pi) \quad \text{for } q \in R^{m+n}, \pi \in G(m + n, m).$$

With the obvious extension of terminology, F^p is an integrand with constant coefficients.

3. **Elliptic integrands.** In the parametric existence and regularity theory it is natural to impose the condition of ellipticity on the integrand. One says that an integrand $F: R^{m+n} \times G(m + n, m) \to R^+$ is *elliptic* (with respect to the boundary operator ∂ under consideration) if and only if there is a positive function $\gamma: R^{m+n} \to$

R^+ which is locally bounded away from 0 such that for each $p \in R^{m+n}$ and each (flat) m-disk D in R^{m+n},

$$\boldsymbol{F}^p(S) - \boldsymbol{F}^p(D) \geqq \gamma(p)[\mathscr{H}^m(S) - \mathscr{H}^m(D)]$$

whenever S is an m-dimensional surface in R^{m+n} with $\boldsymbol{\partial} S = \boldsymbol{\partial} D$. In particular for the boundary ∂D, D itself is the unique $\boldsymbol{F}^p$ minimal surface. For each $p \in R^{m+n}$, the number $\gamma(p)$ is called the *ellipticity bound* for F at p. In case the codimension n equals 1, the ellipticity of $\boldsymbol{F}$ (with respect to any reasonable boundary operator ∂) is equivalent to the uniform convexity of each F^p. The set of elliptic integrands contains a (computable) convex neighborhood of the m-dimensional area integrand in the C^2 topology. Also, if $f: R^{m+n} \to R^{m+n}$ is a diffeomorphism, then the image integrand $f^{\#}F$ is elliptic if and only if F is. Finally, the ellipticity of F implies that the various Euler equations associated with F are (nonlinear) elliptic systems of partial differential equations, and, "in the small," the ellipticity of F is equivalent to the ellipticity of these systems.

4. **Rectifiable sets.** A set S in R^{m+n} is called *m-rectifiable* if and only if for each $\varepsilon > 0$ there exists a compact C^1 m-dimensional submanifold M of R^{m+n} such that

$$\mathscr{H}^m[(S \sim M) \cup (M \sim S)] < \varepsilon.$$

Solutions to elliptic variational problems in the context of geometric measure theory are usually known to have the structure of compact m-rectifiable sets, possibly with multiplicities. The m-rectifiable sets include all classical m-dimensional geometric objects in R^{m+n} as well as many more general surfaces. One can easily construct, for example, compact m-rectifiable sets which are not locally finitely connected anywhere.

5. **Conditions which imply the regularity almost everywhere of compact m-rectifiable sets.** At the present time there are two main classes of regularity results associated with parametric integrals in the calculus of variations. Stated in the context of m-rectifiable sets a representative formulation of these results is the following:

(1) *First class of results.* Suppose $F: R^{m+n} \times G(m+n, m) \to R^+$ is a C^3 elliptic integrand, $0 \leqq c < \infty$, $S \subset R^{m+n}$ is a compact m-rectifiable set such that for each $p \in S \sim \partial S$, and for each sufficiently small $r > 0$,

$$\boldsymbol{F}^p(S \cap \{q: |q-p| \leqq r\}) \leqq (1 + cr)\boldsymbol{F}^p(T_r)$$

whenever $T_r \subset \{q: |q-p| \leqq 2r\}$ is a compact m-rectifiable set with $\partial T_r = \partial[S \cap \{q: |q-p| \leqq r\}]$. One then can prove that, except possibly for a compact singular set with zero $\mathscr{H}^m$-measure, $S \sim \partial S$ (more precisely, $\operatorname{spt}(\mathscr{H}^m \llcorner S) \sim \partial S$) is a differentiable m-dimensional submanifold of R^{m+n} with locally Hölder continuously turning tangent planes.

(2) *Second class of results.* Let $S \subset R^{m+n}$ be a compact m-rectifiable set. The

initial rate of change of m-dimensional area of S under deformations $f: R \times R^{m+n} \to R^{m+n}$, $f(t, p) = p + tg(p)$, corresponding to smooth vector fields $g: R^{m+n} \to R^{m+n}$ is a continuous linear functional of such vector fields. Hence this initial rate of change of area, or first variation of area, can be regarded in a natural way as a covector valued distribution on R^n, denoted $\delta|S|$. Suppose $\delta|S|$ is $\mathscr{H}^m \llcorner S$ integrable to a power larger than m. One can then prove that, except possibly for a compact singular set with zero $\mathscr{H}^m$-measure S (more precisely $\operatorname{spt}(\mathscr{H}^m \llcorner S)$) is a smooth m-dimensional submanifold of R^{m+n} with locally Hölder continuously turning tangent planes.

These hypotheses are related, although not in an immediately obvious manner. In any case, one can adapt the methods of proof of the second class of results to prove the first class when $F \equiv 1$ and $\partial S = \emptyset$. The optimal regularity results for which one could hope would be those of the second class of results applied to general (variationally) elliptic integrands. Present methods, however, are not adequate to prove such a result. The first class of results was established for $c = 0$ but for variable coefficient integrands in [**AF2**]. The second class of results is established in [**AW**]. For $c > 0$, the first class of results is proved in [**AF4**].

The inequality appearing in the hypotheses of the first class of results is somewhat novel but seems to isolate the basic geometric and analytic ingredient of various regularity techniques and enables one to construct a unified regularity theory for a variety of parametric variational problems. Indeed, suppose $S \subset R^{m+n}$ is a compact m-rectifiable set. Then the hypotheses of the first class of results are satisfied in each of the following cases:

(1) F is a constant coefficient elliptic integrand and S is $\boldsymbol{F}$-minimal with respect to ∂S. Here the inequality is trivially satisfied with $c = 0$ since $F^p = F$ for each $p \in R^{m+n}$.

(2) F is a variable coefficient elliptic integrand and S is $\boldsymbol{F}$-minimal with respect to ∂S. Here one chooses c essentially to be $(\inf F)^{-1}\operatorname{Lip}(F)$.

(3) F is a variable coefficient integrand and S is $\boldsymbol{F}$-minimal with respect to ∂S subject to the constraint that S must avoid certain prescribed C^2 obstacles (the comparison surfaces T_r of the inequality need not avoid the obstacles). By an obstacle one means an open subset $A \subset R^{m+n}$ such that closure$(A) \sim A$ consists of the disjoint union of C^2 submanifolds of R^{m+n}, without boundaries, of various dimensions. The region A can either be bounded (for example, an open ball) or be unbounded (for example, the complement of a submanifold M of R^{m+n}, in which case S is $\boldsymbol{F}$-minimal on M). In addition to the hypotheses of ellipticity for the integrand F one needs to require also that F be *compatible* with the obstacles. This means that if M is the boundary of the obstacle A, $N \subset R^{m+n}$ is a suitable neighborhood of M, and $\rho: N \to M$ is a suitable smooth retraction then, for each $p \in M$ and each surface $T \subset R^{m+n}$,

$$\boldsymbol{F}^p(D\rho(p)(T)) \leqq \boldsymbol{F}^p(T).$$

The m-dimensional area integrand is compatible with any C^2 obstacle and in the space of integrands defined on a fixed C^3 submanifold M of R^{m+n} there is a convex neighborhood of the m-dimensional area integrand in the C^2 topology consisting of integrands on M which can be extended to be elliptic integrands defined in all of R^{m+n} each of which is compatible with the obstacle $R^{m+n} \sim M$. For these problems one essentially takes c to be the sum of $(\inf F)^{-1}\mathrm{Lip}(F)$ with the mth power of the maximum principal curvature of closure$(A) \sim A$ at any point and any normal direction.

(4) F is a variable coefficient elliptic integrand and S is a partitioning hypersurface between regions of prescribed volumes which minimizes an appropriately weighted $\boldsymbol{F}$ integral, the weights depending on the regions bounded. Here it does not seem that c can be estimated *a priori*. However, for each fixed solution hypersurface one can show the existence of an appropriate $c < \infty$.

(5) F is a variable coefficient elliptic integrand and S is a hypersurface (usually) minimizing the sum of $\boldsymbol{F}(S)$ with some bounded volume integral. For $F \equiv 1$, this general class of problems includes the study of surfaces of prescribed mean curvature and surfaces of capillarity.

6. **Reasons for studying parametric variational problems in the context of geometric measure theory.** As we mentioned earlier, the formulation of parametric variational problems in the context of nonparametrized measure theoretic surfaces is a substantial break with the century or so old tradition of studying these problems (largely unsuccessfully in higher dimensions) in the context of functional analysis. The phenomena which arise in parametric problems are quite varied. Indeed, for the case of the constant coefficient prescribed boundary (case (1) above) for $F \equiv 1$ (i.e. the problem of minimizing m-dimensional area—often called *Plateau's problem* in honor of the Belgian physicist, J. Plateau, of the last century who, among other things, studied the geometry of soap films), examples [**AF3**, p. 286] show:

(1) In order to solve the problem of least area—and really achieve the least area—one sometimes has to admit surfaces of infinite topological type into competition (even for two-dimensional surfaces whose boundaries are piecewise smooth simple closed curves).

(2) Complex algebraic varieties are surfaces of least oriented area, so that, in particular, at least all the singularities of complex algebraic varieties occur in solutions to Plateau's problem.

(3) In some cases there are topological obstructions to surfaces of least area being free of singularities.

(4) Sometimes surfaces of least area do not span their boundaries in the sense of algebraic topology.

(5) The realization of certain soap films as mathematical "minimal surfaces" requires that the boundary curves have positive thickness.

One of the most compelling reasons for formulating parametric variational problems in the context of geometric measure theory is that one can thereby treat in a natural way the phenomena of the above examples, while in the functional analysis setting one cannot take account of such phenomena in a natural way. The other principal reasons for the geometric measure theory formulation of the problem are listed in [**AF3**, p. 292].

7. **A property which implies the regularity of an m-rectifiable set in a neighborhood of a point.** The objective of this article is to illustrate the method by which a compact m-rectifiable set S which is suitably well behaved with respect to an elliptic integral $\boldsymbol{F}$ is shown to be a smooth submanifold with Hölder continuously turning tangent planes in a neighborhood of $\mathscr{H}^m \llcorner S$ almost all points. Since S is assumed to be m-rectifiable it follows that at $\mathscr{H}^m \llcorner S$ almost all points $p \in S$, S admits an approximate m-dimensional tangent plane. Furthermore, density ratio estimates enable one to show *a priori* that if S is suitably well behaved with respect to $\boldsymbol{F}$, then an approximate m-dimensional tangent plane to S at p is an actual m-dimensional tangent plane. Such a tangent plane is denoted $\operatorname{Tan}(S, p)$. In the regularity theory, one is primarily concerned with surfaces S which measure theoretically lie very close to being a flat m-disk of suitable radius. This is the basic starting condition of the regularity procedure, and this starting condition is realized at H^m almost all points of S as noted above. We need to make precise some ideas.

TERMINOLOGY. (1) For $k = 1, 2, 3, \ldots$, $x \in R^k$, $0 < r < \infty$, we set $B^k(x, r) = R^k \cap \{z : |z - x| \leqq r\}$, $\partial B^k(x, r) = R^k \cap \{z : |z - x| = r\}$.

(2) We denote by $\mathscr{A}$ the set of all pairs (S, r) such that $0 < r < \infty$ and S is a compact m-rectifiable set with

$$S \subset B^m(0, r) \times R^n \subset R^m \times R^n,$$

$$\partial S \subset \partial B^m(0, r) \times R^n \subset R^m \times R^n$$

(which implies $S \cap \{x\} \times R^n \neq \varnothing$ for each $x \in B^m(0, r)$) and

$$\mathscr{H}^m(S) \leqq \tfrac{3}{2}\mathscr{H}^m(B^m(0, r)).$$

(3) For each $(S, r) \in \mathscr{A}$ we define the *excess* $\operatorname{Exc}(S, r)$ as follows:

$$\operatorname{Exc}(S, r) = r^{-m}[\mathscr{H}^m(S) - \mathscr{H}^m(B^m(0, r))].$$

With this terminology we define the *basic regularity property* in terms of suitable *a priori* constants $0 < a_0 < \infty$, $0 < b_0 < \infty$, $0 < s_0 \leqq 8^{-1}$.

DEFINITION. $(S_0, r_0) \in \mathscr{A}$ is said to possess the *basic regularity property* (with respect to a_0, b_0, s_0) if and only if, for each sequence $x_1, x_2, x_3, \ldots \in B^m(0, 2^{-1})$, there exists a sequence $(S_1, r_1), (S_2, r_2), (S_3, r_3), \ldots \in \mathscr{A}$ with (S_ν, r_ν) depending only on $x_1, x_2, \ldots, x_\nu$ for each ν such that, for each $\nu = 0, 1, 2, \ldots$,

(1) $r_{\nu+1} = s_0 r_\nu = s_0^{\nu+1} r_0$.

(2) $S_\nu \cap B^m(0, 2^{-1}r_\nu) \times R^n \subset B^m(0, 2^{-1}r_\nu) \times B^n(0, a_0 r_\nu \operatorname{Exc}(S_\nu, r_\nu)^{1/(2m)})$.

(3) There exists $y_{\nu+1} \in B^n(0, a_0 r_\nu \operatorname{Exc}(S_\nu, r_\nu)^{1/2})$ and a linear function $l_\nu : R^m \to R^n$

with $|l_\nu| \leqq a_0 \operatorname{Exc}(S_\nu, r_\nu)^{1/2}$ such that

$$S_{\nu+1} = \Phi_\nu(S_\nu) \cap B^m(0, r_{\nu+1}) \times R^n,$$

where

$$\Phi_\nu : R^m \times R^n \to R^m \times R^n,$$

$$\Phi_\nu(x, y) = (x - r_\nu x_{\nu+1}, y - y_{\nu+1} - l_\nu(x - r_\nu x_{\nu+1})).$$

(4) $\operatorname{Exc}(S_{\nu+1}, r_{\nu+1}) \leqq 4^{-1}\operatorname{Exc}(S_\nu, r_\nu) + b_0 r_\nu$.

One then has the following regularity theorem.

THEOREM. *Let $(S_0, r_0) \in \mathscr{A}$ possess the basic regularity property with respect to a_0, b_0, s_0. Then there is a smooth function $f: B^m(0, 2^{-1}r_0) \to R^n$ with Hölder continuous first derivatives such that*

$$S_0 \cap B^m(0, 2^{-1}r_0) \times R^n = \operatorname{graph} f.$$

PROOF. Define $f: B^m(0, 2^{-1}r_0) \to R^n$ by the above condition. f is defined at all points of $B^m(0, 2^{-1}r_0)$ since, as noted earlier, the condition that $\partial S_0 \subset \partial B^m(0, r_0) \times R^n$ implies $S_0 \cap \{x\} \times R^n \neq \varnothing$ for each $x \in B^m(0, r_0)$. One verifies that $\operatorname{card}(S_0 \cap \{x\} \times R^n) = 1$ for each $x \in B^m(0, 2^{-1}r_0)$ by application of (2) of the basic regularity property to the sequence $x_1, x_2, x_3, \ldots \in B^m(0, 2^{-1})$, $x_1 = r_0^{-1}x$, $x_\nu = 0$ for $\nu = 2, 3, 4, \ldots$. To show the differentiability of f at $x \in B^m(0, 2^{-1}r_\nu)$ one chooses $x_1, x_2, x_3, \ldots \in B^m(0, 2^{-1})$ as above, and associates to this sequence the sequence $(S_1, r_1), (S_2, r_2), (S_3, r_3), \ldots \in \mathscr{A}$ and the sequence $l_1, l_2, l_3, \ldots$ of linear mappings $R^m \to R^n$ according to the basic regularity property. From the basic regularity property and elementary properties of geometric series (see the estimates for smoothness to follow) one concludes the existence of $\sum_{\nu=0}^{\infty} l_\nu : R^m \to R^n$ with $|Df(x)| \leqq \sum_{\nu=0}^{\infty} a_0 \operatorname{Exc}(S_\nu, r_\nu)^{1/2}$ such that $Df(x) = \sum_{\nu=0}^{\infty} l_\nu$. To verify the Hölder continuity of Df in $B^m(0, 2^{-1}r_0)$ we suppose $p, q \in B^m(0, 2^{-1}r_0)$ with $0 < |p - q| \leqq \min(2^{-1}, r_2)$ and denote by ν_0 the integer (larger than 1) for which $r_{\nu_0+1} < |p - q| \leqq r_{\nu_0}$. We define sequences $x_1, x_2, x_3, \ldots \in B^m(0, 2^{-1})$, $x_1', x_2', x_3', \ldots \in B^m(0, 2^{-1})$ by setting $x_1 = r_0^{-1}p$, $x_\nu = 0$ for $\nu > 1$ and $x_\nu' = x_\nu$ for $\nu \neq \nu_0$, $x_{\nu_0}' = r_{\nu_0-1}^{-1}(q - p)$. Let $(S_1, r_1), (S_2, r_2), (S_3, r_3), \ldots \in \mathscr{A}$, $(S_1', r_1), (S_2', r_2), (S_3', r_3), \ldots \in \mathscr{A}$ be the associated sequences of elements of $\mathscr{A}$ respectively and $l_1, l_2, l_3, \ldots$ and $l_1', l_2', l_3', \ldots$ be the associated sequences of linear mappings $R^m \to R^n$. Then $(S_\nu, r_\nu) = (S_\nu', r_\nu)$ and $l_\nu = l_\nu'$ for $\nu = 0, 1, \ldots, \nu_0 - 1$. As noted above,

$$Df(p) = \sum_{\nu=0}^{\infty} l_\nu \quad \text{and} \quad Df(q) = \sum_{\nu=0}^{\infty} l_\nu'.$$

Hence

$$Df(p) - Df(q) = \sum_{\nu=0}^{\infty} l_\nu - l_\nu' = \sum_{\nu=\nu_0}^{\infty} l_\nu - l_\nu'.$$

One sets $E_\nu = \max\{\operatorname{Exc}(S_\nu, r_\nu), \operatorname{Exc}(S_\nu', r_\nu)\}$ for each ν and observes

$$|Df(p) - Df(q)| = \left| \sum_{\nu=\nu_0}^{\infty} l_\nu - l'_\nu \right| \leqq \sum_{\nu=\nu_0}^{\infty} |l_\nu| + |l'_\nu|$$

$$\leqq \sum_{\nu=\nu_0}^{\infty} a_0 \operatorname{Exc}(S_\nu, r_\nu)^{1/2} + a_0 \operatorname{Exc}(S'_\nu, r_\nu)^{1/2}$$

$$\leqq 2a_0 \sum_{\nu=\nu_0}^{\infty} E_\nu^{1/2}.$$

The facts that $0 < s_0 \leqq 8^{-1}$ and $E_{\nu+1} \leqq 4^{-1}E_\nu + b_0 r_\nu$ for $\nu = 0, 1, 2, \ldots$ together with elementary summation of geometric series yield

$$E_{\nu+\mu} \leqq 4^{-\mu}(E_\nu + 8b_0 r_\nu)$$

for all nonnegative integers μ, ν. Hence

$$\sum_{\nu=\nu_0}^{\infty} E_\nu^{1/2} = E_{\nu_0}^{1/2} + \sum_{\mu=1}^{\infty} 2^{-\mu}(E_{\nu_0} + 8b_0 r_{\nu_0})^{1/2}$$

$$= E_{\nu_0}^{1/2} + (E_{\nu_0} + 8b_0 r_{\nu_0})^{1/2}$$

$$\leqq [4^{-\nu_0}(E_0 + 8b_0 r_0)]^{1/2} + [4^{-\nu_0}(E_0 + 8b_0 r_0) + 8b_0 r_0]^{1/2}.$$

One sets $\alpha = -(\ln 2)(\ln s_0)^{-1}$ so that $0 < \alpha < 2^{-1}$ and $4s_0^{2\alpha} = 1$. Then

$$s_0^{-1}|p - q|^{2\alpha} > s_0^{-1}|p - q| > r_{\nu_0}$$

and

$$(s_0 r_0)^{-2\alpha}|p - q|^{2\alpha} > (s_0^{\nu_0})^{2\alpha}4^{-\nu_0}.$$

One combines the above inequalities to obtain

$$|Df(p) - Df(q)| \leqq C|p - q|^\alpha$$

where

$$C = 2a_0\{(s_0 r_0)^{-\alpha}(E_0 + 8b_0 r_0)^{1/2} + [(s_0 r_0)^{2\alpha}(E_0 + 8b_0 r_0) + 8b_0 s_0^{-1}]^{1/2}\}$$

with $E_0 = \operatorname{Exc}(S_0, r_0)$ as above.

8. **The basic inequality of regularity theory.** Suppose that $F_0 : R^{m+n} \times G(m+n, m) \to R^+$ is a class 3 elliptic integrand and that $S_0 \subset R^{m+n}$ is a compact m-rectifiable set, $0 \leqq c_0 < \infty$, such that, for each $p \in S_0 \sim \partial S_0$ and each sufficiently small $r > 0$,

$$\boldsymbol{F}_0^p(S_0 \cap \{q : |q - p| \leqq r\}) \leqq (1 + c_0 r)\boldsymbol{F}_0^p(T_r)$$

whenever $T_r \subset \{q : |q - p| \leqq 2r\}$ is a compact m-rectifiable set with $\partial T_r = \partial[S_0 \cap \{q : |q - p| \leqq r\}]$. Suppose also that $p_0 \in S_0 \sim \partial S_0$ and $\operatorname{Tan}(S_0, p_0) \in G(m+n, m)$ exists (with $\mathscr{H}^m$-density one). We wish to show that S_0 is regular in a neighborhood of p_0. It will be necessary to study S_0 in a number of different coordinate systems. A change of coordinates of course changes F_0. We therefore wish to consider the various different forms in which F_0 must be considered. We hence define and fix $\mathscr{F}$ to be set of all constant coefficient integrands $F: R^{m+n} \times G(m+n, m) \to R^+$ of the form

$$F = L^\# \,\theta^\#\, F_0^p,$$

corresponding to all $p \in B^{m+n}(p_0, 1)$, all orthogonal maps θ: $R^{m+n} \to R^{m+n}$ and all linear maps $L: R^m \times R^n \to R^m \times R^n$ of the form $L(x, y) = (x, y + l(x))$ where $l: R^m \to R^n$ is linear with $\|l\| \leqq 1$. It is clear from the definitions that $\mathscr{F}$ is compact in the class 3 topology. Furthermore it can be shown that each $F \in \mathscr{F}$ is elliptic with respect to a uniform ellipticity bound γ [**AF2**, 4.3]. Furthermore with $c = 4c_0$ (for L as above, $\operatorname{Lip}(L) < 2$) one can show that if L, θ, p are as above and $F = L^\# \,\theta^\#\, F_0^p$, $S = L\theta\,(S_0)$ then, for each $q \in S \sim \partial S$ and each sufficiently small $r > 0$,

$$\boldsymbol{F}(S \cap \{z : |z - q| \leqq r\}) \leqq (1 + cr)\boldsymbol{F}(T_r)$$

whenever $T_r \subset \{z : |z - q| \leqq 2r\}$ is a compact m-rectifiable set with $\partial T_r = \partial[S \cap \{z : |z - q| \leqq r\}]$.

We can now state the *basic inequality of regularity theory.*

THEOREM. *There exists* $\varepsilon_0 > 0, b_0 < \infty$ *with the following property: Suppose* $x_* \in B^m(0, 2^{-1}), 0 < r < \varepsilon_0, (S, r) \in \mathscr{A}$ *with* $\operatorname{Exc}(S, r) < \varepsilon_0$. *Suppose also for some* $F \in \mathscr{F}, \boldsymbol{F}(S) \leqq (1 + cr)\boldsymbol{F}(T)$ *whenever* $T \subset B^m(0, r) \times B^n(0, r)$ *is a compact m-rectifiable set with* $\partial S = \partial T$. *Then there exists* $y_* \in B^n(0, a_0 r\operatorname{Exc}(S, r)^{1/2})$ *and a linear function* $l: R^m \to R^n$ *with* $\|l\| \leqq a_0 \operatorname{Exc}(S, r)^{1/2}$ *such that*

$$\operatorname{Exc}(S_*, s_0 r) \leqq 4^{-1}\operatorname{Exc}(S, r) + b_0 r$$

where

$$S_* = \Phi(S) \cap B^m(0, s_0 r) \times R^n,$$

$$\Phi: R^m \times R^n \to R^m \times R^n,$$

$$\Phi(x, y) = (x - rx_*, y - y_* - l(x - rx_*)).$$

It should be clear that if one can prove the basic inequality of regularity theory, then the regularity of S_0 in a neighborhood of p_0 will follow through the use of the regularity theorem of §7. The condition (2) of the basic regularity property of §7 is always satisfied by suitable choice of R^n coordinates [**AF2**, 4.9] and henceforth will not be considered.

9. **The basic inequality of regularity theory is implied by the nonexistence of certain sequences of surfaces in** $\mathscr{A}$. Among the strongest tools available to the geometric measure theorist are compactness theorems for various classes of surfaces. One therefore frequently likes to examine the behavior of suitable limits

of sequences of surfaces. Indeed, the proof of the basic inequality of regularity theory is reduced to showing the nonexistence of certain sequences of surfaces.

THEOREM. *Either there exist $\varepsilon_0 > 0$ and $b_0 < \infty$ with respect to which the basic inequality of regularity theory is valid or there exist $(S_1, r_1), (S_2, r_2), (S_3, r_3), \ldots \in \mathscr{A}$ with the following properties:*

(1) $\lim_\nu r_\nu = 0$;

(2) $\lim_\nu \operatorname{Exc}(S_\nu, r_\nu) = 0$;

(3) $\lim_\nu \operatorname{Exc}(S_\nu, r_\nu)^{-1} r_\nu = 0$;

(4) *for each* $\nu = 1, 2, 3, \ldots$ *there exists* $F_\nu \in \mathscr{F}$ *such that* $\boldsymbol{F}_\nu(S_\nu) \leqq (1 + cr_\nu)\boldsymbol{F}_\nu(T_\nu)$ *whenever* $T_\nu \subset B^m(0, r_\nu) \times B^n(0, r_\nu)$ *with* $\partial T_\nu = \partial S_\nu$;

(5) *for each* $\nu = 1, 2, 3, \ldots$ *there exists some* $x_\nu \in B^m(0, 2^{-1})$ *such that, for each linear function* $l: R^m \to R^n$ *with* $|l| \leqq a_0 \operatorname{Exc}(S_\nu, r_\nu)^{1/2}$,

$$\operatorname{Exc}(S_\nu^*, r_\nu s_0) > 4^{-1}\operatorname{Exc}(S_\nu, r_\nu)$$

where

$$S_\nu^* = \Phi_\nu(S_\nu) \cap B^m(0, r_\nu s_0) \times R^n,$$

$$\Phi_\nu : R^m \times R^n \to R^m \times R^n,$$

$$\Phi_\nu(x, y) = (x - r_\nu x_\nu, y - l(x)).$$

PROOF. The second alternative is implied by the negation of the first.

Henceforth, until §12, we will be concerned with showing the nonexistence of sequences having the properties (1), (2), (3), (4), (5) above. The critical property of these five is (3).

10. **A proof of the basic inequality of regularity theory in a special case.** Our purpose here is to give a fairly complete exposition of the proof of the Theorem 8 in the case when the surfaces under consideration are Lipschitz manifolds. The class of surfaces for which this theorem holds is, of course, much larger. Nonetheless, the proof we give here carries over to the general case.

Whenever $0 < \xi < \infty$ we let $\mathscr{L}_\xi$ be the class of functions f such that domain $f \subset R^m$, image $f \subset R^n$ and $|f(x) - f(\tilde{x})| \leqq \xi|x - \tilde{x}|$ whenever $x, \tilde{x} \in$ domain f. We then let $\mathscr{A}_\xi$ be the subclass of $\mathscr{A}$ consisting of those (S, r) for which $S \subset$ graph f for some $f \in \mathscr{L}_\xi$. Whenever $(S, r) \in \mathscr{A}_\xi$, we set $\partial S = S \cap \partial B^m(0, r) \times R^n$. We also set $\mathscr{A}_{\text{Lip}} = \bigcup \{\mathscr{A}_\xi : 0 < \xi < \infty\}$.

THEOREM. *There exist no sequences* $(S_1, r_1), (S_2, r_2), (S_3, r_3), \ldots \in \mathscr{A}_\xi$ *with the following properties*:

(1) $\lim_\nu r_\nu = 0$;

(2) $\lim_\nu \operatorname{Exc}(S_\nu, r_\nu) = 0$;

(3) $\lim_\nu \operatorname{Exc}(S_\nu, r_\nu)^{-1} r_\nu = 0$;

(4) *for each* $\nu = 1, 2, 3, \ldots$ *there exists* $F_\nu \in \mathscr{F}$ *such that* $\boldsymbol{F}_\nu(S_\nu) \leqq (1 + cr_\nu)\boldsymbol{F}_\nu(T_\nu)$ *whenever* $(T_\nu, r_\nu) \in \mathscr{A}_{\text{Lip}}$, $T_\nu \subset B^m(0, r) \times B^n(0, r)$ *and* $\partial T_\nu = \partial S_\nu$;

(5) *for each* $v = 1, 2, 3, \ldots$ *there exists some* $x_v \in B^m(0, 2^{-1})$ *such that for each linear function* $l: R^m \to R^n$ *with* $\|l\| \leqq a_0 \operatorname{Exc}(S_v, r_v)^{1/2}$ *it is false that*

$$\operatorname{Exc}(L(S_v) \cap D^m(0, s_0 r_v) \times R^n, s_0 r_v) \leqq 4^{-1} \operatorname{Exc}(S_v, r_v)$$

where $L(x, y) = (x - r_v x_v, y - l(x - r_v x_v)), (x, y) \in R^m \times R^n$.

(A) *Preliminaries.* In this section the only measure that occurs is m-dimensional Hausdorff measure $\mathscr{H}^m$ so we write $|A| = \mathscr{H}^m(A)$ and $\int_A f$ or $\int_A f(x)\,dx$ for $\int_A f(x)\,d\mathscr{H}^m x$.

Let $M(n, m)$ be the vector space of $n \times m$ matrices. Whenever $p \in M(n, m)$, p_i^k is the entry in the ith column and kth row, $i = 1, \ldots, m$, $k = 1, \ldots, n$. We set

$$p \cdot q = \sum_{i,k} p_i^k q_i^k, \qquad |p| = \left(\sum_{i,k} (p_i^k)^2 \right)^{1/2}.$$

For each $p \in M(n, m)$,

$$\mu(p) = R^m \times R^n \cap \left\{ (x, y) : y_k = \sum_i p_i^k x_i, k = 1, \ldots, n \right\},$$

$$\omega(p) = |\mu(p) \cap B^m(0, 1) \times R^n| / |B^m(0, 1)|.$$

It is elementary that $|p| \leqq \omega(p)$ and that for each $\varepsilon > 0$ there is a number M_ε with the property that

$$|p|^2 \leqq M_\varepsilon[\omega(p) - 1] \quad \text{if } |p| \leqq \varepsilon,$$

$$\omega(p) \leqq M_\varepsilon[\omega(p) - 1] \quad \text{if } |p| > \varepsilon.$$

We will use the sets $B^m(0, r)$ frequently so we set $B(r) = B^m(0, r)$, $B = B(1)$.

Let $\mathscr{D}$ be the space of smooth compactly supported functions on B with values in R^n with the usual strong topology and let $\mathscr{D}'$ be its dual space; that is, $\mathscr{D}'$ is the space of Schwartz distributions on B with values in R^n.

Suppose $(S, r) \in \mathscr{A}_{\mathrm{Lip}}$ and $S \subset \operatorname{graph} f$ for some $f \in \mathscr{L}_\xi$, $0 < \xi < \infty$. As is well known, f is differentiable at $\mathscr{H}^m$ almost all points of $B(r)$. If x is such a point, we let $Df(x) \in M(n, m)$ be the matrix with entry $(\partial f_k / \partial x_i)(x)$ in the ith column and kth row; let $\mathbf{S}(x) = \mu(Df(x)) \in G(m + n, m)$ be the tangent space to S at $(x, f(x))$ and let $Jf(x) = \omega(Df(x))$. As is well known,

$$|A \cap S| = \int_{\{x : (x, f(x)) \in A\}} Jf \qquad \text{for any Borel set } A \subset R^m \times R^n. \tag{1}$$

In particular,

$$\operatorname{Exc}(S, r) = r^{-m} \int_{B(r)} Jf - 1.$$

For each $\varepsilon > 0$, let

$$A_\varepsilon = B(r) \cap \{x : |Df(x)| \leqq \varepsilon\}, \qquad B_\varepsilon = B(r) \sim A_\varepsilon.$$

We make the basic estimates

$$\begin{aligned}|B_\varepsilon| &\leqq |S \cap B_\varepsilon \times R^n| = \int_{B_\varepsilon} Jf \\ &\leqq M_\varepsilon \int_{B_\varepsilon} Jf - 1 \leqq M_\varepsilon r^m \mathrm{Exc}(S, r);\end{aligned} \tag{2}$$

$$\int_{A_\varepsilon} |Df|^2 \leqq M_\varepsilon \int_{A_\varepsilon} Jf - 1 \leqq M_\varepsilon r^m \mathrm{Exc}(S, r); \tag{3}$$

$$\int_{B_\varepsilon} |Df| \leqq M_\varepsilon \int_{B_\varepsilon} Jf - 1 \leqq M_\varepsilon r^m \mathrm{Exc}(S, r); \tag{4}$$

$$\int_{A_\varepsilon} |Df| \leqq |A_\varepsilon|^{1/2} \left(\int_{A_\varepsilon} |Df|^2 \right)^{1/2} \leqq (M_\varepsilon |B|)^{1/2} r^m \mathrm{Exc}(S, r)^{1/2}. \tag{5}$$

(B) *Associating a nonparametric integrand to a parametric integrand.* Suppose $F : G(m + n, m) \to R$ is C^3. For each $p \in M(n, m)$, let $G(p) = F(\mu(p))\omega(p)$. Let

$$C = G(0); \qquad L(i, k) = \frac{\partial G}{\partial p_i^k}(0), \qquad i = 1, \ldots, m, \qquad k = 1, \ldots, n;$$

$$Q(i, j, k, l) = \frac{\partial^2 G}{\partial p_i^k \, \partial p_j^l}(0), \qquad i, j = 1, \ldots, m; \quad k, l = 1, \ldots, n.$$

Whenever $p, q \in M(n, m)$, let

$$L(p) = \sum_{i,k} L(i, k) p_i^k, \qquad Q(p, q) = \sum_{i,j,k,l} Q(i, j, k, l) p_i^k q_j^l,$$

$$H(p) = G(p) - [C + L(p) + \tfrac{1}{2} Q(p, p)].$$

Using Taylor's theorem we secure a positive real number ε such that

$$|H(p)| \leqq \varepsilon^{-1} |p|^3 \quad \text{if } |p| \leqq \varepsilon, \tag{1}$$

$$|H(p + q) - H(p)| \leqq \varepsilon^{-1}(|p|^2 + |q|^2)|q| \quad \text{if } |p| \leqq \varepsilon, \ |q| \leqq \varepsilon. \tag{2}$$

Finally, we define a second order linear partial differential operator $\mathscr{Q} : \mathscr{D}' \to \mathscr{D}'$ with constant coefficients by setting

$$(\mathscr{Q}u)_l = \sum_{i,j,k} Q(i, j, k, l) \, \partial^2 u_k / \partial x_i \, \partial x_j, \qquad u \in \mathscr{D}', l = 1, \ldots, n;$$

the differentiations here are carried out in the sense of distribution theory.

If F is the m-dimensional parametric area integrand, that is if $F(\Pi) = 1$ for all $\Pi \in G(n, m)$, we have that $G(p) = \omega(p)$ for $p \in M(n, m)$, $C = 1$, $L(p) = 0$ for p

$\in M(n, m)$, $Q(p, q) = p \cdot q$ for $p, q \in M(n, m)$ and $\mathcal{Q}u = \Delta u$, $u \in \mathcal{D}'$, where $(\Delta u)_l = \sum_i \partial^2 u_l/(\partial x_i)^2$ for $u \in \mathcal{D}'$.

(C) LEMMA. *Suppose $F: G(m + n, m) \to R$ is C^2. There is a real number N depending only on F such that*

$$|r^{-m}\operatorname{Exc}(S, r)^{-1}[\boldsymbol{F}(S) - \boldsymbol{F}(B(r))]| \leqq N$$

whenever $(S, r) \in \mathcal{A}_\infty$ and $\partial S = \partial B(r)$.

PROOF. Taking ε as in (B)(1) we write

$$\boldsymbol{F}(S) - \boldsymbol{F}(B(r)) = \int_{A_\varepsilon} C + L(Df) + \tfrac{1}{2}Q(Df, Df) + H(Df)$$

$$+ \int_{B_\varepsilon \times R^n} F(\boldsymbol{S}) - \int_{B(r)} C,$$

where $S = \operatorname{graph} f$, $A_\varepsilon = \{x : |Df(x)| \leqq \varepsilon\}$ and $B_\varepsilon = B(r) \sim A_\varepsilon$. Now

$$\int_{A_\varepsilon} C - \int_{B(r)} C = -\int_{B_\varepsilon} C, \qquad \int_{A_\varepsilon} L(Df) = -\int_{B_\varepsilon} L(Df)$$

because $\partial S = \partial B(r)$. One now applies the estimates (A)(2), (3), (4).

(D) LEMMA. *Suppose $0 < \gamma < \infty$, $F: G(m + n, m) \to R$ is C^2 and $\boldsymbol{F}(S) - \boldsymbol{F}(B) \geqq \gamma(|S| - |B|)$ whenever $S =$ graph φ for some $\varphi \in \mathcal{D}$.*

Then, with Q as in (B),

(1) $$\int_B Q(D\varphi, D\varphi) \geqq \gamma \int_B |D\varphi|^2, \qquad \varphi \in \mathcal{D}.$$

Moreover, (1) *implies*

(2) $$\sum_{i,j,k,l} Q(i, j, k, l)\xi_i\xi_j\eta_k\eta_l \geqq \gamma|\xi|^2|\eta|^2$$

whenever $\xi \in R^m$ and $\eta \in R^n$.

PROOF. Fixing $\varphi \in \mathcal{D}$, set $S_t =$ graph $t\varphi$ for $t \in R \sim \{0\}$ and compute, using (B)(1),

$$t^{-2}[\boldsymbol{F}(S_t) - \boldsymbol{F}(B)] = \int_B \tfrac{1}{2}Q(D\varphi, D\varphi) + t^{-2}H(tD\varphi)$$

$$\to \tfrac{1}{2}\int_B Q(D\varphi, D\varphi), \quad \text{as } t \to 0.$$

For the area integrand we obtain

$$\lim_{t \to 0} t^{-2}[|S_t| - |B|] = \tfrac{1}{2}\int_B |D\varphi|^2,$$

and (1) is proved.

One can deduce (2) from (1) with the help of Parseval's formula (see [**AF2**, 5.4]). See also [**FH1**, 5.1.10] for a different proof of (2).

(E) LEMMA. *Suppose $F_1, F_2, \ldots, F$ are C^3 parametric integrands with* $\lim_\nu F_\nu = F$ *in the C^3 topology*; $(S_1, r_1), (S_2, r_2), \ldots \in \mathscr{A}_{\mathrm{Lip}}$; $S_\nu = \operatorname{graph} f_\nu$, $\nu = 1, 2, \ldots$; $E_1, E_2, \ldots$ *are positive real numbers such that* $\lim_\nu E_\nu = 0$, $\alpha = \limsup_\nu E_\nu^{-1}\operatorname{Exc}(S_\nu, r_\nu) < \infty$, $g_\nu(x) = r_\nu^{-1}E_\nu^{-1/2}f_\nu(r_\nu x)$ *for* $x \in B$. *Then*

(1) *if*

$$\lim_\nu r_\nu^{-m-1}E_\nu^{-1/2}\left|\int_{B(r_\nu)} f_\nu\right| < \infty$$

there is an R^n valued $\mathscr{H}^m \cap B$ measurable function h such that

$$\liminf_\nu \int_B |g_\nu - h| = 0;$$

(2) *for any h as in* (1),

$$\int_B \sum_{i,k} h_k \frac{\partial \varphi_k}{\partial x_i} \leqq M_1\alpha\left(\int_B |\varphi|^2\right)^{1/2}, \qquad \varphi \in \mathscr{D};$$

(3) *for any $\varphi \in \mathscr{D}$,*

$$\liminf_\nu r_\nu^{-m}E_\nu^{-1}[\boldsymbol{F}_\nu(\Phi_\nu(S_\nu)) - \boldsymbol{F}_\nu(S_\nu)] - \mathscr{Q}(h + 2\varphi)(\varphi) = 0$$

where $\Phi_\nu(x, y) = (x, y + r_\nu E_\nu^{1/2}\varphi(r_\nu^{-1}x))$ *for* $(x, y) \in R^m \times R^n$, $\mathscr{Q}$ *is related to F as in* (B) *and h is as in* (1).

PROOF. For each $\varepsilon > 0$ let $A_{\nu,\varepsilon} = B(r_\nu) \cap \{x : |Df_\nu(x)| \leqq \varepsilon\}$, $B_{\nu,\varepsilon} = B(r_\nu) \sim A_{\nu,\varepsilon}$. It is elementary that, for some constant N,

$$\int_B |g_\nu| \leqq \left|\int_B g_\nu\right| + N\int_B |Dg_\nu|, \qquad \nu = 1, 2, \ldots;$$

we have that

$$\lim_\nu \left|\int_B g_\nu\right| = \lim_\nu r_\nu^{-m-1}E_\nu^{-1/2}\left|\int_{B(r_\nu)} f_\nu\right| < \infty$$

and from (A)(4), (5) that

$$\int_B |Dg_\nu| = r_\nu^{-m}E_\nu^{-1/2}\left(\int_{A_{\nu,1}} |Df_\nu| + \int_{B_{\nu,1}} |Df_\nu|\right)$$
$$\leqq E_\nu^{-1/2}[(M_1|B|)^{1/2}\operatorname{Exc}(S_\nu, r_\nu)^{1/2} + M_1\operatorname{Exc}(S_\nu, r_\nu)],$$

$$\lim_\nu \int_B |Dg_\nu| < \infty;$$

well-known arguments now imply the existence of h.

With regard to (2), write

$$\int_B \sum_{i,k} g_{\nu,k} \frac{\partial \varphi_k}{\partial x_i} = -\int_{A_{\nu,1}} \sum_{i,k} \frac{\partial g_{\nu,k}}{\partial x_i} \varphi_k - \int_{B_{\nu,1}} \sum_{i,k} \frac{\partial g_{\nu,k}}{\partial x_i} \varphi_k$$

$$\leqq \left(\int_{A_{\nu,1}} |Dg_\nu|^2\right)^{1/2} \left(\int_B |\varphi|^2\right)^{1/2} + (\sup |\varphi|) \int_{B_{\nu,1}} |Dg_\nu|$$

and apply (A)(3), (4).

To prove (3) we let $\varphi_\nu(x) = r_\nu E_\nu^{1/2} \varphi(r_\nu^{-1} x)$ for $x \in B(r_\nu)$ and $\nu = 1, 2, \ldots$ and write $\boldsymbol{F}_\nu(\Phi_\nu(S_\nu)) - \boldsymbol{F}_\nu(S_\nu)$ as the sum of the terms

(i) $\int_{B(r_\nu)} L_\nu(D\varphi_\nu)$,

(ii) $-\int_{B_{\nu,\varepsilon}} L_\nu(D\varphi_\nu)$,

(iii) $r_\nu^m E_\nu \mathcal{Q}_\nu(g_\nu + 2\varphi)(\varphi)$,

(iv) $-\int_{B_{\nu,\varepsilon}} Q_\nu(Df_\nu + 2D\varphi_\nu, D\varphi_\nu)$,

(v) $\int_{A_{\nu,\varepsilon}} H_\nu(Df_\nu + D\varphi_\nu) - H_\nu(Df_\nu)$,

(vi) $\int_{S_\nu \cap B_{\nu,\varepsilon} \times R^n} (\Phi_{\nu\#} F_\nu - F_\nu)(\boldsymbol{S})$,

where L_ν, Q_ν, H_ν, $\mathcal{Q}_\nu$ are associated with F_ν as in (B) and ε is a number such that (B)(2) is satisfied with H replaced by H_ν for all $\nu = 1, 2, \ldots$. Observe that $\operatorname{Lip} \varphi_\nu = E_\nu^{1/2} \operatorname{Lip} \varphi$; using this fact and (A)(2) to estimate (ii), and (A)(4) to estimate (iv), and (A)(3) to estimate (v), and (A)(2) to estimate (vi) we see that these terms when divided by $r_\nu^m E_\nu$ tend to zero as $\nu \to \infty$. Term (i) vanishes because φ_ν has compact support in $B(r_\nu)$. Finally, it is clear that

$$\liminf_\nu \mathcal{Q}_\nu(g_\nu + 2\varphi)(\varphi) - \mathcal{Q}(h + 2\varphi)(\varphi) = 0.$$

(F) Suppose $(S_1, r_1), (S_2, r_2), (S_3, r_3), \ldots$ is a sequence in A_ξ satisfying conditions (1), (2), (3), (4), of the theorem. Passing to a subsequence if necessary we may assume that $\lim_\nu F_\nu = F \in \mathscr{F}$ in the C^3 topology and that $\lim_\nu x_\nu = \tilde{x} \in B(\frac{1}{2})$. Let $E_\nu = \operatorname{Exc}(S_\nu, r_\nu)$. Suppose f_ν is such that $S_\nu = \operatorname{graph} f_\nu$, $\nu = 1, 2, \ldots$. We may assume without loss of generality that

$$\int_{B(r_\nu)} f_\nu = 0, \qquad \nu = 1, 2, \ldots .$$

Let g_ν, $\nu = 1, 2, \ldots$, be as in (E). Passing again to a subsequence if necessary, we use (E)(1), (2) to secure an R^n valued $\mathscr{H}^m \cap B$ measurable function h such that

(1) $$\lim_\nu \int_B |g_\nu - h| = 0,$$

(2) $$\int_B \sum_{i,k} h_k \frac{\partial \varphi_k}{\partial x_i} \leqq M_1 \left(\int_B |\varphi|^2\right)^{1/2}, \qquad \varphi \in \mathscr{D}.$$

Suppose $t \in R$ and $\psi \in \mathscr{D}$ and let $\varphi = t\psi$. Let Φ_ν be as in (E)(3) and observe that $\partial\Phi_\nu(S_\nu) = \partial S_\nu$ and $\lim_\nu r_\nu E_\nu^{-1} = 0$ so that

$$r_\nu^{-m}E_\nu^{-1}[\boldsymbol{F}_\nu(\Phi_\nu(S_\nu)) - \boldsymbol{F}_\nu(S_\nu)] \geqq -cr_\nu E_\nu^{-1}r_\nu^{-m}\boldsymbol{F}_\nu(\Phi_\nu(S_\nu))$$
$$\to 0 \quad \text{as } \nu \to \infty.$$

By (E)(3), $\mathscr{Q}(h + 2t\psi)(t\psi) \leqq 0$; varying t we have that $\mathscr{Q}h(\psi) = 0$ so that

$$\mathscr{Q}h = 0. \tag{3}$$

Owing to strong ellipticity of $\mathscr{Q}$ as asserted by (D) and owing to the bound (2) we have that h is a real analytic function any of whose derivatives of order greater than zero at a point of B can be estimated *a priori* in terms of the distance of the point to ∂B, the order of the derivative, the ellipticity bound for F and the supremum of the numbers $Q(i, j, k, l)$. See [**FH1**, 5.2].

Let σ be a number with $\beta = 2^{-1} + 3s_0 < \sigma < 1$ and let $\zeta \in \mathscr{D}$ be such that $\zeta(x) = 1$ if $|x| < \sigma$. Let $\theta = \zeta h \in \mathscr{D}$. For each $\nu = 1, 2, \ldots$ we set $\theta_\nu(x) = r_\nu E_\nu^{1/2}\theta(r_\nu^{-1}x)$ for $x \in B(r_\nu)$, $f_\nu^1 = f_\nu - \theta_\nu$, $S_\nu^1 = \operatorname{graph} f_\nu^1$, $g_\nu^1(x) = r_\nu^{-1}E_\nu^{-1/2}f_\nu^1(r_\nu x) = g_\nu(x) - \theta(x)$ $= g_\nu(x) - \zeta(x)h(x)$ for $x \in B$, $h^1(x) = (1 - \zeta(x))h(x)$ for $x \in B$. Evidently,

$$\lim_\nu \int_B |g_\nu^1 - h^1| = 0, \tag{4}$$

$$\lim_\nu \int_{B(\sigma)} |g_\nu^1| = 0. \tag{5}$$

The main assertion of this section (F) is that

$$\lim_\nu r_\nu^{-m}E_\nu^{-1}[|S_\nu^1 \cap B(\beta r_\nu) \times R^n| - |B(\beta r_\nu)|] = 0. \tag{6}$$

We now proceed to prove (6). We apply (E)(3) twice to calculate

$$\lim_\nu r_\nu^{-m}E_\nu^{-1}[\boldsymbol{F}_\nu(S_\nu) - \boldsymbol{F}_\nu(S_\nu^1)] = \mathscr{Q}(h^1 + 2\theta)(\theta), \tag{7}$$

$$\lim_\nu r_\nu^{-m}E_\nu^{-1}[|S_\nu| - |S_\nu^1|] = \Delta(h^1 + 2\theta)(\theta); \tag{8}$$

in particular,

$$\begin{aligned} \lim_\nu E_\nu^{-1}\operatorname{Exc}(S_\nu^1, r_\nu) &= \lim_\nu r_\nu^{-m}E_\nu^{-1}[|S_\nu^1| - |B(r_\nu)|] \\ &= \lim_\nu r_\nu^{-m}E_\nu^{-1}[|S_\nu^1| - |S_\nu|] + r_\nu^{-m}E_\nu^{-1}[|S_\nu| - |B(r_\nu)|] \\ &= -\Delta(h^1 + 2\theta)(\theta) + 1 < \infty. \end{aligned} \tag{9}$$

Owing to the fact that $f_\nu \in \mathscr{L}_\xi$ we have that

$$\lim_\nu \sup\{r_\nu^{-1}|f_\nu(x)| : x \in B(r_\nu)\} = 0 \tag{10}$$

because $\lim_\nu r_\nu^{-m-1}\int_{B(r_\nu)} |f_\nu| = 0$.

We assert that for each sufficiently large ν there are ρ_ν with $\beta < \rho_\nu < \sigma$, $\varepsilon_\nu > 0$ and $(S_\nu^2, r_\nu) \in \mathscr{A}_{\mathrm{Lip}}$ such that if f_ν^2 is such that $S_\nu^2 = \operatorname{graph} f_\nu^2$,

$$f_\nu^2(x) = f_\nu^1(x), \qquad |x| = \rho_\nu r_\nu; \tag{11}$$

$$\partial S_\nu^2 = \partial B(r_\nu); \tag{12}$$

$$\operatorname{Exc}(S_\nu^2, r_\nu) \leqq \varepsilon_\nu E_\nu; \tag{13}$$

$$\int_{B(r_\nu)} |f_\nu^2| \leqq \varepsilon_\nu r_\nu^{m+1} E_\nu^{1/2}; \tag{14}$$

$$\lim_\nu \varepsilon_\nu = 0. \tag{15}$$

This follows from a slightly modified version of [**AF2**, Lemma 4.10] because (9) and (10) hold and $\lim_\nu E_\nu = 0$. This lemma is rather subtle but its proof in our case uses only elementary methods. Passing again to a subsequence if necessary, we may assume that $\lim_\nu \rho_\nu = \rho$ with $\beta \leqq \rho \leqq \sigma$.

We now define functions f_ν^i and members (S_ν^i, r_ν) of $\mathscr{A}_{\mathrm{Lip}}$, $i = 3, 4, 5, 6$, as follows:

$$\begin{aligned}
f_\nu^3(x) &= f_\nu^1(x), && |x| \leqq \rho_\nu r_\nu,\\
&= f_\nu^2(x), && \rho_\nu r_\nu < |x| < r_\nu;\\
f_\nu^4(x) &= f_\nu^2(x), && |x| \leqq \rho_\nu r_\nu,\\
&= f_\nu^1(x), && \rho_\nu r_\nu < |x| < r_\nu;\\
f_\nu^5(x) &= f_\nu^4(x) + \theta_\nu(x), && |x| < r_\nu;\\
f_\nu^6(x) &= 0, && |x| < r_\nu;\\
S_\nu^i &= \operatorname{graph} f_\nu^i, && i = 3, 4, 5, 6.
\end{aligned}$$

Identifying sets with characteristic functions, we have that $S_\nu^3 + S_\nu^4 = S_\nu^1 + S_\nu^2$ so that

$$\boldsymbol{F}_\nu(S_\nu^3) - \boldsymbol{F}_\nu(S_\nu^6) = [\boldsymbol{F}_\nu(S_\nu^1) - \boldsymbol{F}_\nu(S_\nu^4)] + [\boldsymbol{F}_\nu(S_\nu^2) - \boldsymbol{F}_\nu(S_\nu^6)]. \tag{16}$$

Since $\partial S_\nu^5 = \partial S_\nu$, we have $\boldsymbol{F}_\nu(S_\nu) \leqq (1 + c r_\nu)\boldsymbol{F}_\nu(S_\nu^5)$ or

$$0 \leqq [\boldsymbol{F}_\nu(S_\nu^5) - \boldsymbol{F}_\nu(S_\nu)] + c r_\nu \boldsymbol{F}_\nu(S_\nu^5). \tag{17}$$

If γ is an ellipticity bound for the F_ν we have (see (12))

$$\gamma[|S_\nu^3| - |S_\nu^6|] \leqq \boldsymbol{F}_\nu(S_\nu^3) - \boldsymbol{F}_\nu(S_\nu^6). \tag{18}$$

Combining (16), (17), (18) we have that

$$\begin{aligned}
\gamma[|S_\nu^3| - |S_\nu^6|] \leqq{}& [\boldsymbol{F}_\nu(S_\nu^1) - \boldsymbol{F}_\nu(S_\nu)] + [\boldsymbol{F}_\nu(S_\nu^5) - \boldsymbol{F}_\nu(S_\nu^4)]\\
&+ [\boldsymbol{F}_\nu(S_\nu^2) - \boldsymbol{F}_\nu(S_\nu^6)] + c r_\nu \boldsymbol{F}_\nu(S_\nu^5).
\end{aligned} \tag{19}$$

By (9) and (13) we estimate

$$\limsup_{\nu} E_\nu^{-1}\operatorname{Exc}(S_\nu^4, r_\nu) \leqq \lim_{\nu} E_\nu^{-1}[\operatorname{Exc}(S_\nu^2, r_\nu) + \operatorname{Exc}(S_\nu^1, r_\nu)] < \infty;$$

passing again to a subsequence if necessary we may use (E)(1) to secure an R^n valued $\mathscr{H}^m \llcorner B$ measurable function h^4 such that

$$\lim_{\nu} \int_B |g_\nu^4 - h^4| = 0$$

where $g_\nu^4(x) = r_\nu^{-1}E_\nu^{-1/2}f_\nu^4(r_\nu x)$, $x \in B(r_\nu)$. By (5), (14) and the definition of h^4 we see that $h^4 = h^1$. Applying (E)(3), we have that

(20) $$\lim_{\nu} r_\nu^{-m}E_\nu^{-1}[\boldsymbol{F}_\nu(S_\nu^5) - \boldsymbol{F}_\nu(S_\nu^4)] = \mathscr{Q}(h^1 + 2\theta)(\theta).$$

By (13) and (C) we see that

(21) $$\lim_{\nu} r_\nu^{-m}E_\nu^{-1}[\boldsymbol{F}_\nu(S_\nu^2) - \boldsymbol{F}_\nu(S_\nu^6)] = 0.$$

Since it is clear that $\lim_\nu r_\nu^{-m}\boldsymbol{F}_\nu(S_\nu^5) < \infty$, we conclude from (7), (20), (21) and the fact that $\lim_\nu r_\nu E_\nu^{-1} = 0$ that

$$\lim_{\nu} r_\nu^{-m}E_\nu^{-1}[|S_\nu^3| - |S_\nu^6|] = 0$$

which implies (6).

(G) We now complete the proof of the theorem. For each $x \in R^m$, let $l(x) = \sum_i (\partial h/\partial x_i)(\tilde{x})(x_i - \tilde{x}_i) \in R^n$. Choose a function $\psi \in \mathscr{D}$ such that $\psi(x) = 1$ if $|x - \tilde{x}| < 2s_0$, $\psi(x) = 0$ if $|x - \tilde{x}| > 3s_0$ and $|\operatorname{grad} \psi(x)| \leqq 2/s_0$ for $x \in B$. Let

$$\lambda(x) = \beta^{-1}\psi(\beta x)[h(\beta x) - h(\tilde{x}) - l(\beta x - \tilde{x})] \quad \text{for } x \in B(\beta^{-1}),$$

note that $\lambda \in \mathscr{D}$, and observe that, by remarks made in (F) and Taylor's theorem,

(1) $$|D\lambda(x)| \leqq N_1|x - \tilde{x}|, \qquad x \in B,$$

where N_1 is a number depending only on $\mathscr{F}$. Now let

$$\tilde{S}_\nu^1 = S_\nu^1 \cap B(\beta r_\nu) \times R^n,$$

$$\Lambda_\nu(x, y) = (x, y + \beta r_\nu E_\nu^{1/2}\lambda((\beta r_\nu)^{-1}x)), \qquad (x, y) \in R^m \times R^n.$$

In view of (F)(9), (E)(3) and (4), we have

$$\lim_{\nu} (\beta r_\nu)^{-m}E_\nu^{-1}[|\Lambda_\nu(\tilde{S}_\nu^1)| - |\tilde{S}_\nu^1|] = \Delta(\lambda)(\lambda) = \int_B |D\lambda|^2$$

$$\leqq N_1 \int_{B^m(\tilde{x}, 3s_0)} |x|^2\, dx \leqq N_2 s_0^{m+2},$$

where N_2 is a number depending only on $\mathscr{F}$. In view of (F)(6) we see that, for sufficiently large ν,

$$(s_0 r_\nu)^{-m} E_\nu^{-1}[|\Lambda_\nu(\tilde{S}_\nu^1)| - |B(\beta r_\nu)|] \leqq 2N_2 \beta^m s_0^2.$$

Finally, we let

$$A_\nu(x, y) = (x, y - r_\nu E_\nu^{1/2}(h(\tilde{x}) - l(r_\nu^{-1}x))), \qquad (x, y) \in R^m \times R^n$$

and observe that

$$A_\nu(S_\nu) \cap B^m(r_\nu \tilde{x}, 2s_0 r_\nu) \times R^n = \Lambda_\nu(\tilde{S}_\nu^1) \cap B^m(r_\nu \tilde{x}, 2s_0 r_\nu) \times R^n.$$

11. **Coordinate changes for integrands in $\mathscr{F}$ which eliminate the linear components of these integrands.** As we have already noted the proof of the regularity of a surface S_0 in a neighborhood of a point p_0 where $\operatorname{Tan}(S_0, p_0) \in G(m+n, m)$ exists reduces to showing the nonexistence of certain sequences $(S_1, r_1), (S_2, r_2), (S_3, r_3), \ldots \in \mathscr{A}$ together with a corresponding sequence of integrands $F_1, F_2, F_3, \ldots \in \mathscr{F}$. The method of proof closely parallels the argument of the last section showing the nonexistence of certain sequences $(S_1, r_1), (S_2, r_2), (S_3, r_3), \ldots \in \mathscr{A}_\xi$. Indeed, we need to look at the nonparametric integrands $G: M(n, m) \to R^+$ associated with a parametric integrand $F \in \mathscr{F}$. Suppose then, $F \in \mathscr{F}$. Then F can be integrated over the graphs of smooth functions $f: R^m \to R^n$, and this integration can be expressed in terms of a nonparametric integrand $G: M(n, m) \to R^+$ associated with F by the requirement

$$\boldsymbol{F}(\operatorname{graph} f) = \boldsymbol{G}(f) = \int_{B^m(0,1)} G(Df)\, d\mathscr{L}^m$$

for each class 1 function $f: B^n(0, 1) \to R^n$. Since F is of class 3, G will also be of class 3. As observed previously one can write, for $p \in R^{mn}$,

$$G(p) = C + \sum_{i,k} L(i, k) p_i^k + 2^{-1} \sum_{i,j,k,l} Q(i, j, k, l) p_i^k p_j^l + H(p).$$

When one evaluates G on smooth functions f, integration by parts shows that

$$\int_{B^m(0,1)} \sum_{i,k} L(i, k)(\partial f^k/\partial x_i)\, d\mathscr{L}^m$$

$$= \sum_{i,k} L(i, k) \int_{z \in B^{m-1}(0,1)} f^k\left(z_1, \ldots, z_{i-1}, \left(1 - \sum_j z_j^2\right)^{1/2}, z_i, \ldots, z_m\right)$$

$$- f^k\left(z_1, \ldots, z_{i-1}, -\left(1 - \sum_j z_j^2\right)^{1/2}, z_i, \ldots, z_m\right) d\mathscr{L}^{m-1} z$$

so that this integral depends only on the boundary values of f and is unaffected by variations of the form $f + t\phi$ corresponding to $t \in R$ and class ∞ ϕ with $\phi | \partial B^m(0, 1) = 0$. Indeed this fact was utilized in the proof of the theorem of the last section. However, when one must integrate F (or G) over m-rectifiable sets S_0, one does not know *a priori* that integration by parts is justified and one cannot conclude the absence of any contribution from the linear component of G. This is quite serious since for small first derivatives (or small tilts to the tangent planes of S_0 relative to $R^m \times \{0\}$) the linear component of G strongly dominates the quadratic component —so much so that one cannot proceed directly with the proof. At this point one is forced to modify the integrands F in $\mathscr{F}$ by a change of coordinates which eliminates the linear components. Indeed if F, G are as above then the change of coordinates

$$\Phi_F : R^m \times R^n \to R^m \times R^n,$$

$$\Phi_F(x, y) = \left(x_1 + C^{-1} \sum_k L(1, k) y_k, \ldots, x_m + C^{-1} \sum_k L(m, k) y_k, y_1, \ldots, y_n\right)$$

transforms F into a new integrand $J = \Phi_{F\#} F$ having associated nonparametric integrand K expressible

$$K(p) = C + \sum_{i,j,k,l} Q'(i, j, k, l) p_i^k p_j^l + H'(p),$$

i.e. the first order terms in the p_i^k vanish identically. We denote by $\mathscr{J}$ the set of all integrands J obtained from integrands $F \in \mathscr{F}$ by the procedure just described. $\mathscr{J}$ is clearly also compact in the class 3 topology. It is clear also that the set of linear maps Φ_F as above corresponding to all $F \in \mathscr{F}$ is a compact subset of the general linear group. Moreover, one is able to show the following theorem [**AF2**, 4.3, 7.1, 7.2].

THEOREM. *There exists $b_1 < \infty$ such that if $F \in \mathscr{F}$ and $(S, r) \in \mathscr{A}$ with $S \subset B^m(0, r) \times B^n(0, b_1^{-1} r)$, then*

$$\operatorname{Exc}(\psi(S) \cap B^m(0, 7r/8) \times R^n, 7r/8) \leqq b_1 \operatorname{Exc}(S, r)$$

for $\psi = \Phi_F, \Phi_F^{-1}$.

It is not difficult to see then that one can prove the basic inequality of regularity theory provided one can prove this inequality when the hypothesis that $F \in \mathscr{F}$ is replaced by the condition that $F \in \mathscr{J}$, the constant c is suitably rechosen, the conclusion that $\operatorname{Exc}(S_*, s_0 r) \leqq 4^{-1}\operatorname{Exc}(S, r) + b_0 r$ is replaced by the conclusion $\operatorname{Exc}(S_*, s_0 r) \leqq \varepsilon \operatorname{Exc}(S, r) + b_0 r$ for appropriate $\varepsilon > 0$ depending on b_1, and finally the condition that $x_* \in B^m(0, 2^{-1})$ be replaced by the condition that $x_* \in B^m(0, 5/8)$.

12. **Surfaces to which the regularity theory is applicable.** The study of geometric or variational problems in the context of geometric measure theory is based on a correspondence between m-dimensional surfaces in R^{m+n} and Radon measures on appropriate spaces. The most important measure theoretic surfaces are the following:

(1) *m-rectifiable sets.* The m-rectifiable sets have already been defined.

(2) *Variation measures* (the word "variation" here has nothing to do with the calculus of variations). If S is an m-rectifiable subset of R^{m+n}, the variation measure

$\|S\|$ associated with S is given by the formula $\|S\| = \mathscr{H}^m \llcorner S$, i.e. $\|S\|(A) = \mathscr{H}^m(S \cap A)$ for each $A \subset R^{m+n}$. Of course, $\|S\|$ determines $S\mathscr{H}^m$ almost uniquely, but it is difficult to evaluate $\boldsymbol{F}(S)$ from $\|S\|$ alone in the case $F: R^{m+n} \times G(m+n, m) \to R^+$.

(3) *Integral varifolds.* If S is an m-rectifiable subset of R^{m+n}, the integral varifold $|S|$ associated with S is given by the formula $|S| = \varphi_{\#}(\|S\|)$ where $\varphi: R^{m+n} \to R^{m+n} \times G(m+n, m)$ $\varphi(x) = (x, \operatorname{ap}\operatorname{Tan}(S, x))$ for $\|S\|$ almost all $x \in R^{m+n}$. Here $\operatorname{ap}\operatorname{Tan}(S, x)$ denotes the $\mathscr{H}^m$ approximate tangent m-plane to S at x. The general integral varifolds by definition are the finite or convergent infinite sums of integral varifolds corresponding to m-rectifiable sets. Note that if $F: R^{m+n} \times G(m+n, m) \to R^+$, then $\boldsymbol{F}(S) = \int F\, d|S|$.

(4) *Integral currents.* If S is an oriented m-rectifiable set (defined in the obvious way) in R^{m+n}, the current (continuous linear functional on differential m-forms in R^{m+n}) associated with S is given by integration of differential m-forms over S. Such a current can also be regarded as an m-vector valued Radon measure on R^{m+n}. If the current ∂S (defined by exterior differentiation of differential $(m-1)$-forms) is $(m-1)$-rectifiable, then S is an integral current. The general integral currents are by definition the finite or convergent infinite sum of integral currents corresponding to m-rectifiable sets, the boundaries of which are the finite or convergent infinite sum of integral currents corresponding to $(m-1)$-rectifiable sets. An appropriately bounded sequence of integral currents always has a subsequence which converges weakly to an integral current [**FH1**, 4.2.17].

(5) *Integral currents modulo* ν. The integral currents modulo ν bear the same relationship to singular chains with coefficients in the integers modulo ν that the integral currents have to singular chains with coefficients in the integers.

13. **Procedure for finding $\boldsymbol{F}$ minimal surfaces.** Suppose $F: R^{m+n} \times G(m+n, m) \to R^+$ is a suitable integrand and one wishes, say, to find an m-dimensional surface S having a prescribed boundary such that $\boldsymbol{F}(S) \leqq \boldsymbol{F}(T)$ whenever T is an m-dimensional surface with the same boundary. The general procedure is the following. First one takes a minimizing sequence $S_1, S_2, S_3, \ldots$ of m-dimensional surfaces having the prescribed boundary such that $\boldsymbol{F}(S_i)$ approaches its minimum value. Second, one associates with each surface a suitable measure, obtaining, say, the sequence $|S_1|, |S_2|, |S_3|, \ldots$. Third one chooses a subsequence of the sequence of measures which converges weakly to a limit measure V. Weak convergence implies of course that the sequence $\boldsymbol{F}(S_i) = \int F\, d|S_i|$ converges to $\int F\, dV$. In practice it is easy to show the weak convergence of a subsequence of the $|S_i|$, $i = 1, 2, 3, \ldots$, because spaces of measures have strong compactness properties in the weak topology. Finally, one must show that V corresponds to a nice surface, say $V = |S|$. This is where the real work comes. S is the desired $\boldsymbol{F}$-minimal surface.

As the variety of measure theoretic surfaces suggests there are a number of different ways in which the problem of finding an $\boldsymbol{F}$-minimal surface can be formulated. However, in the context of geometric measure theory, usually the problem of showing $V = |S|$ as above has two steps. First one shows that S has the structure of a compact m-rectifiable set (possibly with multiplicities). One then attempts to apply the regularity theory of this article to conclude a further differentiability structure for S.

The proof that S has the structure of an m-rectifiable set is one of the major achievements of geometric measure theory. The usual core of the argument is the following *structure theorem for sets of finite Hausdorff measure* [**FH1**, 3.3.13].

THEOREM. *Let $A \subset R^{m+n}$ be $\mathscr{H}^m$-measurable with $\mathscr{H}^m(A) < \infty$. Then one can write $\mathscr{H}^m$ almost uniquely $A = S \cup Q$, $S \cap Q = \varnothing$ such that S is m-rectifiable and $\mathscr{L}^m(\theta(Q)) = 0$ for almost all orthogonal projections $\theta: R^{m+n} \to R^m$.*

Having shown that a solution to a problem has the structure of a compact m-rectifiable set S, one then wishes to apply the regularity results to S. In general if S is a hypersurface this is always possible. Also it is, in general, always possible in case the solution associated with S has no set of positive measure of higher multiplicities (for some problems this is the case, for other problems it is not).

14. **An existence and regularity theorem for one formulation of the prescribed boundary problem.** Existence and regularity results are known for a number of different variational problems. We now give a careful statement of one such result [**AF2**, 1.4].

DEFINITIONS. (1) A *surface* is a compact m-rectifiable subset of R^{m+n}.

(2) A *boundary* is a compact $(m - 1)$-rectifiable subset of R^{m+n}.

(3) $H_{m-1}(B; G)$ denotes the $(m - 1)$-dimensional Vietoris homology group of B with coefficients in a finitely generated abelian group G. For $\sigma \in H_{m-1}(B; G)$ (intuitively σ is a hole in B), one says that S *spans* σ if and only if $i_*(\sigma) = 0$ where $i_*: H_{m-1}(B; G) \to H_{m-1}(B \cup S; G)$ is induced by the inclusion $i: B \to B \cup S$.

(4) An integrand $F: R^{m+n} \times G(m + n, m) \to R^+$ is called *elliptic with respect to G* if and only if there is a positive function $\gamma: R^{m+n} \to R^+$ which is locally bounded away from 0 such that for each $p \in R^{m+n}$ and each (flat) m-disk D in R^{m+n} and each surface S which spans some nontrivial $\sigma \in H_{m-1}(\partial D; G)$,

$$\boldsymbol{F}^p(S) - \boldsymbol{F}^p(D) \geqq \gamma(p)\,[\mathscr{H}^m(S) - \mathscr{H}^m(D)].$$

THEOREM. *Suppose*

(1) *B is a boundary*;

(2) *G is a finitely generated abelian group*;

(3) $\sigma \in H_{m-1}(B; G)$;

(4) *$F: R^{m+n} \times G(m + n, m) \to R^+$ is an elliptic integrand of class $j \geqq 3$ which is bounded away from* 0.

Then there exists a surface S such that

(1) *S spans σ.*

(2) *$\boldsymbol{F}(S) \leqq \boldsymbol{F}(T)$ whenever T is a surface which spans σ.*

(3) *Except possibly for a compact singular set of zero $\mathcal{H}^m$ measure, S is an m-dimensional submanifold of R^{m+n} of class $j-1$.*

15. **Estimates on singular sets.** Very little is known at the present time about the structure of the singular sets of solutions to general elliptic variational problems (except for their existence). However, for the area integrand there has been substantial progress. For example, we have the following two representative theorems which generalize immediately to manifolds.

THEOREM [**FH2**]. *For every unoriented boundary $B \subset R^n$ of dimension $k-1$ (any k), there exists an unoriented minimal surface (flat chain modulo 2) S with $\partial S = B$ of least k-dimensional area. The interior singular set of S has Hausdorff dimension at most $k-2$. The regular part of S is a real analytic submanifold of R^n.*

THEOREM [**FH2**]. *For every oriented boundary $B \subset R^n$ of dimension $n-2$ there exists an oriented minimal surface (integral current) S with $\partial S = B$ of least $(n-1)$-dimensional area (counting multiplicities). The interior singular set of S has Hausdorff dimension at most $n-8$. In particular, there are no interior singularities if $n \leqq 7$. The regular part of S is a real analytic submanifold of R^n.*

Examples show that both of these results are the best possible (at least in terms of Hausdorff dimension).

EXAMPLE. *The unoriented 2-dimensional surface $S = \{x : x_3 = x_4 = 0$ and $x_1^2 + x_2^2 \leqq 1\} \cup \{x : x_1 = x_2 = 0$ and $x_3^2 + x_4^2 \leqq 1\} \subset R^4$ is of least area among all unoriented 2-dimensional surfaces having boundary $B = \{x : x_3 = x_4 = 0$ and $x_1^2 + x_2^2 = 1\} \cup \{x : x_1 = x_2 = 0$ and $x_3^2 + x_4^2 = 1\}$. The origin O is the singular set of S.*

EXAMPLE [**BDG**]. *Let S be the 7-dimensional oriented cone $O(S^3 \times S^3)$ over $S^3 \times S^3 \subset R^4 \times R^4 = R^8$. Then S has less 7-dimensional area than any other oriented hypersurface T in R^8 with $\partial T = S^3 \times S^3$. The origin O is the singular set of S.*

The results above for oriented minimal hypersurfaces are intimately connected with the possibility of extending Bernsteins's theorem (that a globally defined nonparametric minimal hypersurface must be a hyperplane) to higher dimensions. We have the following:

THEOREM [**FH1**, 5.4.18], [**BDG**]. *If $n = 2, 3, \ldots, 7$ and if the graph of $f: R^n \to R$ is a minimal hypersurface in R^{n+1}, then the graph of f is a hyperplane. On the other hand, for each $n \geqq 8$ there exist functions $g: R^n \to R$ with graphs which are minimal hypersurfaces but which are not hyperplanes.*

16. **A regularity theorem for surfaces whose first variation of area satisfies a variational inequality.** We will discuss in this section the regularity of surfaces satisfying inequalities involving the first variation of their areas. We confine our attention to the first variation of the area integrand because, as far as we know, there

exists as yet no comparable theory for other integrands. On the other hand, our hypotheses will involve only the first variation of area; in no way are these surfaces assumed to satisfy comparison inequalities of the type considered previously. Also, the class of surfaces considered is quite general, even more general than the classes already considered, in that these surfaces are required to satisfy no homological conditions whatsoever.

Naturally formulated analogues for the theorems to be stated here for surfaces in Euclidean space hold for surfaces in any Riemannian manifold.

(1) *The first variation of area.* Suppose $2 \leqq k < n$. We consider k-dimensional surfaces in R^n. Let $G(n, k)$ be the space of k-dimensional linear subspaces of R^n. If U is an open subset of R^n, we let $\mathscr{X}(U)$ be the vector space of smooth R^n-valued functions on R^n with compact support contained in U. Given $g \in \mathscr{X}(U)$ and $(x, \pi) \in R^n \times G(n, k)$, we let $\operatorname{div}_k g(x, \pi)$ be the number $\sum_i (\partial g_i/\partial u_i)(x)$ where $u_1, \ldots, u_n$ are any orthonormal coordinates for R^n such that π is the set of common zeroes of $u_{k+1}, \ldots, u_n$. If F is a smooth diffeomorphism of R^n and $(x, \pi) \in R^n \times G(n, k)$ let $J_k F(x, \pi)$ be the square root of the sum of the squares of the $k \times k$ minors of the matrix with entries $(\partial F_i/\partial u_j)(x), j = 1, \ldots, k, i = 1, \ldots, n$, where $u_1, \ldots, u_n$ are orthonormal coordinates for R^n such that π is the set of common zeroes of $u_{k+1}, \ldots, u_n$. Suppose $g \in \mathscr{X}(R^n)$ and $h_t(x) = x + tg(x)$ for $(t, x) \in R \times R^n$; there is $\varepsilon > 0$ such that $J_k h_t(x, \pi)$ is smooth for $(t, x, \pi) \in (-\varepsilon, \varepsilon) \times R^n \times G(n, k)$ and we have the formula

(a) $$\frac{d}{dt} J_k h_t(x, \pi)\bigg|_{t=0} = \operatorname{div}_k g(x, \pi), \qquad (x, \pi) \in R^n \times G(n, k).$$

This formula can be applied as follows. Suppose S is a compact k-rectifiable subset of R^n with approximate tangent space $\boldsymbol{S}(x) \in G(n, k)$ at $\mathscr{H}^k$ almost all points of S. Let $g \in \mathscr{X}(R^n)$ and let $h_t(x) = x + tg(x)$, $(t, x) \in R \times R^n$. As is well known,

(b) $$\mathscr{H}^k(h_t(S)) = \int_S J_k h_t(x, \boldsymbol{S}(x))\, d\mathscr{H}^k x$$

so that

(c) $$\frac{d}{dt} \mathscr{H}^k(h_t(S))\bigg|_{t=0} = \int_S \operatorname{div}_k g(x, \boldsymbol{S}(x))\, d\mathscr{H}^k x.$$

The right-hand side of (c) can be computed explicitly in case S is also a smooth k-dimensional submanifold of R^n with boundary ∂S; in fact, it is an exercise in advanced calculus to show that

(d) $$\int_S \operatorname{div}_k g(x, \boldsymbol{S}(x))\, d\mathscr{H}^k x = -k \int_S g(x) \cdot H(x)\, d\mathscr{H}^k x + \int_S g(x) \cdot v(x)\, d\mathscr{H}^{k-1} x,$$

where H is the mean curvature vector of S and v is the unit vector normal to ∂S whose negative is tangent to S.

(2) *Varifolds*. These notions generalize as follows. Let $\mathbf{V}_k(R^n)$ be the space of Radon measures on $R^n \times G(n, k)$; the members of $\mathbf{V}_k(R^n)$ are called k-dimensional *varifolds* in R^n. If $V \in \mathbf{V}_k(R^n)$, let $\|V\|$ be the projection of V on R^n and for each $a \in R^n$ let $\Theta^k(\|V\|, a) = \lim_{r \to 0} \|V\| \{x : |x - a| < r\}/\alpha(k)r^k$. Here $\alpha(k) = \mathscr{L}^k(R^k \cap B^k(0, 1))$.

For example, if S is a compact k-rectifiable subset of R^n with approximate tangent space $\mathbf{S}(x)$ at $\mathscr{H}^m$ almost all points x of S, we let

$$|S|(A) = \mathscr{H}^m[R^n \cap \{x : (x, \mathbf{S}(x)) \in A\}] \qquad \text{whenever } A \subset R^n \times G(n, k);$$

clearly, $|S| \in \mathbf{V}_k(R^n)$. We let $\mathbf{R}\mathbf{V}_k(R^n)$ be the set of *k-rectifiable varifolds* in R^n, that is, the set of those $V \in \mathbf{V}_k(\mathbf{R}^n)$ for which there exist positive real numbers $c_1, c_2, \ldots$ and compact k-rectifiable subsets $S_1, S_2, \ldots$ of R^n such that $V = \sum_i c_i|S_i|$; if the c_i can be taken to be positive integers, the varifold is called *integral*, the set of which is denoted $\mathbf{IV}_k(R^n)$.

If $F : R^n \to R^n$ is a smooth diffeomorphism and $V \in \mathbf{V}_k(R^n)$, we generalize (b) by setting

$$F_\# V(A) = \int_{\{(x,\pi):(F(x), DF(x)(\pi)) \in A\}} J_k F(x, \pi)\, dV(x, \pi), \qquad A \subset R^n \times G(n, k);$$

observe that $F_\# V \in \mathbf{V}_k(R^n)$. In fact, if S is a compact k-rectifiable subset of R^n we have $F_\# |S| = |F(S)|$. If $V \in \mathbf{V}_k(R^n)$, $g \in \mathscr{X}(U)$, U is bounded, $h_t(x) = x + tg(x)$ for $(t, x) \in R^n \times G(n, k)$ we use (a) to calculate

$$\text{(e)} \qquad \frac{d}{dt} \|h_{t\#} V\|(U)\Big|_{t=0} = \int \operatorname{div}_k g(x, \pi)\, dV(x, \pi).$$

Given $V \in \mathbf{V}_k(R^n)$ we define a linear mapping

$$\delta V : \mathscr{X}(R^n) \to R$$

by setting $\delta V(g)$ equal to the expression in the right-hand side of (e) for $g \in \mathscr{X}(R^n)$. We call δV the first variation distribution of G. We let $\|\delta V\|$ be the total variation of δV; that is, $\|\delta V\|$ is the largest Borel regular measure such that

$$\|\delta V\|(U) = \sup\{\delta V(g) : g \in \mathscr{X}(U) \text{ and } \sup|g| \leqq 1\}$$

for any open subset U of R^n. For example, if S is as in (d),

$$\|\delta|S|\,\| = \mathscr{H}^k \llcorner k|H| + \mathscr{H}^{k-1} \llcorner \partial S.$$

Thus if $V \in \mathbf{V}_k(R^n)$ and $\|\delta V\|$ is a Radon measure on R^n (that is, $\|\delta V\|$ is finite on bounded sets), it seems reasonable to call the absolutely continuous part of δV with respect to $\|V\|$ the mean curvature of V and the singular part of δV with respect to $\|V\|$ the (mean curvature) boundary of V. The reader should bear this in mind when considering the examples.

EXAMPLE 1. Consider a typical soap bubble as below giving a compact 2-rectifiable set S in R^3. Let $V = |S| \in \boldsymbol{IV}_0(R^3)$. It can be shown that for some number c (depending on V), $\|\delta V\| \leqq c\|V\|$.

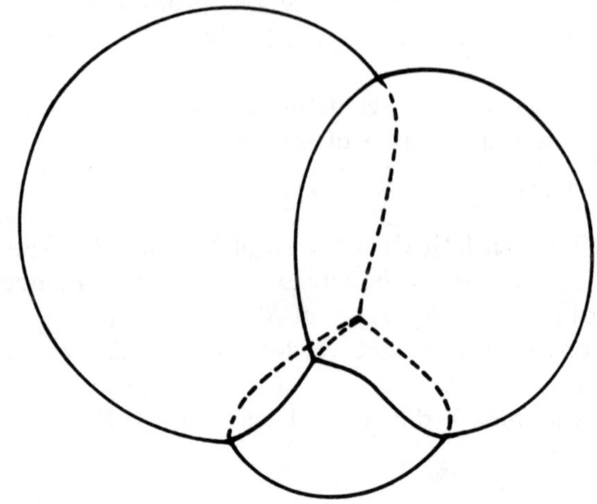

EXAMPLE 2. Consider the one-dimensional complex in R^2 sketched below. The angles ABC, CDE, EDF and FGA are $2\pi/3$ and the angles GOA and AOB are $2\pi/16$. Let S_1 be the outer configuration of 24 segments and let S_2 be the remaining 24 segments. Let $S = S_1 \cup S_2$. There is a unique number d depending on S such that if $V = |S_1| + d|S_2| \in \boldsymbol{RV}_1(R^2)$ then support $\|\delta V\|$ is a set of 16 points; these points are indicated by little circles in the sketch. Clearly, $0 < d < 1$. One may apply the formula (d) to see this readily.

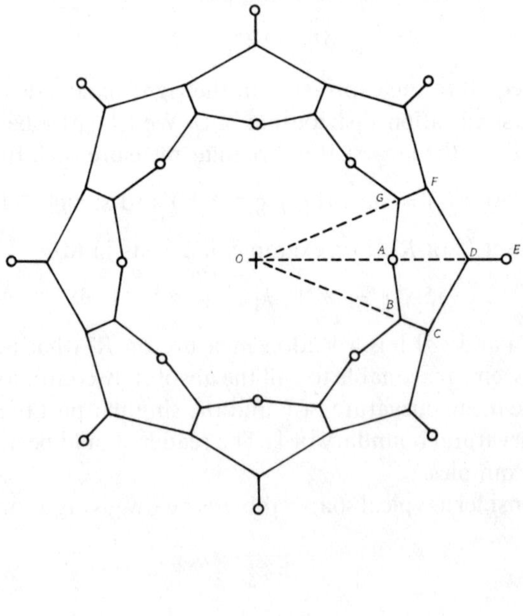

Now consider the infinite one-dimensional complex sketched below which is the union of homotheties $S = S^0, S^1, \ldots$ of S; for each $i = 0, 1, 2, \ldots$ let V^i be the homothetic image of V whose support is equal to S^i. There is a unique number θ with $0 < d < 1$ such that

$$V = \sum_{i=0}^{\infty} d^i V^i \in \boldsymbol{RV}_1(R^2)$$

and support $\|\delta V\|$ consists of 8 points; these points are indicated by little circles in the picture.

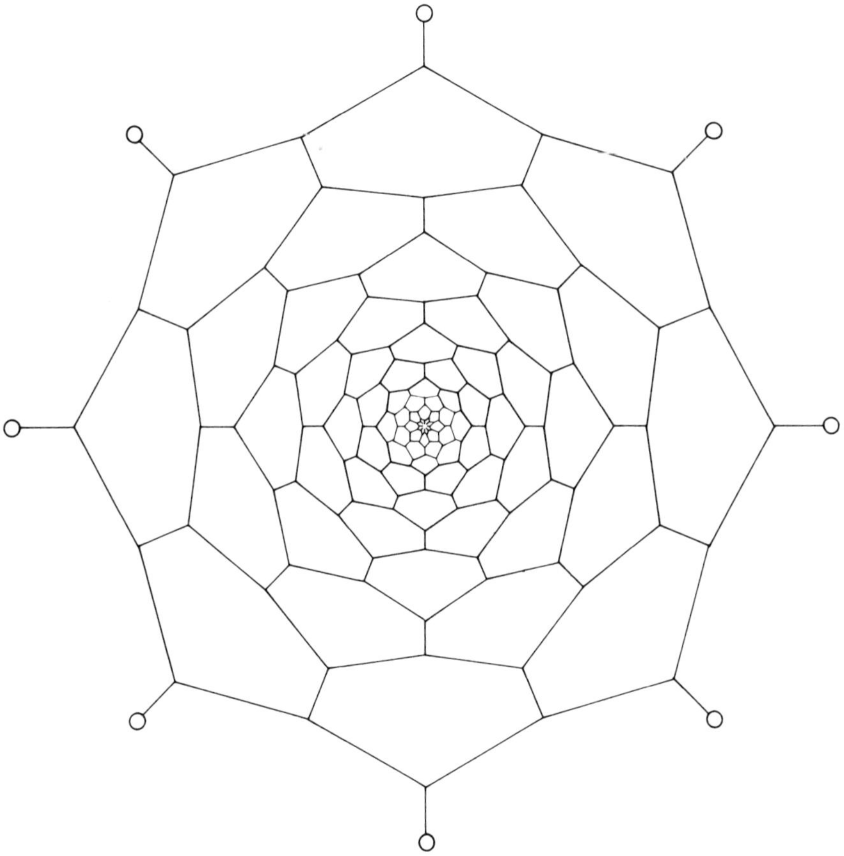

EXAMPLE 3. Suppose M is a smooth compact m-dimensional manifold. Whenever $0 \leqq k \leqq m$, let $S_k = \boldsymbol{IV}_k(M) \cap \{V : V \neq 0 \text{ and } \delta V = 0\}$. In other words, S_k is the set of nonzero stationary integral varifolds in M of dimension k. It is proved in [**AF1**] that

(I) $S_k \neq \emptyset$;
(II) for any $B < \infty$, $S_k \cap \{V : \|V\|(M) \leqq B\}$ is weakly compact.
The regularity question for such varifolds is partly settled by the regularity theorem of this section.

For example, if M is the unit sphere in R^{m+1} the great k-spheres are members of S_k and they can be characterized by the property that they minimize area in S_k.

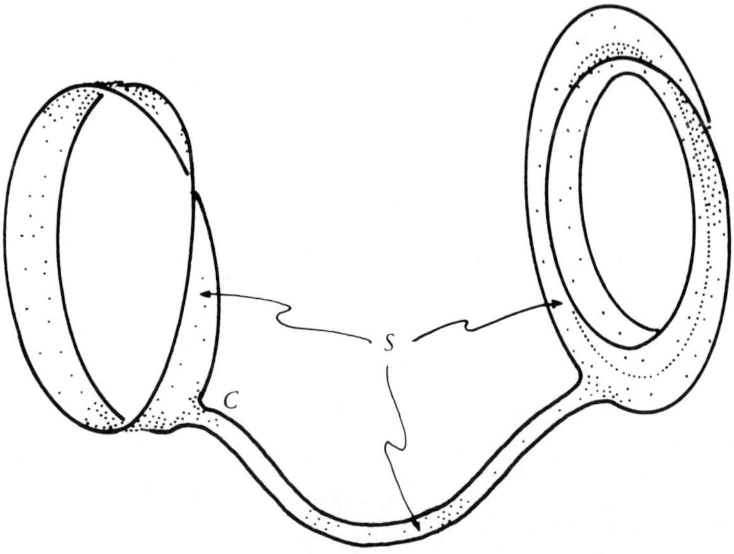

EXAMPLE 4. Were one to take a wire shaped as the curve C above and dip it in a soap solution one might obtain, upon drawing the wire out of the solution and puncturing two regions, a soap film S looking like a Möbius band near the left-hand part of the curve, like a strip near the middle, and, near the right-hand part of the curve, a surface with cross section exhibited on the far right; notice that this last surface contains a singular circle. We would have that $\|\delta|S|\| = \mathscr{H}^1 \llcorner C$. In fact, there exists a retraction of the pair (S, C) onto (C, C). Thus in no way is $\partial S = C$ in the sense of algebraic topology. It is in the sense of mean curvature that $\partial S = C$.

EXAMPLE 5. Suppose $k \leqq p < \infty$ and $q = p/(p-1)$. For each r with $0 < r < \infty$, let $S_r = R^n \cap \{x : |x| = r \text{ and } x_i = 0, i = k+1, \ldots, n\}$. It is easy to see that

$$\sup\left\{\delta|S_r|(g) : g \in \mathscr{X}(R^n) \text{ and } \left(\int_{S_r} |g|^q d\mathscr{H}^k\right)^{1/q} \leqq 1\right\}$$
$$= k^{(p+1)/p}\alpha(k)^{1/p} r^{(k-p)/p} = k^{(k+1)/k}\alpha(k)^{1/k} \quad \text{if } p = k,$$
$$\to \infty \quad \text{as } r \to \infty \quad \text{if } p > k.$$

This illustrates in part the nature of hypothesis (III) in Theorem (f) below.

EXAMPLE 6. For each $t \in R$ consider the complex analytic submanifolds

$$S_t = C^2 \cap \{(z, w) : w = tz^2 \text{ and } |z|^2 + |w|^2 < 1\}.$$

It is easily seen by direct calculation that the mean curvature vector of each of these submanifolds is zero so that $\|\delta|S_t|\|\{(z, w) : |z|^2 + |w|^2 < 1\} = 0$ for all $t \in R$. Clearly,

$$\limsup_{t \to \infty} \{|z| : (z, w) \in S_t\} = 0 \quad \text{and} \quad \lim_{t \to \infty} |S_t| = 2|C^2 \cap \{(z, w) : |w| < 1, z = 0\}|$$

in the weak topology of $V_2(R^4)$. Thus (II) of Theorem (f) below can not be omitted.

(f) REGULARITY THEOREM. *Corresponding to each p with $k < p < \infty$ and each ε with $0 < \varepsilon < 1$ there is $\eta > 0$ with the following property:*
Suppose

(I) $V \in V_k(R^n)$, $0 \in \text{support } \|V\|$ *and*

$$\limsup_{r \downarrow 0} \|V\|B^n(a, r)/\alpha(k)r^k \geqq 1 \quad \textit{for } \|V\| \textit{ almost all } a \in B^n(0, 1);$$

(II) $\|V\|B^n(0, 1) \leqq (1 + \eta)\alpha(k)$;

(III) $\delta V(g) \leqq \eta(\int |g|^q \, d\|V\|)^{1/q}$ *whenever* $g \in \mathscr{X}(B^n(0, 1))$, *where* $q = p/(p - 1)$.

Then there are a linear isometry $\theta : R^n \to R^n$ and continuously differentiable functions $f_j : R^k \to R$, $j = 1, \ldots, n - k$, such that

$$B^n(0, 1-\varepsilon) \cap \theta(\text{support } \|V\|) = B^n(0, 1-\varepsilon) \cap \{x : x_{k+j} = f_j(x_1, \ldots, x_k), j = 1, \ldots, n-k\}$$

and such that $(\partial f_j/\partial x_i)(0) = 0$, $i = 1, \ldots, k$, $j = 1, \ldots, n - k$,

$$\left| \frac{\partial f_j}{\partial x_i}(u) - \frac{\partial f_j}{\partial x_i}(v) \right| \leqq \varepsilon |u - v|^{1 - k/p}, \qquad u, v \in R^k.$$

Moreover, $\Theta^k(\|V\|, a)$ is a real number for all $a \in B^n(0, 1)$ and, if $S = B^n(0, 1 - \varepsilon)$ $\cap$ support $\|V\|$,

$$V(A) = \int_A \Theta^k(\|V\|, x) \, d|S|(x, \pi) \quad \textit{for } A \subset B^n(0, 1 - \varepsilon) \times G(n, k).$$

We point out that this theorem gives nontrivial information about smooth k-dimensional submanifolds M of R^n which are closed relative to $B^n(0, 1)$ and pass through 0; condition (II) requires that the area of M inside $B^n(0, 1)$ is $\leqq (1 + \eta)\alpha(k)$ and condition (III) requires that

$$k\left(\int_M |H|^p\right)^{1/p} \leqq \eta$$

where H is the mean curvature vector of M. One then concludes that $M \cap B^n(0, 1 - \varepsilon)$ is close in the $C^1_{1-k/p}$ topology to a disc.

Modulo translations and homothetic transformations the hypotheses of the regularity theorem will be satisfied at a dense set of points in $U \cap$ support $\|V\|$ wherever U is an open subset of R^n and $\rho > 0$ is such that $\Theta^k(\|V\|, x) > \rho$ for $\|V\|$, almost all $x \in U$ and $\delta V(g) \leqq \rho^{-1}(\int |g|^q \, d\|V\|)^{1/q}$ for $g \in \mathscr{X}(U)$. One would like to apply the regularity theorem at almost all points of a stationary ($\delta V = 0$) varifold for which the density is bounded away from zero, but it is not known whether this is possible.

References

For a more complete set of references to geometric measure theory one should consult the bibliography of [**FH1**].

[**AW**] W. K. Allard, *On the first variation of a varifold*, Ann. of Math. **95** (1972), 417–491.

[**AF1**] F. J. Almgren, Jr., *The theory of varifolds. A variational calculus in the large for the k dimensional area integrand*, Princeton University, Princeton, N.J., 1965 (multilithed notes).

[**AF2**] ———, *Existence and regularity almost everywhere of solutions to elliptic variational problems among surfaces of varying topological type and singularity structure*, Ann. of Math. (2) **87** (1968), 321–391. MR **37** #837.

[**AF3**] ———, *Measure theoretic geometry and elliptic variational problems*, Bull. Amer. Math. Soc. **75** (1969), 285–304. MR **39** #2034.

[**AF4**] ———, *Existence and regularity almost everywhere of solutions to parametric elliptic variational problems with constraints* (in preparation).

[**BDG**] E. Bombieri, E. De Giorgi and E. Giusti, *Minimal cones and the Bernstein problem*, Invent. Math. **7** (1969), 243–268. MR **40** #3445.

[**FH1**] H. Federer, *Geometric measure theory*, Die Grundlehren der math. Wissenschaften, Band 153, Springer-Verlag, Berlin and New York, 1969. MR **41** #1976.

[**FH2**] ———, *The singular sets of area minimizing rectifiable currents with codimension one and of area minimizing flat chains modulo two with arbitrary codimension*, Bull. Amer. Math. Soc. **76** (1970), 767–771. MR **41** #5601.

PRINCETON UNIVERSITY

TWO MINIMAX PROBLEMS IN THE CALCULUS OF VARIATIONS

MELVYN S. BERGER

The solution of many nonlinear problems of differential geometry and mathematical physics involves a proof of the existence of smooth critical points of some functional F over a sufficiently broad admissible class of functions C. Such problems are often difficult to solve because either (i) the given problem admits some trivial solution and one must insure that the critical point obtained does not coincide with this trivial one, or (ii) the minimum of the functional F may not be attained in C. In either case one is led to study the critical points of F which do not render F a (relative or absolute) minimum, i.e. minimax points. In this article we study two specific instances of these basic problems that are indicative of new general theories yet to be explored.

In the first problem (the structure of stationary states for nonlinear wave equations) the difficulty of type (i) occurs. Here the study of the minimax points is simplified by introducing a constraint in the admissible class C, and so reducing the problem to a more geometric one. Nonetheless the resulting isoperimetric problem (π) in the calculus of variations is still not easily studied by general techniques, since it does not satisfy either the Palais-Smale (compactness) "condition C" or its variants. The basic reason for this lack of compactness is that the relevant functionals are defined over the whole space $\boldsymbol{R}^N$ (and not over some bounded subdomain of $\boldsymbol{R}^N$). Yet special considerations yield a rather complete solution for (π) since, for example, one can obtain simple necessary and sufficient conditions for the existence of nontrivial solutions of (π).

In the second problem considered, namely the construction of Hermitian metrics on complex analytic manifolds M^N with prescribed Hermitian scalar curvature, attainment of the minimum of the functional is in question, so that the difficulty of

---

AMS 1970 *subject classifications*. Primary 58E15, 35J60, 53C55; Secondary 53A30, 35J50.

type (ii) mentioned above prevails. This difficulty is surmounted by breaking up the variational problem (P) into two parts (P_1) and (P_2), so that the minimum problem for (P_1) has a solution u_c unique up to an additive constant c, while the minimum problem for (P_2) is solvable provided a certain coefficient (depending on u_c) is sufficiently large.

1. **Stationary states for nonlinear wave equations.** We consider the nonlinear Schrödinger and Lorentz-invariant Klein-Gordon equations defined on $\boldsymbol{R}^{N+1}$

$$-iu_t = \Delta u + f(|u|^2)u, \tag{1}$$

$$u_{tt} = (\Delta - m^2)u + f(|u|^2)u. \tag{2}$$

We seek solutions of (1) and (2) of the form $u(x, t) = e^{i\lambda t}v(x)$ with λ a real number and $v(x)$ a real-valued function in $L_2(\boldsymbol{R}^N)$. Thus v satisfies an equation of the form

$$\Delta v - \beta v + f(|v|^2)v = 0 \tag{3}$$

where $\beta = \lambda$ for (1) and $\beta = (m^2 - \lambda^2)$ for (2). In the sequel we set $f(|v|^2) = g|v|^\sigma$ (as is usual in contemporary work) where g is a real constant. Clearly this problem can be viewed as a nonlinear generalization of the spectral problem for linear elliptic operators. We prove the following sharp result.

THEOREM 1. *Suppose* $N > 2, f(|v|^2) = g|v|^\sigma$ *with* $\sigma \geqq 0$ *and* $\beta \neq 0$, *then* (3) *has nontrivial solutions decaying to zero at* ∞ *if and only if* $0 < \sigma < 4/(N-2), \beta > 0$, *and* $g > 0$. *Furthermore such nontrivial solutions of* (3) *decay exponentially at* ∞.

PROOF OF NECESSITY. First we note that, for $\beta \neq 0$, all solutions of (3) decay to zero exponentially as $|x| \to \infty$ if they tend to zero at all as $|x| \to \infty$. This result is proved via the results of Kato [**5**] mentioned above, and is the special feature of nonlinear problems which we call the "amplification of decay at infinity." Thus if $v(x)$ is a solution of (3) in $L_2(\boldsymbol{R}^N)$, then $v(x)$ is a solution of the linear equation

$$\Delta v + [p(x) - \beta]v = 0,$$

where $p(x) = g|v|^\sigma$ has a certain decay at ∞, since $v(x)$ does; this implies, by (3), that $v(x)$ has a much stronger decay at ∞ than merely $v \in L_2(\boldsymbol{R}^N)$. This argument can be iterated to prove exponential decay.

The case $\beta < 0$ for arbitrary g, σ is an immediate consequence of Kato [**5**]. For $\beta > 0$ and $g \leqq 0$ one finds that if v is a desired solution,

$$\int_{\boldsymbol{R}^N} (|\nabla v|^2 + \beta v^2 + |g|\,|v|^{\sigma+2})\,dx = 0$$

so that $v \equiv 0$. For the remaining cases we prove

LEMMA. *Suppose* $v(x) \to 0$ *exponentially at* ∞ *and satisfies* $\Delta v + G'(v) = 0$ *in* $\boldsymbol{R}^N$, *then*

$$2N \int_{\boldsymbol{R}^N} G(v)\, dx = (N-2) \int_{\boldsymbol{R}^N} vG'(v)\, dx.$$

PROOF. The function $v(x)$ is a critical point of $I(u(x)) = \int_{\boldsymbol{R}^N}[\frac{1}{2}|\nabla u|^2 - G(u)]\, dx$. Thus as a function of c, $dI(v(cx))/dc|_{c=1} = 0$. Hence, after a simple calculation and a change of variables, we find

$$I(v(x)) = \tfrac{1}{2}c^{2-N} \int_{\boldsymbol{R}^N} |\nabla v|^2 - c^{-N} \int_{\boldsymbol{R}^N} G(v)$$

so

$$(N-2) \int_{\boldsymbol{R}^N} |\nabla v|^2 = 2N \int_{\boldsymbol{R}^N} G(v).$$

Since also $\int_{\boldsymbol{R}^N}|\nabla v|^2 = \int_{\boldsymbol{R}^N} G'(v)v$, the result follows.

To use this lemma for (3), set $g(v) = -\beta v + g|v|^\sigma v$ so that $G(v) = -\frac{1}{2}\beta v^2 + g|v|^{\sigma+2}/(\sigma+2)$. Thus if v is a decaying solution of (3), and $N > 2$,

$$2\beta \int_{\boldsymbol{R}^N} v^2 = g\left\{\frac{2N}{(\sigma+2)(N-2)} - 1\right\} \int_{\boldsymbol{R}^N} |v|^{\sigma+2}.$$

Consequently if $g, \beta > 0$ for $v \not\equiv 0$, then the last equation implies that $\sigma < 4/(N-2)$; whereas if $\beta = 0$, $\sigma = 4/(N-2)$ and an application of the Kelvin transformation and the unique continuation theorem show that the equation $\Delta v + |v|^\sigma v = 0$ has no exponentially decaying solutions $\not\equiv 0$.

PROOF OF SUFFICIENCY. The problem can be reduced by restricting attention to radially symmetric solutions $v(x) = v(|x|)$. Setting $|x| = r$ and $r^{(N-1)/2}v(r) = w(r)$, w satisfies

$$w_{rr} - p(r)w + gr^{-\sigma}|w|^\sigma w = 0 \quad \text{where } p(r) = \beta + \tfrac{1}{4}r^{-2}(N-3)(N-1),$$

$$w(0) = w(\infty) = 0. \tag{4}$$

To study (4) the following simple result is useful.

LEMMA. *The linear map L with domain $\dot{W}_{1,2}(0,\infty)$ and range $L_{\sigma+2}(0,\infty)$ defined by $Lu = u/r^{\sigma/(\sigma+2)}$ is compact provided $0 < \sigma < 4/(N-2)$. In addition, for $u \in \dot{W}_{1,2}(0,\infty)$, the following inequality holds*:

$$\frac{1}{4}\int_0^\infty (u^2/r^2)\, dr \leqq \int_0^\infty u_r^2\, dr.$$

Now after scaling by a constant factor, we find that the solutions of (4) are in (1-1) correspondence with the critical points of the isoperimetric variational problem (π): Find the critical points of the functional $J(w) = \int gr^{-\sigma}|w|^{\sigma+2}\, dr$ subject to the constraint $G(w) = \int_0^\infty (w_r^2 + p(r)w^2)\, dr$ = a positive constant (R say).

The above lemma implies that $J(w)$ is weakly continuous when restricted to the sphere-like set $\partial\Sigma_R = \{w|G(w) = R\}$ in the $W_{1,2}(0, \infty)$ topology, so that (π) satisfies the "compactness" hypotheses necessary for the application of the general critical point theory of Ljusternik-Schnirelmann (see [**6**, p. 117] for example). Consequently, using the antipodal symmetry of (π) one finds

COROLLARY 1. *The equation* (3) *has a countably infinite number of distinct solutions decaying to zero exponentially at* ∞, *provided the necessary conditions of Theorem* 1 *are satisfied.*

REMARKS. (i) Analogues of Theorem 1 hold for $\boldsymbol{R}^N$ $(N = 1, 2)$ in which case σ is unrestricted. Note however that in all cases the results obtained here are of a *strictly nonlinear* nature, since in the linear case $\sigma = 0$ and no nontrivial solutions exist.

(ii) Clearly Corollary 1 can be obtained by a direct analysis of (4) independent of critical point theory. However we wanted to emphasize the fact that *once one has obtained a compactness lemma, the standard results of modern critical point theory become applicable.*

2. **Hermitian manifolds with prescribed Hermitian scalar curvature.** Let M be a compact complex N-dimensional manifold with Kähler metric g. In this section, we attempt to find a Hermitian metric $\tilde{g}$ on M, conformally equivalent to g, with prescribed Hermitian scalar curvature f. This problem is equivalent to finding a smooth solution of a certain semilinear elliptic partial differential equation defined on (M, g). Indeed, following Chern [**4**] and Chavel [**3**] with $\tilde{g} = e^{2u}g$, this equation can be written

$$N\Delta u - R(x) + fe^{2u} = 0 \tag{5}$$

where Δ denotes the Laplace-Beltrami operator on the real $2N$-dimensional Riemannian manifold associated with (M, g), and $R(x)$ denotes the Hermitian scalar curvature relative to (M, g). Here we prove

THEOREM 2. *Suppose (M, g) is a given Kähler manifold whose mean scalar curvature over M is <0 (i.e. $\int_M R(x)\,dV < 0$), and let f be any negative function defined on M. Then there is a Hermitian metric $\tilde{g}$, conformally equivalent to g, such that the Hermitian scalar curvature of (M, g) is $f(x)$. Conversely, if $\int f(x)\,dV \geqq 0$ and g is a Kähler metric of negative scalar curvature, no Hermitian metric (conformal to g) with curvature $f(x)$ exists.*

REMARK. If $N = 1$, this result was proved in [**1**] and [**7**]. However the isoperimetric variational method used to study (5) in [**1**] does not extend to the case $N > 1$, since the extensions of the Sobolev imbedding theorems used there fail if $N > 1$.

Idea of the proof. In order to find a solution to (5) it is natural to minimize the functional

$$I(u) = \int_M \{\tfrac{1}{2}N|\nabla u|^2 + R(x)u - \tfrac{1}{2}fe^{2u}\}\, dV$$

over the admissible class $C = W_{1,2}(M, g)$. Under the hypothesis of Theorem 2, one cannot easily demonstrate that $\inf_C I(u)$ is attained. We resolve this difficulty by adopting an idea in our paper [**2**]. We split (5) into two parts by considering the system

$$\tfrac{1}{2}N\Delta v - R(x) + \bar{R} = 0 \quad \text{where } \bar{R} = [\mu(M)]^{-1}\int_M R(x)\, dV, \tag{6}$$

$$\tfrac{1}{2}N\Delta w - \bar{R} + fe^v e^w = 0. \tag{7}$$

If the pair (v, w) satisfies (6), (7), then clearly $u = \frac{1}{2}(v + w)$ satisfies (5). Now since $\int_M(\bar{R} - R(x))\, dV = 0$, (6) is uniquely solvable apart from an arbitrary constant c. On the other hand, in order to solve (7) we try to solve the variational problem:

(P_2) Minimize the functional

$$\tilde{I}(w) = \int_M (\tfrac{1}{4}N|\nabla w|^2 + \bar{R}w - fe^v e^w)\, dV \tag{8}$$

over the class $X = W_{1,2}(M, g)$ where v is any solution of (6). We shall show that (P_2) can be solved by choosing $\min_M v(x)$ sufficiently large.

PROOF OF THEOREM 2. For brevity, suppose f is strictly negative on M. We first demonstrate

LEMMA A. (P_2) *has a solution for* $v = v_0$, *provided* $\min_M e^{v_0}$ *is sufficiently large.*

PROOF. Let $X = \{w | w \in W_{1,2}(M, g)\}$ and $\min_M e^{v_0} = e_0$. Then $c(X) = \inf_X \tilde{I}(w) \geqq \inf_X \int(\bar{R}w + \beta e^v e^w)\, dV$ where $\beta = -\inf_M f(x)$. Now by hypothesis, $\bar{R} < 0$, the minimum of $g(x) = \bar{R}x + \beta e_0 e^x$ is attained at $x_0 = \log(|\bar{R}|(e_0\beta)^{-1}) > -\infty$. Thus $c(X) \geqq \mu(M)g(x_0)$.

We now show that $c(X)$ is attained by some function $w \in X$. Indeed let $\{w_n\}$ be a sequence in X with $\tilde{I}(w_n) \to c(X)$ and $\tilde{I}(w_n) \leqq c(X) + 1$, so

$$\tfrac{1}{4}N\int_M |\nabla w_n|^2 \leqq c(X) + 1 + |f(x_0)|\mu(M), \tag{9}$$

$$\int_M \bar{R}w_n - \int_M fe^v e^w \leqq c(X) + 1. \tag{10}$$

To show that $\|w_n\|^2_{W_{1,2}}(M, g) = \int_M (|\nabla w_n|^2 + w_n^2)\, dV$ is uniformly bounded, by virtue of (9) and a well-known inequality of Friedrichs, it suffices to show that the mean value over M of $w_n, \bar{w}_n$, is uniformly bounded. Furthermore, suppose $e_0 = \min_M e^{v_0} > \beta^{-1}|\bar{R}|$. Then as $e^x \geqq 1 + x$ for all x, we obtain from (10) that $\int \bar{R}w_n + \int \beta e_0(1 + w_n) \leqq c(X) + 1$ and $(-|\bar{R}| + \beta e_0)\mu(M)\bar{w}_n + \int_M \beta e_0 \leqq c(X) + 1$. Thus $\bar{w}_n$ cannot tend to ∞, as $n \to \infty$. On the other hand, $\bar{w}_n$ cannot tend to $-\infty$, because inequality (10) and our hypothesis that $f < 0$ imply that $-|\bar{R}|\bar{w}_n \mu(M) \leqq c(X) + 1$. Thus $\|w_n\|_{W_{1,2}}(M, g)$ are uniformly bounded, and so $\{w_n\}$ contains a weakly convergent sequence in $W_{1,2}(M, g)$ with weak limit $\bar{w}$.

Now we show that $\tilde{I}(w) = c(X)$, so that $\bar{w}$ is the desired solution of (P_2). As $w_n \to \bar{w}$ weakly in $W_{1,2}(M, g)$, $w_n \to \bar{w}$ strongly in $L_2(M, g)$; and since we may suppose that $w_n \to w$ almost everywhere, Fatou's lemma implies that $\int_M(-f)e^v e^{\bar{w}} \leqq \liminf \int_M(-f)e^v e^{w_n}$ whereas it is well known that $\int_M|\nabla\bar{w}|^2 \leqq \liminf \int_M|\nabla w_n|^2$. Thus $\tilde{I}(w_n) \geqq \tilde{I}(\bar{w})$, and consequently if $n \to \infty$ we conclude that $\tilde{I}(\bar{w}) = c(X)$.

The proof of Theorem 2 can now be concluded by showing that the function $\bar{w}$ (shown to exist in the above lemma) is smooth so that $\bar{w}$ satisfies (7). Indeed, by virtue of the remarks made just after the statement of Theorem 2, one can find a smooth function v_0 satisfying both equation (6) and the conditions of Lemma A.

To prove that $\bar{w}$ is smooth, we first prove that $\bar{w}$ can be chosen to be bounded. To this end, we prove

LEMMA B. *There is a number $K > 0$ such that if $\tilde{I}(u) = \int_M F(x, u, \nabla u)\, dV$, then for all $u > K$ and $x \in M$, $F(x, u, 0) \geqq F(x, K, 0)$, while for all $u \leqq -K$, $F(x, u, 0) \geqq F(x, -K, 0)$.*

Once Lemma B is established, it is clear that we may choose the minimizing sequence w_n of Lemma A such that $|w_n| \leqq K$, so that $|\bar{w}| \leqq K$ a.e.; then the classical regularity theory yields the fact that $\bar{w} \in C^\infty(M)$ after a possible redefinition on a set of measure zero. Indeed $\bar{w}$ satisfies an equation of the type $\Delta w = f(\bar{w})$, $f(\bar{w}) \in L_\infty(M^N)$, so that by the L_p regularity theory $\bar{w} \in W_{2,p}(M, g)$ for all finite p; thus for some sufficiently large p, $\Delta\bar{w} = f(\bar{w})$ where $f(\bar{w}) \in C^1(M, g)$ so that $\bar{w} \in C^2(M, g)$ by the Schauder regularity theory. By iterating this argument we obtain $\bar{w} \in C^\infty(M, g)$.

PROOF OF LEMMA B. In proving Lemma A we showed that if $F(x, u, 0) = \bar{R}w - fe^{v_0}e^w$ then $F(x, u, 0) \geqq F(x, u_0, 0)$ where $u_0 = \log(|\bar{R}|(e_0\beta)^{-1}) < 0$. Thus choosing $K = -u_0$, it remains to prove that $F(x, u, 0) \geqq F(x, K, 0)$ for $u > K$. This last inequality follows simply from a consideration of the real-valued function $f(x, s) = \bar{R}s + |f|e^v e^s$. Indeed, for $s > K$,

$$f(x, s) - f(x, K) = \int_K^s \frac{\partial f}{\partial s}\, ds = \int_K^s (\bar{R} + |f|e^v e^s)\, ds.$$

Now $f_s(x, s) = \bar{R} + |f|e^v e^s > 0$ for $s > \log(|\bar{R}|(|f|e^v)^{-1})$ [and since $\bar{R}/|f|e^v < \bar{R}/\beta e_0$] for $s > 0$.

To prove the final part of Theorem 2, suppose u is a solution of equation (5), where $R \leqq 0$ and $\int f \geqq 0$. Then multiplying by e^{-2u} and integrating over M we find

$$\int_M e^{-2u}\{N\Delta u - R(x)\}\, dV + \int_M f\, dV = 0.$$

Thus

$$2N\int_M e^{-2u}|\nabla u|^2\,dV - \int_M R(x)e^{-2u}\,dV + \int_M f\,dV = 0.$$

Consequently the assumption that $\int_M f \geqq 0$ leads to a contradiction.

ADDED IN PROOF. Recent work of J. Kazdan and F. Warner indicates that Theorem 2 can be sharpened by letting f be any Hölder continuous function, strictly negative at some point of M, provided conformal equivalence of metrics is understood in a sense more general than the pointwise notion used here.

BIBLIOGRAPHY

1. M. S. Berger, *Riemannian structures of prescribed Gaussian curvature for 2-manifolds*, J. Differential Geometry **5** (1971), 325–332.

2. ———, *On the conformal equivalence of compact 2-dimensional manifolds*, J. Math. Mech. **19** (1969/70), 13–18. MR **39** #6219.

3. I. Chavel, *Two variational problems in Hermitian geometry*, Indiana Univ. Math. J. **20** (1970/71), 175–183. MR **41** #6127.

4. S. S. Chern, *On holomorphic mappings of hermitian manifolds of the same dimension*, Proc. Sympos. Pure Math., vol. 19, Amer. Math. Soc., Providence, R.I., 1968, pp. 157–170. MR **38** #2714.

5. T. Kato, *Growth properties of solutions of the reduced equation with a variable coefficient*, Comm. Pure Appl. Math. **12** (1959), 403–425. MR **21** #7349.

6. M. M. Vainberg, *Variational methods for the study of non-linear operators*, GITTL, Moscow, 1956; English transl., Holden-Day, San Francisco, Calif., 1964. MR **19**, 567; MR **31** #638.

7. J. Kazdan and F. Warner, *Curvature functions for 2-manifolds*. I (to appear).

YESHIVA UNIVERSITY, BELFER GRADUATE SCHOOL

EXISTENCE THEORY FOR BOUNDARY VALUE PROBLEMS FOR QUASILINEAR ELLIPTIC SYSTEMS WITH STRONGLY NONLINEAR LOWER ORDER TERMS

FELIX E. BROWDER

Let Ω be a bounded open subset of the Euclidean n-space R^n, $n \geqq 1$, such that the Sobolev Imbedding Theorem holds on Ω. We consider a quasilinear elliptic system of order $2m$ on Ω ($m \geqq 1$) of the form

$$A(u) + B(u) = f \tag{1}$$

where

$$A(u) = \sum_{|\alpha| \leqq m} (-1)^{|\alpha|} D^\alpha A_\alpha(x,u,\ldots,D^m u) \tag{2}$$

and

$$B(u) = \sum_{|\beta| \leqq m-1} (-1)^{|\beta|} D^\beta B_\beta(x,u,\ldots,D^{m-1} u). \tag{3}$$

We use the familiar conventions that u, the unknown, is an r-vector function on Ω ($r \geqq 1$) with real values, $D^\alpha = \prod_{j=1}^n (\partial/\partial x_j)^{\alpha_j}$ is the monomial differential operator whose order is denoted by $|\alpha| = \sum_{j=1}^n \alpha_j$.

In the present paper, we develop an existence theory for solutions of boundary value problems of variational type for the partial differential system (1) on a closed subspace V of the Sobolev space $W^{m,p}(\Omega)$ under the assumption that $A(u)$ defines a regular strongly elliptic problem on V as described in Browder [7] (and in particular, its coefficient functions A_α satisfy some polynomial growth conditions in terms of the derivatives of u) while the lower order term $B(u)$ is strongly nonlinear so that it does not satisfy the corresponding growth conditions on its coefficient functions and does not define a mapping from the space V to its conjugate space V^*. This new existence theory is based upon the consideration of some new classes of mappings in Banach spaces which are not everywhere defined and

AMS 1970 *subject classifications*. Primary 35J60, 47H15; Secondary 47H05.

the construction of a corresponding abstract existence theory for these nonlinear mappings.

To formulate the hypotheses which we impose upon the operators $A(u)$ and $B(u)$, we introduce the vector space R^{s_m} whose elements are $\xi = \{\xi_\alpha : |\alpha| \leqq m\}$ where each component ξ_α is an r-vector. Similarly, $\eta = \{\eta_\beta : |\beta| \leqq m - 1\} \in R^{s_{m-1}}$. Each ξ in R^{s_m} can be written in an unique way as a pair (ζ,η) where η, the lower order part of ξ, lies in $R^{s_{m-1}}$ and $\zeta = \{\xi_\alpha : |\alpha| = m\}$ is the pure mth order part of ξ. In terms of this notation, each coefficient function A_α is a function from $\Omega \times R^{s_m}$ into R^r while each B_β is a function from $\Omega \times R^{s_{m-1}}$ into R^r. Upon these functions, we impose the following hypotheses:

(A) *$A_\alpha(x,\xi)$ is measurable in x on Ω for each fixed ξ in R^{s_m} and continuous in ξ for almost all x in Ω. In addition:*

(1) *Let $b_{m,n,p}$ be the greatest integer less than $m - n/p$, $\xi_b = \{\xi_\alpha : |\alpha| \leqq b_{m,n,p}\}$. There exist continuous functions c_α and c_1 from R^{s_b} to $L^{p_\alpha}(\Omega)$ and R^1, respectively, such that the following inequalities hold:*

$$|A_\alpha(x,\xi)| \leqq c_\alpha(\xi_b)(x) + c_1(\xi_b) \sum_{m - n/p \leqq |\beta| \leqq m} |\xi_\beta|^{p_{\alpha\beta}},$$

with the exponents p_α and $p_{\alpha\beta}$ satisfying the inequalities:

$$\begin{aligned} p_\alpha &= p', && \textit{for } |\alpha| = m; \\ p_\alpha &> s'_\alpha, && \textit{for } m - n/p \leqq |\alpha| < m,\ s_\alpha^{-1} = p^{-1} - n^{-1}(m - |\alpha|); \\ p_\alpha &= 1, && \textit{for } |\alpha| < m - n/p; \end{aligned}$$

and

$$\begin{aligned} p_{\alpha\beta} &\leqq p - 1, && \textit{for } |\alpha| = |\beta| = m; \\ p_{\alpha\beta} &< s_\beta(s'_\alpha)^{-1}, && \textit{for } m - n/p \leqq |\alpha| \leqq m,\ |\beta| < m; \\ p_{\alpha\beta} &\leqq s_\beta, && \textit{for } |\alpha| < m - n/p. \end{aligned}$$

Under the assumption (A)(1) which we have just stated, it follows immediately from the Sobolev Imbedding Theorem for the given p with $1 < p < +\infty$ and a closed subspace V of $W^{m,p}(\Omega)$, that for any pair of elements u and v of V we may define the generalized Dirichlet form $a(u,v)$ for the operator $A(u)$ as given by the formula:

$$a(u,v) = \sum_{|\alpha| \leqq m} (A_\alpha(\xi(u)), D^\alpha v) \tag{4}$$

where $A_\alpha(\xi(u))$ is the r-vector function on Ω given by

$$A_\alpha(\xi(u))(x) = A_\alpha(x,\xi(u)(x)); \qquad \xi(u)(x) = \{D^\alpha u(x) : |\alpha| \leqq m\}$$

and (,) denotes the inner product for r-vector functions on Ω with respect to Lebesgue n-measure. It follows from Holder's inequality that there exists a bounded continuous mapping T_0 of the Banach space V into its conjugate space V^* such that, for each u and v of V,

(5) $$a(u,v) = (T_0(u),v),$$

where $(T_0(u),v)$ denotes as usual the action of the functional $T_0(u)$ as an element of V^* upon the element v of V.

If we introduce the additional two assumptions of the following (A)(2), (3) (due originally to Leray and Lions [**19**]), the mapping T_0 satisfies the following condition of *pseudomonotonicity*: *If* $\{u_j\}$ *is an infinite sequence in* V *which converges weakly to* u *in* V, *and if* $\limsup (T_0(u_j),u_j - u) < 0$, *then* $T_0(u_j)$ *converges weakly to* $T_0(u)$ *in* V^* *and* $(T_0(u_j),u_j)$ *converges to* $(T_0(u),u)$. (This is a slight variant of the definition originally due to Brézis [**1**], cf. also Browder [**8**].) Operators A of this type can be thought of as the *regular* elliptic case for our present purposes, so that the existence theory for $B(u) \equiv 0$ will be the standard theory of pseudomonotone mappings from a reflexive Banach space V to its conjugate space V^*. The precise form of the additional assumptions is:

(2) *For each* x *in* Ω, η *in* $R^{s_{m-1}}$ *and each pair of distinct elements* ζ *and* ζ' *of pure mth order jets,*

$$\sum_{|\alpha|=m} (A_\alpha(x,\zeta,\eta) - A_\alpha(x,\zeta',\eta)) \cdot (\zeta_\alpha - \zeta'_\alpha) > 0.$$

(3) *For each pair of elements* c *and* c' *in* R^r,

$$\sum_{|\alpha|=m} (A_\alpha(x,\zeta,\eta) - c_\alpha) \cdot (\zeta_\alpha - c'_\alpha) \to +\infty$$

as $|\zeta| \to +\infty$, *uniformly for* $|\eta|$ *bounded.*

The assumptions which we make upon the strongly nonlinear perturbing term $B(u)$ are of quite a different character.

(B) *For each* β *with* $|\beta| \leqq m - 1$, $B_\beta(x,\eta)$ *is measurable in* x *for fixed* η, *continuous in* η *for almost all fixed* x, *and essentially bounded for* $|\eta|$ *bounded. Let*

$$\Psi(x,\eta) = \sum_{|\beta| \leqq m-1} B_\beta(x,\eta) \cdot \eta_\beta \geqq 0,$$

$$\begin{aligned} \varphi_\alpha(r) &= K_\alpha r^{s_\alpha} \quad \text{for } s_\beta^{-1} > p^{-1} - n^{-1}(m - |\alpha|) \geqq 0, \\ &= \text{an arbitrary continuous function, if } |\alpha| < m - n/p. \end{aligned}$$

Then

(1) *There exists a function* ε *from* R^+ *to* R^+ *with* $\lim_{r\to 0}\varepsilon(r) = 0$, *such that*

(6) $$|B_\beta(x,\eta)| \leqq \varepsilon(|\eta|)\Psi(x,\eta) + \sum_{|\alpha| \leqq m-1} \varphi_\alpha(|\eta_\alpha|)$$

for all β *and* η *and almost all* x *in* Ω.

(2) *For each* $\varepsilon > 0$, *there exists a constant* $K_\varepsilon > 0$ *such that, for all* β *and* x *and any pair* η,η',

(7) $$\begin{aligned} B_\beta(x,\eta) \cdot \eta'_\beta &\leqq \varepsilon\psi(x,\eta') \\ &\quad + K_\varepsilon\Big\{1 + \Psi(x,\eta) + \sum_{|\alpha| \leqq m-1} [\varphi_\alpha(|\eta'_\alpha|) + \varphi_\alpha(|\eta_\alpha|)]\Big\}. \end{aligned}$$

Condition (B) of $B(u)$ is a positivity condition imposing no restriction on the growth of $B_\beta(x,\eta)$ for large η.

The simplest example for this condition is that in which $r = 1$, $B(u) = \varphi(u)$ where for $|u| \geqq k_0 > 0$, $\varphi(u)$ has the same sign as u and such that for each $\varepsilon > 0$, there exist $\rho_\varepsilon, u_\varepsilon, k_\varepsilon > 0$ such that

$$|\varphi(u)| \leqq \varepsilon|\varphi(\rho u)| + k_\varepsilon \qquad (\rho \geqq \rho_\varepsilon, |u| \geqq u_\varepsilon).$$

We impose the following very mild restriction upon the space V (which is always trivially satisfied for the Dirichlet problem where $V = W_0^{m,p}(\Omega)$).

(V) *$V \cap C^m(\bar{\Omega})$ is dense in V. We set $V_0 = V \cap C^{m-1}(\bar{\Omega})$.*

For u,v in V_0, we set $b(u,v) = \sum_{|\beta| \leqq m-1}(B_\beta(\xi(u)),D^\beta v)$.

DEFINITION 1. *The subset V_1 of V is defined by*

$$\begin{aligned} V_1 = \{u | u \in V;\ &\text{for } |\beta| \leqq m-1, \text{ we have } B_\beta(\xi(u)) \in L^1(\Omega); \\ &\Psi(x,\xi(u)(x)) \in L^1(\Omega); \text{ there exists } u^* \text{ in } V^* \\ &\text{such that } (u^*,v) = \textstyle\sum_{|\beta| \leqq m-1}(B_\beta(\xi(u)),D^\beta v) \text{ for all } v \text{ in } V_0\}. \end{aligned}$$

DEFINITION 2. *For u in V_1, we set*

$$T_1(u) = u^*, \qquad b(u,v) = (T_1(u),v) = (u^*,v) \quad \text{for } v \text{ in } V_0.$$

As in the regular case, the solution u of the boundary value problem in whose existence we are interested is defined as follows:

DEFINITION 3. *Let f be a given element of V^*. We seek an element u of V_1 such that for each v in V_0,*

$$a(u,v) + b(u,v) = (f,v).$$

If we introduce an operator T_1 as in Definition 2 whose domain consists of the subset V_1 of V with $(T_1(u),v) = b(u,v)$ for each u in V_1, v in V_0, then the solution u of our given boundary value problem is simply the solution u of the equation

$$T(u) = f \tag{8}$$

where

$$\begin{aligned} T(u) &= T_0(u) + T_1(u), \\ \text{domain } (T) &= \text{domain } (T_1). \end{aligned} \tag{9}$$

We can now state our basic results on the existence of solutions of elliptic boundary value problems for systems of the form (1).

THEOREM 1. *Let Ω be a bounded open set in R^n for which the Sobolev Imbedding Theorem is valid, V a closed subspace of $W^{m,p}(\Omega)$ for which the hypothesis* (V) *holds. Let $A(u)$ and $B(u)$ be differential operators which satisfy the hypotheses* (A)(1),(2),(3) *and* (B), *respectively. Suppose that*

$$\{a(u,u) + b(u,u)\}\|u\|_{m,p}^{-1} \to \infty \tag{10}$$

as $\|u\|_{m,p} \to \infty$ *in* V_0.

Then for each f in V^*, *there exists u in* V_1 *such that* $T(u) = f$, *i.e. u is a solution of the boundary value problem for* $A(u) + B(u) = f$.

THEOREM 2. *Suppose that in Theorem* 1, *we replace the hypothesis* (10) *by the weaker hypothesis that* $\|u\|_{m,p}^{-1}\{a(u,u) + b(u,u)\} \geqq -k$, *and*

$$\|u\|_{m,p}^{-1}\{a(u,u) + b(u,u)\} + \|T(u)\|_{V^*} \to \infty \tag{11}$$

as $\|u\|_{m,p} \to \infty$ *for u in* V_0. *Then the existence of a solution u of* $T(u) = f$ *still holds for each f in* V^*.

As a specialization of Theorem 2, we obtain

THEOREM 3. *Let* Ω *be a bounded open set in* R^n *for which the Sobolev Imbedding Theorem holds, V a closed subspace of* $W^{m,p}(\Omega)$ *for which the hypothesis* (V) *holds. Let* $A(u)$ *be a quasilinear elliptic operator which satisfies the conditions* (A)(1),(2),(3), *B a strongly nonlinear lower order operator which satisfies the hypothesis* (B). *Suppose that the following two conditions hold*:

(a) *There exists a constant k such that for all u in* V_0,

$$a(u,u) + b(u,u) \geqq -k\|u\|_{m,p}.$$

(b) *There exists a continuous function* φ *from* R^+ *to* R^+ *such that if u is a solution of the boundary value problem for* $(A + B)(u) = w$, *then*

$$\|u\|_{m,p} \leqq \varphi(\|w\|_{V^*}).$$

Then a solution u in V_1 *of the equation* $A(u) + B(u) = f$ *with respect to the boundary conditions imposed by the space V exists for each fixed element f of* V^*.

THEOREM 4. *Let* Ω *be a bounded open subset of* R^n *for which the Sobolev Imbedding Theorem holds, V a closed subspace of* $W^{m,p}(\Omega)$ *for which the hypothesis* (V) *holds. Suppose that we are given one parameter families* $\{A_t(u)\}$ *and* $\{B_t(u)\}$ *of differential operators on* Ω *which satisfy the conditions of* (A)(1),(2),(3) *and* (B) *uniformly in t in* $[0,1]$ *with* $A_{\alpha,t}(x,\xi)$ *and* $B_{\beta,t}(x,\xi)$ *continuous in t, uniformly in x on* Ω *and* ξ *bounded. Suppose that* $A_{\alpha,1}(x,\xi)$ *and* $B_{\beta,1}(x,\xi)$ *are odd functions of* ξ. *Suppose further that there exists a continuous function* φ *from* R^+ *to* R^+ *such that if for any t in* $[0,1]$ *and any u in* V_1 *with*

$$A_t(u) + B_t(u) = w$$

in the sense of Definition 3 *with respect to the space V, we have* $\|u\|_{m,p} \leqq \varphi(\|w\|_{V^*})$.

Then for each f in V^*, *there exists a solution u in* V_1 *of the boundary value problem with respect to V of the equation* $A_0(u) + B_0(u) = f$.

As we remarked above, the proofs of Theorems 1 through 4 are obtained by showing that the mapping T, the realization of the given nonlinear boundary value problem as a map from a subset of V to V^*, lies in a suitable class of such mappings for which corresponding abstract existence theorems can be proved. The class of mappings which we consider is given in the following definition:

DEFINITION 4. *Let V be a Banach space, V' a dense linear subspace of V, T a mapping of a subset $D(T)$ of V to V^*. Then T is said to be pseudo-monotone (in the extended sense) with respect to V' if the following conditions hold:*

(a) *V' is contained in $D(T)$ and T is continuous from each finite-dimensional subspace of V' to the weak topology of V^*.*

(b) *Suppose that $\{u_j\}$ is an infinite sequence in V', u an element of V, w an element of V^*. Suppose that u_j converges weakly to u in V, that*

$$(T(u_j),v) \to (w,v)$$

for all v in V', and that

$$\limsup\, (T(u_j),u_j) \leqq (w,u).$$

Then u lies in $D(T)$, $T(u) = w$, and $(T(u_j),u_j) \to (w,u)$.

A related definition which we apply in the subsequent discussion is the corresponding extension of the concept of a mapping T which satisfies condition $(S)_+$ or (S) (as defined in Browder [7], [**8**], [**10**]).

DEFINITION 5. *Let V be a Banach space, V' a dense linear subspace of V, T a mapping of a subset $D(T)$ of V into V^*. Then T is said to satisfy condition $(S)_+$ with respect to V' if the following conditions hold:*

(a) *V' is contained in $D(T)$ and T is continuous from each finite-dimensional subspace of V' to the weak topology of V^*.*

(b) *Suppose that $\{u_j\}$ is an infinite sequence in V', u an element of V, w an element of V^*. Suppose that u_j converges weakly to u in V, that for each v in V, we have $(T(u_j),v) \to (w,v)$, while $\limsup\, (T(u_j),u_j) \leqq (w,u)$. Then u lies in $D(T)$, $T(u) = w$, $\lim\, (T(u_j),u_j) = (w,u)$, and in addition u_j converges strongly to u in V.*

The corresponding definition for condition (S) replaces part (b) of Definition 5 by the following:

(b') *Suppose that $\{u_j\}$ is an infinite sequence in V_0, u an element of V, w an element of V^*. Suppose that u_j converges weakly to u in V, that $(T(u_j),v)$ converges to (w,v) for each v in V_0, and that $\lim\, (T(u_j),u_j) = (w,v)$. Then u lies in $D(T)$, $T(u) = w$, and u_j converges strongly to u in V.*

The basic analytic facts about our given class of nonlinear elliptic boundary value problems which make Definitions 4 and 5 useful for the proof of the existence theorems stated above are contained in the following two theorems.

THEOREM 5. *Let Ω be a bounded open subset of R^n for which the Sobolev Imbedding Theorem is valid, V a closed subspace of $W^{m,p}(\Omega)$ for some p with $1 < p < +\infty$ which satisfies condition (V) with $V_0 = V \cap C^{m-1}(\Omega)$. Suppose that $A(u)$ is a quasi-linear elliptic differential operator of order $2m$ of the form (2) which satisfies the conditions (A)(1),(2),(3), and $B(u)$ a lower order differential operator of the form (3) which satisfies the condition (B). Let $a(u,v)$ be the generalized Dirichlet form for the* *operator $A(u)$, T_0 the mapping of V into V^* given by $(T_0(u),v) = a(u,v)$ for u and v in V. Let T_1 be the mapping of V_1 into V^* corresponding to the differential operator $B(u)$ with V_1 given by Definition 1 above and T_1 by Definition 2 above. Let $T = T_0 + T_1$.*

Then, T is pseudo-monotone with respect to V' for any dense subspace V' of V_0.

THEOREM 6. *Suppose that the hypotheses of Theorem 5 hold and that in addition the condition (A)(3) is replaced by the stronger condition: There exist continuous functions c_0 and c from $R^{s_{m-1}}$ to R^+ with $c_0(\eta) > 0$ for each η such that*

$$\sum_{|\alpha|=m} A_\alpha(x,\zeta,\eta)\cdot\zeta_\alpha \geqq c_0(\eta)|\zeta|^p - c(\eta)$$

for all x in Ω and all η and ζ.

Then T satisfies condition $(\mathrm{S})_+$ with respect to V_0.

We now give the detailed proofs of Theorems 5 and 6, and after completing these proofs, we turn to the detailed development of the existence theory for equations of the form $T(u) = f$ for mappings T which are pseudo-monotone with respect to V'.

PROOF OF THEOREM 5. It follows by standard arguments that T_0 is a bounded continuous pseudo-monotone mapping V into V^*. We must therefore analyze T_1 in detail and consider its properties in relation to those of the sum $T = T_0 + T_1$.

For each β with $|\beta| \leqq m-1$, $B_\beta(\xi(u))$ defines an element of $L^\infty(\Omega)$ for each u in $C^{m-1}(\bar{\Omega})$ and the mapping $u \to B_\beta(\xi(u))$ yields a continuous mapping of $C^{m-1}(\bar{\Omega})$ into $L^s(\Omega)$ for any s, which maps bounded sets into bounded sets. Hence for any u and v in V_0, we may define the generalized Dirichlet form for $B(u)$ by

$$b(u,v) = \sum_{|\beta|\leqq m-1} (B_\beta(\xi(u)),D^\beta v)$$

and $b(u,v)$ satisfies an inequality of the form

$$|b(u,v)| \leqq c_1(\|u\|_{C^{m-1}})\|v\|_V$$

for a suitably constructed continuous function c_1 from R^+ to R^+. It follows that for each u in V_0, there exists a uniquely defined element $T_1(u)$ of V^* such that

$$(T_1(u),v) = b(u,v) \qquad (v \in V_0),$$

where the unique definition of T_1 follows from the assumption that $V \cap C^m(\bar{\Omega})$ is dense in V. Moreover, T_1 is continuous from V_0 endowed with its C^{m-1}-norm to V^* and maps sets X bounded in the C^{m-1}-norm in V_0 into bounded subsets of V^*. It follows that $T = T_0 + T_1$ is continuous from finite-dimensional subsets of V_0 to the weak topology of V^*.

By definition $V_1 = \{u | u \in V;$ for each β with $|\beta| \leqq m-1, B_\beta(\xi(u)) \in L^1(\Omega)$; $\sum_{|\beta|\leqq m-1} B_\beta(\xi(u))(x)\cdot D^\beta u(x) \in L^1(\Omega)$; $\exists u^*$ in V^* such that for each v in V_0, $(u^*,v) = \sum_{|\beta|\leqq m-1}(B_\beta(\xi(u)),D^\beta v)\}$. For each u in V_1 and each v in V_0, we may define the generalized Dirichlet form $b(u,v)$ for $B(u)$ by setting

$$b(u,v) = \sum_{|\beta| \leq m-1} (B_\beta(\xi(u))D^\beta v),$$

and we set

$$T_1(u) = u^*$$

for u in V_1, where u^* as given in the Definition 1 of V_1 is uniquely defined by the density of V_0 in V. Then

$$b(u,v) = (T_1(u),v) \qquad (v \in V_0, u \in V_1)$$

and the mapping T_1 is extended to its domain $V_1 = D(T_1) = D(T)$.

We now prove that T is pseudo-monotone with respect to V_0. Let $\{u_j\}$ be an infinite sequence in V_0 such that u_j converges weakly in V to an element u of V, and suppose that, for a given element w of V^*, $\limsup\, (T(u_j),u_j) \leq (w,u)$ while for all v in a dense subspace V' of V_0, $(T(u_j),v) \to (w,v)$.

We wish to prove that u lies in $D(T)$, $T(u) = w$, and $\lim\, (T(u_j),u_j) = (w,u)$. We note first that we can pick an arbitrary infinite subsequence of the original sequence $\{u_j\}$ and carry out the proof for this subsequence rather than the original sequence. That this suffices for the proof of the property that T is pseudo-monotone with respect to V_0 follows without argument for the conclusions that $u \in D(T)$ and $T(u) = w$. On the other hand, if for any given infinite sequence $\{u_j\}$ which satisfies the above conditions, we can pick an infinite subsequence for which $(T(u_{j_k}),u_{j_k}) \to (w,u)$ as $k \to \infty$, then it follows immediately that $(T(u_j),u_j) \to (w,u)$ for the original sequence. (Indeed, otherwise we could replace the original sequence by an infinite subsequence $\{u_j\}$ for which $|(T(u_j),u_j) - (w,u)| > \delta > 0$ for all j, and for no subsequence of this last sequence, could the asserted convergence hold.)

Since the injection mapping $W^{m,p}(\Omega)$ into $W^{m-1,p}(\Omega)$ is compact by the boundedness of Ω and the validity of the Sobolev Imbedding Theorem, it follows that u_j converges strongly to u in $W^{m-1,p}(\Omega)$. Hence, we may find an infinite subsequence which we again denote by $\{u_j\}$ such that for each β with $|\beta| \leq m-1, D^\beta u_j(x)$ converges almost everywhere to $D^\beta u(x)$ as $j \to \infty$. We assume without loss of generality that this holds for our original sequence. By definition

$$\Psi(x,\eta) = \sum_{|\beta| \leq m-1} B_\beta(x,\eta)\eta_\beta$$

is a nonnegative function of x and η for which there holds the inequality

$$|B_\beta(x,\eta)| \leq \varepsilon(|\eta|)\Psi(x,\eta) + \sum_\alpha \varphi_\alpha(|\eta_\alpha|), \tag{6}$$

for all x,η and all β with $|\beta| \leq m-1$. For each v in V_0, we note that

$$b(v,v) = \int_\Omega \Psi(x,\xi(v)(x))\, dx.$$

By hypothesis, $b(u_j,u_j) = (T_1(u_j),u_j) = (T(u_j),u_j) - (T_0(u_j),u_j)$ has the property that

$$\limsup b(u_j,u_j) \leqq (w,u) + K_0 = K_1$$

since T_0 is a bounded mapping of V into V^*. If we apply Fatou's lemma to the sequence of nonnegative functions $\{\Psi(x,\xi(u_j)(x))\}$ which converges almost everywhere in Ω to $\Psi(x,\xi(u)(x))$, it follows that

$$\int_\Omega \Psi(x,\xi(u)(x))\,dx \leqq \liminf \int_\Omega \Psi(x,\xi(u_j)(x))\,dx \leqq K_1.$$

Hence

$$\Psi(x,\xi(u)(x)) = \sum_{|\beta| \leqq m-1} B_\beta(\xi(u))(x) \cdot D^\beta u(x) \in L^1(\Omega).$$

Let β be a multi-index with $|\beta| \leqq m - 1$. By the inequality (6), there exists a constant K such that, for almost all x in Ω,

$$|B_\beta(\xi(u))(x)| \leqq K\Psi(x,\xi(u)(x)) + \sum_{|\alpha| \leqq m-1} \varphi_\alpha(|D^\alpha u(x)|).$$

By the Sobolev Imbedding Theorem and the preceding results, it follows that $B_\beta(\xi(u))$ lies in $L^1(\Omega)$.

For β fixed, there exists for each $\delta > 0$ a constant $K(\delta) > 0$ such that, for any integer j and almost all x in Ω, either

$$|D^\beta u_j(x)| \leqq K(\delta)$$

or

$$|B_\beta(\xi(u_j))(x)| \leqq \delta\Psi(x,\xi(u_j)(x)) + \sum_{|\alpha| \leqq m-1} \varphi_\alpha(|D^\alpha u_j(x)|).$$

Since u_j converges weakly to u in $W^{m,p}(\Omega)$, it follows from the Sobolev Imbedding Theorem that the functions $\varphi_\alpha(|D^\alpha u_j(x)|)$ are uniformly equi-integrable for all j. Applying the similar inequality for $|B_\beta(\xi(u))(x)|$, it follows that, for any j and any measurable subset Ω_0 of Ω,

$$\int_{\Omega_0} B_\beta(\xi(u_j))(x) - B_\beta(\xi(u)(x))|\,dx$$
$$\leqq 2\max\left(K(\delta)\operatorname{meas}(\Omega_0),\delta K_1 + \int_{\Omega_0} [\varphi_\alpha(|D^\alpha u_j(x)|) + \varphi_\alpha(|D^\alpha u(x)|)]\,dx\right).$$

Thus, the sequence of functions $\{|B_\beta(\xi(u))(x) - B_\beta(\xi(u_j))(x)|\}$ which converges to zero almost everywhere in Ω is uniformly equi-integrable. By the Vitali version of the dominated convergence theorem, it follows that $B_\beta(\xi(u_j)) \to B_\beta(\xi(u))$ strongly in $L^1(\Omega)$ as $j \to \Omega$. Hence, for any v in V_0, we see that

$$(T_1(u_j),v) = b(u_j,v) = \sum_{|\beta| \leqq m-1} (B_\beta(\xi(u_j)),D_\beta v) \sum_{|\beta| \leqq m-1} (B_\beta(\xi(u)),D^\beta v).$$

By hypothesis, $(T(u_j),v)$ converges to (w,v) for any v in V_0. Since T_0 maps bounded sets in V into bounded subsets of V^* and since V is reflexive, we may assume without loss of generality by passing to an infinite subsequence that $T_0(u_j)$ converges weakly in V^* to an element y of V^*. Hence, for any v in V_0,

$$(T_1(u_j),v) = (T(u_j),v) - (T_0(u_j),v) \to (w - y,v).$$

In particular, it follows that

$$(w - y,v) = \sum_{|\beta| \leqq m-1} (B_\beta(\xi(u)),D^\beta v) \qquad (v \in V_0).$$

Thus, we have the fact that u lies in $V_1 = D(T)$, $T_1(u) = w - y$.

Since $T_1(u)$ is an element of V^* and since u_j converges weakly to u in V, we know that

$$(T_1(u),u) = \lim_j (T_1(u),u_j) = \lim_j \sum_{|\beta| \leqq m-1} (B_\beta(\xi(u)),D^\beta u_j).$$

For each β, $B_\beta(\xi(u))(x) \cdot D^\beta u_j(x)$ converges almost everywhere in Ω to $B_\beta(\xi(u))(x) \cdot D^\beta u(x)$, while by the inequality (7) of condition (B) on $B(u)$, its nonnegative part is dominated from above by

$$\varepsilon\Psi(x,\xi(u_j)(x)) + K_\varepsilon\Big(1 + \Psi(x,\xi(u)(x)) + \sum_{|\alpha| \leqq m-1} [\varphi_\alpha(|D_u^\alpha(x)|) + \varphi_\alpha(|D_{u_j}^\alpha(x)|)]\Big)$$

and is uniformly equi-integrable on Ω for all j. Hence

$$(T_1(u),u) \leqq \int_\Omega \sum_{|\beta| \leqq m-1} B_\beta(\xi(u))(x) \cdot D^\beta u(x)\,dx = \int_\Omega \Psi(x,\xi(u)(x))\,dx.$$

By our previous discussion, we know that

$$\int_\Omega \Psi(x,\xi(u)(x))\,dx \leqq \liminf_j \int_\Omega \Psi(x,\xi(u_j)(x))\,dx \leqq \liminf (T_1(u_j),u_j).$$

Combining these inequalities, we see that

$$(T_1(u),u) \leqq \lim (T_1(u_j),u_j).$$

On the other hand,

$$\begin{aligned}\limsup (T_0(u_j),u_j) &\leqq \limsup (T(u_j),u_j) - \liminf (T_1(u_j),u_j)\\ &\leqq (w,u) - (T_1(u),u) = (y,u).\end{aligned}$$

Since u_j converges weakly to u in V, $T_0(u_j)$ converges weakly to y in V^*, and T_0 is pseudo-monotone, it follows that $T_0(u) = y$. Hence

$$T(u) = T_0(u) + T_1(u) = y + (w - y) = w.$$

Since T_0 is pseudo-monotone, it also follows that $(T_0(u_j),u_j)$ converges to (y,u). Hence

$$\begin{aligned}\liminf (T(u_j),u_j) &= \liminf (T_1(u_j),u_j) + \lim (T_0(u_j),u_j)\\ &\geqq (w - y,u) + (y,u) \geqq (w,u).\end{aligned}$$

Since $\limsup (T(u_j),u_j) \leqq (w,u)$, it follows that $(T(u_j),u_j) \to (w,u)$.

Thus the proof of Theorem 5 is complete. Q.E.D.

PROOF OF THEOREM 6. Under the hypothesis of Theorem 6, T_0 satisfies condition $(S)_+$, i.e. given a sequence $\{u_j\}$ in V converging weakly to u in V for which $\limsup (T_0(u_j),u_j - u) \leqq 0$, then u_j converges strongly to u and $T_0(u_j)$ converges weakly to $T_0(u)$. By the proof of Theorem 5, we have u_j converging weakly to u, $T_0(u_j)$ converging weakly to y, and $\limsup (T_0(u_j),u_j) \leqq (y,u)$. Hence

$$\limsup (T_0(u_j),u_j - u) = \limsup (T_0(u_j),u_j) - (y,u) \leqq 0.$$

Therefore, u_j converges strongly to u in V. Q.E.D.

We now establish the first of the abstract theorems from which Theorems 1 through 4 will be derived.

THEOREM 7. *Let V be a reflexive Banach space, V_0 a separable dense linear subspace of V, T a mapping from $D(T)$ to V^* such that for any dense subspace V' of V_0, T is pseudo-monotone with respect to V'. Suppose that T is coercive with respect to the V-norm, i.e. $(T(u),u)\|u\|_V^{-1} \to \infty$ as $\|u\|_V \to \infty$ $(u \in V_0)$.*

Then, the range of T is all of V^*.

PROOF OF THEOREM 1 FROM THEOREM 7. We apply Theorem 5 with $V_0 = V \cap C^{m-1}(\bar{\Omega})$.

PROOF OF THEOREM 7. We choose an increasing sequence $\{V_n\}$ of finite-dimensional subspaces of V_0 whose union is dense in V. Let $V' = \bigcup_n V_n$. For each n, let γ_n be the injection mapping of V_n into V, γ_n^* the dual projection of V^* onto V_n^*, and let $T_n = \gamma_n^* T \gamma_n : V_n \to V_n^*$ be the nth Galerkin approximant to T. By the condition (a) in the definition of pseudo-monotonicity with respect to V_0, each T_n is continuous.

Since for a given f in V^*, we can replace T by T_f given by $T_f(u) = T(u) - f$ and T_f satisfies the same hypotheses as T, it suffices to prove that 0 lies in $R(T)$. By the coercivity of T, there exists $R > 0$ such that for all u in V with $\|u\| = R$, we have $(T(u),u) > 0$. Since $(T_n(u),v) = (T(u),v)$ for all u and v in V_n, it follows that for u in $V_n \cap S_R(0,V)$, $(T_n(u),u) > 0$. It follows from a standard finite-dimensional degree argument that there exists u_n in $V_n \cap B_R(0,V)$ such that $T_n(u_n) = 0$.

Since V is reflexive and the sequence $\{u_n\}$ is bounded, we can find an infinite subsequence $\{u_{n_j}\}$ which converges weakly in V to an element u of V. Let $u_j = u_{n_j}$. Let v be any element of V'. Then v lies in one of the spaces V_n. For $n_j \geqq n$, we have

$$(T(u_j),v) = (T_{n_j}(u_{n_j}),v) = 0.$$

Hence, $(T(u_j),v) \to 0$ for any v in V_0. Moreover, since u_j itself lies in V_{n_j},

$$(T(u_j),u_j) = (T_{n_j}(u_{n_j}),u_{n_j}) = 0.$$

Hence,

$$\lim (T(u_j),u_j) = 0 = (0,u).$$

Since T is pseudo-monotone with respect to V', it follows that u lies in $D(T)$ and $T(u) = 0$. Q.E.D.

THEOREM 8. *Let V be a reflexive Banach space, V_0 a separable dense linear space of V, T a mapping from $D(T)$ to V^* such that for any dense subspace V' of V_0, T is pseudo-monotone with respect to V'. Suppose that $\|u\|^{-1}(T(u),u) \geqq -k$, and*

$$\|T(u)\| + (T(u),u)/\|u\| \to \infty$$

as $\|u\| \to \infty$ in $V_0^{\#}$.

Then the range of T is all of V^*.

Theorem 2 follows from Theorem 8 by a direct specialization using Theorem 5.

PROOF OF THEOREM 8. If we replace T by T_f with $T_f(u) = T(u) - f$ for a given element f of V^*, we note that the hypotheses are invariant under the change. Hence to prove that $R(T) = V^*$, it suffices to prove that $R(T)$ contains 0. To continue the proof, we apply the result of the following proposition:

PROPOSITION 1. *Let V be a reflexive Banach space, V_0 a dense linear subspace of V, T a mapping of a subset $D(T)$ of V into V^* with the property that T is pseudo-monotone with respect to V_0. Then*

(1) *There exists an equivalent norm on V and a duality mapping S from V to V^* with respect to this norm such that S is a bounded continuous mapping of V into V^* which satisfies condition $(S)_+$, $\|S(u)\| = \|u\|$ for all u in V, and $(S(u),u) = \|u\|^2$.*

(2) *For each t with $0 < t \leqq 1$, $T_t = (1-t)T + tS$ satisfies condition $(S)_+$ with respect to V_0.*

PROOF OF PROPOSITION 1. For the proof of part (1), we refer to §17 of Browder [**8**], together with a recent result of Trojanski that each reflexive Banach space V has an equivalent locally uniformly convex norm.

We turn therefore to the proof of part (2) of our assertion. Let $\{u_j\}$ be a sequence in V_0 with u_j converging weakly to u in V. Suppose that for a given t, with $0 < t < 1$, and a given w in V^* that

$$\limsup\,(T_t(u_j),u_j) \leqq (w,u)$$

while, for all v in V_0,

$$\lim\,(T(u_j),v) = (w,v).$$

Since S is a bounded mapping, we may assume that $S(u_j)$ converges weakly to y in V. Let $s = t(1-t)^{-1}$. Suppose first that $\limsup\,(S(u_j),u_j) > (y,u)$. Then, if we pass to an infinite subsequence, we find that

$$\limsup\,(T(u_j),u_j) < ((1-t)^{-1}w - sy,u).$$

For each v in V_0,

$$(T(u_j),v) = (1-t)^{-1}(T_t(u_j),v) - s(S(u_j),v) \to ((1-t)^{-1}w - sy,v).$$

It follows from the pseudo-monotonicity of T with respect to V_0 that u lies in $D(T)$, $T(u) = (1 - t)^{-1}w - sy$, and

$$\lim (T(u_j),u_j) = ((1 - t)^{-1}w - sy,u).$$

This contradicts the strict inequality above, and shows that $\limsup (S(u_j),u_j) \leqq (y,u)$.

Hence, we may assume that $\limsup (S(u_j),u_j) \leqq (y,u)$. Since S satisfies condition $(S)_+$, it follows that u_j converges strongly to $u, S(u) = y$, and that $\lim (S(u_j),u_j) = (y,u)$. Hence

$$\begin{aligned}\limsup (T(u_j),u_j) &= (1 - t)^{-1} \limsup (T_t(u_j),y_j) - s \lim (S(u_j),u_j) \\ &\leqq ((1 - t)^{-1}w - sy,u).\end{aligned}$$

Since

$$(T(u_j),v) \to ((1 - t)^{-1}w - sy,v)$$

for every v in V_0, it follows from the pseudo-monotonicity of T with respect to V_0 that u lies in $D(T)$, that $T(u) = (1 - t)^{-1}w - sy$, and that

$$\lim (T(u_j),u_j) = ((1 - t)^{-1}w - sy,u).$$

Finally, it follows from the preceding paragraph that u lies in $D(T_t) = D(T)$, that $T_t(u) = w$, and that $(T_t(u_j), u_j) \to (w,u)$. Hence T_t satisfies condition $(S)_+$ with respect to V_0 since we have already shown that u_j converges strongly to u in V. Q.E.D.

PROOF OF THEOREM 8 COMPLETED. We may choose $R > 0$ so large that for u in $S_R(0,V)$, we have

$$\|T(u)\| + (T(u),u)/\|u\| > 0.$$

We choose an increasing sequence $\{V_n\}$ of finite-dimensional subspaces of V_0 with dense union in V, and we let $V' = \bigcup_n V_n$. By the hypothesis of Theorem 8, T is pseudo-monotone with respect to V_0.

We form the Galerkin approximants T_n for T with respect to the spaces V_n as in the proof of Theorem 7. For each t in $[0,1]$, we let $T_t = (1 - t)T + tS$ where S is the mapping described in Proposition 1. If S_n and $T_{t,n}$ are the Galerkin approximants of S and T_t, respectively, we see that

$$T_{t,n} = (1 - t)T_n + tS_n.$$

By the proof of Theorem 7, it follows that it suffices to show the existence of an integer N such that for each $n \geqq N$, there exists u_n in $B_R(0,V) \cap V_n$ such that $T_n(u_n) = 0$. We assert that this will follow if we can show that for $n \geqq N$, for any t in $[0,1]$ and any u in $S_R(0,V) \cap V_n$ that

$$T_{t,n}(u) \neq 0.$$

Indeed, if this latter assertion holds, we may identify each V_n with the conjugate space V_n^* by a Hilbert space structure and consider the Brouwer degree $\deg(T_{t,n},S_R \cap V_n,0)$. This degree is defined for each $n \geqq N$ and independent of t in $[0,1]$. For $t = 1$, the fact that $(S_n(u),u) = (S(u),u) > 0$ for nonzero u in V_n implies by a linear homotopy to the identity map of V_n that

$$\deg(T_{1,n},S_R \cap V_n,0) = \deg(S_n,S_R \cap V_n,0) = +1.$$

Hence

$$\deg(T_n,S_R \cap V_n,0) = +1$$

and there will exist a solution u_n in $B_R \cap V_n$ of the equation $T_n(u_n) = 0$.

In the contrary case, we can find a sequence of integers $\{n_j\}$ going to infinity, a sequence $\{t_j\}$ in $[0,1]$ and for each j, an element u_j in $S_R \cap V_{n_j}$ such that $T_{t_j,n_j}(u_j) = 0$. We may assume without loss of generality that t_j converges to t_0 as $j \to \infty$. Since $(T(u),u) \geqq -k\|u\|$, it follows that $t_0 < 1$. If $t_0 = 0$,

$$T_{n_j}(u_j) = (1 - t_j)^{-1}T_{t_j,n_j}(u_j) - t_j(1 - t_j)^{-1}S_{n_j}(u_j),$$

so that

$$\|T_{n_j}(u_j)\|_{V_{n_j}^*} \to 0 \qquad (j \to \infty).$$

It follows that

$$(T(u_j),u_j) = (T_{n_j}(u_j),u_j) \to 0 \qquad (j \to \infty),$$

and that, for each v in V',

$$(T(u_j),v) = (T_{n_j}(u_j),v) \to 0$$

for $n_j \geqq n$ where v lies in V_n. It follows from the pseudo-monotonicity of T with respect to V' that for any weak sequential limit u of a subsequence of $\{u_j\}$ that u lies in $B_R(0,V)$ and $T(u) = 0$. Hence, we need only consider the case in which $t_0 \neq 0$.

For $0 < t_0 < 1$, we note that $T_{t_0,n_j}(u_j)$ satisfies the condition

$$\|T_{t_0,n_j}(u_j)\|_{V_{n_j}^*} \to 0.$$

Applying the argument just used for the case in which $t_0 = 0$, we may assume that u_j converges weakly to u in V, and note that $(T_{t_0}(u_j),u_j) \to 0$ and that for any v in V_0, $(T_{t_0}(u_j),v) \to 0$. By Proposition 1, T_{t_0} satisfies condition $(S)_+$ with respect to V_0. Hence, it follows that u_j converges strongly to u and $T_{t_0}(u) = 0$. Since all the u_j lie in S_R, if follows that u lies in S_R.

We note finally that this latter case is excluded by the hypothesis, which we have not previously applied. Indeed, if $T_{t_0}(u) = 0$ and if $s_0 = t_0(1 - t_0)^{-1}$, it follows that

$$T(u) = -s_0S(u).$$

Hence

$$\|T(u)\| = s_0\|u\| \quad \text{and} \quad (T(u),u) = -s_0\|u\|^2.$$

Thus

$$0 < \|T(u)\| + (T(u),u)/\|u\| = 0,$$

which is a contradiction.

This contradiction, which excludes the last case, establishes the conclusion of the theorem. Q.E.D.

THEOREM 9. *Let V be a reflexive Banach space, V_0 a dense separable linear subspace of V, $\{T_t: 0 \leqq t \leqq 1\}$ a one parameter family of mappings from $D(T_t)$ into V^* each of which is pseudo-monotone with respect to any dense subspace V' of V_0.*

Suppose that the following conditions hold:

(1) *There exists $R_1 > 0$ such that $T_1(-u) = -T_1(u)$ for u outside $B_{R_1}(0,V) \cap V_0$.*

(2) *$T_t(u)$ is continuous in t, uniformly in u on bounded subsets of V_0.*

(3) *There exists a continuous function φ from R^+ to R^+ such that if $T_t(u) = w$ for some t in $[0,1]$, then $\|u\| \leqq \varphi(\|w\|)$.*

Then the range of each T_t is the whole space V^*.

Theorem 4 follows from a stronger version of Theorem 9 using the concept of a pseudo-monotone homotopy. A detailed treatment will be given in another paper. We content ourselves here with the detailed proof of Theorem 9.

PROOF OF THEOREM 9. Let $T_{t,n}$ be the nth Galerkin approximant of T_t with respect to an increasing sequence $\{V_n\}$ of finite-dimensional subspaces of V_0 whose union V' is dense in V. Each of the mappings $T_{1,n}$ is odd outside of $B_{R_1}(0,V) \cap V_n$. Let $R_0 > 0$ be given. Then we can choose $R > R_1$ such that for any u in $S_R(0,V)$ and any t in $[0,1]$, $\|T_t(u)\| > R_0 + 1$. To show that $R(T_0)$ is the whole of V^*, it suffices to show that a given element f of $B_{R_0}(0,V^*)$ lies in $R(T_0)$. For each $\varepsilon > 0$, let

$$T_{t,\varepsilon} = T_t + \varepsilon S$$

where S is the mapping discussed in Proposition 1. For $\varepsilon > 0$ and sufficiently small, we know that $T_{t,\varepsilon}(S_R)$ never hits $B_{R_0}(0,V^*)$. Suppose this is true for all ε such that $0 \leqq \varepsilon \leqq \varepsilon_0$ ($\varepsilon_0 > 0$).

Suppose first that, for each ε with $0 < \varepsilon \leqq \varepsilon_0$, there exists an integer N_ε such that for each $n \geqq N_\varepsilon$, each u in $S_R(0,V) \cap V_n$, each ξ in $[0,1]$, and each t in $[0,1]$ we have

$$T_{t,\varepsilon,n}(u) \neq \xi\gamma_n^*(f).$$

We may then define the Brouwer degree

$$\deg(T_{t,\varepsilon,n}, B_R(0,V) \cap V_n, \xi\gamma_n^*(f)) \qquad (n \geqq N_\varepsilon),$$

and this degree is independent of t and ξ in $[0,1]$. For $t = 1$, $\xi = 0$, it follows from the fact that T_1 is odd on $S_R(0,V)$ that $T_{1,\varepsilon,n}$ is odd on $S_R(0,V) \cap V_n$. By the Borsuk-

Ulam theorem, the above degree is therefore nonnull. Hence

$$\deg(T_{0,\varepsilon,n}, B_R(0,V) \cap V_n, \gamma_n^*(f)) \neq 0$$

for $n \geqq N$, $0 < \varepsilon < \varepsilon_0$. Hence there will exist a solution $u_{\varepsilon,n}$ in $B_R(0,V) \cap V_n$ of the equation

$$T_{0,\varepsilon,n}(u_{\varepsilon,n}) = \gamma_n^*(f) \qquad (n \geqq N_\varepsilon).$$

We now choose a sequence $\{\varepsilon_j\}$ converging to 0 and corresponding sequences $\{n_j\}$, $\{u_j\}$ where $n_j \to +\infty$, $u_j \in B_R(0,V) \cap V_{n_j}$ and $T_{0,\varepsilon_j,n_j}(u_j) = \gamma_{n_j}^*(f)$. Let v be any element of V', $v \in V_n$. For $n_j \geqq n$, we see that

$$\begin{aligned}(T_0(u_j),v) &= (T_{0,n_j}(n_j),v)\\ &= (T_{0,\varepsilon_j,n_j}(u_j),v) - \varepsilon_j(S(u_j),v)\\ &= (\gamma_{n_j}^*(f),v) - \varepsilon_j(S(u_j),v)\\ &= (f - \varepsilon_j S(u_j),v) \to (f,v).\end{aligned}$$

Similarly,

$$\begin{aligned}(T_0(u_j),u_j) &= (T_{0,n_j}(u_j),u_j)\\ &= (T_{0,\varepsilon,n_j}(u_j),u_j) - \varepsilon_j(S(u_j),u_j)\\ &= (f - \varepsilon_j S(u_j),u_j) \to (f,u)\end{aligned}$$

if we assume in addition (as we obviously may) that u_j converges weakly to u in V. Applying the pseudo-monotonicity of T_0 with respect to V', it follows that $T_0(u) = f$ with u lying in $B_R(0,V)$. In this case, the conclusion of Theorem 9 holds.

In the contrary case, we may replace T_t by $T_{t,\varepsilon}$ and assume that T_t satisfies condition $(S)_+$ with respect to V' for each t in $[0,1]$. If we consider the contrary case, we obtain there a sequence $\{n_k\}$ of integers going to infinity, two sequences $\{t_k\}$ and $\{\xi_k\}$ in $[0,1]$ and for each k an element u_k of $S_R(0,V) \cap V_{n_k}$ such that $T_{t_k,n_k}(u_k) = \xi_k(\gamma_n^*(f))$. We may assume without loss of generality that ξ_k converges to ξ and that t_k converges to t. Then

$$\| T_t(u_k) - T_{t_k}(u_k)\| \to 0 \qquad (k \to \infty)$$

by the continuity properties of T_t in t. Hence, we obtain

$$\| T_{t,n_k}(u_k) - \xi\gamma_{n_k}^*(f)\|_{V_{n_k}^*} \to 0.$$

It follows from this last relation that if u_k converges weakly to u, then

$$(T_t(u_k),u_k) \to (\xi f,u)$$

and for each v in V_0,

$$(T_t(u_k),v) \to (\xi f,v).$$

Applying the assumption that T_t satisfies condition $(S)_+$ with respect to V', it follows that u_j converges strongly to u and $T_t(u) = \xi f$. This implies that u lies in $S_R(0,V)$ and that $T_t(\mathrm{u})$ lies in $B_{R_0}(0,V^*)$. This contradicts our previous assumptions, and this contradiction to the contrary case establishes the conclusion of the theorem. Q.E.D.

REMARKS. (1) The proofs of Theorems 7,8, and 9 are more naturally considered as a part of the theory of the generalized degree for mappings T satisfying the condition of pseudo-monotonicity with respect to dense subspaces V_0. Such a theory including the present case but of a much more general character (which extends the generalized degree theory for A-proper mappings given in Browder-Petryshyn [**15**]) is given by the writer in [**12**].

(2) Nonlinear eigenvalue problems for elliptic operators of the type treated here have been considered by the writer in [**13**], using Galerkin methods along the lines of [**10**] (see also [**7**] and [**11**]).

(3) Extensions of the elliptic existence theory to the corresponding classes of variational inequalities, parabolic problems, and nonlinear wave equations will be treated in a subsequent paper.

(4) Brézis and Strauss in a forthcoming paper [**2**] have treated the existence and regularity theory of the equation $-\Delta u + \beta(u) = f$ where β is a maximal monotone function or graph. Brezis has remarked to the writer in connection with a discussion of the results of the present paper that some of the technical devices used in the proof of Theorem 5 (especially the use of Fatou's lemma) have points in common with a forthcoming paper of Strauss [**27**] on weak solutions of strongly nonlinear wave equations of Klein-Gordon type. Results on $-\Delta u + p(u) = f$ have also been obtained by J. P. Gossez as an outgrowth of his results on monotone mappings in nonreflexive Banach spaces.

(5) The existence result of Theorem 8 sharpens existence theorems for pseudo-monotone mappings under subcoercivity assumptions obtained in Browder-Hess [**14**], Petryshyn [**24**] and Wille [**29**]. Existence theorems of the Borsuk-Ulam type and other related forms of the nonlinear Fredholm alternative have been obtained in recent years by Pohozaev [**26**], Browder [**7**], [**8**], [**9**], Hess [**18**], Necas [**22**], Petryshyn [**24**], and Fitzpatrick [**17**].

BIBLIOGRAPHY

1. H. Brézis, *Équations et linequations non-linéaires dans les espaces vectoriels en dualité*, Ann. Inst. Fourier (Grenoble) **18** (1968), fasc. 1, 115–175. MR **42** #5113.

2. H. Brézis and W. Strauss, (in preparation).

3. F. E. Browder, *Nonlinear elliptic boundary value problems*, Bull. Amer. Math. Soc. **69** (1963), 862–874. MR **27** #6048.

4. ———, *Nonlinear elliptic boundary value problems*. II, Trans. Amer. Math. Soc. **117** (1965), 530–550. MR **40** #4054.

5. ———, *Existence and uniqueness theorems for solutions of nonlinear boundary value problems*, Proc. Sympos. Appl. Math., vol. 17, Amer. Math. Soc., Providence, R.I., 1965, pp. 24–49. MR **33** #6092.

6. ———, *Problèmes non linéaires*, Séminaire de Mathématiques Supérieures, no. 15, Les Presses de l'Université de Montréal, Montréal, Que., 1966. MR **40** #3380.

7. ———, *Existence theorems for nonlinear partial differential equations*, Proc. Sympos. Pure Math., vol. 16, Amer. Math. Soc., Providence, R.I., 1970, pp. 1–60. MR **42** #4855.

8. ———, *Nonlinear operators and nonlinear equations of evolution in Banach spaces*, Proc. Sympos. Pure Math., vol. 18, part 2, Amer. Math. Soc., Providence, R.I. (to appear).

9. ———, *Nonlinear elliptic boundary value problems and the generalized topological degree*, Bull. Amer. Math. Soc. **76** (1970), 999–1005. MR **41** #8818.

10. ———, *Nonlinear eigenvalue problems and Galerkin approximations*, Bull. Amer. Math. Soc. **74** (1968), 651–656. MR **37** #2043.

11. ———, *Nonlinear eigenvalue problems and group invariance*, Functional Analysis and Related Fields, Springer-Verlag, New York, 1970, pp. 1–58.

12. ———, *Generalizations of the topological degree in nonlinear functional analysis* (to appear).

13. ———, *Nonlinear elliptic eigenvalue problems with strong lower-order nonlinearities*, Proc. Rocky Mountain Consortium on Nonlinear Eigenvalue Problems, June 1971 (to appear).

14. F. E. Browder and P. Hess, *Nonlinear mappings of monotone type in Banach spaces*, J. Functional Analysis (to appear).

15. F. E. Browder and W. V. Petryshyn, *Approximation methods and the generalized topological degree for nonlinear mappings in Banach spaces*, J. Functional Analysis **3** (1969), 217–245. MR **39** #6126.

16. Ju. A. Dubinskii, *Quasilinear elliptic and aparabolic equations of arbitrary order*, Uspehi Mat. Nauk **23** (1968), no. 1 (139), 45–90 = Russian Math. Surveys **23** (1968), no. 1, 45–90. MR **37** #4405.

17. P. M. Fitzpatrick, *A generalized degree for uniform limits of A-proper mappings*, J. Math. Anal. Appl. (to appear).

18. P. Hess, *On nonlinear mappings of monotone type homotopic to odd operators*, J. Functional Analysis (to appear).

19. J. Leray and J. L. Lions, *Quelques résultats de Visik sur les problèmes elliptiques nonlinéaires par les méthodes de Minty-Browder*, Bull. Soc. Math. France **93** (1965), 97–107. MR **33** #2939.

20. J. L. Lions, *Quelques méthodes de résolution des problèmes aux limites nonlinéaires*, Dunod; Gauthier-Villars, Paris, 1969. MR **41** #4326.

21. G. J. Minty, *On a "monotonicity" method for the solution of nonlinear equations in Banach spaces*, Proc. Nat. Acad. Sci. U.S.A. **50** (1963), 1038–1041. MR **28** #5358.

22. J. Necas, *Sur l'alternative de Fredholm pour les opérateurs nonlinéaires avec applications aux problèmes aux limites*, Ann. Scuola Norm. Sup. Pisa (3) **23** (1969), 331–345. MR **42** #2332.

23. C. B. Morrey, Jr., *Multiple integrals in the calculus of variations*. Die Grundlehren der math. Wissenschaften, Band 130, Springer-Verlag, New York, 1966. MR **34** #2380.

24. W. V. Petryshyn, *Antipodes theorem for A-proper mappings and its applications to mappings of the modified type* (S) *or* $(S)_+$ *and to mappings with the pm property*, J. Functional Analysis **7** (1971), 165–211.

25. ———, *On existence theorems for nonlinear equations involving noncompact mappings*, Proc. Nat. Acad. Sci. U.S.A. **67** (1970), 326–330. MR **42** #3621.

26. S. I. Pohozaev, *The solvability of nonlinear equations with odd operators*, Funkcional. Anal. i Prilozen. **1** (1967), no. 3, 66–73. (Russian) MR **36** #4396.

27. W. Strauss, *On weak solutions of nonlinear hyperbolic equations*, An. Brasil. Acad. Ci. (to appear).

28. M. I. Visik, *Quasi-linear strongly elliptic systems of differential equations of divergence form*, Trudy Moskov. Mat. Obsc. **12** (1963), 125–184 = Trans. Moscow Math. Soc. **1963**, 140–208. MR **27** #6017.

29. F. Wille, *On monotone operators with perturbations*, Arch. Rational Mech. Anal. (to appear).

UNIVERSITY OF CHICAGO

EXISTENCE THEOREMS FOR PROBLEMS OF OPTIMIZATION WITH PARTIAL DIFFERENTIAL EQUATIONS

LAMBERTO CESARI

1. **Existence theorems with state equations in strong form.** We are interested here in problems of optimization in a fixed domain $G \subset E_\nu$ for which the state variable x is an element of a Fréchet L-space S, for which state equations—in either strong or weak forms—and unilateral constraints are expressed in terms of general not necessarily linear operators on S, mapping S into vector-valued L-integrable functions on G and ∂G respectively, and of arbitrary measurable vector-valued control functions u on G and v on ∂G. Note that, in general, S is only a set of elements x for which a concept of convergence has been defined with the usual properties (Fréchet L-space). Nevertheless, in most examples, S is a Banach space on which the weak convergence has been chosen as the concept of convergence.

We denote by Γ a closed subset of ∂G, by μ a suitable measure function on Γ, by T the set of all measurable vector functions $u(t) = (u^1, \dots, u^m)$, $t \in G$ (distributed controls), and by $\mathring{T}$ the set of all μ-measurable vector functions $v(t) = (v^1, \dots, v^{m'})$, $t \in \Gamma$ (boundary controls). We denote by $\mathscr{L}$, $\mathscr{T}$, $\mathscr{M}$, $\mathscr{K}$ given operators $\mathscr{L}: S \to (L_p(G))^r$, $\mathscr{T}: S \to (L_p(\Gamma))^{r'}$, $\mathscr{M}: S \to (L_p(G))^s$, $\mathscr{K}: S \to (L_p(\Gamma))^{s'}$, where $p \geqq 1$, and r, r', s, s' are integers. The images under $\mathscr{L}$, $\mathscr{T}$, $\mathscr{M}$, $\mathscr{K}$ of elements $x \in S$ will be denoted by z, $\mathring{z}$, y, $\mathring{y}$, or $z(t) = (z^1, \dots, z^r) = (\mathscr{L}x)(t)$, $t \in G$; $\mathring{z}(t) = (\mathring{z}^1, \dots, \mathring{z}^{r'}) = (\mathscr{T}x)(t)$, $t \in \Gamma$; $y(t) = (y^1, \dots, y^s) = (\mathscr{M}x)(t)$, $t \in G$; $\mathring{y}(t) = (\mathring{y}^1, \dots, \mathring{y}^{s'}) = (\mathscr{K}x)(t)$, $t \in \Gamma$.

We consider the problem of finding elements $x \in S$, $u \in T$, $v \in \mathring{T}$, so as to minimize a functional of the form

$$(1) \qquad I[x, u, v] = \int_G f_0(t, (\mathscr{M}x)(t), u(t))\, dt + \int_\Gamma g_0(t, (\mathscr{K}x)(t), v(t))\, d\mu,$$

subject to state equations (strong form)

AMS 1970 *subject classifications*. Primary 49A20, 49A50.

$$(2) \qquad (\mathscr{L}x)(t) = f(t, (\mathscr{M}x)(t), u(t)) \qquad \text{a.e. in } G,$$

$$(3) \qquad (\mathscr{T}x)(t) = g(t, (\mathscr{K}x)(t), v(t)) \qquad \mu\text{-a.e. on } \Gamma,$$

and unilateral constraints on the values of u, v, y, $\mathring{y}$ of the forms

$$(4) \qquad u(t) \in U(t, (\mathscr{M}x)(t)) \subset E_m \qquad \text{a.e. in } G,$$

$$(5) \qquad v(t) \in V(t, (\mathscr{K}x)(t)) \subset E_{m'} \qquad \mu\text{-a.e. on } \Gamma,$$

$$(6) \qquad (\mathscr{M}x)(t) \in A(t) \subset E_s \qquad \text{a.e. in } G,$$

$$(7) \qquad (\mathscr{K}x)(t) \in B(t) \subset E_{s'} \qquad \mu\text{-a.e. on } \Gamma.$$

Here we assume that for any $t \in \text{cl } G$ a given subset $A(t)$ of E_s is assigned, and we denote by A the set of all $(t, y) \in E_{\nu+s}$ with $t \in \text{cl } G$, $y \in A(t)$. We assume that for any $(t, y) \in A$ a given subset $U(t, y)$ of E_m is assigned (distributed control space), and we denote by M the set of all (t, y, u) with $(t, y) \in A$, $u \in U(t, y)$. Then, $f_0(t, y, u)$, $f(t, y, u) = (f_1, \ldots, f_r)$ are given functions on M. Analogously, we assume that for any $t \in \Gamma$ a given subset $B(t)$ of $E_{s'}$ is assigned, and we denote by B the set of all $(t, \mathring{y}) \in E_{\nu+s'}$ with $t \in \Gamma$, $\mathring{y} \in B(t)$. We assume that for any $(t, \mathring{y}) \in B$ a given subset $V(t, \mathring{y})$ of $E_{m'}$ is assigned (boundary control space), and we denote by $\mathring{M}$ the set of all $(t, \mathring{y}, v)$ with $(t, \mathring{y}) \in B$, $v \in V(t, \mathring{y})$. Then, $g_0(t, \mathring{y}, v)$, $g(t, \mathring{y}, v) = (g_1, \ldots, g_{r'})$ are given functions on $\mathring{M}$. We denote by $I_1[x, u]$ and $I_2[x, v]$ the two integrals in (1) on G and Γ respectively, whose sum is $I[x, u, v]$.

Usually, S is a Sobolev space in G, state equation (2) represents a system of r partial differential equations in G, and state equation (3) represents either boundary data, or a system of r' partial differential equations on Γ (or on ∂G). But the situation may be quite different, since the operators $\mathscr{L}$, $\mathscr{T}$, $\mathscr{M}$, $\mathscr{K}$ need not be differential operators. All we shall require is a set of axioms (P) in the present setting, as well as in the setting of §2, with state equations in the weak form. In all cases the existence theorems follow in a natural way from uniformly proved closure theorems and lower closure theorems, extending the usual lower semicontinuity theorems for classical free problems of the calculus of variations.

(P_1) *Hypotheses on* $\mathscr{L}$, $\mathscr{T}$, $\mathscr{M}$, $\mathscr{K}$. If $x, x_k \in S$, $k = 1, 2, \ldots$, and $x_k \to x$ in S as $k \to \infty$, then $\mathscr{L}x_k \to \mathscr{L}x$ weakly in $(L_p(G))^r$, $\mathscr{T}x_k \to \mathscr{T}x$ weakly in $(L_p(\Gamma))^{r'}$, $\mathscr{M}x_k \to \mathscr{M}x$ strongly in $(L_p(G))^s$, $\mathscr{K}x_k \to \mathscr{K}x$ strongly in $(L_p(\Gamma))^{s'}$.

(P_2) *Hypotheses on the set* Γ *and measure* μ. We assume that the closed set $\Gamma \subset \partial G$ is the union of finitely many sets $\Gamma_1, \ldots, \Gamma_N$, each Γ_j being the image of a $(\nu - 1)$-dimensional interval I under a transformation T_j of class K (Morrey), say $T_j : I \to \Gamma_j$. To simplify the exposition we assume here that μ is the hyperarea measure defined on Γ by the mappings T_j.

(P_3) *Hypotheses on* f_0, g_0, f, g. We assume that the sets A, B, M, $\mathring{M}$ are closed, that the functions $f_0, f = (f_1, \ldots, f_r)$ are continuous on M, and that the functions $g_0, g = (g_1, \ldots, g_{r'})$ are continuous on $\mathring{M}$. Also, we assume that there is some integrable function $\phi(t) \geqq 0$, $t \in G$, such that $f_0(t, y, u) \geqq -\phi(t)$ for all $(t, y, u) \in M$;

and that there is some μ-integrable function $\psi(t) \geqq 0$, $t \in \Gamma$, such that $g_0(t, \mathring{y}, v) \geqq -\psi(t)$ for all $(t, \mathring{y}, v) \in \mathring{M}$.

These conditions, however, can be relaxed. For instance, f_0 and g_0 may be assumed to be only lower semicontinuous on M and $\mathring{M}$ respectively. Also, the bounds below can be relaxed as follows: For every point $(\bar{t}, \bar{y}) \in A$ there are a neighborhood $N_\delta(\bar{t}, \bar{y})$, real numbers $r, b = (b_1, \ldots, b_r)$, and an L-integrable function $\phi(t) \geqq 0$, $t \in N_\delta(\bar{t}, \bar{y})$, such that $f_0(t, y, u) - r - \sum_j b_j f_j(t, y, u) \geqq -\phi(t)$ for all $(t, y) \in N_\delta(\bar{t}, \bar{y})$ and $u \in U(t, y)$. An analogously relaxed requirement can be formulated for g_0.

Kuratowski's concept of upper semicontinuity of variable sets and modifications. We shall discuss these concepts on the sets $U(t, y)$, but they apply to the sets $V(t, \mathring{y})$ as well, and to the sets $\tilde{\tilde{Q}}(t, y)$ and $\tilde{\tilde{R}}(t, \mathring{y})$ we shall define below. For $(\bar{t}, \bar{y}) \in A$ and $\delta > 0$ let $N_\delta(\bar{t}, \bar{y})$ be the set of all $(t, y) \in A$ at a distance $\leqq \delta$ from $(\bar{t}, \bar{y})$, and let $U(\bar{t}, \bar{y}; \delta)$ be the union of all $U(t, y)$ with $(t, y) \in N_\delta(\bar{t}, \bar{y})$. The sets $U(t, y)$ are said to satisfy property (U) at a point $(\bar{t}, \bar{y}) \in A$ provided $U(\bar{t}, \bar{y}) = \bigcap_\delta \operatorname{cl} U(\bar{t}, \bar{y}; \delta)$. The sets $U(t, y)$ are said to satisfy property (U) in A if they have this property at every $(\bar{t}, \bar{y}) \in A$. If A is closed, then the sets $U(t, y)$ satisfy property (U) in A if and only if M is closed. Property (U) is the original Kuratowski concept of upper semicontinuity.

The sets $U(t, y)$ are said to satisfy property (Q) at a point $(\bar{t}, \bar{y}) \in A$ provided $U(\bar{t}, \bar{y}) = \bigcap_\delta \operatorname{cl} \operatorname{co} U(\bar{t}, \bar{y}; \delta)$. The sets $U(t, y)$ are said to satisfy property (Q) in A if they have this property at every $(\bar{t}, \bar{y}) \in A$.

For every $(t, y) \in A$ we denote by $\tilde{\tilde{Q}}(t, y)$ the set of all points $(z^0, z) \in E_{r+1}$ with $z^0 \geqq f_0(t, y, u)$, $z = f(t, y, u)$, $u \in U(t, y)$. For every $(t, \mathring{y}) \in B$ we denote by $\tilde{\tilde{R}}(t, \mathring{y})$ the set of all points $(z^0, z) \in E_{r'+1}$ with $z^0 \geqq g_0(t, \mathring{y}, v)$, $z = g(t, \mathring{y}, v)$, $v \in V(t, \mathring{y})$. We shall require below properties (U) or (Q) on the sets $\tilde{\tilde{Q}}(t, y)$ and $\tilde{\tilde{R}}(t, y)$. As proved by Cesari, property (Q) for the sets $\tilde{\tilde{Q}}(t, y)$ and $\tilde{\tilde{R}}(t, \mathring{y})$ is the natural extension of Tonelli's and McShane's concept of seminormality for free problems. Also, a great many criteria for property (Q) have been recently proved. For instance, suitable growth conditions imply property (Q). We shall state existence theorems below where properties (U) or (Q) are required. However, there are situations (for instance, where f_0, f, g_0, g possess suitable bounds) where these requirements can be relaxed, or dropped.

A triple (x, u, v) $x \in S$, $u \in T$, $v \in \mathring{T}$, is now said to be admissible provided relations (2), (3), (4), (5), (6), (7) hold, $f_0(t, (\mathscr{M}x)(t), u(t))$ is L-integrable in G, and $g_0(t, (\mathscr{K}x)(t), v(t))$ is μ-integrable on Γ.

We shall consider nonempty classes Ω of admissible triples (x, u, v). Any such class is said to be closed if the following holds: For any sequence (x_k, u_k, v_k) of elements of Ω, and elements $x \in S$, such that $x_k \to x$ weakly in S, $l_1 = \liminf I_1[x_k, u_k] < +\infty$, $l_2 = \liminf I_2[x_k, v_k] < +\infty$, and there are elements $u \in T$, $v \in \mathring{T}$ such that (x, u, v) is admissible, $I_1[x, u] \leqq l_1$, $I_2[x, v] \leqq l_2$, then there is also one of these triples (x, u, v) which also belongs to Ω. This definition of a closed class Ω is justified by lower closure theorems, which precisely guarantee under conditions (P_1), (P_2),

(P_3), that there are elements $u \in T$, $v \in \mathring{T}$ such that (x, u, v) is admissible and $I_1 \leqq l_1, I_2 \leqq l_2$. Given any class Ω of admissible triples (x, u, v), we denote by $\{x\}_\Omega$ the set of all $x \in S$ such that $(x, u, v) \in \Omega$ for some $u \in T$, $v \in \mathring{T}$.

EXISTENCE THEOREM I. *Under hypotheses* (P_1), (P_2), (P_3), *if the sets* $\tilde{\tilde{Q}}(t, y)$ *and* $\tilde{\tilde{R}}(t, \mathring{y})$ *have property* (Q) *in* A *and* B *respectively, if* Ω *is a nonempty closed class of admissible triples* (x, u, v) *such that the set* $\{x\}_\Omega$ *is sequentially relatively compact in* S, *then the functional* (1) *attains its infimum in* Ω.

Condition (Q) above can be relaxed. For instance, it is enough to know that there is a countable decomposition $G = \bigcup_j H_j$ into measurable disjoint sets H_j, such that H_0 has measure zero, and the sets $\tilde{\tilde{Q}}(t, y)$ have property (Q) on each set $A_j = A \cap (H_j \times E_s)$ with respect to $A_j, j = 1, 2, \ldots$. An analogous remark holds for the sets $\tilde{\tilde{R}}(t, \mathring{y})$.

If S is a Sobolev space $(W_p^l(G))^n$ with $p > 1$, $l \geqq 1$, $n \geqq 1$, and norm $\|x\|$, and we choose in S the weak convergence, if G is of class K, and $I[x, u, v] \leqq C$, $(x, u, v) \in \Omega$ implies $\|x\| \leqq c$ for some constant c which may depend on C, then the requirement concerning the set $\{x\}_\Omega$ in Theorem I is certainly trivial. For $p = 1$, this is not the case, and often growth conditions must be required which are suitable generalizations of Tonelli's, Nagumo's, and McShane's analogous growth conditions for free problems. For instance, the following growth condition has been found to be relevant:

(ε) Given $\varepsilon > 0$, there is a function $\phi_\varepsilon \geqq 0$, $\phi_\varepsilon \in L(G)$, such that $|f(t, y, u)| \leqq \phi_\varepsilon(t) + \varepsilon f_0(t, y, u)$ for all $(t, y, u) \in M$.

An analogous growth condition can be expressed in terms of g and g_0.

There are situations where (P_1) holds in a stronger form. We denote by (P_1') the hypothesis analogous to (P_1) where $x_k \to x$ weakly in S implies $\mathcal{T} x_k \to \mathcal{T} x$ strongly in $(L_p(\Gamma))^{r'}$ (instead of weakly as in (P_1)).

EXISTENCE THEOREM II. *Under hypotheses* (P_1'), (P_2), (P_3), *if the sets* $\tilde{\tilde{Q}}(t, y)$ *have property* (Q) *in* A *and the sets* $\tilde{\tilde{R}}(t, y)$ *have property* (U) *in* B, *if* Ω *is a nonempty closed class of admissible triples* (x, u, v) *such that the set* $\{x\}_\Omega$ *is sequentially relatively compact, then the functional* (1) *attains its infimum in* Ω.

If some components of $\mathring{y}_k = \mathcal{T} x_k$ converge weakly and the remaining components converge strongly to the corresponding components of $\mathring{y} = \mathcal{T} x$, then an intermediate existence theorem can be proved, where a suitable intermediate property between (U) and (Q) is used (D. E. Cowles).

2. **State equations in the weak form.** We consider now the case where the functional equations (2), (3) (state equations) are written in weak form, as is customary in partial differential equation theory. To this purpose let W denote a suitable normed space of test functions $w = (w_1, w_2)$, where $w_1 : G \to (L_q(G))^r$ and $w_2 : \Gamma \to (L_q(\Gamma))^{r'}$ are vector-valued functions defined in G and Γ respectively, and

$p^{-1} + q^{-1} = 1$ with the usual conventions. Let W^* be the dual space of W.

We shall use the same general notations as in §1. Instead of the operators $\mathscr{L}, \mathscr{T}$, we shall consider only one operator $\mathscr{F} : S \to W^*$. Instead of the hypothesis (P_1) we shall now require:

(P_1'') If $x, x_k \in S$, $k = 1, 2, \ldots$, and $x_k \to x$ in S, then $\mathscr{F}x_k \to \mathscr{F}x$ in the weak star topology (that is, $(\mathscr{F}x_k)w \to (\mathscr{F}x)w$ for every $w \in W$), and $\mathscr{M}x_k \to \mathscr{M}x$ strongly in $(L_p(G))^s$, $\mathscr{K}x_k \to \mathscr{K}x$ strongly in $(L_p(\Gamma))^{s'}$.

For $x \in S$ we denote by $f_{x,u,v}$ the element of W^* defined by

$$f_{x,u,v}w = \int_G f(t, (\mathscr{M}x)(t), u(t))w_1(t)\,dt + \int_\Gamma g(t, (\mathscr{K}x)(t), v(t))w_2(t)\,d\mu$$

for all $w = (w_1, w_2) \in W$, and where fw_1 and gw_2 are inner products. Instead of the state equations (2), (3), we shall now consider the unique state equation in the weak form $\mathscr{F} = f_{x,u,v}$, or equivalently

$$(\mathscr{F}x)w = f_{x,u,v}w \quad \text{for all } w \in W, \tag{8}$$

or

$$(\mathscr{F}x)(w_1, w_2) = \int_G f(t, (\mathscr{M}x)(t), u(t))w_1(t)\,dt + \int_\Gamma g(t, (\mathscr{K}x)(t), v(t))w_2(t)\,d\mu \tag{9}$$

for all $w = (w_1, w_2) \in W$. We are now interested in the problem of the minimum of a cost functional of the form (1) subject to state equation (8) and unilateral constraints (4), (5), (6), (7).

In the present situation suitable growth conditions must be required, as for instance:

(P_4) If $p = 1$ we assume that f_0, f as well as g_0, g satisfy the (ε) condition above. If $p > 1$ we simply assume that there are functions $\phi \geqq 0$, $\phi \in L(G)$, $\psi \geqq 0$, $\psi \in L(\Gamma)$ and constants $a > 0$, $b > 0$ such that

$$|f(t, y, u)|^p \leqq \phi(t) + af_0(t, y, u) \quad \text{for all } (t, y, u) \in M;$$

and

$$|g(t, \mathring{y}, v)|^p \leqq \psi(t) + bg_0(t, \mathring{y}, v) \quad \text{for all } (t, \mathring{y}, v) \in \mathring{M}.$$

A triple (x, u, v), $x \in S$, $u \in T$, $v \in \mathring{T}$, is now said to be admissible provided relations (4), (5), (6), (7), (8) hold, and $f_0(t, (\mathscr{M}x)(t), u(t))$ is L-integrable in G and $g_0(t, (\mathscr{K}x)(t), v(t))$ is μ-integrable in Γ. We shall then consider nonempty classes Ω of admissible triples (x, u, v), where the definition of closedness is analogous to the one in §1.

EXISTENCE THEOREM III. *Under hypotheses* (P_1''), (P_2), (P_3), (P_4), *if the sets* $\tilde{\tilde{Q}}(t, y)$ *and* $\tilde{\tilde{R}}(t, \mathring{y})$ *have property* (Q) *in* A *and* B *respectively, if* Ω *is a nonempty closed class of admissible triples such that the set* $\{x\}_\Omega$ *is sequentially relatively compact, then the functional* (1) *attains its infimum in* Ω.

Existence theorems analogous to Theorems I, II, III have been also proved, where the operators $\mathscr{L}$, $\mathscr{T}$, $\mathscr{M}$, $\mathscr{K}$, $\mathscr{F}$ depend both on x and suitable components of the controls u and v.

For the results mentioned above, as well as for further ones concerning a number of different aspects of the wide subject, we refer to the work of the author and to the related work of T. S. Angell, R. F. Baum, D. E. Cowles, J. R. La Palm, D. A. Sanchez, and M. B. Suryanarayana.

UNIVERSITY OF MICHIGAN

TOPOLOGICAL METHODS IN THE THEORY OF SHOCK WAVES[1]

CHARLES C. CONLEY AND JOEL A. SMOLLER

1. We consider the hyperbolic system of conservation laws of the form

$$u_t + (f(u))_x = 0, \tag{1}$$

where $u = u(x, t) \in R^n, n > 1, t \geqq 0$ and $-\infty < x < \infty$. Here f is a smooth mapping from an open subset of R^n into R^n, whose Jacobian $df(u)$ has real distinct eigenvalues $\lambda_1(u) < \ldots < \lambda_n(u)$, with corresponding (right) eigenvectors r_i, $i = 1, 2, \ldots, n$. We furthermore assume that the system is genuinely nonlinear, in the sense that $d\lambda_i r_i > 0, i = 1, 2, \ldots, n$. It is well known [**8**] that systems of this type admit discontinuous solutions called shock waves (or simply, shocks); that is, solutions of the type

$$\begin{aligned} u(x,t) &= u_0, \qquad x - st < 0, \\ &= u_1, \qquad x - st > 0, \end{aligned} \tag{2}$$

where u_0 and u_1 are constant vectors and s is a real number called the shock speed. The connection between the system (1) and solution (2) comes from the (Rankine-Hugoniot) jump conditions

$$s(u_0 - u_1) = f(u_0) - f(u_1), \tag{3}$$

which relates the values of the solution on both sides of the discontinuity to the function f.

In order to reject certain extraneous unstable (with respect to small perturbations) solutions of (1), one usually imposes an additional condition, called the

AMS 1970 *subject classifications*. Primary 35F25, 35L60, 35L65, 76L05; Secondary 73D05, 76W05, 34C99.

[1] This research was partially supported by the Air Force Office of Scientific Research Contract No. AFOSR-69-1662.

entropy condition, which solutions are required to satisfy. Thus, there is an index k, $1 \leqq k \leqq n$, such that the following inequalities (the entropy inequalities) hold:

$$\begin{aligned} &\lambda_{k-1}(u_0) < s < \lambda_k(u_0), \\ &\lambda_k(u_1) < s < \lambda_{k+1}(u_1). \end{aligned} \tag{4}$$

Now in order to be able to find a correct class of systems (1) (i.e. a correct class of functions f), in which the Cauchy problem is well-posed, we are motivated by physics into studying associated parabolic systems with "viscosity" of the form

$$u_t + (f(u))_x = \varepsilon u_{xx}, \qquad \varepsilon > 0. \tag{5}$$

Here ε is a small positive parameter and the second derivative terms represent the viscosity. Those systems (1) whose solutions can be obtained as limits of solutions of (5) as the viscosity ε tends to zero will be deemed *acceptable*. Again motivated by the physics, we call the shock wave solution (u_0, u_1, s) of (1) *admissible* if it can be recovered as a limit of progressive wave solutions of (5) as $\varepsilon \to 0$.

By a progressive wave solution of (5), we mean a solution of the form $u = u(\xi)$, $\xi = (x - st)/\varepsilon$. If we put this in (5), we obtain the system of ordinary differential equations

$$u_\xi = -s(u - u_0) + f(u) - f(u_0) \equiv \Phi(u), \tag{6}$$

and the condition that this solution converges to the shock wave gives the boundary conditions

$$\lim_{\xi \to -\infty} u(\xi) = u_0, \qquad \lim_{\xi \to +\infty} u(\xi) = u_1. \tag{7}$$

Thus u_0 and u_1 are critical points of Φ, and this statement, in turn, is equivalent to the jump conditions (3). Our problem therefore is to solve (6) subject to the boundary conditions (7). Thus, in our terminology, if this problem has a solution, the shock wave is admissible.

In [**1**], we studied a rather wide class of hyperbolic systems ($n = 2$) for which the Riemann problem is well-posed for arbitrary jump data [**9**], [**10**], and we showed that all shock wave solutions, of arbitrary strength,[2] are admissible, and that this class of systems is always acceptable. More generally, for this class of systems, we considered associated parabolic systems of the form

$$u_t + (f(u))_x = \varepsilon P u_{xx}, \qquad \varepsilon > 0, \tag{8}$$

where P is a constant matrix having eigenvalues with positive real parts. Let us call P a *suitable* viscosity matrix for a given shock wave solution of (1) if the shock wave can be obtained as a limit of progressive wave solutions of (8). In this general framework, (6) becomes

$$P u_\xi = \Phi(u), \tag{9}$$

[2] The strength of a shock wave (u_0, u_1, s) is $|u_0 - u_1|$.

subject to the same boundary conditions (7). Note that (6) and (9) have exactly the same singular points, and P is suitable means that there is an orbit of (9) connecting the two singular points u_0 and u_1.

In these terms, our work in [**1**] shows that the two sets of suitable matrices for two different shock wave solutions of the same system (1) are generally different. We found, in the case $n = 2$ for shocks of arbitrary strength, that the identity matrix I is always suitable for any shock wave, and that the set of suitable matrices is open. On the other hand, we found examples where diagonal positive definite matrices were not suitable for some admissible shock waves, and we also constructed an example of a system (1) and a positive definite selfadjoint constant matrix P which was unsuitable for every shock wave.

Here in this paper, we shall announce some extensions of these results in two directions. First we shall consider hyperbolic systems (1) in general $n > 1$ dependent variables, for sufficiently weak shocks, where we also allow nonlinear viscosity matrices, i.e., we allow the matrix P to depend on u. Next we shall consider, for the case $n = 2$, more general viscosity matrices than we had previously considered; namely we allow the matrix P to be nonlinear, and also to depend on several parameters.

2. Let us begin by describing the higher dimensional case. Thus, we shall discuss the case of weak shocks in dimensions $n > 1$ (see [**2**], [**8**]) and describe classes of suitable matrices. The purpose of this is to illustrate how to use topological methods from the qualitative theory of flows in this connection, and to provide proofs which are more directly related to insights which allow us to guess at the correct theorems in the first place. The case where $P = I$ was considered by Foy [**5**], who employed a bifurcation technique based on an expansion procedure common in boundary layer theory. We also treat the case where f is a gradient and we obtain, in a rather easy way, a theorem of Kulikovskiĭ [**6**] which states that every positive definite matrix P is suitable for sufficiently weak shocks.

Our method is to combine the notion of index of an isolated invariant set of a flow with the fact that our flows are gradient-like[3] in a region containing the singular points. The main difficulty comes from the fact that this region generally shrinks as the shocks get weaker. However, in the case of gradient f's, this difficulty disappears. For in this case, the system (9) takes the form

$$Pu_\xi = \nabla\Psi(u), \tag{10}$$

and it follows easily that the function Ψ increases along orbits of this flow.

In order to prove these theorems, we first must introduce the idea of an isolating block [**3**], [**4**]. Namely, let V be a vector field in R^n defining a flow α, where we define $x \cdot t = \alpha(x, t)$, $x \in R^n$, $t \in R^1$. Let B be the closure of an open set in R^n and

[3] If the flow is defined by the vector field V, then the flow is called gradient-like if there is a real valued function Ψ satisfying $V \cdot \nabla\Psi > 0$ except at critical points. Thus, Ψ increases along orbits of V.

define three sets b^+, b^- and τ as follows:

$$b^+ = \{p \in \partial B : \exists\, \varepsilon > 0 \ni : p \cdot (-\varepsilon, 0) \cap B = \varnothing\},$$

$$b^- = \{p \in \partial B : \exists\, \varepsilon > 0 \ni : p \cdot (0, \varepsilon) \cap B = \varnothing\},$$

$$\tau = \{p \in \partial B : V \text{ is tangent to } B \text{ at } p\}.$$

(So, for example, if $p \in b^+$, the flow leaves B for small backwards time.) Then B is called an *isolating block* for the flow if $b^+ \cap b^- = \tau$ and $b^+ \cup b^- = \partial B$. It is easy to see that if B is an isolating block, then all of the tangencies to B must be external. It follows that the mapping (defined by the flow) from the set of points in which the orbits enter B to the set of points in which the orbits leave B, is continuous, where-ever this mapping is defined. Indeed, this is perhaps the main point of the notion of an isolating block! Thus, for example, the flow defined by the following picture does not have this property since the set B has an internal tangent.

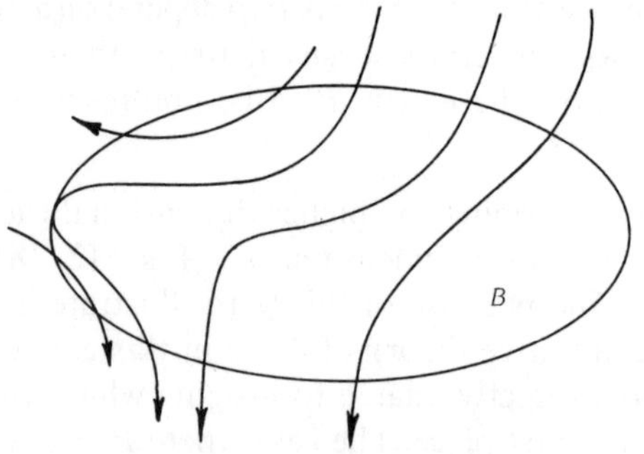

On the other hand, the picture to keep in mind, as an example of an isolating block, is a hyperbolic singular point of a flow.

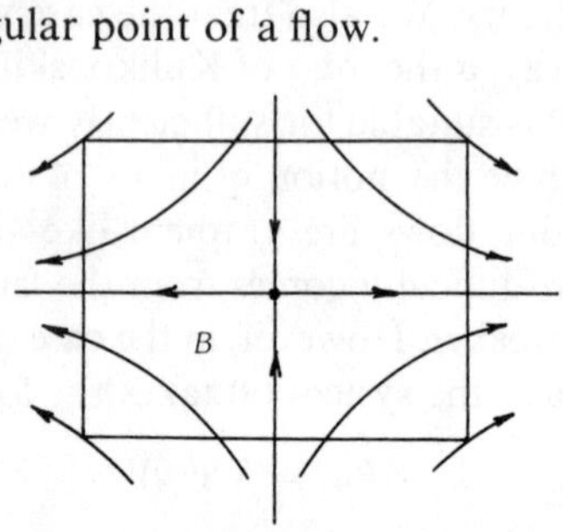

Now, let I be a maximal invariant set contained in a block B. We define the *homotopy index* $h(I)$ of I to be the pair of pointed spaces

$$h(I) = (B/b^+, B/b^-).$$

We also define $h(I) = 0$ to denote the homotopy class where each of B/b^+ and B/b^- has the homotopy type of a pointed point. It turns out that this index is independent of the block B. Furthermore, this notion of index generalizes the usual Morse index in case the set I is a nondegenerate critical point of a gradient vector field. To see this, consider the system of ordinary differential equations $\dot{u} = Au$, where A has p negative eigenvalues, and $n - p$ positive eigenvalues at $u = 0$. We build a block about 0 which is an n-ball B^n (as in the above figure). The stable manifold of the point meets B in a p-ball B^p having boundary a $(p - 1)$-sphere S^{p-1} and the unstable manifold meets B in an $n - p$ ball. Then B can be represented as $B^p \times B^{n-p}$ and b^+ as $\partial B^p \times B^{n-p}$. Thus we have the homotopy equivalence of the pair $(B, b^+) \sim (B^p, \partial B^p)$ and on collapsing the boundary, we find that B/b^+ is a pointed p-sphere. Similarly, B/b^- is a pointed $(n - p)$-sphere.

We next have a summation formula for indices; namely if I_1 and I_2 are disjoint isolated invariant sets, then $I = I_1 \cup I_2$ is an isolated invariant set and choosing disjoint blocks B_1 and B_2 for I_1 and I_2 respectively, we find $B = B_1 \cup B_2$ is a block for I and $B/b^\pm = B_1/b_1^\pm \vee B_2/b_2^\pm$. Here we use the definition for pointed spaces X, Y of $X \vee Y$ which is the pointed space obtained from X and Y by gluing them together at the distinguished points. Hence $h(I) = h(I_1) \vee h(I_2)$. The main lemma in our development can now be proved.

LEMMA. *Let V be a gradient-like vector field defined on an open subset $\mathscr{U}$ of R^n, defining a flow ϕ. Let $B \subset \mathscr{U}$ be an isolating block for ϕ where $B \approx B^n$, $\partial B \approx b^+ \cup b^-$, $b^\pm \approx B^{n-1}$. If V has exactly two nondegenerate critical points u_0 and u_1 in B, then there is an orbit of ϕ connecting u_0 to u_1. (The symbol $\approx$ denotes topological equivalence.)*

In order to prove this lemma, observe that it suffices to prove that there is a third orbit in B. Namely, since V is gradient-like, this orbit "begins" and "ends" in distinct critical points. Next note that if $h(I_1) \neq 0$, then $h(I_1) \vee h(I_2) \neq 0$ for any I_2. Now if there were no third invariant set in B, then $I = u_0 \cup u_1$ would be the maximal invariant set in B, and $h(I) = h(u_0) \vee h(u_1) \neq 0$, by the nondegeneracy condition. However, since $B \approx B^n$ and $b^\pm \approx B^{n-1}$, we have $B/b^\pm \approx$ (pt, pt) and $h(I) = 0$. This contradiction completes the proof of the lemma.

In order to illustrate the use of this lemma, let us consider, for simplicity, the case where the function f in (1) is itself a gradient, $f = \nabla F$. If we assume, without loss of generality, that $u_0 = f(u_0) = 0$, then the system (6) becomes

$$\dot{u} = \nabla\Phi(u), \tag{11}$$

where $\Phi(u) = -s\|u\|^2/2 + F(u)$, so that the flow is gradient-like. Thus, we need only construct an isolating block. Now for sufficiently weak shocks, u_1 is in approximately the r_k direction from u_0 (see [**8**]), and we shall build our block in this direction. To do this, we write (11) as $\dot{u} = (A - sI)u + O_2(u)$, and then expand u in terms of the eigenvectors of A as $u = \sum u_i r_i$, to obtain the system

$$\dot{u}_k = (\lambda_k - s)u_k + O_2(u),$$

$$\dot{u}^+ = (A^+ - sI)u^+ + O_2(u),$$
$$\dot{u}^- = (A^- - sI)u^- + O_2(u),$$

where A^+ and A^- are diagonal matrices with positive and negative entries, respectively.

Now let $\sigma > 0$ and define the sets

$$B(u_k) = \{(\hat{u}_k, u_k) : |u^+| \leqq \sigma, |u^-| \leqq \sigma\},$$
$$b^{\pm}(u_k) = \{(\hat{u}_k, u_k) : |u^{\pm}| = \sigma\}.$$

Then for small $|u|$, we have

$$d|u^-|^2/d\xi = 2\langle \dot{u}^-, u^- \rangle = 2\langle (A^- - sI)u^-, u^- \rangle + O_3(u) < 0,$$

so that the flow comes in along the u^- directions; similarly, $d|u^+|^2/d\xi > 0$. If we let $\delta > 0$ and define

$$B = \bigcup \{B(u_k) : -\delta \leqq u_k \leqq u_k' + \delta\},$$

where u_k' is the kth component of u_1, then along b^+, $|u^-| = \sigma$, $d|u^-|^2/d\xi < 0$ so the u^- component of the flow has norm $> \sigma$ for small negative ξ and thus the flow crosses b^+ transversally and enters B. Similarly, the flow crosses b^- transversally and leaves B. Finally, on the two faces $B(-\delta)$ and $B(u_k' + \delta)$, the entropy inequalities, (4), show, for example, that at 0 the unstable manifold is spanned by $r_k, \ldots, r_n$ so that the flow near 0 is transverse to the plane $B(-\delta)$ for small σ and δ and the orbits leave B for small positive ξ.

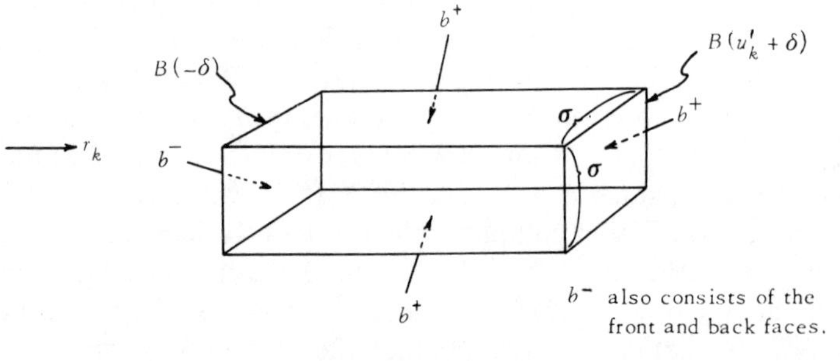

Finally, since $d^2\|u\|^2/d\xi^2 > 0$ on b^+b^-, the tangencies are all external. It follows that B is an isolating block which, moreover, satisfies the hypothesis of our lemma. Thus an application of the lemma shows that there is a connecting orbit. Furthermore, since blocks are "structurally stable," it is not too difficult to show that there is in fact a neighborhood of the identity I which also consists of suitable matrices. Here the matrices can also depend on u.

We remark that in the general case, i.e., the case where f is not a gradient, the construction of the block proceeds in exactly the same way as we showed above.

However, the difficult part of the proof is the construction of the gradient-like function in the block; see [**12**] for details.

3. Next, we turn to the theorem of Kulikovskiĭ [**6**], which is concerned with gradient $f, f = \nabla F$, and arbitrary positive definite viscosity matrices $P = P(u)$. The main point here is that P need not be close to the identity I.

Since the flow is again gradient-like, in this case it suffices (by our lemma in §2) to build the isolating block. Now, however, since P need not be near I, we do not have an obvious distinguished direction in which to build the block (in contrast to the situation of §2 where the r_k direction was distinguished). In order to find such a direction, we examine the nature of the singularities of the two systems

$$\text{(a)} \qquad \dot{u} = \nabla\Phi(u),$$

$$\text{(b)} \qquad P\dot{u} = \nabla\Phi(u),$$

where P is a constant positive definite matrix and $\Phi = -s\|u\|^2/2 + F(u)$. Clearly (a) and (b) have the same rest points. Furthermore, these two systems also have the same dimensional stable and unstable manifolds. To see this, note that since the hessian H of Φ is symmetric and has no zero eigenvalues (from (4)), it follows that $P^{-1}H$ has the same number of eigenvalues with positive real parts as does H. For, if $P^{-1}Hr = ir$, then $0 < \operatorname{Re}(P^{-1}Hr, Hr) = \operatorname{Re} i(r, Hr) = 0$, so $P^{-1}H$ has no purely imaginary eigenvalues. Thus, for each t, $0 \leqq t \leqq 1$, the family

$$[(1 - t)I + tP^{-1}]H$$

has the same number of positive and negative eigenvalues. This proves our statement.

Next, we can show that the linear part, $P^{-1}(H - sI)$, of $P^{-1}\nabla\Phi$ has a real eigenvalue if the shock is sufficiently weak. Observe that if P is symmetric, then this is a trivial consequence of the fact that

$$P^{1/2}[P^{-1}(H - sI)]P^{-1/2} = P^{-1/2}(H - sI)P^{-1/2}$$

is also symmetric. The case of nonsymmetric P is a bit harder. Namely, let the matrix B, where $B^{-1} = B^t$, be such that $B^tHB = D = \operatorname{diag}(\lambda_1, \ldots, \lambda_n)$. Consider the eigenvalue equation $P^{-1}(H - sI)r = \mu r$. Setting $r = Bs$, we obtain $P_1As = \mu s$, where $P_1 = B^tP^{-1}B > 0$, and $A = D - sI = \operatorname{diag}(\lambda_1 - s, \ldots, \lambda_n - s)$. If we let $\tilde{A} = \operatorname{diag}(\lambda_1 - s, \ldots, \lambda_{k-1} - s, 0, \lambda_{k+1} - s, \ldots, \lambda_n - s)$, then $P_1\tilde{A}$ has a zero eigenvalue which turns out to have algebraic multiplicity one. Thus for $\lambda_k - s$ small, P_1A has a real eigenvalue (close to zero), as we asserted. The corresponding eigenvector then picks out the distinguished direction in which to build the isolating block, and the construction then proceeds in a manner analogous to that in §2. This proves the theorem for constant P. The case where $P = P(u)$ follows by a perturbation argument as before.

In the case of gas dynamics, magnetohydrodynamics and several other physical situations, the associated system of ordinary differential equations is of the form (10) where the function Ψ is some sort of generalized entropy—it increases through a shock wave. Thus our results show that sufficiently weak shocks are admissible and each positive definite matrix $P(u)$ is suitable for these shocks. In the physics terminology, sufficiently weak shocks have structure. Thus weak shocks are in reality transition layers of finite thickness, where the thickness decreases as the strength of the shock increases. Now if the shock becomes very strong, the changes in the physical quantities within the shock occur so rapidly, that the concept of thickness becomes meaningless [7]. In particular, there appears to be little physical significance in studying this structure problem for arbitrarily strong shocks. On the other hand, the purely mathematical problem (9) or (10) and (7) of obtaining orbits which connect two singularities, in the large, is an interesting and difficult problem in global analysis.

4. For the case $n = 2$, we study viscosity matrices P which depend on several parameters, and we also consider the case of nonlinear matrices, for shocks of arbitrary strength. (Physically speaking, there are usually several "viscosity parameters." For example in magnetohydrodynamics, there is the ordinary viscosity coefficient, the "second" viscosity coefficient, the magnetic viscosity coefficient, and the thermal conductivity coefficient.) These viscosity parameters tend to zero independently, in contrast to the previously considered cases in which a single viscosity parameter ε tends to zero. In order to motivate our development, let us dwell on a special example for a 2×2 hyperbolic system (1).

First, let us consider the associated parabolic system (5) in which $P = I$. We now look for solutions $u = u(\xi)$, where $\xi = x - st$. If we substitute in (5), we get the system of ordinary differential equations

$$\varepsilon u_\xi = \Phi(u), \tag{12}$$

together with the boundary conditions (7). Let $u(\xi;\varepsilon)$ be the solution to this problem. Now it is rather easy to see that $u(\xi;\varepsilon) = u(\xi/\varepsilon;1)$; that is, as ε varies, the solution curve goes into itself, and only the parametrization of the curve changes. Keeping this in mind, we consider a special viscosity matrix depending on two parameters $\varepsilon_1, \varepsilon_2$ of the form $P = \operatorname{diag}(\varepsilon_1, \varepsilon_2)$, $\varepsilon_1 > 0$, $\varepsilon_2 > 0$. If we again let $U = U(\xi)$, $\xi = x - st$, we obtain a system of ordinary differential equations of the form

$$\begin{aligned} \varepsilon_1 \dot{u}_1 &= \phi_1(u_1, u_2), \\ \varepsilon_2 \dot{u}_2 &= \phi_2(u_1, u_2), \end{aligned} \tag{13}$$

where the dot denotes differentiation with respect to ξ and $(\phi_1, \phi_2) = \Phi$. It is clear that solutions of this system depend only on the ratio $\varepsilon_1/\varepsilon_2$ and are thus independent of ε_1 if $\varepsilon_1 = \varepsilon_2$, as we saw above. We solve the problem (13)—(7) for each fixed $\varepsilon_1 > 0, \varepsilon_2 > 0$, obtaining a solution curve $u = u(\xi;\varepsilon_1, \varepsilon_2)$. We now ask for the behavior of this orbit as $\varepsilon_1^2 + \varepsilon_2^2 \to 0$; in particular, when does this orbit "converge" to the given shock wave? We shall define an appropriate notion of "uniformity" in the parameters which in this case reduces to the fact that for $\varepsilon_1^2 + \varepsilon_2^2$ small, the orbit is near the singular points U_0 and U_1 "most of the time."

Thus, we consider associated parabolic systems of the form

$$u_t + (f(u))_x = \varepsilon(P(u;\mu)u_x)_x, \qquad \varepsilon > 0.$$

Here P is a 2×2 matrix-valued function of U and μ where μ varies over a compact subset of a Euclidean space, and for fixed μ, $P(u;\mu)$ is C^1 in U. In order to be able to use theorems on C^1 perturbations of vector fields, we require that the mapping from parameter space μ into $P(u;\mu)$ is continuous in the topology of uniform convergence on compacta. Then if P^{-1} exists, we call $P(\cdot;\mu)$ suitable if the system $u_\xi = P^{-1}(u;\mu)\Phi(u)$ admits an orbit $u(\xi;\mu)$ connecting the critical points u_0 and u_1. To be more precise, the progressive wave solution converges to the shock wave means the following: if $\gamma(\xi;\mu)$ denotes the connecting orbit with any fixed parametrization, then $\gamma((x-st)/\varepsilon;\mu)$ converges uniformly in x and t to the shock wave solution in any region $|x - st| \geqq \delta > 0$ of the x-t plane.

We next ask what is the dependence of this convergence on the parameter μ? Our definition is as follows:

DEFINITION. Let $P(u;\mu)$ be a μ-parameter family of suitable viscosity matrices. This family is called *uniformly suitable* if parametrizations of the connecting orbits $\gamma(\xi;\mu)$ can be chosen such that for any $\delta > 0$, the functions $\gamma((x - st)/\varepsilon;\mu)$ converge uniformly in x, t and μ to the shock wave solution in the region $(x - st) \geqq \delta > 0$.

If we let N_0 and N_1 be arbitrary neighborhoods of u_0 and u_1, then in the connecting orbit $\gamma(\xi;\mu)$ there are finite subsegments whose complement lies in $N_0 \cup N_1$. For any parametrization, we define the time length of the subsegment to be $|\xi_0 - \xi_1|$, where ξ_i are the parameter values at the endpoints, $i = 1, 2$. For fixed μ, we define $T(\mu;N_0, N_1)$ as the infimum of these lengths. Then we have the theorem that the family $P(u;\mu)$ of viscosity matrices is uniformly suitable if and only if for each choice of neighborhoods N_0 and N_1 of u_0 and u_1 respectively, the function $T(\mu;N_0, N_1)$ is uniformly bounded in μ.

Our main theorem for nonsingular viscosity matrices is that if each element of the (nonsingular) family $P(u;\mu)$ is suitable, then the convergence of the progressive waves to the shock wave is uniform on compact subsets of parameter space, and the progressive wave solutions are uniformly bounded as μ varies over a compact set. The theorem is very nontrivial and again makes use of the notion of isolating block for a flow and some theorems concerning these blocks. In this planar case, however, the proofs of such theorems are greatly simplified.

We then turn to the case where the matrix $P(u;\mu)$ becomes, possibly, singular at certain points u. The system (5) then becomes a combined differential and algebraic system. However, we shall take a different point of view, and treat (5) as a discontinuous vector field, so that we can again apply results from qualitative theory of ordinary differential equations. This requires a new definition of orbit in this general framework (see [**12**] for details). With this definition, we then again have the notion of suitable viscosity matrix, and we can prove the theorem.

THEOREM. *If $P(u)$ is suitable, then all nearby nonsingular viscosity matrices are also suitable.*

Thus the singular orbits are realized as limits of connecting orbits as the relevant parameter tends to zero. These results are then applied to general 2×2 systems of conservation laws which include the well-studied "p-system" arising in gas dynamics: $u_t - v_x = 0, v_t + p(u)_x = 0, p' < 0, p'' > 0$. Here the viscosity matrix is of the form $\mathrm{diag}(0, \alpha(u, v))$, where $\alpha(u, v) > 0$.

It is interesting to note that these singular vector fields arise in other contexts, namely in relaxation oscillations. We can apply these qualitative methods, to prove, for example, the existence of a periodic orbit for the Van der Pol equation.

REFERENCES

1. C. C. Conley and J. A. Smoller, *Viscosity matrices for two dimensional nonlinear hyperbolic systems*, Comm. Pure Appl. Math. **23** (1970), 867–884.

2. ———, *Shock waves as limits of progressive wave solutions of higher order equations*, Comm. Pure Appl. Math. **24** (1971), 459–472.

3. C. C. Conley, *On the continuation of invariant sets of a flow*, Proc. Internat. Congress Math., vol. 2, 1970, pp. 909–913.

4. C. C. Conley and R. W. Easton, *Isolated invariant sets and isolating blocks*, Trans. Amer. Math. Soc. **158** (1971), 35–61.

5. L. R. Foy, *Steady state solutions of hyperbolic systems of conservation laws with viscosity terms*, Comm. Pure Appl. Math. **17** (1964), 177–188. MR **28** #2354.

6. A. G. Kulikovskiĭ, *The structure of shock waves*, Prikl. Mat. Meh. **26** (1962), 631–641 = J. Appl. Math. Mech. **26** (1963), 950–964. MR **26** #3338.

7. L. D. Landau and E. M. Lifschitz, *Mechanics of continuous media*, 2nd ed., GITTL, Moscow, 1953; English transl., part I: *Fluid mechanics*, Course of Theoretical Physics, vol. 6, Pergamon Press, New York; Addison-Wesley, Reading, Mass., 1959. MR **16**, 412; **21** #6839.

8. P. D. Lax, *Hyperbolic systems of conservation laws*. II, Comm. Pure Appl. Math. **10** (1957), 537–566. MR **20** #176.

9. J. A. Smoller, *On the solution of the Riemann problem with general step data for an extended class of hyperbolic systems*, Michigan Math. J. **16** (1969), 201–210. MR **40** #552.

10. ———, *A uniqueness theorem for Riemann problems*, Arch. Rational Mech. Anal. **33** (1969), 110–115. MR **38** #6238.

11. J. A. Smoller and C. C. Conley, *Shock waves as limits of progressive wave solutions of higher order equations*. II, Comm. Pure Appl. Math. **25** (1972), 133–146.

12. ———, *Viscosity matrices for two dimensional nonlinear hyperbolic systems*. II, Amer. J. Math. (to appear).

UNIVERSITY OF MICHIGAN

GENERALIZATIONS OF THE KORTEWEG-DE VRIES EQUATION

THEODORE E. DUSHANE

1. The generalized Korteweg-de Vries (K-dV) equation is

$$u_t - (f(u))_x + u_{xxx} = g_1(u)u_{xx} + g_2(u)(u_x)^2 + p(u) \tag{1.1}$$

where $u = u(x, t)$ and the subscripts indicate partial derivatives. The purpose of this note is to state existence and uniqueness results for global solutions of the initial value problem for (1.1) in $t \geqq 0$ taking data

$$u(x, 0) = u_0(x). \tag{1.2}$$

We further discuss the closely related periodic initial value problem (1.1), (1.2), and

$$u(x, t) = u(x + 2\pi, t) \quad \text{for } t \geqq 0. \tag{1.3}$$

These results are the first of their kind for nonlinear functions g_1 and g_2, and are soon to appear.

The problems (1.1), (1.2) and (1.1), (1.2), (1.3) have been studied most recently by Tsutsumi, Mukasa, and Iino [**10**], for functions f satisfying $f' \geqq 0$ and $\int_0^u f(v)\,dv \geqq 0$ and for $g_1 = 1, g_2 = p = 0$. Since our hypotheses on f are the same, our work is a generalization of theirs to nonlinear second order terms. Previous to this, Sjöberg and Temam have studied the K-dV itself ($f(u) = u^2/2, g_1 = g_2 = p = 0$) by different methods. Kruskal, Miura, Gardner, and Zabusky [**3**], [**5**] have done fundamental work in the discovery of conservation laws for the K-dV and some of its generalizations. They have done other work in the description of solutions to the K-dV. Lax [**4**] has investigated the long term interaction of travelling waves for the K-dV.

AMS 1970 *subject classifications*. Primary 35Q99, 35G25; Secondary 35B20.

The difficult part of our development, as in all previous existence and uniqueness proofs for the K-dV and its generalizations, is in obtaining a priori estimates for the Sobolev norms of solutions to the problem. As an important example, the conservation laws of Kruskal, Miura, et al. make available a priori estimates for the K-dV and their generalizations of it.

We will actually obtain the solution to the problem (1.1), (1.2) as a limit of solutions to the perturbed problem

$$(1.4) \quad u_t - (f(u))_x + u_{xxx} = g_1(u)u_{xx} + g_2(u)(u_x)^2 + p(u) - \varepsilon(u_{xxxx} - (f(u))_{xx})$$

with data (1.2). The perturbation $-\varepsilon(u_{xxxx} - (f(u))_{xx})$ comes from two sources. The fourth order perturbation is needed to obtain local existence to a perturbed equation in §2; the second order perturbation is an unimportant consideration at this stage. The second order perturbation makes the a priori estimates of §3 go through with relatively light restrictions on f.

REMARK. Temam [**9**] was the first to use the fourth order perturbation $-\varepsilon u_{xxxx}$ to solve the K-dV itself, though his techniques were different from ours. Tsutsumi [**10**] also used this perturbation to solve the $g_1 = g_2 = p = 0$ problem.

Definitions and useful lemmas. For $f \in L^2(R^1)$, let $\|f\|^2 = \int f^2(x)\,dx$. If $f(x)$ is measurable and bounded, let $\|f\|_\infty = \operatorname{ess\,sup}|f(x)|$ $(-\infty < x < \infty)$. For s a nonnegative integer, let H^s be the space of functions u in $L^2(R^1) = L^2$ having weak L^2-derivatives $D^k u$ of orders $k = 1, \dots, s$. For $u(x) \in H^s$, let $\|u\|_s^2 = \sum_{k \le s} \|D^k u\|^2$. Let C^k be the space of functions k times continuously differentiable on R^1. Let $C^\infty = \bigcap_{k=1}^{\infty} C^k$. We will say $u(x,t) \in L^\infty(0, T; H^s)$ if u, as a function of x, is in H^s for each t, $0 \leqq t \leqq T$ and $\sup_{0 \leqq t \leqq T} \|u(\cdot, t)\|_s < \infty$. Finally, for f integrable, let $If(u) = \int_0^u f(v)\,dv$.

LEMMA 1.1. *If $u(x) \in H^s$, $s \geqq 1$, then for $p \leqq s - 1$, $\alpha = (2p+1)/2s$,*

$$(1.5) \qquad \|D^p u\|_\infty \leqq c_1(\|u\|^{1-\alpha}\|D^s u\|^\alpha + \|u\|) \leqq c_2\|u\|_s,$$

where c_1, c_2 are independent of u.

LEMMA 1.2. *Let $k \geqq 1$ and let $f(u) \in C^k$, $f(0) = 0$, $u(x,t) \in L^\infty(0, T; H^k)$, then $f(u(x,t)) \in L^\infty(0, T; H^k)$ and we have*

$$(1.6) \qquad \|f(u(t))\|_1 \leqq cM_1(f, b)\|u(t)\|_1,$$

$$(1.7) \qquad \|f(u(t))\|_k \leqq cM_k(f, b)(1 + \|u(t)\|_{k-1}^{k-1})\|u(t)\|_k,$$

where $M_k(f, b) = \max_s \sup_v |D^s f(v)|$ $(s = 1, \dots, k, |v| \leqq b)$ for $b = \sup_{0 \leqq \tau \leqq t} \|u(\tau)\|_\infty$.

REMARK ON THE PROOF OF LEMMA 1.1. We prove (1.5) first for functions v having compact support using a Sobolev type result from Agmon [**1**, p. 32]. Fix x_0; using an appropriate test function ϕ of compact support so that $\phi \equiv 1$ in a neighborhood of x_0, we apply the result to $v = \phi u$ from which the lemma follows.

2. **Local existence for the perturbed initial value problem.** We have

THEOREM 2.1. *Let $u_0 \in H^{s-1}$ for $s \geqq 3$ and let $f \in C^{s-1}$, $p, g_1, g_2 \in C^{s-2}$, then (1.4), (1.2) has a solution $u(x, t) \in L^\infty(0, T^s; H^s)$ for some $T^s > 0$. For $s \geqq 9$, u is a genuine pointwise solution.*

OUTLINE OF THE PROOF. Let ϕ be a solution to the linear problem

$$\phi_t + \phi_{xxx} = -\varepsilon\phi_{xxxx}, \qquad \phi(x, 0) = u_0(x). \tag{2.1}$$

For $n = 1, 2, \ldots$, let the sequence $\{u^n\}$ be defined by the linearized problem

$$\begin{aligned} u_t^n - f(v^{n-1})_x + u_{xxx}^n &= g_1(v^{n-1})v_{xx}^{n-1} + g_2(v^{n-1})(v_x^{n-1})^2 \\ &\quad + p(v^{n-1}) - \varepsilon(u_{xxxx} - (f(v^{n-1}))_{xx}), \\ u^n(x, 0) &= 0, \end{aligned} \tag{2.2}$$

where $u^0 \equiv 0$ and $v^{n-1} = \phi + u^{n-1}$.

Notice that by adding (2.1) and (2.2) that $\lim_n v^n$ taken in some appropriate H^s sense should be a solution of (1.4), (1.2). To show this is in fact the case, two essential facts are used:

(A) If $u_0 \in H^s$,

$$\|D^k\phi(t)\| \leqq \|D^k u_0\| \quad \text{for } k = 0, \ldots, s \text{ and } t \geqq 0. \tag{2.3}$$

(B) Let $w(x, t)$ be a solution to $w_t + w_{xxx} = -\varepsilon w_{xxxx} + a(x, t)$ taking data $w(x, 0) = 0$. Then if $a(x, t) \in L^\infty(0, T; H^{s-3})$, then $w(x, t) \in L^\infty(0, T; H^s)$ and satisfies

$$\|D^k w(t)\| \leqq c(\varepsilon)(t + t^{3/4} + t^{1/2} + t^{1/4}) \sup_{0 \leqq \tau \leqq t} \|a(\tau)\|_{k-3} \tag{2.4}$$

for $c(\varepsilon) > 0$, $k = 3, 4, \ldots, s$.

We first show that $\|u^n(t)\|_k \leqq c(\varepsilon, k, t)$ for $0 \leqq t \leqq t_0$, $k = 3, 4, \ldots, s$ independent of n if $u_0 \in H^{s-1}$. This is proved by induction on k then on n using (2.3) and (2.4) as well as Lemmas 2.1 and 2.2. To complete the proof of Theorem 2.1, we show that, for some T^s,

$$\sup_{0 \leqq \tau \leqq T^s} \|u^{n+1}(\tau) - u^n(\tau)\|_s < r \sup_{0 \leqq \tau \leqq T^s} \|u^n(\tau) - u^{n-1}(\tau)\|_s$$

for some r, $0 < r < 1$. This is proved by subtracting (2.2) for n from (2.2) for $n + 1$ and applying (2.4) to $w = u^{n+1} - u^n$, making use of (2.3).

REMARK. The technique used thus far is local in time and $c(\varepsilon)$ appearing in (2.4) is like $\varepsilon^{-3/4}$ for ε near zero. Hence, estimates independent of ε must be found to pass from a local to a global solution in time.

3. A priori estimates and global existence for the perturbed problem.

THEOREM 3.1. *Let $u \in L^\infty(0, T^k; H^k)$ for some $T^k > 0$, let $k = m + 5$ and let $f \in C^{m+1}$, $g_1, g_2, p \in C^m$ satisfy If $fu \geqq 0$, $f(0) = 0$, $f'(u) \geqq 0$, $p'(u) \leqq 0$, $up(u) \leqq 0$, $f(u)p(u) \leqq 0$, and $g_1(u) + ug_1'(u) \geqq ug_2(u)$, $g_1(u) \geqq 0$, $g_2'(u) \leqq 0$, $f'(u)g_1(u) + f(u)g_1'(u)$*

$\geqq f(u)g_2(u)$ *and* $g_2'(u) \leqq g_1''(u)$. *Then we have* $u(x,t) \in L^\infty(0, T; H^m)$ *for all* $T > 0$ *and*

$$\|u(t)\|_n \leqq c(T; \|u_0\|_n) \quad \text{for } 0 \leqq t \leqq T, n = 0, \ldots, m, \tag{3.1}$$

$$\varepsilon \int_0^t \|u(\tau)\|_{n+2}^2 \, d\tau \leqq c(T; \|u_0\|_n) \quad \text{for } 0 \leqq t \leqq T, n = 0, 2, 3, \ldots, m. \tag{3.2}$$

COROLLARY 3.1. *We have Theorem* 3.1 *for* $f(u) = u^{2n+1}$, $g_1(u) = u^{2m}$, $g_2(u) = -u^{2r+1}$, $p(u) = -u^{2s+1}$ *for* n, m, r, s *nonnegative integers.*

REMARKS ON THE PROOF OF THEOREM 3.1. From (3.1) and (3.2) we may easily obtain global existence for (1.4), (1.2). To this end, we show, by induction on n, for each fixed $T > 0$,

$$\frac{d}{dt}\left(\|D^n u\|^2 + \int F(u, \ldots, D^n u) \, dx \right) \leqq c(T)(\|D^n u\|^2 + 1 - \varepsilon \|D^{n+2} u\|^2), \tag{3.3}$$

for properly chosen F's. Here are the actual choices: for $n = 1$,

$$F(u) = I(f(u));$$

for $n = 2$,

$$F(u) = -\tfrac{1}{3}I(2g_2(u) + g_1'(u))u_x u_{xx} + \tfrac{1}{3}I(\tfrac{5}{2}f''(u) - (2g_2(u) + g_1'(u)))g_1(u_x)^2;$$

for $n = 3$,

$$F(u) = -2I(3g_1'(u) + 5g_2(u))u_{xx}u_{xxx}.$$

We make use of Lemma 1.1 to prove these cases. For $n \geqq 4$, Lemma 1.1 proves (3.3) for $F = 0$.

Note. We may modify the proofs of Theorems 2.1 and 3.1 to obtain similar results for $f(u) = u^2/2$, $p(u) = 0$, the original K-dV equation with nonlinear second order terms added. We use the simpler perturbation $-\varepsilon u_{xxxx}$ and sufficient hypotheses on g_1, g_2 are: $g_1(u) + ug_1'(u) \geqq ug_2(u)$, $g_1(u) \geqq 0$, $g_2'(u) \leqq 0$, and $2ug_1(u) + u^2 g_1'(u) \geqq u^2 g_2'(u)$.

4. **Existence for the unperturbed problem by letting ε tend to zero.** We have

THEOREM 4.1. *If* $u_0 \in H^s$ *and* $f, g_1, g_2 \in C^{s+3}$ *and satisfy the hypotheses of Theorem* 3.1, *then* (1.1), (1.2) *has a solution* $u(x,t) \in L^\infty(0, T; H^s)$ *for each* $T > 0$.

REMARKS ON THE PROOF. Form a sequence $u_{0,n} \to u_0$ in H^s, $u_{0,n} \in C^\infty$. Then let u_n be the solution to (1.4), (1.2) for data $u_{0,n}$ and $\varepsilon = 1/n$. Then we find a subsequence of $\{u_n\}$, call it $\{u_n\}$ which converges on a dense set of linear $\{t_m\}$ in $t \geqq 0$. Then along an arbitrary $t > 0$, we show u_n is Cauchy in H^{s-3} of compact sets in space by a technique similar to that of Oleinik [7, p. 122]. Using Banach space arguments, we show the limit u actually lies in H^s for $s \geqq 6$.

As an easy consequence of the methods of §3, we have

THEOREM 4.2. *Let* $u \in L^\infty(0, T; H^3)$ *be a solution to* (1.2), (1.1) *or* (1.4), (1.1) *and let* $g_1(u) \geqq 0$ *and* $f, g_1, g_2, p \in C^2$, *then u is a unique solution.*

5. **Periodic problems.** The periodic problems are solved in much the same way as the nonperiodic. In the local existence proof we replace Fourier integrals by Fourier series and the same techniques work. The rest of the development of the existence theory may be readily imitated to prove existence and uniqueness for (1.1), (1.2), (1.3).

REFERENCES

1. S. Agmon, *Lectures on elliptic boundary value problems*, Van Nostrand Math. Studies, no. 2, Van Nostrand, Princeton, N.J., 1965. MR **31** #2504.

2. Y. Kametaka, *Korteweg-de Vries equation.* I–IV, Proc. Japan Acad. **45** (1969), 552–558, 656–665. MR **40** #6043; #6044; **31** #7311; #7312.

3. M. D. Kruskal, et al., *Korteweg-de Vries equation and generalizations.* V. *Uniqueness and non-existence of polynomial conservation laws*, J. Mathematical Phys. **11** (1970), 952–960. MR **42** #6410.

4. P. D. Lax, *Integrals of nonlinear equations of evolution and solitary waves*, Comm. Pure Appl. Math. **21** (1968), 467–490. MR **38** #3620.

5. R. M. Miura, C. S. Gardner and M. D. Kruskal, *Korteweg-de Vries equation and generalizations.* II. *Existence of conservation laws and constants of motion*, J. Mathematical Phys. **9** (1968), 1204–1209. MR **40** #6042b.

6. T. Mukasa and R. Iino, *On the global solution for the simplest generalized Korteweg-de Vries equation*, Math. Japon. **14** (1969), 75–83. MR **41** #7313.

7. O. A. Oleĭnik, *Discontinuous solutions of non-linear differential equations*, Uspehi Mat. Nauk **12** (1957), no. 3 (75), 3–73; English transl., Amer. Math. Soc. Transl. (2) **26** (1963), 95–172. MR **20** #1055; **27** #1721.

8. A. Sjöberg, *On the Korteweg-de Vries equation, existence and uniqueness*, Dept. of Comp. Sci., Uppsala University, Uppsala, Sweden, 1967.

9. R. Temam, *Sur un problème non-linéaire*, J. Math. Pures Appl. (9) **48** (1969), 159–172. MR **41** #5799.

10. M. Tsutsumi, T. Mukasa and R. Iino, *Parabolic regularizations for the generalized Korteweg-de Vries equation*, Proc. Japan Acad. **46** (1970), 921–925.

UNIVERSITY OF CALIFORNIA, BERKELEY

GENERAL RELATIVITY, PARTIAL DIFFERENTIAL EQUATIONS, AND DYNAMICAL SYSTEMS

ARTHUR E. FISCHER AND JERROLD E. MARSDEN

0. **Introduction.** In this paper we study two aspects of the Einstein equations of evolution for an empty spacetime. In the first part (§§1–3) we give a simple direct proof that the differentiability of the Cauchy data is maintained for short time. In the second part (§§4–5) we sketch how, on a suitable configuration space, Einstein's equations can be considered as forced geodesics modified by terms which reflect a moving coordinate system equipped with its own system of clocks. Both of these topics will be presented in more detail elsewhere [**15**], [**12**].

Let Ω be a bounded open domain in $\mathbf{R}^3$, let $\mathring{g}_{\mu\nu}(x^i)$, $(x^i) \in \Omega \subset \mathbf{R}^3$, $0 \leqq \mu, \nu \leqq 3$, $1 \leqq i \leqq 3$, be a Lorentz metric of signature $(-, +, +, +)$, and let $\mathring{k}_{\mu\nu}(x^i)$ be a symmetric 2-covariant tensor field on Ω. The first proof that Cauchy data $(\mathring{g}_{\mu\nu}(x^i), \mathring{k}_{\mu\nu}(x^i))$ of Sobolev class (H^s, H^{s-1}), $s \geqq 4$, evolves for small time into a Ricci zero ($R_{\mu\nu} = 0$) spacetime $g_{\mu\nu}(t, x^i)$ which is also of class H^s was given by Choquet-Bruhat [**3**], [**4**], based on earlier work by herself [**2**], and Lichnerowicz [**21**]. Her method of proof is to normalize the Ricci tensor by using harmonic coordinates so that the resulting system is a quasilinear strictly hyperbolic system (no multiple characteristics). The result then follows by quoting a theorem of Leray [**20**, p. 230] about quasilinear strictly hyperbolic systems, as modified by Dionne [7, p. 82]. Leray's original version of the theorem loses a derivative (i.e., H^s Cauchy data only has an H^{s-1} evolution), but Dionne remedies this defect.

Our method of proof is based on a simple observation; namely, the Ricci tensor in harmonic coordinates can be reduced to a quasilinear symmetric hyperbolic *first* order system of the form

$$A^0(u)\partial u/\partial t = A^i(u)\partial u/\partial x^i + B(u),$$

where u and $B(u)$ are 50 component column vectors, $A^0(u)$ and $A^i(u)$ are symmetric,

AMS 1970 *subject classifications*. Primary 83C99, 83C10, 35L45, 35L60; Secondary 83C30, 83C20, 83C45, 35L05.

and $A^0(u)$ is positive-definite. This observation is inspired by the well-known fact that any single second order hyperbolic equation can be reduced to a first order symmetric hyperbolic system.

Let $g_{\alpha\beta}(t, x)$ be an H^s-solution of $R_{\mu\nu} = 0$ with given Cauchy data. Because of the form invariance of the system $R_{\mu\nu} = 0$, $\bar{g}_{\mu\nu} = (\partial x^\alpha/\partial \bar{x}^\mu)(\partial x^\beta/\partial \bar{x}^\nu)g_{\alpha\beta}$ is also an H^s-solution of $R_{\mu\nu} = 0$, if $\bar{x}^\alpha(x^\mu)$ is an H^{s+1}-coordinate transformation. By arranging so that $\bar{x}^\alpha(x^\mu)$ is the identity in a neighborhood of the spacelike hypersurface $t = 0$, $\bar{g}_{\mu\nu}$ has the same Cauchy data as $g_{\mu\nu}$. Hence solutions to the Cauchy problem cannot be functionally unique. However, we prove a uniqueness theorem that says the evolution is unique up to the H^{s+1}-isometry class of the spacetime. This result sharpens, by one degree of differentiability, the uniqueness theorem stated in [**3**].

In [**15**] we shall show how our existence and uniqueness theorems can be obtained intrinsically for arbitrary manifolds, not necessarily compact, in the class of metrics for which the space-manifold is complete, and which satisfies suitable asymptotic conditions.

In §§4–5 we consider in what sense Einstein's equations in 3-dimensional form are a Lagrangian system of the classical form kinetic energy minus potential energy. We show how on a suitable configuration space (the manifold $\mathscr{D} \times \mathscr{M}$), the evolution equations are a degenerate dynamical system. Various terms in the Einstein system are given a geometrical explanation. In particular, the central role played by certain Lie derivative terms in the presence of a shift vector field is shown to be analogous to the space-body transitions of hydrodynamics (see Ebin-Marsden [**9**]) or the rigid body (see Marsden-Abraham [**23**]). These results are a geometrical reinterpretation of the basic work of Arnowitt-Deser-Misner [**1**], DeWitt [**6**] and Wheeler [**29**].

We thank D. Ebin, H. Lewy, P. Lax, M. Protter, R. Sachs, and A. Taub for a variety of helpful suggestions.

1. **Existence of an H^s-spacetime for (H^s, H^{s-1}) Cauchy data.** For empty space relativity, one searches for a Lorentz metric $g_{\mu\nu}(t, x^i)$ whose Ricci curvature $R_{\mu\nu}$ is zero; i.e., $g_{\mu\nu}(t, x^i)$ must satisfy the system

$$\begin{aligned} R_{\mu\nu}\left(t, x^i, g_{\mu\nu}, \frac{\partial g_{\mu\nu}}{\partial x^\alpha}, \frac{\partial^2 g_{\mu\nu}}{\partial x^\alpha \partial x^\beta}\right) &= -\frac{1}{2} g^{\alpha\beta} \frac{\partial^2 g_{\mu\nu}}{\partial x^\alpha \partial x^\beta} - \frac{1}{2} g^{\alpha\beta} \frac{\partial^2 g_{\alpha\beta}}{\partial x^\mu \partial x^\nu} + \frac{1}{2} g^{\alpha\beta} \frac{\partial^2 g_{\alpha\nu}}{\partial x^\beta \partial x^\mu} \\ &\quad + \frac{1}{2} g^{\alpha\beta} \frac{\partial^2 g_{\alpha\mu}}{\partial x^\beta \partial x^\nu} + H_{\mu\nu}\left(g_{\mu\nu}, \frac{\partial g_{\mu\nu}}{\partial x^\alpha}\right) \\ &= 0 \end{aligned}$$

where $H_{\mu\nu}(g_{\mu\nu}, \partial g_{\mu\nu}/\partial x^\alpha)$ is a rational combination of $g_{\mu\nu}$ and $\partial g_{\mu\nu}/\partial x^\alpha$ with denominator $\det g_{\mu\nu} \neq 0$. Note that the contravariant tensor $g^{\mu\nu}$ is a rational combination of the $g_{\mu\nu}$'s with denominator $\det g_{\mu\nu} \neq 0$.

Let $G_{\mu\nu} = R_{\mu\nu} - \frac{1}{2}g_{\mu\nu}R$ be the Einstein tensor, where $R = g^{\alpha\beta}R_{\alpha\beta}$ is the scalar curvature. Then, as is well known, G^0_μ contains only first order time derivatives of $g_{\mu\nu}$. Thus $G^0_\mu(0, x^i)$ can be computed from the Cauchy data $g_{\mu\nu}(0, x^i)$ and $\partial g_{\mu\nu}(0, x^i)/\partial t$ alone, and therefore $G^0_\mu(0, x^i) = 0$ is a necessary condition on the Cauchy data in order that a spacetime $g_{\mu\nu}(t, x^i)$ have the given Cauchy data and satisfy $G_{\mu\nu} = 0$, which is equivalent to $R_{\mu\nu} = 0$.

The existence part of the Cauchy problem for the system $R_{\mu\nu} = 0$ is as follows: *Let $(\mathring{g}_{\mu\nu}(x^i), \mathring{k}_{\mu\nu}(x^i))$ be Cauchy data of class $(H^s(\Omega), H^{s-1}(\Omega))$, $s \geqq 4$, such that $\mathring{G}^0_\mu(x^i) = 0$. Let Ω_0 be a proper subdomain, $\overline{\Omega}_0 \subset \Omega$. Find an $\varepsilon > 0$ and a spacetime $g_{\mu\nu}(t, x^i)$, $|t| < \varepsilon$, $(x^i) \in \Omega_0 \subset \Omega$ such that*

(a) *$g_{\mu\nu}(t, x^i)$ is H^s jointly in $(t, x^i) \in (-\varepsilon, \varepsilon) \times \Omega_0$,*

(b) *$(g_{\mu\nu}(0, x^i), \partial g_{\mu\nu}(0, x^i)/\partial t) = (\mathring{g}_{\mu\nu}(x^i), \mathring{k}_{\mu\nu}(x^i))$, and*

(c) *$g_{\mu\nu}(t, x^i)$ has zero Ricci curvature.*

The system $R_{\mu\nu} = 0$ is a quasilinear system of ten second order partial differential equations for which the highest order terms involve mixing of the components of the system. As it stands, there are no known theorems about partial differential equations which can be applied to resolve the Cauchy problem. However, as was first noted in 1922 by Lanczos [**18**] (and in fact in 1916 by Einstein himself for the linearized equations [**10**]) the Ricci tensor simplifies considerably in harmonic coordinates, i.e., in a coordinate system (x^α) for which the contracted Christoffel symbols vanish, $\Gamma^\mu = g^{\alpha\beta}\Gamma^\mu_{\alpha\beta} = 0$. (For the existence of such a coordinate system for an arbitrary Lorentz metric see Theorem 3.3.) In fact, an algebraic computation shows that

$$R_{\mu\nu} = -\frac{1}{2}g^{\alpha\beta}\frac{\partial^2 g_{\mu\nu}}{\partial x^\alpha\,\partial x^\beta} + \frac{1}{2}g_{\mu\alpha}\frac{\partial\Gamma^\alpha}{\partial x^\nu} + \frac{1}{2}g_{\nu\alpha}\frac{\partial\Gamma^\alpha}{\partial x^\mu} + H_{\mu\nu}$$

so that in a coordinate system for which $\Gamma^\mu = 0$,

$$R_{\mu\nu} = R^{(h)}_{\mu\nu} = -\tfrac{1}{2}g^{\alpha\beta}(\partial^2 g_{\mu\nu}/\partial x^\alpha\,\partial x^\beta) + H_{\mu\nu}.$$

The operator $-\frac{1}{2}g^{\alpha\beta}(\partial^2/\partial x^\alpha\,\partial x^\beta)$ operates the same way on each component of the system $g_{\mu\nu}$ so that there is no mixing in the highest order derivatives. Thus the normalized system $R^{(h)}_{\mu\nu} = 0$ is considerably simpler than the full system. In fact, the system $R^{(h)}_{\mu\nu} = 0$ has only simple characteristics so that $R^{(h)}_{\mu\nu} = 0$ is a strictly hyperbolic system.

The importance of the use of harmonic coordinates and of the system $R^{(h)}_{\mu\nu} = 0$ is based on the fact that it is sufficient to solve the Cauchy problem for $R^{(h)}_{\mu\nu} = 0$; this remarkable fact discovered by Fourès-Bruhat [2] is based on the observation that the condition $\mathring{\Gamma}^\mu(x^i) \equiv \mathring{g}^{\alpha\beta}(x^i)\mathring{\Gamma}^\mu_{\alpha\beta}(x^i) = 0$ is propagated off the hypersurface $t = 0$ for solutions $g_{\mu\nu}$ of $R^{(h)}_{\mu\nu} = 0$. This is established in the next lemma.

1.1. LEMMA. *Let $(\mathring{g}_{\mu\nu}(x^i), \mathring{k}_{\mu\nu}(x^i))$ be of Sobolev class (H^s, H^{s-1}) on Ω, $s > \frac{1}{2}n + 2$, $n = 3$, and suppose that $(\mathring{g}_{\mu\nu}(x^i), \mathring{k}_{\mu\nu}(x^i))$ satisfies*

(a) $\mathring{\Gamma}^\mu(x^i) = 0$,
(b) $\mathring{G}^0_\mu(x^i) = 0$.

If $g_{\mu\nu}(t, x)$, $|t| < \varepsilon$, $x \in \Omega_0$, Ω_0 *a proper subdomain*, $\overline{\Omega}_0 \subset \Omega$, *is an* H^s*-solution of*

$$(g_{\mu\nu}(0, x), \partial g_{\mu\nu}(0, x)/\partial t) = (\mathring{g}_{\mu\nu}(x^i), \mathring{k}_{\mu\nu}(x^i)),$$
$$R^{(h)}_{\mu\nu} = -\tfrac{1}{2}g^{\alpha\beta}(\partial^2 g_{\mu\nu}/\partial x^\alpha\, \partial x^\beta) + H_{\mu\nu} = 0,$$

then $\Gamma^\mu(t, x^i) = 0$ *for* $|t| < \varepsilon$, $x \in \Omega_0$.

PROOF. The case $s > \frac{1}{2}n + 2$ is treated in [**14**]; here we assume $s > \frac{1}{2}n + 3$. Let $g_{\mu\nu}(t, x^i)$ satisfy (a), (b) and $R^{(h)}_{\mu\nu} = 0$. Then a straightforward computation shows that $\Gamma^\mu(t, x^i) = g^{\alpha\beta}(t, x^i)\Gamma^\mu_{\alpha\beta}(t, x^i)$ satisfies $\partial\Gamma^\mu(0, x^i)/\partial t = 0$. From $G^{\mu\nu}{}_{|\nu} = 0$ (where $_{|\nu}$ means covariant derivative) and $R^{(h)}_{\mu\nu} = 0$, Γ^μ is shown to satisfy the system of linear equations

$$g^{\alpha\beta}\frac{\partial^2\Gamma^\mu}{\partial x^\alpha\, \partial x^\beta} + A^{\beta\mu}_\alpha\left(g_{\mu\nu}, g^{\mu\nu}, \frac{\partial g_{\mu\nu}}{\partial x^\lambda}\right)\frac{\partial\Gamma^\alpha}{\partial x^\beta} = 0.$$

This linear system can be reduced to a linear first order symmetric hyperbolic system for which a uniqueness and existence theorem holds. This is exactly analogous to Theorems 1.2 and 3.1 below. But from the uniqueness for this system, $\Gamma^\mu(0, x^i) = 0$ and $\partial\Gamma^\mu(0, x^i)/\partial t = 0$ imply $\Gamma^\mu(t, x^i) = 0$. ∎

According to the lemma, an H^s-solution of $R^{(h)}_{\mu\nu} = 0$ with prescribed Cauchy data is also a solution of $R_{\mu\nu} = 0$ (since $\Gamma^\mu(t, x) = 0 \Rightarrow R^{(h)}_{\mu\nu} = R_{\mu\nu}$), provided that the Cauchy data satisfies (a) $\mathring{\Gamma}^\mu = 0$ and (b) $\mathring{G}^0_\mu = 0$. As mentioned above (b) is a necessary condition on the Cauchy data for a solution $g_{\mu\nu}(t, x)$ to satisfy $R_{\mu\nu} = 0$. If (a) is not satisfied, then a set of Cauchy data can be found whose evolution under $R^{(h)}_{\mu\nu} = 0$ leads to an H^s-spacetime which by an H^{s+1}-coordinate transformation gives rise to a spacetime with the original Cauchy data (see Fischer-Marsden [**14**]).

From the theorem of Dionne concerning quasilinear strictly hyperbolic systems [**7**] and Lemma 1.1, Choquet-Bruhat [**3**], [**4**] concludes that Cauchy data of class (H^s, H^{s-1}) has an H^s-time evolution. We prove this result directly by reducing the strictly hyperbolic system $R^{(h)}_{\mu\nu} = 0$ to a quasilinear symmetric hyperbolic first order system. Aside from the practical nature of putting the Einstein evolution equations in this form (systems of this type are very well understood; the Cauchy problem has a simple resolution in this form), there is the aesthetic value of bringing relativity into a form which uniformly governs most other equations of mathematical physics, such as Maxwell's equations, the Dirac equation, the Lundquist equations of magnetohydrodynamics, Euler's equations for a compressible fluid, and the equations describing the motion of elastic bodies.

1.2. THEOREM. *Let* Ω *be an open bounded domain in* $\mathbf{R}^3$ *with* Ω_0 *a proper subdomain*, $\overline{\Omega}_0 \subset \Omega$, *and let* $(\mathring{g}_{\mu\nu}(x), \mathring{k}_{\mu\nu}(x))$, $(x^i) \in \Omega$, $0 \leqq \mu, \nu \leqq 3$, $1 \leqq i \leqq 3$, *be of Sobolev class*

(H^s, H^{s-1}), $s \geqq 4$. *Suppose that* $\mathring{\Gamma}^\mu(x^i) = 0$ *and* $\mathring{G}^0_\mu(x) = 0$. *Then there exists an* $\varepsilon > 0$ *and a unique Lorentz metric* $g_{\mu\nu}(t, x)$, $|t| < \varepsilon$, $(x^i) \in \Omega_0$ *such that*

(1) $g_{\mu\nu}(t, x^i)$ *is jointly of class* H^s,

(2) $R^{(h)}_{\mu\nu}(t, x^i) = 0$,

(3) $(g_{\mu\nu}(0, x^i), \partial g_{\mu\nu}(0, x^i)/\partial t) = (\mathring{g}_{\mu\nu}(x^i), \mathring{k}_{\mu\nu}(x^i))$.

From Lemma 1.1, *this* $g_{\mu\nu}(t, x^i)$ *also satisfies* $R_{\mu\nu}(t, x^i) = 0$. *Moreover,* $g_{\mu\nu}(t, x^i)$ *depends continuously on* $(\mathring{g}_{\mu\nu}(x^i), \mathring{k}_{\mu\nu}(x^i))$ *in the* (H^s, H^{s-1}) *topology. If* $(\mathring{g}_{\mu\nu}(x^i), \mathring{k}_{\mu\nu}(x^i))$ *is of class* (C^∞, C^∞) *on* Ω, *then* $g_{\mu\nu}(t, x^i)$ *is* C^∞ *for all t for which the solution exists.*

Note. The case $s = 4$ is delicate and is treated in [**14**]. Here we assume $s \geqq 5$. In [**14**] we also give a complete discussion for the case of spatial asymptotic conditions.

PROOF. The system $R^{(h)}_{\mu\nu} = 0$ is reduced to a first order system by introducing the ten new unknowns $k_{\mu\nu} = \partial g_{\mu\nu}/\partial t$ and the thirty new unknowns $g_{\mu\nu,i} = \partial g_{\mu\nu}/\partial x^i$ and considering the quasilinear first order system of fifty equations:

$$\partial g_{\mu\nu}/\partial t = k_{\mu\nu},$$

$$\text{(Q)} \qquad g^{ij}(\partial g_{\mu\nu,i}/\partial t) = g^{ij}\frac{\partial k_{\mu\nu}}{\partial x^i},$$

$$-g^{00}\frac{\partial k_{\mu\nu}}{\partial t} = 2g^{0j}\frac{\partial k_{\mu\nu}}{\partial x^j} + g^{ij}\frac{\partial g_{\mu\nu,i}}{\partial x^j} - 2H_{\mu\nu}(g_{\mu\nu}, g_{\mu\nu,i}, k_{\mu\nu}).$$

We are considering $H_{\mu\nu}$ as a polynomial in $g_{\mu\nu,i}$ and $k_{\mu\nu}$ and rational in $g_{\mu\nu}$ with denominator $\det g_{\mu\nu} \neq 0$. At first, we extend our initial data to all of $\mathbf{R}^3$, say to equal the Minkowski metric outside a compact set, and consider the system (Q) on $\mathbf{R}^3$. Note that the Cauchy data need not satisfy the constraints $\mathring{G}^0_\mu = 0$ during the transition.

The matrix g^{ij} has inverse $g_{jk} - (g_{j0}g_{k0}/g_{00})$ (i.e., $g^{ij}(g_{jk} - (g_{j0}g_{k0}/g_{00})) = \delta^i_k$) so that the second set of thirty equations can be inverted to give

$$\text{(1)} \qquad \partial g_{\mu\nu,i}/\partial t = \partial k_{\mu\nu}/\partial x^i.$$

For $g_{\mu\nu}$ of class C^2, (1) implies

$$g_{\mu\nu,i} = \partial g_{\mu\nu}/\partial x^i,$$

so that the system (Q) is equivalent to $R^{(h)}_{\mu\nu} = 0$.

Let

$$u = \begin{pmatrix} g_{\mu\nu} \\ g_{\mu\nu,i} \\ k_{\mu\nu} \end{pmatrix}$$

be a fifty component column vector, where $g_{\mu\nu,i}$ is listed as

$$\begin{pmatrix} g_{00,1} \\ \vdots \\ g_{33,1} \\ \vdots \\ g_{00,3} \\ \vdots \\ g_{33,3} \end{pmatrix},$$

$0^{10} = 10 \times 10$ zero matrix, $I^{10} = 10 \times 10$ identity matrix, and let $A^0(u) = A^0(g_{\mu\nu}, g_{\mu\nu,i}, k_{\mu\nu})$ and $A^j(g_{\mu\nu}, g_{\mu\nu,i}, k_{\mu\nu})$ be the 50×50 matrices given by

$$A^0(g_{\mu\nu}, g_{\mu\nu,i}, k_{\mu\nu}) = \begin{pmatrix} I^{10} & 0^{10} & 0^{10} & 0^{10} & 0^{10} \\ 0^{10} & g^{11}I^{10} & g^{12}I^{10} & g^{13}I^{10} & 0^{10} \\ 0^{10} & g^{12}I^{10} & g^{22}I^{10} & g^{23}I^{10} & 0^{10} \\ 0^{10} & g^{13}I^{10} & g^{23}I^{10} & g^{33}I^{10} & 0^{10} \\ 0^{10} & 0^{10} & 0^{10} & 0^{10} & -g^{00}I^{10} \end{pmatrix},$$

$$A^j(g_{\mu\nu}, g_{\mu\nu,i}, k_{\mu\nu}) = \begin{pmatrix} 0^{10} & 0^{10} & 0^{10} & 0^{10} & 0^{10} \\ 0^{10} & 0^{10} & 0^{10} & 0^{10} & g^{j1}I^{10} \\ 0^{10} & 0^{10} & 0^{10} & 0^{10} & g^{j2}I^{10} \\ 0^{10} & 0^{10} & 0^{10} & 0^{10} & g^{j3}I^{10} \\ 0^{10} & g^{1j}I^{10} & g^{2j}I^{10} & g^{3j}I^{10} & 2g^{j0}I^{10} \end{pmatrix},$$

and let $B(g_{\mu\nu}, g_{\mu\nu,i}, k_{\mu\nu})$ be the fifty component column vector given by

$$B(g_{\mu\nu}, g_{\mu\nu,i}, k_{\mu\nu}) = \begin{pmatrix} k_{\mu\nu} \\ 0^{30} \\ -2H_{\mu\nu}(g_{\mu\nu}, g_{\mu\nu,i}, k_{\mu\nu}) \end{pmatrix}$$

where 0^{30} is the thirty component zero column vector.

Note that $A^0(u)$ and $A^j(u)$ are symmetric, and that $A^0(u)$ is positive-definite if $g_{\mu\nu}$ has Lorentz signature. A direct verification shows that the first-order quasi-linear symmetric hyperbolic system

$$A^0(u)(\partial u/\partial t) = A^j(u)(\partial u/\partial x^j) + B(u)$$

is just the system (Q). From Theorem 2.1 and its generalizations proven below, we conclude that for Cauchy data

$$\mathring{u}(x^i) = \begin{pmatrix} \mathring{g}_{\mu\nu}(x^i) \\ \mathring{g}_{\mu\nu,i}(x^i) \\ \mathring{k}_{\mu\nu}(x^i) \end{pmatrix}$$

of Sobolev class H^{s-1}, $s - 1 > \frac{1}{2}n + 2$, there exists an $\varepsilon > 0$ and a solution

$$u(t, x^i) = \begin{pmatrix} g_{\mu\nu}(t, x^i) \\ g_{\mu\nu,i}(t, x^i) \\ k_{\mu\nu}(t, x^i) \end{pmatrix}$$

of class H^{s-1}. By Sobolev's lemma, $u(t, x^i)$ is also of class C^2, and so, by the second set of equations of (Q), $g_{\mu\nu,i} = \partial g_{\mu\nu}/\partial x^i$. Since $(g_{\mu\nu,i}, k_{\mu\nu}) = (\partial g_{\mu\nu}/\partial x^i, \partial g_{\mu\nu}/\partial t)$ is of class H^{s-1}, $g_{\mu\nu}(t, x^i)$ is in fact of class H^s. The continuous dependence of the solutions on the initial data follows from the general theory below.

To recover the result for the domain Ω from the result for $\mathbf{R}^n$, we can use the standard domain of dependence arguments; see Courant-Hilbert [**5**].

Since Ω is bounded, $(\mathring{g}_{\mu\nu}, \mathring{k}_{\mu\nu})$ of class C^∞ implies that the solution is in the intersection of all the Sobolev spaces and hence is C^∞; again we are using a general regularity result about symmetric hyperbolic systems.

From Lemma 1.1, the $g_{\mu\nu}(t, x^i)$ so found satisfy the field equations $R_{\mu\nu} = 0$. For the case in which the Cauchy data does not satisfy $\mathring{\Gamma}^\mu(x^i) = 0$, see [**14**]. ∎

Although $g_{\mu\nu}(t, x^i)$ is a unique solution of $R^{(h)}_{\mu\nu} = 0$, with prescribed Cauchy data, it is not a unique solution of $R_{\mu\nu} = 0$. We return to this point in §3.

2. **First order quasilinear symmetric hyperbolic systems.** The theory of linear first order symmetric hyperbolic systems is due to Friedrichs [**17**] with some simplifying modifications by Lax [**19**]. The essential ideas for handling quasilinear equations appear in Schauder [**25**], Frankl [**16**], and Petrovskiĭ [**24**]. Friedrichs [**17**, p. 352] mentions that these ideas can be used to prove the unique existence of a solution for a quasilinear first order symmetric hyperbolic system, and it is again mentioned in Courant-Hilbert [**5**, p. 675], but we have been unable to find the details of a complete treatment in the literature. Here we shall outline the methods for $\mathbf{R}^n$ and present an intrinsic version for manifolds elsewhere [**15**]. The basic idea is to find energy type estimates, use the contraction mapping principle to find an H^{s-1}-solution, and then show that this H^{s-1}-solution is in fact H^s.

2.1. THEOREM. *Let $H^s(\mathbf{R}^n, \mathbf{R}^m)$ denote the H^s maps from $\mathbf{R}^n$ to $\mathbf{R}^m$, and let*

$\mathscr{U}^s \subset H^s(\mathbf{R}^n, \mathbf{R}^m)$ be an open subset. Let $\delta > 0$, and for $(t, x, u) \in (-\delta, \delta) \times \mathbf{R}^n \times \mathscr{U}^s$, let $A^i(t, x, u)$ be a symmetric $m \times m$ matrix, and let $B(t, x, u)$ be an m-component column vector. Suppose that $A^i(t, x, u)$, and $B(t, x, u)$ are H^s-functions of (t, x), and are rational functions of u with nonzero denominators. (More generally, one could use Sobolev's "condition T" on compositions of H^s functions [**28**].)

Given $u_0 \in \mathscr{U}^s$, $s > \frac{1}{2}n + 2$, there is an $\varepsilon > 0$, $\varepsilon < \delta$, and a unique $u(t, x)$, $|t| < \varepsilon$, $x \in \mathbf{R}^n$, which is H^s in (t, x) and which satisfies

$$u(t, x) = u_0(x),$$

$$\partial u/\partial t = A^i(t, x, u)(\partial u/\partial x^i) + B(t, x, u).$$

Moreover, the solution $u(t, x)$ depends continuously on u_0 in the H^s-topology. If $A^i(t, x, u)$ and $B(t, x, u)$ are functions of (t, x) in $\bigcap_{s > n/2+2} H^s((-\varepsilon, \varepsilon) \times \mathbf{R}^n)$ and u_0 is in $\bigcap_{s > n/2+2} H^s(\mathbf{R}^n)$, then $u(t, x)$ is in $\bigcap_{s > n/2+2} H^s((-\varepsilon, \varepsilon) \times \mathbf{R}^n)$.

Note. For the above applications, we should replace $\partial u/\partial t$ by $A^0(t, x, u)\partial u/\partial t$ where A^0 is a symmetric positive-definite matrix. Here we consider the case $s > \frac{1}{2}n + 2$; the case $s > \frac{1}{2}n + 1$ is obtained in [**14**]. Moreover, 2.1 can be generalized to the case in which the coefficients and the Cauchy data satisfy asymptotic conditions. This case is more delicate and is discussed following the proof of the present theorem.

PROOF. Let $\| \ \|_s$ denote the H^s-norm for functions $u: \mathbf{R}^n \to \mathbf{R}^m$ in $\mathscr{U}^s$. Let E denote the set of continuous curves $\omega: [-\delta, \delta] \to \mathscr{U}^s$ such that $\omega(0) = u_0$ and $\|\omega(t) - u_0\|_s \leqq 1$, $-\delta \leqq t \leqq \delta$. Thus E is a complete metric space, and we want to define a map $f: E \to E$ by

$$f(\omega)(t) = u_0 + \int_0^t A^i(s, x, \omega(s, x)) \frac{\partial}{\partial x^i} f(\omega)(s, x)\, ds + \int_0^t B(s, x, \omega(s, x))\, ds$$

where the integration is done as a curve in H^{s-1}. From the linear theory of first order symmetric hyperbolic systems, it follows that for δ sufficiently small there is a unique such mapping $f: E \to E$. Indeed, for $\omega \in E$, the unique solution of the linear system

$$\text{(I)} \qquad \begin{aligned} u(0) &= u_0 \\ \partial u/\partial t &= A^i(t, x, \omega)(\partial u/\partial x^i) + B(t, x, \omega) \end{aligned}$$

is exactly $f(\omega)$. Moreover, from the usual Leray estimates of the linear theory (see Courant-Hilbert [**5**, p. 671]), it is easy to show that for $\omega \in E$ and for u satisfying (I), there is a constant β independent of δ, ω, and u such that $\|u(t)\|_s \leqq e^{\beta|t|}\|u_0\|_s$. Thus f maps E to E.

Let F denote the completion of E in the H^{s-1}-norm. We remark that F is, by the Rellich-Garding theorem, a compact set, although we shall not need this fact. Next we note that for δ sufficiently small, $f: E \to E$ is a contraction in the H^{s-1}-norm, i.e., for $\omega_1, \omega_2 \in E$, there exists a k, $0 < k < 1$, such that

$$\|f(\omega_1) - f(\omega_2)\|_{s-1} \leqq k\|\omega_1 - \omega_2\|_{s-1}. \tag{1}$$

This follows from the estimates

$$\begin{aligned}\|f(\omega_1) - f(\omega_2)\|_{s-1} \leqq \int_0^t \Big\| & A^i(s, x, \omega_1(s, x)) \frac{\partial}{\partial x^i} f(\omega_1)(s, x) \\ & - A^i(s, x, \omega_2(s, x)) \frac{\partial}{\partial x^i} f(\omega_2)(s, x)\Big\|_{s-1} ds \\ & + \int_0^t \|B(s, x, f(\omega_1)(s, x)) - B(s, x, f(\omega_2)(s, x))\|_{s-1}\, ds\end{aligned}$$

and

$$\left\|A^i(s, x, \omega_1) \frac{\partial}{\partial x^i} f(\omega_1) - A^i(s, x, \omega_2) \frac{\partial}{\partial x^i} f(\omega_2)\right\|_{s-1}$$

$$\begin{aligned} & \leqq \left\|A^i(t, x, \omega_1) \frac{\partial}{\partial x^i} [f(\omega_1) - f(\omega_2)]\right\|_{s-1} \\ & + \left\|A^i(t, x, \omega_1) \frac{\partial}{\partial x^i} f(\omega_2) - A^i(t, x, \omega_2) \frac{\partial}{\partial x^i} f(\omega_2)\right\|_{s-1}. \end{aligned} \tag{2}$$

The first term in (2) is handled as in the proof of the energy estimates; the second term is bounded by $c_1\|\omega_1 - \omega_2\|_{s-1}\|f(\omega_2)\|_s$ (c_1, c_2, c_3 are constants). Thus

$$\begin{aligned}\sup_{-\delta < t < \delta} \|f(\omega_1) - f(\omega_2)\|_{s-1} \leqq c_2 & \sup_{-\delta < t < \delta} \|f(\omega_1) - f(\omega_2)\|_{s-1} \\ & + \delta c_3 \sup_{-\delta < t < \delta} \|\omega_1 - \omega_2\|_{s-1}\end{aligned}$$

from which (1) follows.

Thus f extends to a contraction on the complete metric space F so by the contraction mapping principle f has a unique fixed point, a solution in H^{s-1} to the quasilinear system we are studying. Since f depends continuously on u_0, so does the fixed point. We remark that the original argument of Schauder used compactness of F and the Schauder fixed point theorem. P. Lax has pointed out that the above contraction argument can be replaced by the extraction of a weakly convergent sequence in L_2, and using H^s-boundedness, to deduce convergence in H^k, $k < s$. This again yields a solution in H^{s-1}.

Finally we show that the solution $u(t, x)$ in H^{s-1} is in fact in H^s. The trick is to look at the differential equation satisfied by the second spatial derivative of u, the solution found in H^{s-1}. Now

$$\partial u/\partial t = A^i(t, x, u)(\partial u/\partial x^i) + B(t, x, u)$$

so if Du is the first differential of u,

$$\frac{\partial}{\partial t}(Du) = D_2A^i(t, x, u)\frac{\partial u}{\partial x^i} + D_3A^i(t, x, u) \cdot Du \cdot \frac{\partial u}{\partial x^i} \tag{3}$$
$$+ A^i(t, x, u)\frac{\partial}{\partial x^i}Du + DB(t, x, u)$$

where D_2A^i and D_3A^i are the partial derivatives with respect to the second and third variables, respectively. If we consider (3) as a linear equation in the unknown $v = Du$ of the form

$$\partial v/\partial t = A^i(\partial v/\partial x^i) + C \cdot v + D$$

then we must treat $D_3A^i(t, x, u)(\partial u/\partial x^i)$ as a coefficient. However, since u is only H^{s-1}, $\partial u/\partial x^i$ is only H^{s-2}. However, if we differentiate again it is easy to see that $w = D^2u$ satisfies

$$\partial w/\partial t = A^i(\partial w/\partial x^i) + \tilde{C} \cdot w + \tilde{D}$$

where now A^i, $\tilde{C}$, $\tilde{D}$ are H^{s-2}-functions ($\partial Du/\partial x^i$ is, for example, taken to be part of D^2u). The reason is just that second derivatives do not occur multiplied together as the first ones did. Now if $s > \frac{1}{2}n + 3$, the coefficients are in H^r, $r = s - 2 > \frac{1}{2}n + 1$, so by the linear theory w which is initially in H^{s-2} remains in H^{s-2}. Hence u remains in H^s. (However one only needs $s > \frac{1}{2}n + 2$ here by using the fact that only the lower order terms are affected and $r > \frac{1}{2}n$; see below.

This argument also shows that the map $u_0 \mapsto u_t$ is continuous in H^s and not merely in H^{s-1}. Moreover, if the coefficients $A^i(t, x, u)$, and $B(t, x, u)$ are smooth, then the same argument shows that if we have a solution in H^s whose initial condition is in H^{s+1} then in fact the solution is in H^{s+1} (as long as it is defined in H^s). Hence smooth initial conditions remain smooth. ∎

Now in our application, we are considering a system of the form

$$A^0(t, x, u)(\partial u/\partial t) = A^i(t, x, u)(\partial u/\partial x^i) + B(t, x, u).$$

This case may be handled as follows. We assume as above that A^0, A^i and B are H^s-functions. By using the technique above, we are led to consider first the linear case. We proceed as in [**5**] to reduce to the case $A^0 = \text{Id}$, by writing $A^0 = TT^*$ and letting $v = T^{-1}u$. However, in the case that A^0 depends on t, x in an H^s manner, this modifies the B term by replacing it by an H^{s-1} term.

Without further conditions on the $A^i(t, x)$, Cauchy data u_0 of class H^s need only have a time evolution $u(t, x)$ of class H^{s-1}. For example, if $A^i(t, x) = 0$, $B(t, x) = B(x)$, then $\partial u/\partial t = B(x)u$ can be integrated explicitly to give $u(t, x) = e^{tB(x)}u_0(x)$. For $B(x)$ of class H^{s-1}, $u(t, x)$ need only be of class H^{s-1}.

The appropriate condition on the matrices $A^i(t, x)$ can be found from the following standard lemma from perturbation theory [**30**].

2.2. Lemma. *Let F be a Banach space, $D_A \subset F$ a dense domain, and $A: D_A \subset F \to F$*

a linear operator which is a generator. Let $B: F \to F$ be a bounded operator. Then $A + B: D_A \subset F \to F$ is a generator whose domain is exactly D_A.

Thinking of F as H^{s-1}-functions, D_A as H^s-functions, and $B: F \to F$ as multiplication by an H^{s-1} matrix, we see that D_A will be the domain of the closure of the operator $A = A^i(\partial/\partial x^i)$ from H^s to H^{s-1}. In concrete examples like 3.1 below, the domain of this operator is not hard to work out. Thus in the symmetric hyperbolic case, where we know that A is a generator, we know that solutions to the full system with a B term which is H^{s-1} will remain in D_A if they start out in that set.

The same remarks remain valid in the quasilinear case; that is, if A^i is H^s and B is H^{s-1} then solutions which start out in D_A will remain there. In both this and the linear case, it is important to realize that the H^s energy estimates fail, and so the proof of Theorem 2.1 as given breaks down. This failure occurs because in estimating derivatives of order s of the B term, one runs into H^r for $r < \frac{1}{2}n$ and the requisite ring structure of H^k is no longer available. However, one can use the H^{s-1}-estimates, and the regularity argument of Theorem 2.1 together with the linear theory Lemma 2.2. In addition to applications of Lemma 2.2 to quasilinear systems, in the next section we shall also need to consider the linear first order symmetric hyperbolic systems $A^0(t, x)(\partial u/\partial t) = A^i(t, x)(\partial u/\partial x^i) + B(t, x) \cdot u$ where $A^0(t, x)$ and $A^i(t, x)$ are of class H^s in (t, x) but $B(t, x)$ is only of class H^{s-1}.

3. **Uniqueness for the Einstein equations.** In this section we show that any two H^s-spacetimes which are Ricci flat and which have the same Cauchy data are related by an H^{s+1}-coordinate transformation. The key idea is to show that any H^s-spacetime when expressed in harmonic coordinates is also of class H^s. This in turn is based on an old result of Sobolev [**28**]; namely, that solutions to the wave equation with (H^s, H^{s-1}) coefficients preserve (H^{s+1}, H^s) Cauchy data. We can give an easy proof of this result by using Lemma 2.2 and the well-known result that any single second order hyperbolic equation can be reduced to a system of symmetric hyperbolic equations.

3.1. THEOREM. *Let $(\psi_0(x), \dot\psi_0(x))$ be of Sobolev class (H^{s+1}, H^s) on $\mathbf{R}^3$. Then there exists a unique $\psi(t, x)$ of class H^{s+1} that satisfies*

$$(\psi(0, x), \partial\psi(0, x)/\partial t) = (\psi_0(x), \dot\psi_0(x)),$$

$$g^{\mu\nu}(t, x)(\partial^2\psi/\partial x^\mu\, \partial x^\nu) + b^\mu(t, x)(\partial\psi/\partial x^\mu) + c(t, x)\psi = 0$$

where $g^{\mu\nu}(t, x)$ is a Lorentz metric of class H^s, $b^\mu(t, x)$ a vector field of class H^{s-1}, and $c(t, x)$ is of class H^{s-1}.

PROOF. As in the proof of Theorem 1.2, this single equation can be reduced to a first order symmetric hyperbolic system

$$A^0(t, x)(\partial u/\partial t) = A^i(t, x)(\partial u/\partial x^i) + B(t, x) \cdot u$$

where $A^0(t, x)$ and $A^i(t, x)$ are of class H^s, $B(t, x)$ is of class H^{s-1}, and u is the 5 component column vector

$$u = \begin{pmatrix} \psi \\ \psi_{,i} \\ \psi_{,0} \end{pmatrix}.$$

In fact the A^0 and A^i are exactly as in Theorem 1.2 with $0^{10} \to 0$ and $I^{10} \to 1$ and

$$B(t, x) = \begin{pmatrix} 0 & 0 & 0 & 0 & 1 \\ 0 & 0 & 0 & 0 & 0 \\ 0 & 0 & 0 & 0 & 0 \\ 0 & 0 & 0 & 0 & 0 \\ c(t, x) & b^1(t, x) & b^2(t, x) & b^3(t, x) & b^0(t, x) \end{pmatrix}.$$

Regard $A = A^i\, \partial/\partial x^i$ as a densely defined operator on H^{s-1} with domain H^s. A simple check using positive-definiteness of g_{ij} (ellipticity of the operator $g^{ij}(\partial^2/\partial x^i\, \partial x^j)$) shows that the conditions of Lemma 2.2 on the A^i are met with the domain of $\bar{A} = \mathrm{cl}(A^i \partial/\partial x^i)$ in H^{s-1} being at least as large as $H^{s-1} \oplus H^s \oplus H^{s-1}$ on the three blocks of u. Thus if u_0 is in D_A, then

$$u = \begin{pmatrix} \psi \\ \psi_{,i} \\ \psi_{,0} \end{pmatrix}$$

remains in D_A, which means that ψ is H^{s+1} in x and H^s in t. From the differential equation itself we see that in fact ψ is also H^{s+1} in t. ∎

We remark that this proof "works" because we use the symmetric hyperbolic system in u and thus the coefficients need only be of class H^s, H^{s-1}.

From Theorem 3.1, we can now prove that when one transforms an H^s-spacetime to harmonic coordinates, it stays H^s.

3.2. THEOREM. *Let $g_{\mu\nu}(t, x)$ be an H^s-spacetime. Then there exists an H^{s+1}-coordinate transformation $\bar{x}^\lambda(x^\mu)$ such that*

$$\bar{g}_{\mu\nu}(\bar{x}^\lambda) = \frac{\partial x^\alpha}{\partial \bar{x}^\mu}(\bar{x}^\lambda)\frac{\partial x^\beta}{\partial \bar{x}^\nu}(\bar{x}^\lambda) g_{\alpha\beta}(x^\mu(\bar{x}^\lambda))$$

is an H^s-spacetime with $\bar{\Gamma}^\mu(\bar{t}, \bar{x}) = \bar{g}^{\alpha\beta}\bar{\Gamma}^\mu_{\alpha\beta}(\bar{t}, \bar{x}) = 0$.

PROOF. To find $\bar{x}^\lambda(x^\mu)$ consider the wave equation

$$\Box\psi = -g^{\alpha\beta}(\partial^2\psi/\partial x^\alpha\, \partial x^\beta) + g^{\alpha\beta}\Gamma^\mu_{\alpha\beta}(\partial\psi/\partial x^\mu) = 0,$$

and let $\bar{t}(t, x)$ be the unique solution of the wave equation with Cauchy data $t(0, x) = 0$, $\partial\bar{t}(0, x)/\partial t = 1$, and let $\bar{x}^i(t, x)$ be the unique solution of the wave equation with Cauchy data

$$\bar{x}^i(0, x) = x^i, \qquad \partial\bar{x}^i(0, x)/\partial t = 0.$$

For $g_{\mu\nu}$ of class H^s, Γ^μ is of class H^{s-1}, so by Theorem 3.1, $\bar{t}(t, x)$ and $\bar{x}(t, x)$ are H^{s+1}-functions and in fact by the inverse function theorem for H^s-functions, $(\bar{t}(t, x), \bar{x}(t, x))$ is an H^{s+1} diffeomorphism in a neighborhood of $t = 0$.

Since $\Box\bar{x}^\mu(t, x) = 0$ is an invariant equation,

$$\bar{\Box}\bar{x}^\mu = -\bar{g}^{\alpha\beta}\frac{\partial^2\bar{x}^\mu}{\partial\bar{x}^\alpha\,\partial\bar{x}^\beta} + \bar{g}^{\alpha\beta}\bar{\Gamma}^\nu_{\alpha\beta}\frac{\partial\bar{x}^\mu}{\partial\bar{x}^\nu} = \bar{g}^{\alpha\beta}\bar{\Gamma}^\mu_{\alpha\beta} = 0$$

in the barred coordinate system, so $\bar{x}^\mu$ is a system of harmonic coordinates. ∎

As a simple consequence of Theorem 3.2 we have the following uniqueness result for the Einstein equations:

3.3. THEOREM. *Let $g_{\mu\nu}(t, x)$ and $\bar{g}_{\mu\nu}(t, x)$ be two H^s-spacetimes with zero Ricci tensor and such that $(g_{\mu\nu}(0, x), \partial g_{\mu\nu}(0, x)/\partial t) = (\bar{g}_{\mu\nu}(0, x), \partial\bar{g}_{\mu\nu}(0, x)/\partial t)$. Then $g_{\mu\nu}(t, x)$ and $\bar{g}_{\mu\nu}(t, x)$ are related by an H^{s+1}-coordinate change in a neighborhood of $t = 0$.*

PROOF. From Theorem 3.2 there exist H^{s+1}-coordinate transformations $y^\mu(x^\alpha)$ and $\bar{y}^\mu(x^\alpha)$ such that the transformed metrics $(\partial x^\alpha/\partial y^\mu)(\partial x^\beta/\partial y^\nu)g_{\alpha\beta}$ and $(\partial x^\alpha/\partial\bar{y}^\mu)(\partial x^\beta/\partial\bar{y}^\nu)\bar{g}_{\alpha\beta}$ satisfy $R^{(h)}_{\mu\nu} = 0$. Since the Cauchy data for $g_{\mu\nu}$ and $\bar{g}_{\mu\nu}$ are equal the transformed metrics also have the same Cauchy data. By the uniqueness part of Theorem 1.2, $(\partial x^\alpha/\partial y^\mu)(\partial x^\beta/\partial y^\nu)g_{\alpha\beta} = (\partial x^\alpha/\partial\bar{y}^\mu)(\partial x^\beta/\partial\bar{y}^\nu)\bar{g}_{\alpha\beta}$. Since the composition of H^{s+1}-coordinate changes is also H^{s+1}, $\bar{g}_{\alpha\beta}$ is related to $g_{\alpha\beta}$ by an H^{s+1}-coordinate change in a neighborhood of $t = 0$. ∎

4. **The Einstein system on the manifold $\mathcal{M}$.** We now consider a dynamical formulation of general relativity from the 3-dimensional point of view of Arnowitt, Deser and Misner [**1**], DeWitt [**6**], and Wheeler [**29**]. All tensor fields, such as g, k, X are referred to a fixed oriented smooth 3-dimensional manifold M.

Let

$\mathcal{M} = \mathrm{Riem}(M)$ = manifold of all smooth Riemannian metrics (positive-definite) on M;

$S_2(M)$ = the linear space of all smooth symmetric 2-covariant tensor fields on M; and

$\mathcal{D} = \mathrm{Diff}(M)$ = the group of smooth orientation-preserving diffeomorphisms of M.

In the dynamical formulation of general relativity, one is concerned with the evolution of initial Cauchy data $(\mathring{g}, \mathring{h}) \in \mathcal{M} \times S_2(M)$ on some 3-dimensional hypersurface M of a yet to be constructed Ricci-flat (vacuum) spacetime V_4.

As one is interested in finding the evolution g_t of Riemannian metrics only up to the isometry class $\{(\eta_t^{-1})^* g_t | \eta_t \in \mathcal{D}\}$ of g_t (here $(\eta_t^{-1})^*$ is the "push forward" of covariant tensor fields), the evolution is determined only up to an arbitrary curve $\eta_t \in \mathcal{D}$ with $\eta_0 =$ the identity diffeomorphism. In other words, only the orbit class or geometry of g_t is determined (see DeWitt [**6**] and Fischer [**11**] for the structure of the orbit space $\mathcal{M}/\mathcal{D}$). Moreover, one is free to specify on M an arbitrary system of clock rates.

These degeneracies are reflected in the evolution equations as follows:

THE EINSTEIN SYSTEM. *Let* $(\mathring{g}, \mathring{k}) \in \mathcal{M} \times S_2(M)$ *satisfy the constraints:*

$$\delta(\mathring{k} - \mathring{g}(\operatorname{Tr} \mathring{k})) = 0, \qquad \tfrac{1}{2}((\operatorname{Tr} \mathring{k})^2 - \mathring{k} \cdot \mathring{k}) + 2R(\mathring{g}) = 0.$$

Let X_t be an arbitrary time-dependent vector field on M (the shift vector field) and N_t an arbitrary time-dependent scalar field on M (the lapse function) such that

$$N_t(m) > 0,$$
$$N_0^2 - \|X_0\|^2 > 0 \quad \textit{for all } (t, m) \in R \times M.$$

The problem is to find a curve $(g_t, k_t) \in \mathcal{M} \times S_2(M)$ which satisfies the evolution equations

$$\partial g_t / \partial t = N_t k_t - L_{X_t} g_t,$$
$$\partial k_t / \partial t = N_t S_{g_t}(k_t) - 2N_t \operatorname{Ric}(g_t) + 2 \operatorname{Hess}(N_t) - L_{X_t} k_t,$$

and which has initial conditions $(g_0, k_0) = (\mathring{g}, \mathring{k})$.

Our notation is the following:

$\delta k =$ divergence of $k = (\delta k)_i = -k^j_{i|j}$,
$\operatorname{Tr} k =$ Trace $k = g^{ij} k_{ij} = k^i_i$,
$k \cdot k =$ dot product for symmetric tensors $= k_{ij} k^{ij}$,
$k \times k =$ cross product for symmetric tensors $= k_{il} k^l_j$,
$S_g(k) = k \times k - \frac{1}{2}(\operatorname{Tr} k) k = k_{il} k^l_j - \frac{1}{2}(g^{mn} k_{mn}) k_{ij}$,
$\|X\|^2 =$ norm of $X = g_{ij} X^i X^j$,
$L_{X_t} g_t = \{$Lie derivative of g_t with respect to the time-dependent vector field $X_t\}$ $= X_{i|j} + X_{i|j}$ ($_{|j}$ = covariant derivative with respect to the time-dependent metric),
$L_{X_t} k_t =$ Lie derivative of $k_t = X^l k_{ij|l} + k_{il} X^l{}_{|j} + k_{jl} X^l{}_{|i}$,
$\operatorname{Ric}(g_t) =$ (Ricci curvature tensor formed from g_t) $= R_{ij} = \Gamma^k_{ij,k} - \Gamma^k_{ki,j} + \Gamma^k_{ij} \Gamma^l_{kl} - \Gamma^l_{ik} \Gamma^k_{lj}$,
$R(g_t) =$ scalar curvature $= R^k_k$,
$\operatorname{Hess}(N) =$ Hessian of $N =$ double covariant derivative $= N_{|i|j}$.

In the case that we choose $N_t = 1$ and $X_t = 0$, the proper configuration space for the Einstein system is the manifold $\mathcal{M}$. We equip $\mathcal{M}$ with a metric $\mathcal{G}$, referred to as the *DeWitt metric*, by setting for $g \in \mathcal{M}$,

$$\mathcal{G}_g : T_g\mathcal{M} \times T_g\mathcal{M} \approx S_2(M) \times S_2(M) \to \mathbf{R},$$

$$\mathcal{G}_g(h, k) = \int_M ((\operatorname{Tr} h)(\operatorname{Tr} k) - h \cdot k)\mu_g$$

where $\mu_g = (\det g)^{1/2}\, dx^1 \wedge dx^2 \wedge dx^3$ is the usual volume element.
The following is a straightforward computation (see [**12**]):

4.1. PROPOSITION. *The Lagrangian* $L_0(g, h) = \frac{1}{2}\,\mathcal{G}_g(h, h)$ *is nondegenerate and the associated Lagrangian vector field exists and is given by the second order system*

$$\text{(Z)} \qquad \begin{aligned} \partial g_t/\partial t &= k_t, \\ \partial k_t/\partial t &= k_t \times k_t - \tfrac{1}{2}(\operatorname{Tr} k_t)k_t - \tfrac{1}{8}(((\operatorname{Tr} k_t)^2 - k_t \cdot k_t)g). \end{aligned}$$

For each $(\mathring{g}, \mathring{k}) \in \mathcal{M} \times S_2(M)$ *there exists a unique smooth curve* $(g_t, k_t) \in \mathcal{M} \times S_2(M)$ *defined for short time with initial conditions* $(g_0, k_0) = (\mathring{g}, \mathring{k})$ *and which satisfies* (Z).

Now one adds a potential term to L_0; set

$$L(g, k) = \tfrac{1}{2}\,\mathcal{G}_g(k, k) - 2\int_M R(g)\mu_g$$

where $R(g)$ is the scalar curvature of g. Adding this potential term adds a gradient term to the equations of motion. For the potential $V = 2\int_M N R(g)\mu_g$ (where N is a positive scalar on M included for later use), a computation gives

$$-\operatorname{grad} V = -2N(\operatorname{Ric}(g) - \tfrac{1}{4}R(g)g) + 2\operatorname{Hess}(N),$$

where the gradient has been computed with respect to the DeWitt metric. Using the pointwise conservation law $\frac{1}{2}((\operatorname{Tr} k)^2 - k \cdot k) + 2R(g) = 0$ (see [**12**]), we have

4.2. PROPOSITION. $(g_t, k_t) \in \mathcal{M} \times S_2(M)$ *is an integral curve of the second order system determined by* $L = \frac{1}{2}\mathcal{G}_g(k, k) - 2\int_M R(g)\mu_g$ *iff*

$$\begin{aligned} \partial g_t/\partial t &= k_t, \\ \partial g_t/\partial t &= S_{g_t}(k_t) - 2\operatorname{Ric}(g_t). \end{aligned}$$

Eardley, Liang, and Sachs [**8**] have given conditions for which the velocity terms $S_g(k)$ dominate the $\operatorname{Ric}(g)$ term (for example near a singular hypersurface) so that the latter can with some justification be neglected. In this case the integral curves can be given explicitly, and are the geodesics of $\mathcal{M}$ with respect to the metric $\mathcal{G}$.

5. **The evolution equations with a shift vector field and space-body transitions.** Now suppose we consider the equations with an arbitrary shift vector field X_t. We assert that there is a simple method for solving these equations if the solution for $X = 0$ is known.

5.1. PROPOSITION. *Let* g_t, k_t *be a solution of the Einstein system with* $N = 1, X = 0$.

Then given X_t we construct its flow η_t. Then the solution of the Einstein system with $N = 1$, shift X_t, and the same initial conditions g_0, k_0 is given by

$$\bar{g}_t = (\eta_t^{-1})^* g_t, \qquad \bar{k}_t = (\eta_t^{-1})^* k_t.$$

PROOF. The extra terms involving the Lie derivatives are picked up as follows:

$$\begin{aligned}\partial \bar{k}_t/\partial t &= (\eta_t^{-1})^*(\partial k_t/\partial t) - L_{X_t}(\eta_t^{-1})^* k_t \\ &= (\eta_t^{-1})^*(S_{g_t}(k_t) - 2\,\mathrm{Ric}(g_t)) - L_{X_t}\bar{k}_t \\ &= S_{\bar{g}_t}(\bar{k}_t) - 2\,\mathrm{Ric}(\bar{g}_t) - L_{X_t}\bar{k}_t,\end{aligned}$$

where we have used the fact that $\partial(\eta_t^{-1})^* k/\partial t = -L_{X_t}(\eta_t^{-1})^* k$. ∎

Proposition 5.1 shows that even though the evolution equations with a shift involve extra terms which are nonlinear and involve derivatives, the more general system can be solved merely by solving an ordinary differential equation; namely, by finding the flow of X_t.

In order to take into account the shift vector field X_t we enlarge the configuration space $\mathscr{M}$ to $\mathscr{D} \times \mathscr{M}$. For $\eta \in \mathscr{D}$, it is easy to see that $T_\eta \mathscr{D}$ is the set of maps $X \circ \eta$ where X is a vector field on M. The Lagrangian of the preceding section is transferred to $\mathscr{D} \times \mathscr{M}$ by setting, for $(\eta, g) \in \mathscr{D} \times \mathscr{M}$,

$$\bar{L}: T_\eta \mathscr{D} \times T_g \mathscr{M} \to \mathbf{R},$$

$$(X \circ \eta, h) \to \tfrac{1}{2}\mathscr{G}_g(h + L_X g, h + L_X g) - 2\int_M R(g)\mu_g.$$

We observe that for $\lambda \in \mathbf{R}, \lambda \neq 0, \bar{L}(\lambda X \circ \eta, \lambda h) = \lambda^2 \bar{L}(X \circ \eta, h)$ so that $\bar{L}$ is quadratic in the velocities $(X \circ \eta, h)$. On $T\mathscr{M}$, of course, this is not true.

On $T(\mathscr{D} \times \mathscr{M})$, $\bar{L}$ is, roughly speaking, degenerate in the direction of $\mathscr{D}$. This degeneracy has the effect of introducing some ambiguity into the equations of motion, which is, however, precisely removed by the specification of a curve $\eta_t \in \mathscr{D}$. A direct computation proves the following:

5.2. PROPOSITION. *For any smooth curve η_t with generator the vector field $X_t = (d\eta_t/dt) \circ \eta_t^{-1}$, a possible Lagrangian vector field for the degenerate Lagrangian $\bar{L}(\eta, g; X \circ \eta, h) = L(g, h + L_X g)$ is given by the equations*

$$\begin{aligned}\partial g_t/\partial t &= k_t - L_{X_t} g_t, \\ \partial k_t/\partial t &= S_{g_t}(k_t) - 2\,\mathrm{Ric}(g_t) - L_{X_t} k_t.\end{aligned}$$

There is a natural action of the group $\mathscr{D}$ on $\mathscr{D} \times \mathscr{M}$ given by

$$\Phi_\eta: \mathscr{D} \times \mathscr{M} \to \mathscr{D} \times \mathscr{M}$$

$$(\zeta, g) \to (\eta \circ \zeta, (\eta^{-1})^* g).$$

This action leads to a natural symmetry and consequent conservation laws for our system.

5.3. PROPOSITION. *Let $\Phi_\eta : \mathscr{D} \times \mathscr{M} \to \mathscr{D} \times \mathscr{M}$ be as above with tangent action $T\Phi_\eta : T(\mathscr{D} \times \mathscr{M}) \to T(\mathscr{D} \times \mathscr{M})$. Then $\bar{L} \circ T\Phi_\eta = \bar{L}$ for each $n \in \mathscr{D}$ and $\delta(k - g(\mathrm{Tr}\, k)) \otimes \mu_g$ ($\otimes$ = tensor product) is a constant of the motion.*

We remark that the standard conservation theorems (cf. [**22**], [**23**]), used to prove this proposition, have to be modified to take into account the degeneracy of $\bar{L}$. The infinite dimensionality of the symmetry group leads to a differential rather than an integral identity; see [**12**] for details.

The Lie derivative terms that appear in 5.2 have a natural geometric interpretation related to changing from space to body coordinates in a manner similar to that of the rigid body and hydrodynamics (cf. Marsden-Abraham [**23**] and Ebin-Marsden [**9**]). More specifically we consider the manifold M to be the body, and the flow η_t of the shift vector field X_t as being a rotation of M. We then make the convention that an observer is in *body coordinates* if he is *on* the manifold, and is in *space coordinates* if he is *off* the manifold.

Now let g_t be a time-dependent metric field on M. We assume that the field is rigidly attached to M as it moves so that we set $g_t = g_{\mathrm{body}}$. An observer in body coordinates then finds $(\partial g_{\mathrm{body}}/\partial t) = k_{\mathrm{body}}$ as the "velocity" of the metric. An observer in space coordinates sees the metric field g_{body} as it is dragged past him by the moving manifold; he sees the metric field $g_{\mathrm{space}} = (\eta_t^{-1})^* g_{\mathrm{body}}$ and computes

(1) $\partial g_{\mathrm{space}}/\partial t = k_{\mathrm{space}} - L_X g_{\mathrm{space}}$,

(2) $\partial k_{\mathrm{space}}/\partial t = S_{g_{\mathrm{space}}}(k_{\mathrm{space}}) - 2\,\mathrm{Ric}(g_{\mathrm{space}}) - L_X k_{\mathrm{space}}$,

where $k_{\mathrm{space}} = (\eta_t^{-1})^* k_{\mathrm{body}}$. But (1) and (2) are just the evolution equations (with $N_t = 1$).

Finally, we remark that the Hessian term in the evolution equations can be accounted for by introducing the general relativistic time translation group $\mathscr{T} = C^\infty(M; \mathbf{R})$ and defining an extended Lagrangian on $\mathscr{T} \times \mathscr{D} \times \mathscr{M}$. The pointwise conservation of the Hamiltonian is closely related to the invariance of the Lagrangian under the action of the group of general relativistic time translations $\mathscr{T}$. Moreover, as with the shift vector field, there is associated with an arbitrary lapse function and a solution of the evolution equations with $N = 1$ and with given Cauchy data, a proper time function τ and an intrinsic shift vector field Y. From the integration of Y, together with τ, we determine from the solution for $N = 1$ the solution with the arbitrary lapse function and with the same Cauchy data. See [**12**] for a detailed analysis of these topics.

BIBLIOGRAPHY

1. R. Arnowitt, S. Deser and C. W. Misner, *The dynamics of general relativity*, Gravitation: An Introduction to Current Research, Wiley, New York, 1962. MR **26** #1182.

2. Y. Fourès-Bruhat, *Théorème d'existence pour certains systèmes d'équations aux dérivées partielles non linéaires*, Acta Math. **88** (1952), 141–225. MR **14**, 756.

3. Y. Choquet-Bruhat, *Espaces-temps einsteiniens généraux chocs gravitationnels*, Ann. Inst. H. Poincaré Sect. A **8** (1968), 327–338. MR **38** #1897.

4. ———, *Solutions C^∞ d'équations hyperboliques non linéaires*, C. R. Acad. Sci. Paris **272** (1971), 386–388.

5. R. Courant and D. Hilbert, *Methods of mathematical physics*. Vol. II: *Partial differential equations* (Vol. II by R. Courant), Interscience, New York, 1962. MR **25** #4216.

6. B. DeWitt, *Quantum theory of gravity*. I. *The canonical theory*, Phys. Rev. **160** (1967), 1113–1148.

7. P. A. Dionne, *Sur les problèmes de Cauchy hyperboliques bien posés*, J. Analyse Math. **10** (1962/63), 1–90. MR **27** #472.

8. D. Eardley, E. Liang and R. Sachs, *Velocity dominated singularities in irrotational dust cosmologies*, J. Mathematical Phys. **13** (1972), 99–107.

9. D. G. Ebin and J. Marsden, *Groups of diffeomorphisms and the motion of an incompressible fluid*, Ann. of Math. (2) **92** (1970), 102–163. MR **42** #6865.

10. A. Einstein, *Näherungsweise Integration der Feldgleichungen der Gravitation*, S.-B. Preuss. Akad. Wiss. **1916**, 688–696.

11. A. Fischer, "The theory of superspace," in *Relativity*, M. Carmeli, S. Fickler and L. Witten (Editors), Plenum Press, New York, 1970, 303–357.

12. A. Fischer and J. Marsden, *The Einstein equations of evolution—a geometric approach*, J. Mathematical Phys. **13** (1972), 546–568.

13. ———, *General relativity as a dynamical system on the manifold $\mathscr{A}$ of a riemannian metrics which cover diffeomorphisms*, in *Methods of local and global differential geometry in general relativity*, Lecture Notes in Physics, vol. 14, Springer-Verlag, Berlin, 1972, 176–188.

14. ———, *The Einstein evolution equations as a first-order quasi-linear symmetric hyperbolic system*. I, Comm. Math. Phys. **28** (1972), 1–38.

15. ———, *First -order symmetric hyperbolic systems on non-compact manifolds and general relativity* (to appear).

16. F. Frankl, *Über das Anfangswertproblem für lineare und nichtlineare hyperbolische partielle Differentialgleichungen zweiter Ordnung*, Mat. Sb. **2 (44)** (1937), 814–868.

17. K. O. Friedrichs, *Symmetric hyperbolic linear differential equations*, Comm. Pure Appl. Math. **7** (1954), 345–392. MR **16**, 44.

18. C. Lanczos, *Ein vereinfachendes Koordinatensystem für die Einsteinschen Gravitationsgleichungen*, Phys. Z. **23** (1922), 537–539.

19. P. Lax, *On Cauchy's problem for hyperbolic equations and the differentiability of solutions of elliptic equations*, Comm. Pure Appl. Math. **8** (1955), 615–633. MR **17**, 1212.

20. J. Leray, *Lectures on hyperbolic equations with variable coefficients*, Institute for Advanced Studies, Princeton, N.J., 1952.

21. A. Lichnerowicz, *Relativistic hydrodynamics and magnetohydrodynamics*, Benjamin, New York, 1967.

22. J. E. Marsden, *Hamiltonian one parameter groups*: *A mathematical exposition of infinite dimensional Hamiltonian systems with applications in classical and quantum mechanics*, Arch. Rational Mech. Anal. **28** (1967/68), 362–396. MR **37** #1735.

23. J. Marsden and R. Abraham, *Hamiltonian mechanics on Lie groups and hydrodynamics*, Proc. Sympos. Pure Math., vol. 16, Amer. Math. Soc., Providence, R.I., 1968, pp. 237–244.

24. I. Petrovskiĭ, *Über das Cauchysche Problem für lineare and nichtlineare hyperbolische partielle Differentialgleichunges*, Mat. Sb. **2 (44)** (1937), 814–868.

25. J. Schauder, *Das Anfangswertproblem einer quasilinearen hyperbolischen Differentialgleichungen weiter Ordnung in beliebiger Anzahl von unabhangigen Veränderlichen*, Fund. Math. **24** (1935), 213–246.

26. S. S. Sobolev, *Méthode nouvelle à résourdre le problème de Cauchy pour les équations linéaires*

hyperboliques normales, Mat. Sb. **1** (**43**) (1936), 39–72.

27. ———, *Sur la théorie des équations hyperboliques aux dérivées partielles*, Mat. Sb. **5** (**47**) (1939), 71–99. (Russian) MR **1**, 237.

28. ———, *Applications of functional analysis in mathematical physics*, Izdat. Leningrad. Gos. Univ., Leningrad, 1950; English transl., Transl. Math. Monographs, vol. 7, Amer. Math. Soc., Providence, R.I., 1963. MR **14**, 565; MR **29** #2624.

29. J. A. Wheeler, *Geometrodynamics and the issue of the final state*, Relativité, Groupes et Topologie (Lectures, Les Houches, 1963 Summer School of Theoret. Phys., Univ. Grenoble), Gordon and Breach, New York, 1964, pp. 315–520. MR **29** #5596.

30. K. Yosida, *Functional analysis*, Die Grundlehren der math. Wissenschaften, Band 123, Academic Press, New York; Springer-Verlag, Berlin, 1965. MR **31** #5054.

UNIVERSITY OF CALIFORNIA, SANTA CRUZ AND BERKELEY

ELLIPTIC EQUATIONS ON MINIMAL SURFACES

ENRICO GIUSTI

1. A main motivation for the study of elliptic equations on manifolds is that information on solutions of elliptic equations on a manifold M can be translated into information on the topological structure of M itself.

When, in addition, the manifold in question is a minimal hypersurface (of codimension one) a further motivation is the search for *a priori* estimates independent of the particular minimal surface.

An estimate of this kind has been established by Bombieri, De Giorgi and Miranda [**2**]:

$$|Du(0)| \leqq c_1 \exp\{c_2(1 + u(0)/R)\} \tag{1}$$

for a nonparametric minimal hypersurface

$$M = \{(x, y) \in \boldsymbol{R}^n \times \boldsymbol{R};\ |x| < R,\ y = u(x)\},$$

where $u(x)$ is a positive function in the ball $\{|x| < R\}$. The estimate (1) does not depend on M in the sense that the constants c_1 and c_2 appearing in it depend only on the dimension n.

In the following I want to discuss some results, obtained by Bombieri and myself [**3**], regarding solutions and supersolutions of elliptic equations on minimal hypersurfaces. For the solutions we obtain a Harnack inequality which, in particular, implies that every positive harmonic function on a minimal hypersurface in $\boldsymbol{R}^{n+1}$ is constant; for supersolutions we get weaker estimates which can be applied to the problem of the behavior of complete minimal graphs.

2. We will describe briefly the De Giorgi formalism for minimal hypersurfaces.

We shall consider open sets $A \subset \boldsymbol{R}^{n+1}$ with the following property: A contains every point $x_0 \in \boldsymbol{R}^{n+1}$ such that

AMS 1970 *subject classifications*. Primary 49F10; Secondary 35B45, 35J20.

$$\operatorname{meas}\{B(x_0, r) \cap (\boldsymbol{R}^{n+1} \backslash A)\} = 0 \quad \text{for some } r > 0.$$

$[B(x_0, r) = \{x \in \boldsymbol{R}^{n+1};\ |x - x_0| < r\}.]$

DEFINITION 1. Let A be an open set in $\boldsymbol{R}^{n+1}$ as before and let ϕ_A be the characteristic function of A. We say that the boundary, ∂A, of A is an (oriented) hypersurface in Ω, Ω being an open set, if the derivatives $D_i\phi_A$ are measures in Ω (i.e., if $\phi_A \in BV_{\text{loc}}(\Omega)$). If ∂A is a hypersurface, let $|D\phi_A|$ be the total variation of the vector valued measure $D\phi_A = (D_1\phi_A, \ldots, D_{n+1}\phi_A)$. Let $v(x)$ be the (vector) density of $D\phi_A$ with respect to $|D\phi_A|$ at the point x:

$$v(x) = \lim_{r \to 0^+} D\phi_A(B(x, r))/|D\phi_A|(B(x, r))$$

The limit at the right-hand side exists $|D\phi_A|$-almost everywhere and we have $|v(x)| = 1$ $|D\phi_A|$-a.e. With the aid of v we define differential operators

$$\delta_i = D_i - v_i \sum_{h=1}^{n+1} v_h D_h \qquad (i = 1, \ldots, n + 1).$$

If the boundary of A is a hypersurface of class C^1 in $B(x_0, r)$, then $|D\phi_A|(B(x_0, r))$ is the area of the portion of ∂A lying in $B(x_0, r)$ and $v(x)$ coincides with the inner normal vector so that the operators δ_i are tangential derivatives. In particular if $x(w)$ $(w = (w_1, \ldots, w_n))$ is a parametric representation of $\partial A \cap B(x_0, r)$ then it is easily seen that

$$\delta_i = \sum_{\alpha,\beta=1}^{n} \frac{\partial x_i}{\partial w_\alpha} g^{\alpha\beta} \frac{\partial}{\partial w_\beta}$$

($g^{\alpha\beta}$ is the inverse matrix of $g_{\alpha\beta} = \sum_{i=1}^{n+1} (\partial x_i/\partial w_\alpha)(\partial x_i/\partial w_\beta)$) so that the second order operator $\sum_{i=1}^{n+1} \delta_i\delta_i$ is the Laplace operator on ∂A.

For those who are familiar with the formalism of Federer and Fleming [**6**], [**8**] we include a dictionary. The correspondence becomes clear if one considers that, once the coordinates and the metric are chosen in $\boldsymbol{R}^{n+1}$, there exists a canonical isomorphism between $(n + 1)$-forms and functions (and between n-forms and vectors).

Federer and Fleming	*De Giorgi*
$(n + 1)$-current	distribution
n-current	vector valued distribution
$T \in \boldsymbol{N}_{n+1}^{\text{loc}}$	$g \in BV_{\text{loc}}$
$M_K(T)$	$\int_K g(x)\,dx$
∂T	$-Dg$
$M_K(\partial T)$	$\|Dg\|(K)$
$\|\|T\|\|$	$g(x)\,dx$
$\|\|\partial T\|\|$	$\|Dg\|$
$\boldsymbol{\partial T}(z)$	$-v(z)$
$\langle \partial T, \omega\rangle = \langle T, d\omega\rangle$	$-\int\langle \boldsymbol{a}, v\rangle\|Dg\| = \int g \operatorname{div} \boldsymbol{a}\,dx$
$T = \boldsymbol{E}^{n+1} \llcorner A$	ϕ_A

DEFINITION 2. Let Ω be an open set in $\boldsymbol{R}^{n+1}$. We say that ∂A is a minimal hypersurface in Ω if for every $g \in BV(\Omega)$ with compact support $K \subset \Omega$ we have

$$\int_K |D\phi_A| \leqq \int_K |D\phi_A + Dg|. \tag{2}$$

It is known (De Giorgi [**4**], Reifenberg [**14**], Federer [7]) that if ∂A is a minimal hypersurface in Ω, then $\partial A \cap \Omega$ is a real analytic hypersurface, except possibly for a closed set N whose Hausdorff dimension does not exceed $n - 7$.

In addition, from the minimality of ∂A in Ω it follows that

$$\int \delta_i \psi |D\phi_A| = 0 \tag{3}$$

for every $\psi \in C_0^1(\Omega)$ $(i = 1, \ldots, n + 1)$, so that integration by parts is allowed on ∂A.

In particular we can take $\psi = v_i \eta$ (if $\partial A \cap \operatorname{spt} \eta$ is a regular surface) and sum over i. Considering that from the definition of δ_i, it follows that $\sum_{i=1}^{n+1} v_i \delta_i = 0$, we get easily

$$\sum_{i=1}^{n+1} \delta_i v_i = 0$$

whenever the sum at the left-hand side (which gives the sum of principal curvatures of ∂A) exists.

Equation (3) enables us to consider (weak) solutions and supersolutions of elliptic partial differential operators on $\Sigma = \partial A \cap \Omega$ in divergence form; by a solution of the equation

$$Eu = \sum_{i,j=1}^{n+1} \delta_i(a_{ij}(x)\delta_j u) = 0 \tag{4}$$

where the a_{ij} are bounded and $|D\phi_A|$ measurable functions, we mean a function $u(x) \in W^{1,2}_{\mathrm{loc}}(\Sigma)$ such that for every ψ with compact support in Ω we have

$$\sum_{i,j=1}^{n+1} \int a_{ij}(x) \delta_j u \delta_i \psi |D\phi_A| = 0. \tag{5}$$

Equation (4) is elliptic in the sense that there exists a constant L such that

$$|\xi|^2 \leqq \sum_{i,j=1}^{n+1} a_{ij}(x)\xi_i\xi_j \leqq L|\xi|^2$$

for every $x \in \Sigma$ and $\xi \in \boldsymbol{R}^{n+1}$. We suppose also that $a_{ij} = a_{ji}$.

In a similar way we call a supersolution (of the operator E) a function $u(x)$ such that

$$\sum_{i,j=1}^{n+1} \int a_{ij}(x)\delta_j u \delta_i \psi |D\phi_A| \geqq 0 \tag{6}$$

for every $\psi \in C_0^1(\Omega)$, $\psi \geqq 0$.

3. In the following we will put $\Sigma_R = \partial A \cap B_R$, where B_R is the open $(n+1)$-ball of radius R. We have

THEOREM 1. *Let ∂A be a minimal hypersurface in B_R and let u be a positive solution of the equation* (4) *on Σ_R. Then we have*

$$\sup_{\Sigma_r} u(x) \leqq \exp(c_3 L^{1/2}) \inf_{\Sigma_r} u(x) \tag{7}$$

for $r \leqq \beta' R$, where β' and c_3 depend only on the dimension $n+1$.

REMARKS. If Σ lies on the hyperplane $x_{n+1} = 0$, we have $\delta_\alpha = D_\alpha$ $(\alpha = 1, \ldots, n)$ and $\delta_{n+1} = 0$ so that equation (4) reduces to a usual elliptic partial differential equation. If in addition Σ is connected, one may deduce easily the classical Harnack inequality (see Moser [**12**]):

$$\sup_{\Sigma'} u(x) \leqq c_4(\Sigma, \Sigma')^{L^{1/2}} \inf u(x) \tag{8}$$

for $\Sigma' \subset\subset \Sigma$. If Σ is convex, one can take

$$c_4(\Sigma, \Sigma') = \left\{ \frac{\operatorname{diam} \Sigma}{\operatorname{dist}(\Sigma', \partial\Sigma)} \right\}^{c_5}$$

with c_5 depending only on n.

The example $\partial^2 u/\partial x^2 + L\partial^2 u/\partial y^2 = 0$ with solution $u(x, y) = \cos y \exp(L^{1/2}x)$ shows that the dependence on L in (8) cannot be improved.

A similar result has been recently obtained by Moser [**13**] for parabolic equations in $\boldsymbol{R}^n$.

A simple consequence of Theorem 1 is the following:

THEOREM 2. *If ∂A is a minimal hypersurface in $\boldsymbol{R}^{n+1}$, then every solution of equation* (4) *which is bounded on one side is constant. In particular, every positive harmonic function on ∂A is constant.*

REMARK. From Theorem 2 it follows at once that a minimal hypersurface in $\boldsymbol{R}^{n+1}$ is connected. This result is due to Almgren and De Giorgi (unpublished) and is a prerequisite more than a consequence of Theorem 2, since it is needed in the proof of Theorem 1.

The following results are corollaries of Theorem 2.

COROLLARY 1 (MIRANDA [**11**]). *A minimal hypersurface in $\boldsymbol{R}^{n+1}$ which is contained in a half-space is a hyperplane.*

PROOF. This follows from the fact that the coordinate functions are harmonic on ∂A; i.e., $\sum_{i=1}^{n+1} \delta_i \delta_i x_s = 0$ $(s = 1, \ldots, n+1)$.

If x_0 is a regular point of ∂A one can represent $\Sigma_r(x_0)$ as the zero set of a real analytic function $\phi(x)$, with $|D\phi| \neq 0$ on Σ_r. Then one has, changing the sign of ϕ if necessary,

$$\nu(x) = D\phi/|D\phi|$$

and the following relations are easily proved:

$$\delta_i \nu_j = \delta_j \nu_i \qquad (i, j = 1, \ldots, n+1), \tag{9}$$

$$\sum_{i=1}^{n+1} \delta_i \delta_i \nu_h + c^2 \nu_h = 0 \qquad (h = 1, \ldots, n+1), \tag{10}$$

where $c^2 = \sum_{i,h=1}^{n+1} (\delta_i \nu_h)^2$ is the sum of the squares of the principal curvatures of ∂A at x.

COROLLARY 2 (BERNSTEIN'S THEOREM). *A complete minimal graph over $\boldsymbol{R}^2$ is a plane.*

A complete minimal graph over $\boldsymbol{R}^n$ is a minimal hypersurface ∂A, where

$$A = \{(x, y) \in \boldsymbol{R}^n \times \boldsymbol{R};\ y < u(x)\}$$

and u is a function in BV_{loc}. We note that if ∂A is a minimal graph then the function u defining A is in fact real analytic (see e.g. Giusti [**10**]). For a minimal graph we have

$$\nu_i = -D_i u/(1 + |Du|^2)^{1/2} \qquad (i = 1, \ldots, n),$$

$$\nu_{n+1} = (1 + |Du|^2)^{-1/2}$$

so that $\nu_{n+1} > 0$.

PROOF OF COROLLARY 2. As remarked originally by Bernstein, the functions

$$g_h = \operatorname{arctg} \nu_h/\nu_3 \qquad (h = 1, 2)$$

are harmonic on ∂A (this is easily seen using (9), (10) and $\sum_{i=1}^{n+1} \delta_i \nu_i = 0$) and bounded. From Theorem 2 it follows that g_h, and hence $D_h u = -\nu_h/\nu_3$, are constant.

The next result is a generalization of a theorem of Moser [**12**].

COROLLARY 3. *Let ∂A be a complete minimal graph over $\boldsymbol{R}^n$ and suppose that the function u which defines A has all the partial derivatives bounded in $\boldsymbol{R}^n$, with the possible exception of one of them. Then u is linear.*

REMARK. The above result improves Moser's which assumes that all the derivatives of u are bounded.

We note that using the methods of Fleming [**9**] and De Giorgi [**5**] and the result of Simons [**15**] one can prove that u is linear assuming only the boundedness of $n - 7$ derivatives.

Starting from the example given in [**1**] it is not difficult to show that the boundedness of $n - 8$ partial derivatives is not sufficient for getting the same conclusion.

We actually prove that the conclusion of Corollary 3 holds if

$$|Du| \leqq K\left(1 + \left|\sum_{h=1}^{n+1} a_h D_h u\right|\right) \tag{11}$$

for some constants K, a_h and for every $x \in \boldsymbol{R}^n$.

PROOF OF COROLLARY 3. The function $g = \operatorname{arctg}(\langle a, v\rangle / v_{n+1})$ is a bounded solution of the equation

$$\sum_{i=1}^{n+1} \delta_i\{[\langle a, v\rangle^2 + v_{n+1}^2]\delta_i g\} = 0$$

so that if

$$\langle a, v\rangle^2 + v_{n+1}^2 \geqq A > 0 \tag{12}$$

(this condition is equivalent to (11)), the equation is elliptic and the function g is constant.

Hence $\langle a, v\rangle = Bv_{n+1}$ and, by (12), $(1 + B^2)v_{n+1}^2 \geqq A$, whence

$$1 + |Du|^2 \leqq (1 + B^2)/A$$

and by Moser's theorem u is linear.

4. A result similar to Theorem 1 is valid for supersolutions.

THEOREM 3. *Let ∂A be a minimal hypersurface in B_R and let $u(x)$ be a positive supersolution of the operator*

$$E = \sum_{i,j=1}^{n+1} \delta_i(a_{ij}(x)\delta_j).$$

Then we have, for $0 < p < n/(n-2)$,

$$\left\{\frac{1}{|D\phi_A|(B_r)} \int_{B_r} u^p |D\phi_A|\right\}^{1/p} \leqq c_6(p)^{L^{1/2}} \inf_{\Sigma_r} u(x)$$

for $r < \beta' R$; c_6 and β' being absolute constants.

As for the case of solutions this result is known in $\boldsymbol{R}^n$ (Trudinger [**16**]) except for the dependence on the ellipticity constant L.

Theorem 3 has an interesting application to minimal graphs.

THEOREM 4. *If ∂A is a complete minimal graph we have, for every $r > 0$,*

$$\frac{1}{|D\phi_A|(B_r)} \int_{B_r} v_{n+1} |D\phi_A| \leqq c_7 \inf_{\Sigma_r} v_{n+1}. \tag{13}$$

In fact $v_{n+1} > 0$ *satisfies*

$$\sum_{i=1}^{n+1} \delta_i \delta_i v_{n+1} = -c^2 v_{n+1} \leqq 0.$$

If we call Π_R the projection of Σ_R on $\boldsymbol{R}^n$ and remember that

$$|D\phi_A|(B_R) = \int_{\Pi_R} (1 + |Du|^2)^{1/2}\,dx, \qquad \int_{B_R} v_{n+1}|D\phi_A| = \text{meas}\,\Pi_R,$$

we obtain the estimate

$$\sup_{\Pi_R}(1 + |Du|^2)^{1/2} \leqq c_8(\text{meas}\,\Pi_R)^{-1} \int_{\Pi_R} (1 + |Du|^2)^{1/2}\,dx \tag{14}$$

which is equivalent to

$$\sup_{\Pi_R}(1 + |Du|^2)^{1/2} \leqq c_9 R^n (\text{meas}\,\Pi_R)^{-1} \tag{15}$$

because

$$\omega_n R^n \leqq |D\phi_A|(B_R) \leqq \tfrac{1}{2}(n+1)\omega_{n+1} R^n.$$

COROLLARY 4. *Let* ∂A *be a complete minimal graph. For every n-ball* B_r *of radius r we have*

$$\sup_{B_r} |Du| \leqq c_{10}\left(1 + (1/r)\sup_{B_r} |u|\right)^n. \tag{16}$$

PROOF. Let $R^2 = r^2 + \sup_{B_r} u^2$. Then we have $B_r \subset \Pi_R$ so that meas $\Pi_R \geqq \omega_n r^n$ and (16) follows from (15).

REMARK. The estimates (14), (15) and (16) are related with the growth at infinity of complete minimal graphs. In fact a bound for the gradient

$$\sup_{B_r} |Du| \leqq c_{11}\left(1 + (1/r)\sup_{B_r} |u|\right) \tag{17}$$

implies the polynomial growth

$$|u(x)| \leqq K(1 + |x|)^{c_{12}}$$

for some constants K, c_{12}.

The proof of Corollary 4 makes use of the weak lower estimate

$$\text{meas}\,\Pi_R \geqq c_{13} r^n$$

while one would like an estimate

$$\text{meas}\,\Pi_R \geqq c_{14} R^{n-1} r.$$

However it does not seem easy to settle this point, so that the polynomial growth of complete minimal graphs remains to be proved.

REFERENCES

1. E. Bombieri, E. De Giorgi and E. Giusti, *Minimal cones and the Bernstein problem*, Invent. Math. **7** (1969), 243–268. MR **40** #3445.

2. E. Bombieri, E. De Giorgi and M. Miranda, *Una maggiorazione a priori relativa alle ipersuperfici minimali non parametriche*, Arch. Rational Mech. Anal. **32** (1969), 255–267. MR **40** #1898.

3. E. Bombieri and E. Giusti, *Harnack's inequality for elliptic differential equations on minimal surfaces*, Invent. Math. **15** (1972), 24–46.

4. E. De Giorgi, *Frontiere orientate di misura minima*, Seminario di Matematica della Scuola Normale Superiore de Pisa, 1960/61, Editrice Tecnico Scientifica, Pisa, 1961. MR **31** #3897.

5. ———, *Una estensione del teorema di Bernstein*, Ann. Scuola Norm. Sup. Pisa (3) **19** (1965), 79–85. MR **31** #2643.

6. H. Federer, *Geometric measure theory*, Die Grundlehren der math. Wissenschaften, Band 153, Springer-Verlag, New York, 1969. MR **41** #1976.

7. ———, *The singular set of area minimizing rectifiable currents with codimension one and of area minimizing flat chains modulo two with arbitrary codimension*, Bull. Amer. Math. Soc. **76** (1970), 767–771. MR **41** #5601.

8. H. Federer and W. H. Fleming, *Normal and integral currents*, Ann. of Math. (2) **72** (1960), 458–520. MR **23** #A588.

9. W. H. Fleming, *On the oriented plateau problem*, Rend. Circ. Mat. Palermo (2) **11** (1962), 69–90. MR **28** #499.

10. E. Giusti, *Superfici cartesiane di area minima*, Rend. Sem. Mat. Fis. Milano **40** (1970), 3–31.

11. M. Miranda, *Frontiere minimali con ostacoli*, Ann. Univ. Ferrara **16** (1971), 29–37.

12. J. Moser, *On Harnack's theorem for elliptic differential equations*, Comm. Pure Appl. Math. **14** (1961), 577–591. MR **28** #2356.

13. ———, *On a pointwise estimate for parabolic differential equations*, Comm. Pure Appl. Math. **24** (1971), 727–740.

14. E. R. Reifenberg, *Solution of the plateau problem for m-dimensional surfaces of varying topological type*, Acta Math. **104** (1960), 1–92. MR **22** #4972.

15. J. Simons, *Minimal varieties in riemannian manifolds*, Ann. of Math. (2) **88** (1968), 62–105. MR **38** #1617.

16. N. S. Trudinger, *On Harnack type inequalities and their application to quasilinear elliptic equations*, Comm. Pure Appl. Math. **20** (1967), 721–747. MR **37** #1788.

UNIVERSITY OF PISA, ITALY

UNIVERSITY OF CALIFORNIA, BERKELEY

JUSTIFICATION OF MATCHED ASYMPTOTIC EXPANSION SOLUTIONS FOR SOME SINGULAR PERTURBATION PROBLEMS

FRANK HOPPENSTEADT[1]

Problems involving differential equations with small parameters appearing in the coefficients have some interesting features. Often the coefficients depend on a parameter in a regular way while the solution of the problem is quite singular in its dependence. Such problems are frequently called singular perturbation problems.

Singular perturbation problems arise naturally in descriptions of various physical phenomena and in studies into the structure of solutions of differential equations. In many investigations, matching or boundary layer methods have been used to formally construct solutions. These utilize the dependence of solutions on several variables obtained from the original variables of the problem through scale changes. Certain features common to several investigations of singular perturbation problems are described here. In the course of this, an approach is outlined which justifies the use of matched asymptotic expansions.

A point of view similar to the one illustrated here was taken in the article [**1**] to study initial and boundary value problems for systems of ordinary differential equations with several small parameters. Extensions of our results to similarly general problems involving systems of partial differential equations, while not discussed in this article, can be carried out through straightforward modifications of the methods developed in [**1**].

The general method is described in the example of §1. This is treated in a formal and imprecise way, and it is used only to illustrate the approach. Later specific applications are made to some problems involving partial differential equations.

1. **Example.** Consider the initial value problem

AMS 1970 *subject classifications*. Primary 35B25; Secondary 35F20.

[1] This research was supported in part by the Army Research Office at Durham under Grant no. DA-ARO-D-31-124-71-G108.

(P_ε) $$\varepsilon(du/dt) = F(t,u,\varepsilon), \qquad u|_{t=0} = \mathring{u}(\varepsilon),$$

where u takes values in some Banach space, say E, $\varepsilon > 0$ is a small parameter, t is a real variable and F is some mapping (possibly unbounded) acting in E. The data F and $\mathring{u}$ are required to satisfy some additional smoothness conditions which are specified in the particular applications. Typically, these involve smooth dependence on t,ε and some condition ensuring local existence of solutions for each ε.

Formally setting $\varepsilon = 0$ in (P_ε) gives

$$0 = F(t,u,0), \qquad u|_{t=0} = \mathring{u}(0),$$

a problem which is usually overdetermined. We elect to discard the initial condition and define the reduced problem to be

(P_0) $$F(t,u,0) = 0$$

in which t occurs as a parameter. We assume

(A) *There is a smooth (as a function of t) solution $u = u_0(t)$ of (P_0) on some interval $0 \leqq t \leqq T$.*

The solution of the full problem (P_ε) is to be studied relative to the solution of the reduced problem (P_0). This is accomplished by replacing (P_ε) with two auxiliary problems whose solutions depend regularly on ε.

The first of these auxiliary problems, called the *outer problem*, is formulated as follows: Find a solution $u = u^*$ of the equation

$$\varepsilon(du^*/dt) = F(t,u^*,\varepsilon)$$

which is a smooth function of ε at $\varepsilon = 0$ for $0 \leqq t \leqq T$ such that $u^*(t,0) \equiv u_0(t)$ for $0 \leqq t \leqq T$.

This problem, denoted by (O_ε), is set with the goal of determining the part of the solution of (P_ε) which depends smoothly on t and ε.

If (O_ε) has a solution, Taylor's theorem may be applied with the result

$$u^*(t,\varepsilon) = u_0(t) + \sum_{r=1}^{N} u_r^*(t)\varepsilon^r + O(\varepsilon^{N+1})$$

where N is some number related to the degree of smoothness required of the data. A formal calculation shows the coefficients of this Taylor expansion satisfy equations of the form

$$F_u(t,u_0(t),0)u_r^* = P_r(t,u_0,\ldots,u_{r-1}^*)$$

for $r = 1, 2, \ldots$. Here $F_u(t,u_0(t),0)w$ denotes the linear approximation to $F(t,u_0(t) + w,0)$ for w with small norm. This is usually some derivative of $F(t,u,0)$.

We assume

(B) *The linear operator $F_u(t,u_0(t),0)$ is invertible.*

If (B) holds, then the functions u_r can be determined successively. This gives some evidence that a solution, $u^*(t,\varepsilon)$, of (O_ε) can be found.

It is desirable to have $O(\varepsilon^{N+1})$ in the expansion of $u^*(t,\varepsilon)$ hold uniformly for $0 \leqq t \leqq T$. If T is finite, this is usually not a problem. However, if $T = \infty$, the form of the expansion may be inadequate. Either suitable stability conditions must be imposed or a different technique for expanding the solution must be used; e.g., a multi-time method.

Suppose now that (O_ε) has been solved. New variables are introduced in (P_ε) by $U = u - u^*$, $\tau = t/\varepsilon$. The result is the second auxiliary problem

$$(\mathrm{S}_\varepsilon) \qquad (dU/d\tau) = \tilde{F}(\varepsilon\tau,U,\varepsilon), \qquad U|_{\tau=0} = \mathring{u}(\varepsilon) - u^*(0,\varepsilon),$$

where $\tilde{F}(\varepsilon\tau,U,\varepsilon) = F(\varepsilon\tau,U + u^*,\varepsilon) - F(\varepsilon\tau,u^*,\varepsilon)$.

The problem (S_ε) appears to be a regular perturbation problem, but it must be considered on the growing intervals $0 \leqq \tau \leq T/\varepsilon$. A straightforward scheme leads to an expansion of the form

$$U = \sum_{r=0}^{N} U_r(\tau)\varepsilon^r + O(\varepsilon^{N+1}).$$

Some sort of stability condition is usually imposed to ensure that $O(\varepsilon^{N+1})$ holds uniformly for $0 \leqq \tau \leqq T/\varepsilon$. This is often done by specifying "matching conditions" of the form $\lim_{\tau\to\infty} U_r(\tau) = 0$ for $r = 0,1,2,\ldots$. On the other hand, these conditions can be guaranteed through a restriction on the spectrum of the operator $F_u(0,u_0(0),0)$, which appears in the linear part of the equation used to determine the functions U_r.

We assume

(C) *The spectrum of* $F_u(t,u_0(t),0)$ *lies in the left half*, $\mathrm{Re}\,\lambda < 0$.

Under conditions like (A), (B), (C), and with sufficient smoothness of the data, we expect (P_ε) to have a solution $u(t,\varepsilon)$ which may be written as

$$u(t,\varepsilon) = u^*(t,\varepsilon) + U(\tau,\varepsilon)$$

where u^* is a smooth function of (t,ε) and U is a smooth function of (τ,ε) which decays as $\tau \to \infty$. In various applications of this method, u^* is called the outer solution of (P_ε) and U is called the boundary layer solution of (P_ε). In fact, U is significant only for t/ε near zero; i.e., for t in some $o(1)$ neighborhood of the initial time, $t = 0$. Thus, the solution of (P_ε) is described by u^* outside a boundary (or initial) layer.

In many singular perturbation problems arising in applications either condition (B) or (C) fails to be satisfied. For example, $F_u(t,u_0(t),0)$ may be invertible on some interval $0 \leqq t < t_1$, but not at $t = t_1$. This is similar to turning point problems in the theory of linear ordinary differential equations. Examples of this arise in various forms in problems involving partial differential equations; e.g., Eckhaus

[2], Višik-Ljusternik [3] where (P_ε) is a linear elliptic system and Keller-Kogelman [4] where (P_ε) is a nonlinear hyperbolic system.

On the other hand, the null space of $F_u(t,u_0(t),0)$ may have fixed positive dimension over $[0,T]$. Certain cases of this were studied by Trenogin [5]. There the solution of (O_ε) proceeds in a way similar to degenerate forms of the implicit function theorem.

In [4], the stability condition (C) fails and solutions are constructed by a two-time procedure. In [5], (C) is replaced by conditions on the nonlinearity in (P_ε).

2. **Application to a quasilinear parabolic problem.** In [6], the method of §1 was applied to the following initial boundary value problem. Let Ω be a smooth, bounded domain in E^n. Consider the problem

$$\varepsilon(\partial u/\partial t) - \sum_{i,j=1}^{n} a_{ij}(x,t,u,\nabla u,\varepsilon)(\partial^2 u/\partial x_i \partial x_j) = f(x,t,u,\nabla u,\varepsilon),$$

$$u|_{\partial\Omega} = 0 \quad \text{for } 0 \leqq t \leqq T, \qquad u|_{t=0} = \mathring{u}(x,\varepsilon) \quad \text{for } x \in \Omega,$$

where ∇u denotes the gradient of u and the matrix (a_{ij}) is positive definite uniformly in its arguments. Conditions are given in [6] (similar to (A), (B), (C)) under which the program outlined in §1 can be executed. The result is that this problem has a solution, $u(x,t,\varepsilon)$, on $\overline{\Omega} \times [0,T]$ for each small $\varepsilon > 0$, and there is a corresponding decomposition of u:

$$u(x,t,\varepsilon) = u^*(x,t,\varepsilon) + U(x,\tau,\varepsilon)$$

where u^* and U are smooth functions of ε at $\varepsilon = 0$. Moreover, the order relations $O(\varepsilon^{N+1})$ occurring in the Taylor expansions of u^* and U are shown to hold in the $C^1(\Omega)$-norm uniformly in t. The function U was shown to decay exponentially as $\tau \to \infty$.

3. **Application to an abstract parabolic problem.** Consider the initial value problem

$$\varepsilon(du/dt) + A(t,\varepsilon)u = f(t,u,\varepsilon), \qquad u|_{t=0} = \mathring{u}(\varepsilon).$$

Here $u \in E$, a Banach space, and f has the form

$$f = \sum_{i,j,k=1}^{\infty} \varepsilon^i t^j f_{ijk} u^k$$

where f_{ijk} is a k-linear bounded operator acting in E. The operator $A(t,\varepsilon)$ is a closed, linear operator with domain of definition D everywhere dense in E and independent of (t,ε) for $0 \leqq t \leqq T$, $0 \leqq \varepsilon \leqq \varepsilon_0$.

This equation is said to be of parabolic type if for each (t,ε), $-A(t,\varepsilon)$ is the infinitesimal generator of an analytic semigroup of bounded operators acting in E.

Conditions similar to (A), (B), (C) of §1 along with smoothness conditions on the data were imposed on this problem in the article [7]. The result of that analysis is the existence of a solution, $u(t,\varepsilon)$, for each small $\varepsilon > 0$ and a decomposition of u as

$$u(t,\varepsilon) = u^*(t,\varepsilon) + U(\tau,\varepsilon)$$

where u^* and U depend smoothly on ε near $\varepsilon = 0$. Again, the order relations $O(\varepsilon^{N+1})$ appearing in the Taylor expansions of u^* and U were shown to hold uniformly for $0 \leqq t \leqq T$, and U was shown to decay exponentially as $\tau \to \infty$.*

BIBLIOGRAPHY

1. F. Hoppensteadt, *Properties of solutions of ordinary differential equations with small parameters*, Comm. Pure Appl. Math. **24** (1971), 807–840.

2. W. Eckhaus, *Boundary layers in linear elliptic singular perturbation problems*, SIAM Rev. (to appear).

3. M. I. Višik and L. A. Ljusternik, *The asymptotic behavior of solutions of linear differential equations with large or quickly changing coefficients and boundary conditions*, Uspehi Mat. Nauk **15** (1960), no. 4, (94), 23–91 = Russian Math. Surveys **15** (1960), no. 4, 23–91. MR **23** #A1919.

4. J. B. Keller and S. Kogelman, *Asymptotic solutions of initial value problems for nonlinear partial differential equations*, SIAM J. Appl. Math. **18** (1970), 748–758. MR **41** #7258.

5. V. A. Trenogin, *Asymptotic behavior and existence of a solution of the Cauchy problem for a first order differential equation in a Banach space*, Dokl. Akad. Nauk SSSR **152** (1963), 63–66 = Soviet Math. Dokl. **4** (1963), 1261–1265. MR **29** #1409.

6. F. Hoppensteadt, *On quasilinear parabolic equations with a small parameter*, Comm. Pure Appl. Math. **24** (1971), 17–38. MR **42** #4863.

7. ———, *Asymptotic series solutions of some nonlinear parabolic equations with a small parameter*, Arch. Rational Mech. Anal. **35** (1969), 284–298. MR **40** #1694.

COURANT INSTITUTE OF MATHEMATICAL SCIENCES, NEW YORK UNIVERSITY

---

* A stronger version of the results in [7] has been obtained. In fact, the restriction that the operators f_{ijk} be bounded operators can be replaced by the weaker restriction that, for each k,

$$f_{ijk}(A^{-1}(t,0)u)^k$$

is a bounded k-linear operator acting in E. This brings the quasilinear problem described in §2 into the abstract setting.

DEFORMATIONS LEAVING A HYPERSURFACE FIXED

HOWARD JACOBOWITZ[1]

1. **Introduction.** Classically the asymptotic lines of a surface in Euclidean three space were called "lines of bending." This reflected the fact that if $U \subset \boldsymbol{R}^2$ and $u_0 : U \to \boldsymbol{R}^3$ gives an analytic surface $u_0(x,y)$ which admits a family of isometric deformations $u_t(x,y)$, analytic in (x,y) for each t, such that all the surfaces coincide on some curve $\Gamma \subset U$, then $u_0(\Gamma)$ is an asymptotic line of the surface $u_0(x,y)$ and therefore of each surface $u_t(x,y)$. Naturally one must assume that u_t is not identically equal to u_0 in any neighborhood of Γ.

We here study the analogous problem for submanifolds of Euclidean space of arbitrary dimension and codimension. We introduce a class of hypersurfaces which generalizes the idea of an asymptotic curve—roughly, H^{n-1} is an asymptotic hypersurface for $U^n \subset E^N$ if $H^{n-1} \subset U^n$ and if at each of its points H^{n-1} has, with respect to some normal to U^n, second order contact with the tangent plane to U^n. For "low" codimension, $N \leqq \frac{1}{2}n(n+1)$, U^n has an isometric deformation leaving some hypersurface H^{n-1} fixed only if H^{n-1} is asymptotic. Similar results hold also if H is of codimension greater than one in U. On the other hand, for "high" codimension, $N \geqq \frac{1}{2}n(n+3)+1$, U^n has such deformations, at least locally, for arbitrary hypersurfaces provided U^n is nondegenerate and H^{n-1} and U^n are analytic. Thus isometric deformations leaving a hypersurface fixed are rare for low codimension and common for high. Note that for $N = \frac{1}{2}n(n+1)$ the relevant system of equations is a determined one, since N is then the number of components of the metric tensor of an n-dimensional manifold. The result for low codimension is obtained as a consequence of the Holmgren uniqueness theorem and provides a variation of the classical statement: We take $u_0(x,y)$

AMS 1970 *subject classifications.* Primary 35F25, 53A05; Secondary 53A30.

[1] During the period of this research the author was supported by NSF grants GP-29697 and GP-18961.

analytic and assume $u_t(x,y)$ is analytic either in t or in (x,y) and is C^1 in the other. Also no analyticity is required if the original surface has negative Gaussian curvature. Note that the classical theorem is true as long as the family $\{u_t\}$ contains three distinct analytic surfaces [**1**, p. 336]. The high codimension result uses the well-known device of Nash to invert the linearized problem and Newton's method to then construct the family of deformations. The results one may anticipate for codimensions satisfying $\frac{1}{2}n(n+1) < N < \frac{1}{2}n(n+3)+1$ are discussed with reference to the special case $n = 2$, where such results can be obtained using a simple argument.

We also study conformal deformations for $N < \frac{1}{2}n(n+1)$. Our result is that such deformations holding H fixed are possible, for H not an asymptotic hypersurface, only if on H the second fundamental form, evaluated at some normal vector, agrees with the original metric.

2. **Definitions and preparatory remarks.** Let $U \subset \boldsymbol{R}^n$, $u: U \to E^N$; we use E^N in place of $\boldsymbol{R}^N$ to emphasize the Euclidean metric. Denote the metric induced by u by $F(u)$. In terms of local coordinates $(x_1, \ldots, x_n) \in U$ and $(u^1(x), \ldots, u^N(x)) \in E^N$, one has

$$F_{ij}(u) = \sum_{\nu=1}^{N} \frac{\partial u^\nu}{\partial x_i}\frac{\partial u^\nu}{\partial x_j}$$

where F_{ij} are the components of the induced metric relative to these coordinates. A C^1 *isometric deformation* $u_t(x)$ of u is a C^1 function of (x,t) with $u_0(x) = u(x)$ and $F(u_t) = F(u)$; a $C^{1,\omega}$ *isometric deformation* is in addition analytic in either x or t. We exclude the trivial deformation $u_t(x) = u(x)$ but do not exclude deformations which arise as ambient isometries. When we speak of deformations u_t leaving a hypersurface fixed it is to be understood that u_t does not coincide with u on any neighborhood of the hypersurface. We shall use "deformation" to always mean "isometric deformation."

An *infinitesimal deformation* $v(x)$ of u is a C^1 function satisfying $dF(u)v = 0$. In local coordinates $dF(u)$ is given by

$$(dF(u)v)_{ij} = \sum_{\nu=1}^{N} \left\{\frac{\partial u^\nu}{\partial x_i}\frac{\partial v^\nu}{\partial x_j} + \frac{\partial u^\nu}{\partial x_j}\frac{\partial v^\nu}{\partial x_i}\right\}.$$

If $u_t(x)$ is a deformation, then $\partial u_t(x)/\partial t|_{t=0}$ is an infinitesimal deformation. Thus the existence of a deformation with this partial derivative not identically zero and leaving a hypersurface fixed implies a nonuniqueness result for the initial value problem for the partial differential operator $dF(u_0)$. Hence the hypersurface is characteristic for this operator. But so are all hypersurfaces. Thus in the next section we shall seek to replace $dF(u_0)$ by an equivalent operator and then to investigate geometric properties of characteristic surfaces of this second operator.

A map $u: U \to \boldsymbol{R}^{n+K}$, for U an open set of $\boldsymbol{R}^n$, is called *nondegenerate* if at each point $p \in U$ the osculating space $\mathcal{O}(p)$ is of maximal rank. The osculating space is the linear space generated by the first and second derivatives of u. That is, one introduces local coordinates $\{x_1, \ldots, x_n\}$ on U and defines $\mathcal{O}(p)$ as the linear span of $\{u_{x_1}(p), \ldots, u_{x_n}(p), u_{x_1x_1}(p), \ldots, u_{x_ix_j}(p), \ldots, u_{x_nx_n}(p)\}$. This linear space is independent of the choice of coordinates. If u is nondegenerate then $\dim \mathcal{O}(p) = \min(n + K, \frac{1}{2}n(n + 3))$.

Let $H \subset \boldsymbol{R}^q$, $q < n$, and $h: H \to U$ be a C^1 embedding. We often identify H and $h(H)$. If $u: U \to E^{n+K}$ then denote by $\Pi|_H$ the pullback (or restriction) to H of the second fundamental form on U. Thus $\Pi|_H: T_H \otimes T_H \otimes N \to \boldsymbol{R}^1$ where T_H is the tangent bundle of H and N is the normal bundle of $u(U)$ in E^{n+K}. Equivalently $\Pi|_H: T_H \otimes T_H \to \operatorname{Hom}(N, \boldsymbol{R}^1)$. In low codimensions, $K \leqq \frac{1}{2}n(n - 1)$, call H *asymptotic at* $p \in H$ if, in the fibre above p, $\Pi|_H$ annihilates some normal vector and *asymptotic* if this is true for each $p \in H$.

Now let H be a C^2 submanifold of U. Define the augmented osculating space $\mathcal{O}^*(p)$ to be the space spanned by the osculating space of $u|_H$ at p together with the tangent space to $u(U)$ at p. In terms of local coordinates such that H is given by $x_1 = x_2 = \ldots = x_{n-q} = 0$ we have $\mathcal{O}^*(p) =$ linear span of $\{u_{x_i}(p), u_{x_Ix_J}(p) | 1 \leqq i \leqq n, n - q < I \leqq J \leqq n\}$. So $\dim \mathcal{O}^*(p) \leqq n + \frac{1}{2}q(q + 1)$. H is clearly asymptotic at p precisely if one has strict inequality. We now take this to be the definition of an asymptotic surface when $K > \frac{1}{2}n(n + 1)$. Alternatively for H a C^1 surface and any value of K, H is not asymptotic if it can be approximated arbitrarily well in C^1 by a smooth surface whose augmented osculating space is everywhere of maximum dimension. If H is of codimension one and is asymptotic, then we say H is an *asymptotic hypersurface*; this is the case we are most interested in. Note that we are speaking of a hypersurface in U not in the larger space E^{n+K}. For $n + K \leqq \frac{1}{2}n(n + 1)$, H is an asymptotic hypersurface if and only if there exists a normal to the osculating space of H which is also normal to U. We note for later use the following lemma which may be considered a partial generalization of Meusnier's theorem.

LEMMA 1. *Two hypersurfaces of $U \subset E^N$ which are tangent at some point have the same augmented osculating space there.*

For $n = 2$ and $K = 1$ an asymptotic hypersurface H is just an asymptotic curve in the classical sense—the principle curvature normal of H lies in the tangent plane of the surface U. Hence if the induced metric on U has positive curvature no asymptotic hypersurfaces exist. It is not clear if a similar intrinsic nonexistence criteria exists for $n > 2$ and $K > 1$.

EXAMPLE. Let $s: S^n \to E^{n+1}$ be the standard map and $x_1, \ldots, x_n$ local coordinates on S^n. The map $s' = s \oplus \varepsilon x_n^2$, in a neighborhood of the origin, defines a nondegenerate n-submanifold of E^{n+2} with positive sectional curvatures. The hypersurface $\{x | x_n = 0\}$ is asymptotic.

NONEXAMPLE. The author does not know of any submanifold $U^n \subset E^N$, $N = \frac{1}{2}n(n + 1)$ and $n > 2$, which is free of asymptotic hypersurfaces. For $n = 3$, a simple argument shows no such submanifold exists.

Let $H \subset U \subset E^{n+K}$ and introduce orthogonal (with respect to the induced metric) coordinates $\{x_1,\ldots,x_n\}$ on U such that $H = \{x|x_1 = 0\}$ and orthogonal vectors $\{u_{x_1},\ldots,u_{x_n},e_1,\ldots,e_K\}$ in E^{n+K}. Choose some ordering for the set $\{(I,J), 2 \leqq I \leqq J \leqq n\}$ and let (I,J) denote the value of the ordering at this pair. Thus $1 \leqq (I,J) \leqq \frac{1}{2}n(n-1)$. We claim that if H is not an asymptotic hypersurface then the $\frac{1}{2}n(n-1)$ by K matrix B given by $B_{(I,J),k} = u_{x_I x_J} \cdot e_k$ (scalar product in E^{n+K}) has maximal rank. For $K \leqq \frac{1}{2}n(n-1)$, this means that B has a left inverse. That B has maximal rank is a consequence of the following lemma from linear algebra. The proof is simple. Here u_i, v_j, and w_k are all vectors in $\boldsymbol{R}^Q$. A set of Q' such vectors has maximal rank if $Q' \leqq Q$ and the set is linearly independent or $Q' \geqq Q$ and the set spans $\boldsymbol{R}^Q$.

LEMMA 2. *Let $\{u_1,\ldots,u_Q\}$ be a basis for $\boldsymbol{R}^Q$ and $\{w_1,\ldots,w_P,u_{N+1},\ldots,u_Q\}$ a set of maximal rank. Then the matrix $B_{ij} = u_i \cdot w_j$, $1 \leqq i \leqq N$, $1 \leqq j \leqq P$, is of maximal rank.*

3. **Low codimension.**

THEOREM. *Let $u: U^n \to E^{n+K}, K \leqq \frac{1}{2}n(n-1)$, be an analytic map admitting a $C^{1,\omega}$ isometric deformation which leaves some C^2 hypersurface fixed. Then this hypersurface is necessarily asymptotic.*

REMARK. For $n = 2$ and $K = 1$, this theorem is a generalization of the classical result.

In this section we prove the following theorem.

THEOREM. *Let $u: U^n \to E^{n+K}, K \leqq \frac{1}{2}n(n-1)$, be analytic and H be a given C^2 hypersurface which is not asymptotic. Then the initial value problem*

$$dF(u)v = 0, \qquad v = 0 \quad \text{on } H$$

has only the trivial solution.

This theorem implies the first one. For let us assume the given $C^{1,\omega}$ deformation is analytic in x. If H is not asymptotic for u_0 then it is also not asymptotic for u_t, $|t|$ small. Apply the above theorem for $u = u_t$, t in this interval, to obtain $\partial u_t/\partial t = 0$. Thus $u_t(x) \equiv u(x)$ and u does not admit any nontrivial deformation leaving H fixed. Now assume that the $C^{1,\omega}$ deformation is analytic in t. For $u = u_0$ we determine a neighborhood of H for which the conclusion of the second theorem is valid, i.e. $v = 0$ in this neighborhood for any v solving the initial value problem. We can inductively establish that all partial derivatives of u_t with respect to t vanish at $t = 0$. That is, $u_t(x) \equiv u(x)$.

We now prove this second theorem. Let us first assume H is an analytic hypersurface. Introduce an analytic change of coordinates so that in a neighborhood of the origin H is given by $\{(x_1,\ldots,x_n)|x_1 = 0\}$ and the metric g has the form $g_{ii} = A^i \delta_{ij}$. In this section we do not employ the summation convention with regard to repeated indices. Rather the summation sign and summed index are explicitly indicated. The range of indices used shall be $1 \leqq i \leqq j \leqq n, 2 \leqq I \leqq J \leqq n$, $1 \leqq k \leqq K$, $1 \leqq \nu \leqq n + K$, $1 \leqq l \leqq n$. The system of equations we are investigating becomes

$$\sum_{\nu}\left(\frac{\partial u^{\nu}}{\partial x_i}\frac{\partial v^{\nu}}{\partial x_j} + \frac{\partial u^{\nu}}{\partial x_j}\frac{\partial v^{\nu}}{\partial x_i}\right) = 0$$

or, more briefly

$$u_{x_i} \cdot v_{x_j} + u_{x_j} \cdot v_{x_i} = 0.$$

For each point $x \in U$ we extend the orthogonal set $\{u_{x_1}, \ldots, u_{x_n}\}$ to an orthogonal basis for E^{n+K}, say $\{u_{x_1}, \ldots, u_{x_n}, e_1, \ldots, e_K\}$ and in such a way that each $e_k(x)$ is also an analytic function of x. If we write

$$v(x) = \sum_i \theta^i u_{x_i} + \sum_k \phi^k e_k$$

and now consider the system of equations for the unknowns θ_i, $i = 1, \ldots, n$, and ϕ_k, $k = 1, \ldots, K$, we see that this system is of the form

$$A^i\theta^i_{x_1} + A^1\theta^1_{x_i} + \sum_l \theta^l \frac{\partial A^1}{\partial x_l}\delta_{i1} + 2\sum_k \phi^k e_{kx_i} \cdot u_{x_1} = 0,$$

$$2\sum_k \phi^k e_{kx_I} \cdot u_{x_J} + A^I\theta^I_{x_J} + A^J\theta^J_{x_I} + \sum_l \theta^l \frac{\partial A^J}{\partial x_l}\delta_{IJ} = 0.$$

We have made use of $u_{x_i} \cdot u_{x_j} = A^i\delta_{ij}$ and $u_{x_i} \cdot e_k = 0$. Note that we have divided the system of $\frac{1}{2}n(n + 1)$ equations into n equations containing differentiation with respect to x_1 and $\frac{1}{2}n(n - 1)$ equations involving only differentiations with respect to $x_I, I \geqq 2$. We assume a solution θ^i, ϕ^k exists. In particular,

$$-2B\phi = \sum_l \theta^l \frac{\partial A^J}{\partial x_l}\delta_{IJ} + (A^I\theta^I_{x_J} + A^J\theta^J_{x_I})$$

where the $\frac{1}{2}n(n - 1) \times K$ matrix B is given by

$$B^k_{IJ} = e_{kx_I} \cdot u_{x_J} = -e_k \cdot u_{x_I x_J}.$$

If H is not an asymptotic hypersurface at the origin then by Lemma 2, B has maximal rank in a U neighborhood of H. Since $n + K \leqq \frac{1}{2}n(n + 1)$ this means B has rank K. Thus ϕ is necessarily a linear function of the right-hand side and we may write

$$\phi = \sum_{I,J} C_{IJ}\theta^I_{x_J} + \sum_i C_i\theta^i.$$

Returning to the first n equations in our decomposition, we have

$$D\theta_{x_1} + \sum_J D_J\theta_{x_J} + \sum_i D'_i\theta^i = 0.$$

Each of the matrices D, D_J, D'_i is analytic and D is given by $D_{ij} = A^i\delta_{ij}$. Hence D is nonsingular. The initial condition $v = 0$ on H means $\theta_i(0,x_2,\ldots,x_n) = \phi_k(0,x_2,\ldots,x_n) = 0$. The hyperplane $x_1 = 0$ is clearly not a characteristic for the above system and applying the Holmgren uniqueness theorem for analytic systems one obtains $\theta \equiv 0$ and so also $\phi \equiv 0$.

Recall that in the Holmgren theorem the initial surface need not be analytic; this allows us to eliminate the assumption that H is analytic. Two derivatives are needed to define an asymptotic hypersurface so let us assume H is only C^2 and is not an asymptotic hypersurface. One can introduce (analytic) coordinates $\{x_1,\ldots,x_n\}$ so that H is tangent to $\{x|x_1 = 0\}$ at the origin. By Lemma 1, this latter hypersurface is also not asymptotic near the origin.[2] Now we can again reduce the analytic system $dF(u)v = 0$ to a determined system. However, the hypersurface H is noncharacteristic for this system since the hypersurface $\{x|x_1 = 0\}$ is. Thus the Holmgren theorem again applies.

REMARKS. (1) The theorem also holds if the standard metric on E^{n+K} is replaced by any other analytic Riemannian metric.

(2) Returning to the classical two-dimensional case we see that if the Gaussian curvature is negative then the following is true: A C^3 surface admits a C^1 deformation holding a C^1 curve fixed only if this curve is an asymptote; no other asymptote intersecting this one can also be fixed. This is a consequence of the uniqueness theorems for the Cauchy and characteristic initial value problems for a strictly hyperbolic system. For in the case $n = 2$, $K = 1$, one can identify the characteristics of the relevant system of two equations with the asymptotes of the original surface. Thus negative curvature problems are hyperbolic and positive ones are elliptic.

(3) One can consider deformations preserving structures other than the metric and investigate the geometric properties of their fixed hypersurfaces. For instance let U admit a conformal deformation, $F(u_t) = \sigma_t F(u_0)$ with $\sigma_t \in C^1(t)$. In the statement of our result we think of the second fundamental form as a mapping of the normal bundle of U to the space of symmetric bilinear forms on the tangent bundle.

THEOREM. *If there exists a $C^{1,\omega}$ conformal deformation of $U^n \subset E^{n+K}$, $K < \frac{1}{2}n(n-1)$, $n > 2$, holding H^{n-1} fixed then the image of the unit normal bundle under the second fundamental form contains a metric on H conformal to $F(u_0)$; for $n = 2$ such a conformal deformation never exists.*

Note that this theorem includes the case where H is asymptotic if we take that metric which is identically zero to be conformal to all other metrics.

In a similar manner we can study the existence of deformations leaving fixed lower dimensional submanifolds.

---

[2] A simple modification of Lemma 1 shows that the above theorems are valid for C^1 hypersurfaces. This is used in Remark (2).

THEOREM. *Let H be a submanifold of $U^n \subset E^N$. Deformations of U^n holding H fixed cannot exist if H satisfies each of the following conditions:*

(i) $\mathcal{O}^*(H) = E^N$,

(ii) $\dim \mathcal{O}(H) \cap T_U \geqq n - 1$.

The proof of this theorem is an immediate consequence of the following lemmas.

LEMMA. *Let $H = \{x|x^1 = \ldots = x^l = 0\}$ satisfy* (i) *and* (ii). *For any solution v of $dF(u)v = 0$ one has that $\theta^1, \ldots, \theta^{l-1}, \phi^1, \ldots, \phi^K$ can be written as a linear combination of θ^r and $\theta^j_{x_i}$ for $r \geqq l$ and $i, j \geqq l + 1$.*

LEMMA. *Let the conclusion of the previous lemma hold. If the solution v vanishes on $\{x|x_1 = \ldots = x_p = 0\}$ then it also vanishes on $\{x|x_1 = \ldots = x_{p-1} = 0\}$ for each $p \leqq l$.*

4. **High codimension.** When the codimension K satisfies $K \geqq \frac{1}{2}n(n + 1) + 1$ we have the following result.

THEOREM. *Let $U^n \subset E^{n+K}$ be an analytic, nondegenerate submanifold. Then for any analytic hypersurface $H \subset U^n$ there is an analytic deformation, defined for some neighborhood of H in U^n, which leaves H fixed.*

By the definition of nondegeneracy given in §2 and the above inequality on K we see that the osculating space has everywhere dimension $\frac{1}{2}n(n + 3)$. In particular this dimension is a constant and one can find a unit vector $n(p)$ orthogonal to the osculating space and varying analytically. Fix some analytic hypersurface H in U. It is clear that U has an infinitesimal deformation v which vanishes on H. For $dF(u)n = 0$ and also $dF(u)\phi n = 0$ for all scalar functions ϕ (see below). So set $v = \phi n$ for a proper choice of ϕ. Indeed let us choose ϕ so that its gradient also vanishes on H.

We now construct a deformation $u(p,t)$ with $u(p,0) = u(p)$ for $p \in U$ and $u(p,t) = u(p)$ for $p \in H$ and $t \in (-1,1)$. Also $u(p,t)$ will be analytic in each variable. We need to work with complex analytic functions. Thinking of $U \subset \boldsymbol{R}^n$ and $u: U \to \boldsymbol{R}^{n+K}$ we can extend u to a complex analytic function mapping some ball $B(r_0) \subset \boldsymbol{C}^n$, of radius r_0 and center $0 \in U \subset \boldsymbol{R}^n \subset \boldsymbol{C}^n$, into $\boldsymbol{C}^{n+K}$. We may likewise take v complex analytic on the same ball. A nondegenerate map will now be a complex analytic map w such that $\{w_{z_i}, w_{z_i z_j}, 1 \leqq i \leqq j \leqq n\}$ spans a linear space, over $\boldsymbol{C}^1$, of dimension $\frac{1}{2}n(n + 3)$. And a metric (in local coordinates) will be a complex analytic, symmetric matrix valued function which is positive definite when restricted to $\boldsymbol{R}^n \cap B(r_0)$.

For a complex analytic function, or mapping, set

$$|f|_r = \sup\{|f(z)|;\ z \in B(r)\} \quad \text{and} \quad |f|_{r,k} = \max_I \{|D^I f|_r;\ I = (I_1, \ldots, I_n), |I| \leqq k\}.$$

If f also depends on t then for each fixed t one can define a norm $|f|_{r,k}$ and make sense of inequalities $|f|_{r,k} \leqq \varepsilon(t)$.

We have the following lemma which is due to Nash and in this form can be found in [**2**]. The lemma asserts the existence of a right inverse for $dF(w)$ as long as w is C^2 close to u. Let $u,v,w,\ldots$ denote maps of $B(r)$ into $\boldsymbol{C}^{n+K}$ and $g,h,\ldots$ metrics on $B(r)$.

LEMMA. *Let $u: U \to \boldsymbol{R}^{n+K}$, $K \geqq \frac{1}{2}n(n+1)$, be nondegenerate on $B(r_0)$. There exists an $\varepsilon > 0$ and a map $\eta(v,g)$ of $B(r)$ into $\boldsymbol{C}^{n+K}$ such that if $|w-u|_{r,2} < \varepsilon$ for some $r < r_0$ then for all h, $dF(w)\eta(w,h) = h$. Further $|\eta(w,h)|_r < C|h|_r$ for some constant C.*

The proof of the lemma is based on the observation that a smooth solution to the linear algebraic system

$$\sum_v \frac{\partial x^v}{\partial x_i} v^v = 0, \qquad \sum_v \frac{\partial^2 w^v}{\partial x_i \partial x_j} v^v = -\frac{h_{ij}}{2}$$

of $\frac{1}{2}n(n+3)$ equations is also a solution to the system of partial differential equations $dF(w)v = h$. This also makes clear the fact that if n is orthogonal to the osculating space then $dF(u)\phi n = 0$ for all scalars ϕ.

We now construct the desired deformation. Let $u_0 = u + tv$, v an infinitesimal deformation which vanishes, along with its first derivatives, on H. Define inductively $u_i = u_{i-1} + \eta(u_{i-1}, F(u) - F(u_{i-1}))$. Fix some $s > 2$. We shall show there exist two positive numbers Δ and λ, independent of i, such that u_i is well defined and

$$|u_i - u_{i-1}|_{\delta_i} \leqq \lambda 2^{-is}|t| \quad \text{for } |t| < \Delta.$$

Here $\delta_i = r_0(\frac{1}{2} + (\frac{1}{2})^i)$. Δ also depends on the infinitesimal deformation v.

Assume u_i is defined and the above estimate holds for $i = 1,2,\ldots,I$ and some Δ. We can assume that Δ is so small that u_0 is nondegenerate for $|t| < \Delta$. Indeed we restrict Δ further and assume that the above lemma applied to u_0 instead of u yields some ε independent of t in this range.

For u_{I+1} to be defined on δ_{I+1}, one needs

$$|u_I - u_0|_{\delta_{I+1},2} < \varepsilon$$

and this follows from writing $u_I - u_0 = (u_I - u_{I-1}) + \ldots + (u_1 - u_0)$ and then using the Cauchy estimates and the above inductive estimate to bound the derivatives. We use $s > 2$ and obtain λ as a function of s and ε. See [**2**] for details here and below.

Now

$$\begin{aligned} |u_{I+1} - u_I|_{\delta_{I+1}} &= |\eta(u_I, F(u) - F(u_I))|_{\delta_{I+1}} \leqq C|F(u) - F(u_I)|_{\delta_{I+1}} \\ &\leqq C|F(u) - (F(u_{I-1}) + dF(u_{I-1})(u_I - u_{I-1}) + F(u_I - u_{I-1}))|_{\delta_{I+1}} \\ &\leqq C|F(u_I - u_{I-1})|_{\delta_{I+1}} \leqq C_1(|u_I - u_{I-1}|_{\delta_{I+1},1})^2 \\ &\leqq C_2\{(\delta_{I+1} - \delta_I)^{-1}|u_I - u_{I-1}|_{\delta_I}\}^2 \leqq \lambda 2^{-(I+1)s}|t|, \end{aligned}$$

assuming another smallness restriction on λ and also $\Delta < 1$.

Thus u_i converges to some u_∞ uniformly on $B(\frac{1}{2}r_0) \times (-\Delta, \Delta)$ provided

$$|u_1 - u_0|_{r_0} \leqq \lambda 2^{-s}|t|.$$

Since $F(u) - F(u + tv) = tdF(u)v + t^2F(v)$ and $dF(u) = 0$, one has $|u_1 - u_0|_{r_0} < ct^2$ and a smallness condition on Δ gives $|u_1 - u_0|_{r_0} \leqq \lambda 2^{-s}|t|$. Thus u_∞ exists and is analytic on $B(\frac{1}{2}r_0) \times (-\Delta, \Delta)$. Clearly $F(u_\infty) = F(u)$ and $u_\infty(p,0) = u(p)$. Further from the fact that v is zero on H and has a zero derivative normal to H, it follows that $F(u_0) = F(u)$ on H and our iteration process has no need to modify u_0 on H and therefore does not. Thus $u_\infty(p,t) = u_0(p) = u(p)$ for $p \in H$. Also u_∞ is not trivial since $du_\infty/dt = v \not\equiv 0$. This proves the theorem.

One can drop the assumption of analyticity. This necessitates the construction of a different inverse to the linearized operator and a modification of the iteration procedure. Details will appear elsewhere. Also, at least in the analytic case, a global version of the above theorem is valid. We take H closed and find an appropriate global function φ using Theorem A of Cartan. The iterates are globally defined and, assuming U is compact, again converge to an isometric deformation leaving H fixed.

We have not treated the case where the codimension is in the range $\frac{1}{2}n(n - 1) < K \leqq \frac{1}{2}n(n + 1)$. Here we expect that deformations exist for most hypersurfaces: perhaps for those that are nowhere asymptotic. This is so for $n = 2$.

THEOREM. *Given a nondegenerate two-dimensional surface in $E^N, N > 3$, and a curve on this surface which is nowhere asymptotic there exists a deformation leaving this curve fixed.*

The proof is based on the observation that any such surface in E^4 can be projected onto some hyperplane so that the image has negative curvature. Hence the proof reduces to showing that given an embedding u, metrics $g(t)$ with $F(u) = g(0)$, and a curve γ whose arclength does not depend on t and such that $u(\gamma)$ is nowhere asymptotic, one can find embeddings $u(t)$, agreeing on γ, and satisfying $F(u(t)) = g(t)$, $u(0) = u$. Since the curvature is negative this can be reduced, in several different ways, to an easily handled problem in nonlinear hyperbolic equations.

REFERENCES

1. L. Eisenhart, *A treatise on the differential geometry of curves and surfaces*, Ginn, Boston, Mass., 1909.

2. H. Jacobowitz, *Implicit function theorems and isometric embeddings*, Ann. of Math. **95** (1972), 191–225.

RICE UNIVERSITY

THE REGULARITY OF THE SOLUTION TO A CERTAIN VARIATIONAL INEQUALITY[1]

DAVID KINDERLEHRER

1. **Introduction.** The object of this paper is to demonstrate that the solution to a special type of variational inequality has Lipschitz continuous first derivatives. The problem especially concerns minimal surfaces and the proof is valid only for two dimensions. Let us first state the problem under consideration, then the nature of the proof we shall present, and finally what has been previously known about the solution.

Let Ω be a strictly convex domain in the $z = x_1 + ix_2$ domain with smooth, say $C^{2,\alpha}$, boundary $\partial\Omega$. Let $\psi \in C^2(\overline{\Omega})$ satisfy $\max_\Omega \psi > 0$ in Ω and $\psi < 0$ on $\partial\Omega$ and let us set

$$\mathscr{K} = \{v \in C^{0,1}(\overline{\Omega}) : v \geqq \psi \text{ in } \Omega \text{ and } v = 0 \text{ on } \partial\Omega\}.$$

Let $a(p) = (a_1(p), a_2(p))$, $p = (p_1, p_2)$, denote a locally coercive C^1 vector field, by which we mean that, for each compact $C \subset R^2$, there exists a $\nu = \nu(C) > 0$ such that the C^1 function $a(p)$ satisfies

$$(a(p) - a(q))(p - q) \geqq \nu|p - q|^2 \quad \text{for } p, q \in C.$$

We consider the unique $u \in \mathscr{K}$ such that

$$\int_\Omega a_j(Du)(v - u)_{x_j}\, dx \geqq 0 \quad \text{for } v \in \mathscr{K}. \tag{1}$$

In the above, Du denotes the gradient of u. The function $u(z)$ is referred to as the solution of the variational inequality (1) with obstacle ψ. Our intention is to show that $Du \in C^{0,1}(\overline{\Omega})$ provided that the obstacle is concave and that $a(p)$ satisfies a

AMS 1970 *subject classifications*. Primary 35J20; Secondary 53A10.

[1] This research was supported in part by the Air Force Office of Scientific Research grant no. AFOSR 71–2098.

certain conformal mapping condition. This mapping condition is the statement that the complex gradient of an harmonic is analytic for the problem $a_j(p) = p_j$ which arises in the study of the Dirichlet integral. For the problem involving minimal surfaces this condition expresses the conformality of the Gauss mapping of a minimal surface.

In addition, if the obstacle is strictly concave, we are able to show that the Gauss curvature of the solution surface,

$$K = W^{-4}(u_{x_1x_1}u_{x_2x_2} - (u_{x_1x_2})^2), \qquad W^2 = 1 + |Du|^2,$$

satisfies

$$(2) \quad K(z) \leqq -K_0 < 0, \qquad z \in \Omega - I, \qquad I = \{z : u(z) = \psi(z)\}, \quad \text{where } K_0 > 0.$$

If the obstacle is concave, but only Lipschitz, and $a(p)$ is an arbitrary locally coercive $C^{1,\alpha}$ vector field then it is possible to show that $K(z) < 0$ in $\Omega - I$, but precise conditions necessary for the validity of an estimate of the form (2) are not known. The proof here is somewhat different than [**9**, Theorem 5].

Let us consider briefly the case $a_j(p) = p_j$. The solution $u(z)$ is harmonic in $\Omega - I$, $I = \{z : u(z) = \psi(z)\}$ so that $f(z) = u_{x_1}(z) - iu_{x_2}(z)$ is analytic there. Therefore, for a fixed $z \in \Omega - I$, the modulus of the difference quotient

$$(3) \qquad \left|\frac{f(t) - f(z)}{t - z}\right|, \qquad t \in \Omega - I,$$

is a subharmonic whose maximum can be determined by its values on $\partial I \cup \partial\Omega$. To show that this modulus is bounded independently of z, t we are assisted by the knowledge that $f(z) = f^*(z) = \psi_{x_1}(z) - i\psi_{x_2}(z)$ for $z \in \partial I$ and that each $z \in \partial I$ is the limit of a sequence $z_n \in \text{Interior } I$ provided that ψ is strictly concave. For this enables us to approximate (3) when $z \in \partial I$ by the continuous subharmonics

$$w_n(t) = \left|\frac{f(t) - f^*(z_n)}{t - z_n}\right|, \qquad t \in \Omega - I.$$

With some accounting, the Lipschitz constant of $f(z)$ is shown to depend only on various a priori quantities so that we may pass to the limit and obtain that $f(z) \in C^{0,1}(\bar{\Omega})$ if ψ is concave and C^2. More general cases are reduced to ones concerning analytic functions by use of the classical theorem of Lichtenstein. The proof of (2) is almost identical.

In their 1969 paper [**10**], Lewy and Stampacchia proved that the solution to the variational inequality

$$u \in \mathscr{K}: \quad \int_\Omega a_{jk}(z)u_{x_j}(z)v_{x_k}(z)\,dx \geqq 0, \quad \text{for } v \in \mathscr{K},$$

satisfies $u \in C^{1,\alpha}(\Omega)$ under appropriate assumptions about a_{jk} and for any dimension. In the special case $a_{jk} = \delta_{jk}$ and in two dimensions for a strictly concave real analytic $\psi(z)$ they proved that ∂I is an analytic Jordan curve. Since this implies that only certain cusps are singularities of the curve ∂I, a type of growth estimate is available for $Du(z)$. In view of [**9**] it is also known that these cusps do not occur, from which it follows for this case that $u(z)$ is extendible as an harmonic function of z into a neighborhood of ∂I.

Existence of the solution to (1) has been proven, in the form stated here, by Lewy and Stampacchia [**11**], where the solution is shown to be in the space $H^{2,q}(\Omega)$, any q, $1 < q < \infty$, and hence in $C^{1,\alpha}(\Omega)$ for any α, $0 < \alpha < 1$. Existence and regularity under different hypotheses have been shown by Brézis and Stampacchia [**2**] and Stampacchia [**13**] to which we refer for a more complete bibliography.

Recently Brézis has studied the general question of regularity for this and also nonhomogeneous problems [**3**], [**4**]. In all of the work just noted, smoothness is a consequence of integral inequalities. But since, in general, the solution to the problem (1) cannot belong to a Sobolev space $H^{m,q}(\Omega)$ for $m \geqq 3$, it might be true that $C^{1,\alpha}$ smoothness is the most available from these techniques for second order problems. At the end of this paper we illustrate an application of our method to a simple case of the problem treated by Brézis [**4**]. Although it is a rather limited example, it does provide hope that some conclusions are valid for the solutions of nonhomogeneous problems.

The questions related more directly to minimal surfaces and surfaces of constant mean curvature have been considered by M. Miranda [**7**] in parametric form and Giaquinta and Pepe [**6**]. We refer to the article by Miranda in these Proceedings [**8**], where the interesting work of Massari and Santi is also discussed. In addition, we ought to note the important papers of Tomi [**14**], [**15**].

2. **The solution to a certain Beltrami equation.** We study here a Beltrami equation which we shall apply to our situation. For any function $v \in C^{0,\alpha}(\Omega)$ let us denote $H(v) = \sup_\Omega\{|v(z) - v(z')|/|z - z'|^\alpha\}$. Let $I \subset \Omega$ be a closed set with the property

(*) the point set I is the closure of its interior and is connected.

THEOREM 1. *Let $I \subset \Omega$ satisfy* (*) *and suppose that $I \subset \Omega_0 \subset \Omega$ for a compact Ω_0. Let $\mu(z) \in C^{0,\alpha}(\bar{\Omega})$ for some α, $0 < \alpha < 1$, and satisfy $|\mu(z)| \leqq \mu_0 < 1$ in $\bar{\Omega}$. Suppose that $h(z) \in C^0(\bar{\Omega}) \cap C^1(\bar{\Omega} - I)$ satisfies*

$$h_{\bar{z}} = \mu h_z \quad \text{in } \Omega - I,$$
$$h = h^* \quad \text{in } I,$$

where $h^* \in C^1(\bar{\Omega})$. Then $h \in C^{0,1}(\bar{\Omega})$ and

$$|h(z) - h(z')| \leqq c|z - z'| \quad \text{for } z, z' \text{ in } \Omega \tag{4}$$

with $c = c(\Omega, \mu_0, H(\mu), \Omega_0, \sup_{\partial\Omega}|Dh|, \sup_I|Dh^*|)$.

PROOF. We proceed to use the Lichtenstein Theorem ([**5**, Chapter IV], for example) to reduce this problem to one involving analytic functions. There exists a

$C^{1,\alpha}$ homeomorphism φ of $\bar{\Omega}$ onto itself with the $C^{1,\alpha}$ inverse φ^{-1} leaving fixed two points $z_0 \in \Omega$ and $z_0' \in \partial\Omega$ which satisfies

$$\begin{aligned} &\varphi_{\bar{z}} = \mu\varphi_z \quad \text{in } \Omega, \\ (5) \qquad &|D\varphi| \leqq \operatorname{const}(\Omega, \mu_0, H(\mu)) = K, \\ &|D\varphi^{-1}| \leqq \operatorname{const}(\Omega, \mu_0, H(\mu)) = K'. \end{aligned}$$

In the set $\Omega - G$, $G = \varphi(I)$, there is a complex analytic $f(\zeta)$, continuous in $\overline{\Omega - G}$ satisfying

$$h(z) = f(\varphi(z)), \qquad z \in \Omega - G.$$

The function $f^*(\zeta) = h^*(\varphi^{-1}(\zeta))$ is C^1 in $\bar{\Omega}$ and is equal to $f(\zeta)$ on ∂G. We observe that the restriction of f to $\partial\Omega$ has a Lipschitz constant bounded by $K' \sup_{\partial\Omega}|Dh(z)|$. We pass to a lemma.

LEMMA 1. *Let $G \subset \Omega$ satisfy condition (*). Suppose that $f \in C^0(\bar{\Omega})$ is analytic in $\Omega - G$ and satisfies*

$$f = f^* \qquad \textit{for } \zeta \in G,$$
$$|f(\zeta) - f(\zeta')| \leqq k|\zeta - \zeta'| \quad \textit{for } \zeta, \zeta' \in \partial\Omega,$$

where $f^* \in C^1(\Omega)$. Then

$$f \in C^{0,1}(\overline{\Omega - G}) \quad \textit{and} \quad |f(\zeta) - f(\zeta')| \leqq C|\zeta - \zeta'|$$

for $\zeta, \zeta' \in \Omega - G$ where $C = \max\{k, \sup_G|Df^*|, 2\sup|f|(\operatorname{dist}(G, \partial\Omega))^{-1}\}$.

PROOF. For $\zeta_0 \in \partial\Omega \cup \partial G$ let us estimate the modulus

$$w(t) = w(t, \zeta_0) = |f(\zeta_0) - f(t)|/|\zeta_0 - t|, \qquad t \in \Omega - G.$$

We first suppose that $\zeta_0 \in \partial G$ and select $\zeta_n \in$ Interior G, $\zeta_n \to \zeta_0$ and set

$$w_n(t) = |f(t) - f^*(\zeta_n)|/|t - \zeta_n|, \qquad t \in \Omega - G.$$

The functions w_n, subharmonic in $\Omega - G$, may be continuously extended to $\overline{\Omega - G}$ since $\zeta_n \in$ Interior G and $f \in C^0(\bar{\Omega})$. Hence there is a point $\zeta \in \partial(\Omega - G)$ such that $w_n(\zeta) = \max w_n(t)$. If $\zeta \in \partial G$ then, since $f = f^*$ on G,

$$w_n(t) \leqq \left|\frac{f^*(\zeta_n) - f^*(\zeta)}{\zeta_n - \zeta}\right| \leqq \sup_G|Df^*|.$$

On the other hand, if $\zeta \in \partial\Omega$, then

$$w_n(t) \leqq \frac{2}{\operatorname{dist}(G, \partial\Omega)} \sup|f|.$$

For each $t \in \Omega - G$,

$$w(t, \zeta_0) = \lim_{n\to\infty} w_n(t),$$

so that

$$w(t, \zeta_0) \leqq \max\left\{\sup_G |Df^*|, 2 \sup|f|(\operatorname{dist}(G, \partial\Omega))^{-1}\right\}.$$

Let us now suppose that $\zeta_0 \in \partial\Omega$. Then it is easy to see that

$$w(t, \zeta_0) \leqq \max\{k, 2 \sup|f|(\operatorname{dist}(G, \partial\Omega))^{-1}\}.$$

For any $t \in \Omega - G$,

$$w(t, \zeta) \leqq \max_{\zeta \in \partial(\Omega - G)} w(t, \zeta) = w(t, \zeta_0) \quad \text{for some } \zeta_0 \in \partial(\Omega - G).$$

Hence by the foregoing,

$$w(t, \zeta) \leqq \max\left\{\sup_G |Df^*|, k, 2 \sup|f|(\operatorname{dist}(G, \partial\Omega))^{-1}\right\} = C.$$

To complete the proof of Theorem 1, one notices that $|h(z) - h(z')| \leqq C|\varphi(z) - \varphi(z')| \leqq CK|z - z'|$ for $z, z' \in \Omega - I$. The set G satisfies (*) inasmuch as it is the homeomorphic image of I. One might note further that $\operatorname{dist}(G, \partial\Omega)$ is a function of Ω and μ_0 and Ω_0. This is a consequence of well-known properties of quasi-conformal mappings. To suppose the contrary would lead one to suppose the existence of homeomorphisms φ_n, functions $\mu_n, |\mu_n| \leqq \mu_0 < 1$, and a sequence $z_n \in \Omega_0$, $z_n \to z$, such that $\varphi_n(z_n) \to \zeta \in \partial\Omega$ and $\varphi_{n\bar{z}} = \mu_n \varphi_{nz}$. This allows one to determine a homeomorphism φ by the uniform Hölder continuity of $\{\varphi_n\}$ such that $\varphi(z) = \zeta \in \partial\Omega$. But this is not possible for a homeomorphism since $\partial\Omega \cap \Omega_0 = \emptyset$. We refer to [**1**, Chapter 6].

3. **Lipschitz continuity of first derivatives.** The result of the previous paragraph will be applied to variational inequalities with the vector fields

$$a_j(p, t) = (1 + t^2 p^2)^{-1/2} p_j, \qquad 0 < t \leqq 1.$$

From this example the reader will readily see how to employ Theorem 1 in more general situations. Let $u_t \in C^{1,\alpha}(\Omega) \cap \mathscr{K}$ denote the solution to

$$(\mathrm{P}_t) \qquad \int_\Omega a_j(Du, t)(v - u)_{x_j}\, dx \geqq 0 \quad \text{for all } v \in \mathscr{K}.$$

We exhibit the function $h(z)$ appropriate for the application of Theorem 1. Let $G \subset R^2$ be any open set and $w \in C^2(G)$ be a solution of the minimal surface equation

$$(6) \qquad -\frac{\partial}{\partial x_j}\left(\frac{w_{x_j}}{W}\right) = 0, \qquad W^2 = 1 + |Dw|^2, \quad \text{in } G.$$

It is well known, and may be shown simply by using the transformation introduced by Radó [**12**], for example, that the function

$$h(z) = (w_{x_1} - iw_{x_2})/(1 + W), \qquad z \in G,$$

satisfies the Beltrami equation

$$(7) \qquad h_{\bar z} = \mu h_z, \qquad \mu = \bar h^2, z \in G.$$

This is the statement that the Gauss mapping of a minimal surface is conformal. Since u_t is a solution to

$$-\partial a_j(Du_t, t)/\partial x_j = 0, \qquad z \in \Omega - I_t, \qquad I_t = \{z : u_t(z) = \psi(z)\},$$

the function $w(\zeta) = u_t(t\zeta)$ is a solution to the minimal surface equation (6) in the domain $G = \{\zeta : t\zeta \in \Omega - I_t\}$ as a function of $\zeta = \xi_1 + i\xi_2$. Hence $h(\zeta)$ formed from w satisfies (7), whence

$$(8) \qquad h_t(z) = \frac{u_{tx_1}(z) - iu_{tx_2}(z)}{1 + (1 + t^2|Du_t|^2)^{1/2}}, \qquad 0 < t \leqq 1,$$

satisfies the equation

$$\frac{\partial}{\partial \bar z} h_t = \mu_t \frac{\partial}{\partial z} h_t, \qquad z \in \Omega - I_t,$$

where

$$\mu_t = t^2 \frac{(u_{tx_1} + iu_{tx_2})^2}{(1 + (1 + t^2|Du_t|^2)^{1/2})^2}.$$

We set $h^*(z) = (\psi_{x_1} - i\psi_{x_2})/(1 + (1 + t^2|D\psi|^2)^{1/2})$.

THEOREM 2. *Let ψ be a concave C^2 obstacle. Then the solution u_t to the problem* (P_t) *is in* $C^{1,1}(\bar\Omega)$. *Moreover,*

$$|Du_t(z) - Du_t(z')| \leqq c_0|z - z'|, \qquad z, z' \in \Omega,$$

where c_0 depends on ψ and $\dot\Omega$.

PROOF. Let us first assume that ψ is strictly concave. To verify the hypotheses of Theorem 1, we need only note that I_t satisfies (*) and that $\sup_{\partial\Omega}|Dh_t|$ is bounded. That I_t satisfies (*) follows from [**9**], for the proof given there is valid for any analytic vector field, indeed for any C^1 vector field with slight modifications. On the other hand, $w(\zeta) = u_t(t\zeta)$ is easily seen to be the solution of (P_1) with the integration extended over $\{\zeta : t\zeta \in \Omega\}$ and obstacle $\psi(t\zeta)$ so that [**9**] may be applied directly.

The general theory of elliptic equations insures us that

$$\sup_{\partial\Omega}|Dh_t| \leqq \text{const} \sup(\textstyle\sum|u_{x_jx_k}|) \leqq \text{const},$$

where the last constant depends on Ω and ψ but not on t (cf. [**1**, Chapter 5.6]).

We choose $\Omega_0 = \{z \in \Omega : \psi(z) > 0\} \subset \Omega$. Evidently, $I_t \subset \Omega_0$ for all $t, 0 < t \leqq 1$. In view of the simple dependence of $a_j(p, t)$ on t, it is elementary to verify that the remaining parameters occurring in Theorem 1 are independent of t. For example, the Hölder constant of μ_t, a function of Du_t, is estimated by the norm in $H^{2,q}(\Omega)$ of u_t. That this norm is independent of t follows from [**11**, Equation 4.7] and the estimate

$$|Du_t(z)| \leqq \sup_{\Omega}|D\psi(z)| \quad \text{in } \Omega,$$

the latter of which insures uniform ellipticity of the equations

$$-\partial a_j(Du_t, t)/\partial x_j = f \quad \text{in } \Omega.$$

Hence, the assertion of the theorem is valid provided ψ is strictly concave since h_t is a smooth function of Du_t and vice versa.

To pass to the limit in the case when ψ is only concave, we approximate to $\psi(z)$ by $\psi_\varepsilon(z) = \psi(z) - \varepsilon|z - z_0|^2$ with z_0 a point where ψ assumes its mayimum. For $\varepsilon > 0$ sufficiently small the solutions u_t^ε of (P_t) with obstacle ψ_ε have uniformly bounded $H^{2,q}(\Omega)$ norms. Furthermore; it is known that $u_t^\varepsilon \to u_t$ uniformly as $\varepsilon \to 0$ (t fixed), cf. [**11**], and hence we see that for a subsequence $\{\varepsilon'\}$, $Du_t^{\varepsilon'} \to Du_t$ uniformly. By the preceding, it is clear that the Du_t^ε have Lipschitz constants independent of t, so the theorem follows.

4. **Sufficient conditions for the estimation of the curvature.** We describe here a sufficient condition for the validity of the estimate (2) in the case of the problem (P_t). We begin with a theorem to replace Theorem 1.

THEOREM 3. *Let $I \subset \Omega$ be compact and $\mu \in C^{0,\alpha}(\overline{\Omega})$ for some α, $0 < \alpha < 1$, satisfy $|\mu(z)| \leqq \mu_0 < 1$. Suppose that $h \in C^1(\overline{\Omega} - I) \cap C^0(\overline{\Omega})$ is a homeomorphism of $\Omega - I$ such that*

$$h_{\bar{z}} = \mu h_z \qquad \text{in } \Omega - I,$$

$$|h(z) - h(z')| \geqq c_0|z - z'| \quad \text{for } z, z' \in \partial\Omega \text{ and some } c_0 > 0.$$

Let $h^*(z) \in C^1(\overline{\Omega})$ be a homeomorphism of Ω such that

$$(9) \qquad \begin{aligned} &(h^*)^{-1} \text{ is continuously differentiable} \\ &h = h^* \quad \text{in } I, \\ &h(\Omega - I) \cap h^*(\Omega - I) = \varnothing. \end{aligned}$$

Then there is a constant $c > 0$ such that

$$|h(z) - h(z')| \geqq c|z - z'| \quad \text{for } z, z' \in \Omega - I.$$

PROOF. The proof follows the same arguments as Theorem 1 and its lemma, except that the condition (*) is replaced by the conditions (9). We denote by φ the $C^{1,\alpha}$ homeomorphism of Ω onto itself satisfying (5) and set $f(\zeta) = h(\varphi^{-1}(\zeta))$, a continuous function in Ω which is complex analytic in $\Omega - G$, $G = \varphi(I)$, and $f^*(\zeta) = h^*(\varphi^{-1}(\zeta)) \in C^1(\overline{\Omega})$.

Since $|D\varphi|$ is bounded, the hypothesis (9) implies that $(f^*)^{-1}$ is a Lipschitz function. Hence, there exists a constant $c_1 > 0$ such that

$$|f^*(\zeta) - f^*(\zeta')| \geqq c_1|\zeta - \zeta'|, \qquad \zeta, \zeta' \in \overline{\Omega}. \tag{10}$$

Since h is a homeomorphism of $\Omega - I$, the modulus

$$v(t, \zeta) = \left|\frac{t - \zeta}{f(t) - f(\zeta)}\right|, \qquad t, \zeta \in \Omega - G,$$

is a continuous subharmonic which may be continuously extended to $\overline{\Omega - G}$ for fixed $t \in \Omega - G$. It assumes its maximum at some $\zeta \in \partial\Omega \cup \partial G$. If $\zeta \in \partial G$, we may select $\zeta_n \in \Omega - G$, $\zeta_n \to \zeta$, which exist since $\Omega - G$ is open, and consider

$$v_n(t) = \left|\frac{t - \zeta_n}{f(t) - f^*(\zeta_n)}\right|.$$

The $\{v_n(t)\}$ are continuous in $\overline{\Omega - G}$ by the condition (9). In view of (10), we may estimate their modulus. We can consider the case $\zeta \in \partial\Omega$ in a manner identical to Lemma 1. It follows that there exists a constant $c_2 > 0$ such that

$$c_2|\zeta - \zeta'| \leqq |f(\zeta) - f(\zeta')| \quad \text{for } \zeta, \zeta' \in \Omega - G.$$

Since φ^{-1} is Lipschitz, the theorem follows.

Let us apply Theorem 3. We shall assume that $\psi(z)$ is strictly concave, which insures that the mapping $\psi_{x_1} - i\psi_{x_2}$ is a homeomorphism of Ω. Since the mapping

$$p_1 - ip_2 \longrightarrow \frac{p_1 - ip_2}{1 + (1 + t^2p^2)^{1/2}}$$

is a 1:1 smooth mapping of the plane we may verify the conditions that h_t be a homeomorphism of $\Omega - I_t$ and $h_t(\Omega - I_t) \cap h^*(\Omega - I_t)$ by [**9**, Theorem 2, Lemma 1.3], since those statements assert the identical facts for the complex gradient of u_t. As we have noted in the proof of Theorem 1, we may assume the validity of these arguments for the solution to (P_t). To show that $(h^*)^{-1}$ is continuously differentiable with

$$h^*(z) = h_t^*(z) = h_1^*(z) - ih_2^*(z) = \frac{\psi_{x_1}(z) - i\psi_{x_2}(z)}{1 + (1 + t^2|D\psi(z)|^2)^{1/2}}$$

we shall show that $\det(\partial h_j^*/\partial x_k)$ is positive in $\overline{\Omega}$. Indeed,

$$\begin{aligned}\det\left(\frac{\partial h_j^*}{\partial x_k}\right) &= \frac{1}{W_t(1 + W_t)^2}(\psi_{x_1x_1}\psi_{x_2x_2} - (\psi_{x_1x_2})^2), \\ &= \frac{W^4}{W_t(1 + W_t)^2}K_\psi > 0,\end{aligned} \qquad W_t^2 = 1 + t^2|D\psi|^2,$$

where K_ψ denotes the Gauss curvature of ψ. Hence the strict concavity is necessary to insure both that $(h^*)^{-1}$ exist and that it be differentiable. All the conditions of (9) are satisfied.

We shall find a simple estimate for $K(z)$. The estimate (13) corrects that given in [**9**, Lemma 5.3].

LEMMA 2. *Let $u \in C^2(G)$, $G \subset R^2$ an open set, satisfy*

$$-\frac{\partial}{\partial x_j} a_j(Du, t) = 0 \quad \text{in } G. \tag{11}$$

Then the Gauss curvature of the surface defined by $u(z)$ satisfies

$$K \leqq -\frac{1}{W^4(1 + W_t^2)}\{u_{x_1x_1}^2 + 2u_{x_1x_2}^2 + u_{x_2x_2}^2\} \tag{12}$$

for each $z \in G$, where $W^2 = 1 + |Du|^2$ and $W_t^2 = 1 + t^2|Du|^2$. Let $G = \Omega - I_t$, ϑ the angle between the inward normal of $\partial\Omega$ and the positive x_1-axis, and s the arclength of $\partial\Omega$. Then

$$K \leqq -\frac{W^2 - 1}{W^4(1 + W_t^2)}\left(\frac{d\vartheta}{ds}\right)^2 \quad \text{on } \partial\Omega. \tag{13}$$

PROOF. The estimate (12) is a simple calculation based on (11). To obtain it we note that

$$-(1 + t^2u_{x_2}^2)(u_{x_1x_1}u_{x_2x_2} - u_{x_1x_2}^2) = u_{x_2x_2}^2 + u_{x_1x_2}^2 + t^2(u_{x_1}u_{x_2x_2} - u_{x_2}u_{x_1x_2})^2,$$
$$-(1 + t^2u_{x_1}^2)(u_{x_1x_1}u_{x_2x_2} - u_{x_1x_2}^2) = u_{x_1x_1}^2 + u_{x_2x_2}^2 + t^2(u_{x_1}u_{x_1x_2} - u_{x_2}u_{x_1x_1})^2.$$

Summing the above we see that

$$\begin{aligned} -(1 + W_t^2)(u_{x_1x_1}u_{x_2x_2} - u_{x_1x_2}^2) &= u_{x_1x_1}^2 + 2u_{x_1x_2}^2 + u_{x_2x_2}^2 + t^2(u_{x_1}u_{x_2x_2} - u_{x_2}u_{x_1x_2})^2 \\ &\quad + t^2(u_{x_1}u_{x_1x_2} - u_{x_2}u_{x_1x_1})^2 \\ &\geqq u_{x_1x_1}^2 + 2u_{x_1x_2}^2 + u_{x_2x_2}^2. \end{aligned}$$

The estimate follows.

Let us note that if λ is any eigenvalue of the matrix $(u_{x_jx_k})$, then $\lambda^2 \leqq u_{x_1x_1}^2 + 2u_{x_1x_2}^2 + u_{x_2x_2}^2$. Hence it suffices to estimate a λ to verify (13). We follow [**9**, p. 238]. If g is any function of class C^1 in a neighborhood of $\partial\Omega$,

$$\partial g/\partial s = -\sin\vartheta g_{x_1} + \cos\vartheta g_{x_2}.$$

Hence, $0 = \partial^2u/\partial s^2 = -u_\nu d\vartheta/ds + (\sin\vartheta)^2u_{x_1x_1} - 2\sin\vartheta\cos\vartheta u_{x_1x_2} + (\cos\vartheta)^2u_{x_2x_2}$, where u_ν denotes the interior normal. By the maximum principle, $u = 0$ on $\partial\Omega$ and $u > 0$ in Ω implies that $u_\nu > 0$ on $\partial\Omega$. Hence, $u_\nu(d\vartheta/ds) > 0$ on $\partial\Omega$. Note that the choice of interior normal implies that $d\vartheta/ds > 0$ when $\partial\Omega$ is traversed in the usual counterclockwise sense. Hence denoting by $\lambda(z)$ the positive eigenvalue of $(u_{x_jx_k}(z))$

at $z \in \partial\Omega$, we obtain that $\lambda(z) \geqq u_\nu(z)\, d\vartheta/ds > 0$.

Therefore,

$$K \leqq -\frac{1}{W^4(1 + W_t^2)}\lambda^2 \leqq -\frac{W^2 - 1}{W^4(1 + W_t^2)}\left(\frac{d\vartheta}{ds}\right)^2, \qquad z \in \partial\Omega.$$

From Lemma 2, we observe that h_t^{-1} is Lipschitz near $h_t(\partial\Omega)$, for one computes that $\det(\partial h_j/\partial x_k) = KW^4/W_t(1 + W_t^2)$. It follows that for some positive constant $c_0 > 0$,

$$|h_t(z) - h_t(z')| \geqq c_0|z - z'| \quad \text{for } z, z' \in \partial\Omega.$$

The conclusion of Theorem 3 holds, therefore, for h_t, $0 < t \leqq 1$. In view of the relation

$$u_{x_1} - iu_{x_2} = 2h_t/(1 - t^2|h_t|^2)$$

and noting that $|h_t| < 1$, we may conclude that there is a constant $\gamma > 0$ such that

$$\gamma < (u_{x_1x_1})^2 + (u_{x_1x_2})^2 + (u_{x_2x_2})^2 \quad \text{for } z \in \Omega - I_t.$$

Applying (12) to the above, we may conclude with

THEOREM 4. *Let ψ be a C^2 strictly concave obstacle. Then the curvature $K(z)$ of the solution u_t to the problem* (P_t) *satisfies*

$$K(z) \leqq -K_0 < 0 \quad \textit{for } z \in \Omega - I_t$$

where $K_0 > 0$ may depend on t, ψ, Ω.

5. **A remark about a nonhomogeneous problem.** We close with a simple remark. For a fixed constant λ, let $u \in \mathscr{K}$ be the solution to the variational inequality

$$(14) \qquad u \in \mathscr{K}: \quad \int_\Omega u_{x_j}(v - u)_{x_j}\, dx \geqq \lambda \int_\Omega (v - u)\, dx \quad \text{for } v \in \mathscr{K}.$$

Let $\varphi \in H^{2,p}(\Omega) \cap C^\infty(\Omega)$ denote the solution to the problem $\Delta\varphi = \lambda$ in Ω and $\varphi = 0$ on $\partial\Omega$. It is clear, then, that $U(z) = u(z) - \varphi(z)$ is the solution to the variational inequality (1) for obstacle $\psi - \varphi$ and $a_j(p) = p_j$. Hence if $\psi - \varphi$ is strictly concave and $\max_\Omega(\psi - \varphi) > 0$, the function U, hence u, satisfies the conclusion of Theorem 2. On the other hand, if $\psi - \varphi < 0$ in Ω, then φ is easily seen to be the solution of (14). The difficulty, therefore, lies in satisfying the constraint that $\psi - \varphi$ be concave. Miss Silvia Mazzone, in a yet unpublished paper, has studied the existence of solutions for the problem (14) for general $\lambda = \lambda(x, u)$ and $a(p)$ locally coercive, where the λ is placed inside the integral sign, of course.

REFERENCES

1. L. Bers, F. John and M. Schecter, *Partial differential equations*, Proc. Summer Seminar (Boulder, Col., 1957), Lectures in Appl. Math., vol. 3, Interscience, New York, 1962. MR **29** #346.

2. H. R. Brézis and G. Stampacchia, *Sur la régularité de la solution d'inéquations elliptiques*, Bull. Soc. Math. France **96** (1968), 153–180. MR **39** #659.

3. H. Brézis, *Seuil de régularité pour certains problèmes unilatéraux*, C. R. Acad. Sci. Paris (to appear).

4. ———, *Nouveaux théorèmes de régularité pour les problèmes unilatéraux* (to appear).

5. R. Courant and D. Hilbert, *Methods of mathematical physics*. Vol. II: *Partial differential equations*, Interscience, New York, 1962. MR **25** #4216.

6. M. Giaquinta and L. Pepe, *Esistenza e regolarità per il problema dell'area minima con ostocoli in n variabili*, Ann. Scuola Norm Sup. Pisa **25** (1971), 481–507.

7. M. Miranda, *Frontiere minimali con ostacoli*, Ann. Univ. Ferrara (2) **16** (1971), 29–37.

8. ———, *Existence and regularity of hypersurfaces in R^n with prescribed mean curvature*, Proc. Sympos. Pure Math., vol. 23, Amer. Math. Soc., Providence, R. I., 1972, pp. 1–9.

9. D. Kinderlehrer, *The coincidence set of solutions of certain variational inequalities*, Arch. Rational Mech. Anal. **40** (1970/71), 231–250. MR **42** #6680.

10. H. Lewy and G. Stampacchia, *On the regularity of the solution of a variational inequality*, Comm. Pure Appl. Math. **22** (1969), 153–188. MR **40** #816.

11. ———, *On the existence and smoothness of solutions of some non-coercive variational inequalities*, Arch. Rational Mech. Anal. **41** (1971), 241–253.

12. T. Radó, *Bemerkung über die differentialgleichungen zwei-dimensionaler Variationsprobleme*, Acta. Litt. Sci. Univ. Szeged **1925**, 147–156.

13. G. Stampacchia, *Regularity of solutions of some variational inequalities*, Proc. Sympos. Pure Math., vol. 18, part I, Amer. Math. Soc., Providence, R. I., 1970.

14. F. Tomi, *Ein teilweise friese Randwertproblem für Flächen vorgeschriebener mittlerer Krümmung*, Math. Z. **115** (1970), 104–112. MR **41** #7553.

15. F. Tomi, *Minimal surfaces and surfaces of prescribed mean curvature spanned over obstacles*, Math. Ann. **190** (1970/71), 248-264. MR **42** #6683.

University of Minnesota

Scuola Normale Superiore, Pisa

ASYMPTOTICS OF A NONLINEAR RELATIVISTIC WAVE EQUATION[1]

CATHLEEN S. MORAWETZ AND WALTER A. STRAUSS

1. How can we best describe the asymptotic behavior in time of the solutions of a hyperbolic partial differential equation? We may ask that, for each solution u,

$$\exists u_{\pm} \quad \text{such that } u \sim u_{\pm} \text{ as } t \to \pm\infty \tag{1}$$

where u_+ and u_- are a pair of simpler objects. The problem is to find these simpler objects and to decide in what sense the asymptotic relation holds. Here we consider the Lorentz invariant equation

$$\partial^2 u/\partial t^2 - \Delta u + m^2 u + g u^3 = 0 \tag{2}$$

where m and g are positive constants. It is called conservative because, upon multiplying it by $\partial u/\partial t$ and integrating, we find the energy

$$E = \int [\tfrac{1}{2}(u_t^2 + |\nabla u|^2 + m^2 u^2) + g u^4/4]\, dx$$

to be a constant. The integration is over all three-dimensional space. The solutions we consider have finite energy E and so they are small as $|x| \to \infty$. That m^2 and g are positive has the consequence that each term in E is nonnegative and therefore bounded. Equation (2) is perhaps the simplest relativistic nonlinear equation; cf. the articles of Arthur Wightman and Arthur Jaffe in these Proceedings. It is therefore a prototype of more general semilinear Lorentz invariant hyperbolic systems.

Since there is no decay of the energy E, we may ask whether the solution itself goes to zero. If so, the nonlinear term should be negligible for large $|t|$ and natural candidates for $u_{\pm}$ should be what we call free solutions; that is, solutions of the

AMS 1970 *subject classifications*. Primary 35B40, 35L60, 35P25; Secondary 47H99, 81A48.

[1] This work was supported by Contract no. DA-31-124-ARO-D-365 at the Courant Institute and by NSF grant GP-16919 at Brown University.

linear equation obtained by setting $g = 0$. Condition (1) can now be interpreted to mean that $u - u_\pm$ tends to zero in some norm. To make it nontrivial, we should take a norm in which u, u_+ and u_- do not separately tend to zero. A natural candidate for the norm is available: The *energy norm*

$$\|v(t)\|_e^2 = \int (v_t^2 + |\Delta v|^2 + m^2 v^2)\, dx,$$

which is constant if v is a free solution and does not appear likely to go to zero if v is a solution of (2). This is our interpretation of condition (1).

2. To be more precise, let us put gu^3 on the right side of the equation. Let T_1 and T_2 be two large positive times. Denote by $u_{T_i} = u_{T_i}(x, t)$ the free solution data given by $u = u(x, t)$ at time $t = T_i$ $(i = 1, 2)$. By a standard energy estimate,

$$\|u_{T_1} - u_{T_2}\|_e \leqq \int_{T_1}^{T_2} \|gu^3(\tau)\|_2\, d\tau.$$

($\|\ \|_p$ denotes the L_p norm in space.) But $\|u^3\|_2 \leqq \|u\|_\infty^2 \|u\|_2$, so that

$$\|u_{T_1} - u_{T_2}\|_e \leqq cg \int_{T_1}^{T_2} \|u(\tau)\|_\infty^2\, d\tau$$

where c depends on E. Thus we have the

LEMMA. *$\{u_T\}$ converges as $T \to +\infty$ to a free solution (called u_+) provided*

$$\int^{+\infty} \sup_x |u(x, t)|^2\, dt < \infty.$$

The analogue holds when $T \to -\infty$ and it is easy to see that u_+ and u_- are the required free solutions.

Thus the question is whether the solutions of the nonlinear equation decay uniformly. The answer is yes for the *free* solutions. In fact, Fourier transformation or other methods show that the free solutions decay uniformly like $|t|^{-3/2}$ as $|t| \to \infty$. That this is the maximal possible rate of decay can be seen by the following elementary argument. Consider a free solution v with initial data of compact support, say in $|x| \leqq k$. At time $t > 0$ its support is in $|x| < t + k$. The volume of its support increases like t^3. If it decayed uniformly at any rate faster than $t^{-3/2}$, then $\int v^2\, dx$ would tend to zero. But it is well known that, as one of the terms in the energy, $\int v^2\, dx$ tends to a nonzero constant.

What makes the problem difficult is that u^3 is much larger than u when u is large. What must be exploited is that u^3 is much smaller than u when u is small. In fact, if *either* the initial data of u *or* the coupling constant g is sufficiently small, then the problem has been solved by an iterative procedure by Irving Segal. But if $g = 1$ say and the data are large, it is not clear whether u^3 could swamp all the linear effects. It should be mentioned that the existence and regularity of solutions of (1) are known and that our problem also has an affirmative answer when $m = 0$ by making use of a special invariance property in that case.

3. THEOREM 1. *For solutions of* (2) *with smooth data of compact support there is a constant* c *such that, for all* x, t,

$$|u(x, t)| \leq \frac{c}{1 + |t|^{3/2}}.$$

Hence (1) *follows from the Lemma.*

It turns out that any one of u_-, u and u_+ determine the other two uniquely. We write $u_+ = Su_-$ whenever u_+ and u_- are related as in (1) via some u. Then S is a *nonlinear* operator acting on *free* solutions. What is the domain of S? That is, which free solutions u_- possess u's which possess u_+'s? Let us define the space $\mathscr{F}$ as the limits of the free solutions with smooth data of compact support in the following norm:

$$\|v\|_F^2 = \sup_t \|v(t)\|_e^2 + \sup_t \sup_x |v(x, t)|^2 + \int_{-\infty}^{\infty} \sup_x |v(x, t)|^2 \, dt.$$

The variables run over all space and time. Thus $\mathscr{F}$ is a Banach space of weak solutions of the linear equation. $\mathscr{F}$ is dense in the space of finite energy solutions. The motivation for the last term in the definition of the norm comes from the Lemma.

THEOREM 2. *S is defined on all of $\mathscr{F}$ and is a diffeomorphism of $\mathscr{F}$ onto itself.*

Thus "outgoing waves are just as good as incoming waves".

4. The detailed proofs will appear in *Communications on Pure and Applied Mathematics*. A basic tool is the standard one of inverting the linear part:

$$u(x, t) = u_0(x, t) - \int_0^t \int R(x - y, t - \tau) g u^3(y, \tau) \, dy \, d\tau \tag{3}$$

where the influence (Riemann, Green's, source) function $R(x, t)$ is the free solution with initial data 0, $\delta(x)$. Explicitly, for $t > 0$ the kernel $R(x - y, t - \tau)$ lives on the backward cone $|x - y| < t - \tau$. On the surface of this cone it is the singular function $[4\pi(t - \tau)]^{-1}\delta(|x - y| - t + \tau)$. Inside the cone it is $(4\pi\mu)^{-1} m J_1(m\mu)$, where $\mu^2 = (t - \tau)^2 - |x - y|^2$.

The second main tool uses the sign of the nonlinear term. It is a global estimate of energy type, completely elementary but not obvious. The consequence of it which we need is that $|x|^{-1}[u(x, t)]^4$ is integrable over all space-time. If $u(x, t)$ lives in $|x| < t + k$, we can replace the $|x|^{-1}$ by $(t + k)^{-1}$. Thus

$$\int^{\infty} f(t) \, dt/t < \infty \quad \text{where } f(t) = \int u^4(x, t) \, dx.$$

Fortunately t^{-1} is not integrable. So $\int_I f(t)\,dt$ is arbitrarily small on arbitrarily large time intervals I. That is,

(i) $\liminf_{s\to\infty} \int_{S-T}^{S} \int u^4\,dx\,dt = 0$.

Theorem 1 is proved by a jacking-up procedure starting with (i). The succeeding steps are:

(ii) $u(x,t)$ is arbitrarily small on some arbitrarily long time interval;

(iii) $u(x,t) \to 0$ uniformly;

(iv) $\int^\infty \sup_x |u(x,t)|^2\,dt < \infty$;

(v) $\sup_x |u(x,t)| = O(t^{-3/2})$.

The most interesting step is from (ii) to (iii). Let ε be a small positive number. Let $T = T(\varepsilon)$ be sufficiently large. By step (ii), $|u(x,t)| < \varepsilon$ on some time interval $[t^* - T, t^*]$. Let

$$t^{**} = \sup\{s \mid |u| < \varepsilon \text{ in } [t^* - T, s]\}.$$

If $t^{**} = \infty$ there is nothing to prove. Suppose t^{**} is finite. Take a time t slightly later than t^{**}, $t^{**} \leqq t \leqq t^{**} + \delta$. Break the right side of (3) into four parts. Since $t \geqq t^{**} \geqq T$ is large enough, $|u_0| < \varepsilon/4$. The integral over $[t^{**}, t]$, the tip of the cone, is less than $\varepsilon/4$ if δ is chosen small enough. In $[t - T, t^{**}]$, we have $|u(x,t)| < \varepsilon$. Since u appears in (3) to the third power and ε is small, we can arrange the integral over $[t - T, t^{**}]$ to be less than $\varepsilon/4$. The rest of the integration is over the large base of the cone $[0, t - T]$, where we do not know that u is small. However $t - \tau > T$ in that interval and so $R(x - y, t - \tau)$ is small in some sense. The kernel is actually constant on the hyperboloids $\mu =$ constant, but these hyperboloids bunch together very closely. In any case, from (3) we obtain $|u(x,t)| < 4(\varepsilon/4) = \varepsilon$, which is a contradiction of the definition of t^{**}.

5. Some basic questions remain open. Can S be defined on the whole space of free solutions of finite energy? Are similar results valid in any other dimension? Are they valid for other nonlinear dispersive systems, for example, the Maxwell-Dirac equations? In what way might these results be used for the construction of quantum fields?

COURANT INSTITUTE, NEW YORK UNIVERSITY

BROWN UNIVERSITY

PROPAGATION OF ZEROES OF SOLUTIONS OF P.D.E.'S ALONG LEAVES OF FOLIATIONS[1]

E. C. ZACHMANOGLOU

1. **Introduction.** Let $P(x,D)$ be a partial differential operator of order m with complex-valued coefficients defined and analytic in an open connected set Ω in R^n:

$$P(x,D) = \sum_{|\alpha| \leqq m} a^{\alpha}(x)D^{\alpha}. \tag{1}$$

Here, $x = (x_1, \ldots, x_n), \alpha = (\alpha_1, \ldots, \alpha_n)$ is an n-tuple of nonnegative integers with $|\alpha| = \alpha_1 + \cdots + \alpha_n$ and $D^{\alpha} = D_1^{\alpha_1} \cdots D_n^{\alpha_n}$ with $D_j = \partial/\partial x_j$. The principal part is the homogeneous part of order m,

$$P_m(x,D) = \sum_{|\alpha| = m} a^{\alpha}(x)D^{\alpha}. \tag{2}$$

At a fixed point $x \in \Omega$, the (real) zeroes of $P_m(x,\xi)$ form a cone in R^n which is called the (real) characteristic cone of $P(x,D)$ at x. We will denote by $\mathscr{A}$ the ring of real-valued analytic functions in Ω.

In this paper we consider partial differential operators having the following property: There exist r analytic vector fields, in Ω,

$$A_j = \sum_{i=1}^{n} a_j^i D_i, \qquad j = 1, \ldots, r, \tag{3}$$

with $a_j^i \in \mathscr{A}$, $i = 1, \ldots, n, j = 1, \ldots, r$, such that at each point of Ω the characteristic cone of $P(x,D)$ is orthogonal to every A_j. More precisely, we assume that, for every $x \in \Omega$,

$$P_m(x,\xi) = 0, \xi \in R^n \quad \Rightarrow \quad \sum_{i=1}^{n} a_j^i(x)\xi_i = 0, \qquad j = 1, \ldots, r. \tag{4}$$

AMS 1970 *subject classifications*. Primary 35A05; Secondary 57D30.

[1]Research supported by NSF grant GP-20547.

We will denote by $\mathscr{L}(A_1,\ldots,A_r)$ the Lie algebra generated by $A_1,\ldots,A_r$, i.e. the smallest set of analytic vector fields in Ω which is closed under the operations of taking brackets and linear combinations with coefficients in $\mathscr{A}$.

According to a theorem of Nagano [**1**], which is an extension of the classical theorem of Frobenius, the Lie algebra $\mathscr{L}(A_1,\ldots,A_r)$ defines a unique partition of Ω into maximal integral manifolds of $\mathscr{L}(A_1,\ldots,A_r)$, that is, Ω is the disjoint union of maximal integral manifolds of $\mathscr{L}(A_1,\ldots,A_r)$. This partition is called a foliation and each maximal integral manifold is called a leaf of the foliation.

In this paper we prove that the zeroes of solutions of the equation $P(x,D)u = 0$ propagate along the leaves of the foliation defined by $\mathscr{L}(A_1,\ldots,A_r)$. More precisely let u be a distribution solution of $P(x,D)u = 0$ in Ω and suppose that u vanishes in an open neighborhood of a point $x \in \Omega$. Then u also vanishes in an open neighborhood of every point of the leaf through x of the foliation defined by $\mathscr{L}(A_1,\ldots,A_r)$.

This result includes the well-known result on the propagation of zeroes of solutions of elliptic equations. It includes also the result of Bony [**2**] concerning degenerate elliptic second order equations and the result of Zachmanoglou [**3**] concerning first order equations with complex-valued coefficients.

2. **Some results on foliations.** Let A and B be two vector fields, $A = \sum_{i=1}^n a^i D_i$, $B = \sum_{i=1}^n b^i D_i$. The bracket $[A,B]$ is the commutator

$$[A,B] = AB - BA = \sum_{i=1}^n \left[\sum_{k=1}^n a^k(x) D_k b^i(x) - \sum_{k=1}^n b^k(x) D_k a^i(x) \right] D_i.$$

If A and B have analytic coefficients in Ω, then $[A,B]$ is also a vector field with analytic coefficients in Ω. The (real) vector space of all real analytic vector fields in Ω equipped with the bracket operation is a Lie algebra denoted by $\mathscr{L}(\Omega)$. It is also a module over the ring $\mathscr{A}$ of real analytic functions in Ω. $\mathscr{A}(A_1,\ldots,A_r)$ will denote the smallest $\mathscr{A}$-submodule of $\mathscr{L}(\Omega)$ containing the vector fields $A_1,\ldots,A_r$. A vector subspace of $\mathscr{L}(\Omega)$ which is closed under the bracket operation is a Lie subalgebra of $\mathscr{L}(\Omega)$. $\mathscr{L}(A_1,\ldots,A_r)$ is the smallest $\mathscr{A}$-submodule and Lie subalgebra of $\mathscr{L}(\Omega)$ containing $A_1,\ldots,A_r$.

Let $\mathscr{L}$ be a vector subspace of $\mathscr{L}(\Omega)$. For any $x \in \Omega$ we set

$$\mathscr{L}(x) = \{A(x) : A \in \mathscr{L}\}.$$

$\mathscr{L}(x)$ is a subspace of R^n and is called the integral element of $\mathscr{L}$ at x. An integral manifold N of $\mathscr{L}$ is a connected submanifold of Ω such that for every $x \in N$, the tangent space to N at x is equal to $\mathscr{L}(x)$.

THEOREM 1. *If $\mathscr{L}$ is a Lie subalgebra of $\mathscr{L}(\Omega)$, then through every point $x \in \Omega$ passes a maximal integral manifold L^x of $\mathscr{L}$. Any integral manifold of $\mathscr{L}$ containing x is an open submanifold of L^x.*

According to Theorem 1, $\mathscr{L}$ defines a unique partition of Ω by integral manifolds of $\mathscr{L}$ (that is, Ω is the disjoint union of maximal integral manifolds of $\mathscr{L}$). This partition of Ω will be called the foliation defined by $\mathscr{L}$ and each maximal integral manifold will be called a leaf of the foliation. Note that for every $x \in \Omega$, the dimension of the leaf L^x containing x is equal to the dimension of the integral element $\mathscr{L}(x)$.

With the additional assumption that dim $\mathscr{L}(x)$ is constant in Ω, Theorem 1 is the classical theorem of Frobenius (see Chevalley [**4**]). However, the Frobenius theorem is also valid in the C^∞ case. Theorem 1 was proved by Nagano [**1**] and it is not generally valid in the C^∞ case. (For conditions weaker than analyticity under which Theorem 1 holds, see Hermann [**5**], Matsuda [**6**] and Kumano-Go and Matsuda [**7**].)

We will apply Theorem 1 to the Lie subalgebra $\mathscr{L}(A_1, \ldots, A_r)$. The leaf of its foliation through x will be denoted by $L^x(A_1, \ldots, A_r)$.

An integral curve of an analytic vector field, say A, is a solution $x = x(t)$ of the system

$$dx_i/dt = a^i(x), \qquad i = 1, \ldots, n.$$

A trajectory of a collection $\mathscr{C}$ of analytic vector fields in Ω is a piecewise analytic curve in Ω, each analytic piece of which is an integral curve of a member of $\mathscr{C}$.

By restricting $\mathscr{L}(A_1, \ldots, A_r)$ to the leaf L^x and using the methods described in §3.11 of Bishop and Goldberg [**8**] it is easy to show the following theorem which provides a means for constructing L^x by solving ordinary differential equations.

THEOREM 2. *Let $x \in \Omega$. The set of points of Ω which can be connected to x by trajectories of $\mathscr{L}(A_1, \ldots, A_r)$ is precisely the leaf $L^x(A_1, \ldots, A_r)$.*

However, it is the following theorem which is needed in order to prove our result concerning the propagation of zeroes of solutions of the equation $P(x,D)u = 0$.

THEOREM 3. *Let $x \in \Omega$. The set of points of Ω which can be connected to x by trajectories of $\mathscr{A}(A_1, \ldots, A_r)$ is precisely the leaf $L^x(A_1, \ldots, A_r)$.*

This theorem can be proved by extending the proof of Lemma 2 in [**3**].

3. **The propagation of zeroes.**

LEMMA 1. *Let $A = \sum_{i=1}^n a^i D_i$ be an analytic vector field in Ω so that at each point of Ω the characteristic cone of $P(x,D)$ is orthogonal to A, that is, for every $x \in \Omega$,*

$$P_m(x,\xi) = 0,\ \xi \in R^n \quad \Rightarrow \quad \sum_{i=1}^n a^i(x)\xi_i = 0.$$

Let u be a distribution solution of

$$P(x,D)u = 0 \quad in\ \Omega, \tag{5}$$

and suppose that u vanishes in an open neighborhood of some point x of Ω. *Then u vanishes in an open neighborhood of every point of the integral curve of A passing through x.*

The proof of this lemma consists of locally straightening out the vector field A and applying Theorem 1 of [**9**] concerning the propagation of zeroes of solutions of partial differential equations with flat characteristic cones. Theorem 1 of [**9**] was proved using Holmgren's uniqueness theorem as extended to distribution solutions by Hörmander [**10**] and a method first used by John [**11**].

It follows immediately from Lemma 1 that a solution of $P(x,D)u = 0$ which vanishes in a neighborhood of any point $x \in \Omega$ also vanishes in a neighborhood of every point of a trajectory of $\mathscr{A}(A_1, \ldots, A_r)$ containing x. Combining this with Theorem 3 we obtain the main result of this paper. For convenience we repeat all assumptions mentioned in the introduction.

THEOREM 4. *Let* $P(x,D)$ *be a partial differential operator with analytic coefficients in an open connected set* Ω *of* R^n. *Suppose that there are r analytic vector fields* $A_1, \ldots, A_r$ *in* Ω *such that at each point of* Ω *the characteristic cone of* $P(x,D)$ *is orthogonal to every* A_j, *i.e., for every* $x \in \Omega$,

$$(4) \qquad P_m(x,\xi) = 0, \ \xi \in R^n \ \Rightarrow \ \sum_{i=1}^{n} a_j^i(x)\xi_i = 0, \qquad j = 1, \ldots, r.$$

Let u be a distribution solution of

$$(5) \qquad P(x,D)u = 0 \quad in\ \Omega,$$

and suppose that u vanishes in an open neighborhood of some point $x \in \Omega$. *Then u vanishes in a neighborhood of every point of the leaf* $L^x(A_1, \ldots, A_r)$ *of the foliation defined by* $\mathscr{L}(A_1, \ldots, A_r)$.

COROLLARY 1. *Under the assumption of Theorem 4 and if in addition at every point of* Ω *the dimension of the integral element of* $\mathscr{L}(A_1, \ldots, A_r)$ *is equal to n, then every solution of* $P(x,D)u = 0$ *vanishing in a neighborhood of a point of* Ω *must vanish in the whole of* Ω.

4. **Some examples.** If $P(x,D)$ is an elliptic operator in Ω, then, by definition, for every $x \in \Omega$,

$$P_m(x,\xi) = 0, \ \xi \in R^n \ \Leftrightarrow \ \xi = 0.$$

In this case we can take $A_j = D_j$, $j = 1, \ldots, n$. We have $\mathscr{L}(D_1, \ldots, D_n) = \mathscr{L}(\Omega)$ and its foliation consists of a single leaf, the whole of Ω. Hence any solution of $P(x,D)u = 0$ vanishing in a neighborhood of a point of Ω must vanish in the whole of Ω.

The operators studied by Bony [**2**] are of the form

$$P(x,D) = \sum_{j=1}^{r} A_j^2 + B + c$$

where $A_1, \ldots, A_r$ and B are real first order operators (or vector fields) with analytic coefficients in Ω. At every point of Ω the characteristic cone is orthogonal to the vector fields $A_1, \ldots, A_r$. Under the assumption that the dimension of the integral element of $\mathscr{L}(A_1, \ldots, A_r)$ is equal to n at every point of Ω, Bony showed that any solution of $P(x,D)u = 0$ vanishing in a neighborhood of a point of Ω must vanish in the whole of Ω. This of course is precisely the assertion of Corollary 1.

A first order operator with complex-valued coefficients is of the form

$$P(x,D) = A + iB + c$$

where A and B are real first order operators. The characteristic cone is orthogonal to the vector fields A and B. The propagation of zeroes and uniqueness in the Cauchy problem for this operator were studied in [**3**].

We present now some nontrivial examples of foliations.

In R^2 let

$$A_1 = D_1, \qquad A_2 = x_2 D_2.$$

The foliation defined by $\mathscr{L}(A_1, A_2)$ consists of three leaves,

$$x_2 > 0; \qquad x_2 = 0; \qquad x_2 < 0.$$

In R^3 let

$$A_1 = D_1, \qquad A_2 = D_2, \qquad A_3 = x_1 x_3 D_3.$$

The foliation defined by $\mathscr{L}(A_1, A_2, A_3)$ consists of three leaves,

$$x_3 > 0; \qquad x_3 = 0; \qquad x_3 < 0.$$

In R^3 let

$$A_1 = D_1, \qquad A_2 = (x_2^2 + x_3^2)(D_2 + x_1 D_3).$$

The foliation defined by $\mathscr{L}(A_1, A_2)$ consists of two leaves, the x_1-axis and the complement of the x_1-axis.

Bibliography

1. T. Nagano, *Linear differential systems with singularities and an application to transitive Lie algebras*, J. Math. Soc. Japan **18** (1966), 398–404. MR **33** #8005.

2. Jean-Michel Bony, *Principe du maximum et inégalité de Harnack pour les opérateurs elliptiques degénérés*, Séminaire Brelot-Choquet-Deny, 12e année: 1967/1968, Théorie du potentiel, no. 10.

3. E. C. Zachmanoglou, *Propagation of zeroes and uniqueness in the Cauchy problem for first order partial differential equations*, Arch. Rational Mech. Anal. **38** (1970), 178–188.

4. C. Chevalley, *Theory of Lie groups*. I, Princeton Univ. Press, Princeton, N.J., 1946. MR **18**, 583.

5. Robert Hermann, *The differential geometry of foliations*. II, J. Math. Mech. **11** (1962), 303–315. MR **25** #5524.

6. M. Matsuda, *An integration theorem for completely integrable systems with singularities*, Osaka J. Math. **5** (1968), 279–283. MR **39** #4876.

7. H. Kumano-Go and M. Matsuda, *An analyticity problem and an integration theorem of completely integrable systems with singularities*, Osaka J. Math. **7** (1970), 225–229.

8. R. L. Bishop and S. I. Goldberg, *Tensor analysis on manifolds*, Macmillan, New York, 1968. MR **36** #7057.

9. E. C. Zachmanoglou, *An application of Holmgren's theorem and convexity with respect to differential operators with flat characteristic cones*, Trans. Amer. Math. Soc. **140** (1969), 109–115. MR **39** #1790.

10. L. Hörmander, *Linear partial differential operators*, Academic Press, New York, 1963.

11. F. John, *On linear partial differential equations with analytic coefficients*, Comm. Pure Appl. Math. **2** (1949), 209–253. MR **12**, 185.

PURDUE UNIVERSITY

AFFINE CONNECTIONS WITH ZERO TORSION

BOHUMIL CENKL

A smooth foliation (smooth = $C^r, r = 1, 2, \ldots, \infty$, or real analytic) of codimension q on a smooth manifold M is given by a completely integrable system of linear differential equations—Pfaff's system [**2**].

A classifying space $B\Gamma_q^r$ for such foliations has been defined by Haefliger [**1**]. In fact there is a classifying space $B\Gamma$ associated with any topological groupoid Γ, using the analog of Milnor's joint construction of the classifying space for principal G-bundles. Using the representability theorem of Brown for a homotopy functor from CW-complexes X into the set of homotopy classes of Γ-structures $\Gamma(X)$ on X Haefliger derives the existence of a CW-complex $B\Gamma$ with a canonical Γ-structure on it.

Using some differential geometric constructions one can construct a classifying space $B\Delta_q$ for foliations of codimension q on M with a fixed fundamental group $\pi_1(\mathrm{M})$ = the given group π_1 if the foliations with a flat normal bundles are considered.

By an affine connection on a q-dimensional subbundle N of the tangent bundle $T(M)$ we mean a linear connection on the affine extension $\tilde{N}$ of N. On the principal bundle P associated with N, an affine connection $\tilde{\omega}$ is given by a pair of forms (ω, α), where ω is a 1-form of a linear connection on P and α is an R^q-valued 1-form, zero on vertical vectors. We say that α, or the affine connection, is regular if α considered as a mapping has maximal rank. Suppose that we denote by D the covariant differential of the connection ω, and by $\theta = D\alpha$ the 2-form called the torsion of $\tilde{\omega}$. Now a system of differential equations $\alpha = 0$ such that $\theta = 0$ is a completely integrable system of Pfaff. In fact we have the following known fact.

THEOREM 1. *Let E be a distribution of codimension q on a manifold M and suppose*

AMS 1970 *subject classifications*. Primary 57D30.

that N is a normal bundle to E, i.e. $T(M) = E \oplus N$. The distribution E is integrable if and only if there exists a regular affine connection on N with torsion zero.

Suppose now that we fix the fundamental group $\pi_1(M) = \pi_1$ and assume that the normal bundle N is flat. Then an affine connection $\tilde{\omega} = (\omega, \alpha)$, ω being flat, has zero torsion if and only if the holonomy map from the loop space ΩM into the affine group $A(q)$ (= semidirect product $L(q) \times R^q$, $L(q) = GL(q, R)$) factors through π_1. The integrability conditions for the system $\alpha = 0$ can be formulated in topological terms. Let $A_d(q)$ be $A(q)$ with discrete topology, and $BA_d(q)$ be the classifying space.

THEOREM 2. *A connection $\tilde{\omega} = (\omega, \alpha)$, ω-flat, on a principal bundle $\tilde{P}$, associated to flat N, has zero torsion if and only if either one of the following conditions holds:*
(1) *the structure group $A(q)$ can be reduced to $A_d(q)$;*
(2) *there exists a homomorphism $\lambda: \pi_1 \to A_d(q)$ such that the structure group $A(q)$ can be reduced to $\pi_\lambda = \lambda(\pi_1)$.*

In order to simplify the formulations of the following theorems we introduce some terminology. A flat differential system α_N of codimension q on M is a triple (P, ω, α), where P is the principal $L(q)$-bundle associated to a flat q-dimensional subbundle N of $T(M)$, ω is the flat connection on ω on P and (ω, α) is the affine connection on $\tilde{P}$. α_N is said to be regular if α is regular. Two regular differential systems are I-homotopic if they are homotopic via regular systems. Ordinary homotopy is via 1-parametric family of differential systems.

DEFINITION. $B\Delta_q = \bigcup B\pi_\lambda$; the topological sum over all λ's one from each equivalence class, with respect to an inner automorphism of $A(q)$, of elements in $\mathrm{Hom}(\pi_1, A(q))$ is called the classifying space for the flat differential systems of codimension q.

The space $B\Delta_q$ has the following properties:

THEOREM 3. *There is 1-1 correspondence between the set of homotopy classes of flat differential systems of codimension q on M and the set of homotopy classes of liftings $\tilde{\gamma}: M \to B\Delta_q$ of the classifying map $v: M \to BL(q)$ for the flat q-dimensional subbundles of $T(M)$ with respect to the projection σ in the diagram:*

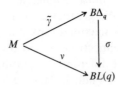

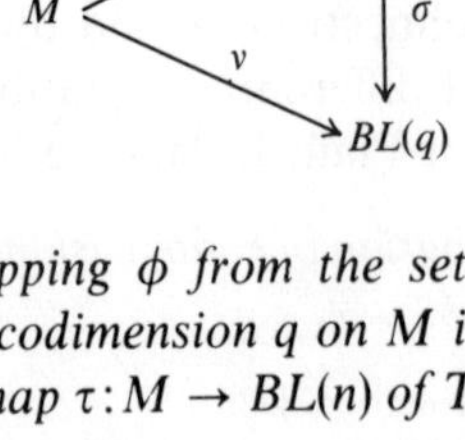

THEOREM 4. *There is a mapping ϕ from the set of I-homotopy classes of flat regular differential systems of codimension q on M into the set of homotopy classes of liftings ρ of the classifying map $\tau: M \to BL(n)$ of $T(M)$ such that*

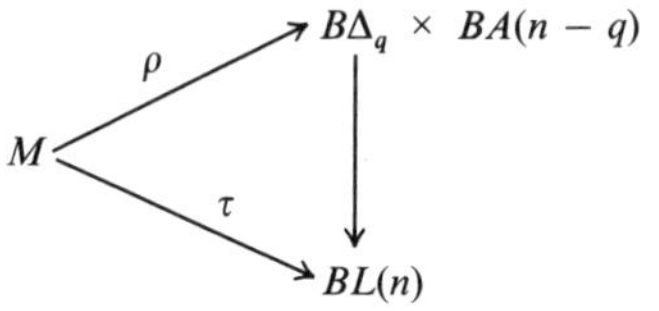

THEOREM 5. *If M is open then ϕ is bijective onto.*

BIBLIOGRAPHY

1. André Haefliger, *Feuilletages sur les variétés ouvertes*, Topology **9** (1970), 183–194. MR **41** #7709.

2. Georges Reeb, *Sur certaines properiétés topologiques des variétés feuilletées*, Actualités Sci. Indust., no. 1183 = Publ. Inst. Math. Univ. Strasbourg 11, Hermann, Paris, 1952, pp. 91–154, 157–158. MR **14**, 1113.

NORTHEASTERN UNIVERSITY

ON THE SPENCER COHOMOLOGY OF A LIE EQUATION

HUBERT GOLDSCHMIDT

The infinitesimal transformations of a continuous pseudogroup, acting on a manifold X, in the sense of Lie and Cartan satisfy a linear partial differential equation on the tangent bundle of X of a special type, which is called a *Lie equation*. As in the case of Lie groups, it is natural in studying pseudogroups to describe the structure of the Lie equation associated to a pseudogroup.

In 1962, D. C. Spencer in his fundamental paper [**10**] elaborated a theory of deformations of pseudogroup structures. It was soon realized that many of the results of this paper were concerned with general systems of overdetermined partial differential equations; the formal theory of differential equations was then developed systematically (see for example the expository articles [**2**] and [**11**] on linear differential equations). Recently, Malgrange and Spencer (see [**7**], [**8**] and [**9**]) using this formal theory of differential equations and the contributions of Guillemin-Sternberg [**6**] gave a new version of Spencer's original work. On the other hand, Guillemin and Sternberg ([**5**] and [**4**]) introduced certain topological Lie algebras, transitive Lie algebras, which correspond to the Lie algebras of transitive pseudogroups.

It is now therefore possible to undertake a systematic study of the structure of Lie equations and the relationship of Lie equations to transitive Lie algebras. In this paper, we examine the correspondence between transitive Lie algebras and Lie equations and the relation between existence theorems for a formally transitive Lie equation, the Spencer cohomology groups of the equation and the transitive Lie algebra determined by the equation. For a study of the structure of formally transitive Lie equations, we refer the reader to [**3**].

1. **The Spencer cohomology of a differential equation.** We begin by recalling various facts from the formal theory of linear differential equations (see [**1**], [**2**],

AMS 1970 *subject classifications*. Primary 58H05; Secondary 58G99, 35N10, 22E65.

or [**11**]) to fix the terminology and the notation which we shall use throughout this paper.

Let X be a differentiable manifold of dimension n; we shall assume that X is connected, for simplicity. If E is a vector bundle over X, we denote by $\mathscr{E}$ the sheaf of germs of C^∞-sections of E and by $J_k(E)$ the bundle of k-jets of sections of E. We denote by $\pi_k: J_{k+l}(E) \to J_k(E)$ the natural projection.

A subbundle $R_k \subset J_k(E)$ will be called a linear (homogeneous) differential equation of order k on E. A section u of E is a solution of R_k if the k-jet $j_k(u)$ of u is a section of R_k. If F is another vector bundle over X, a differential operator $P: \mathscr{E} \to \mathscr{F}$ of order k is said to be associated to R_k if R_k is the kernel of the morphism of vector bundles $p_0(P): J_k(E) \to F$ induced by P. In fact, $u \in \mathscr{E}$ is a solution of R_k if and only if $Pu = 0$. It is clear that there always exists such a differential operator P associated to R_k.

For $l \geqq 0$, we associate a subbundle $R_{k+l} \subset J_{k+l}(E)$ with varying fiber in the following way. Choose a differential operator $P: \mathscr{E} \to \mathscr{F}$ associated to R_k; then R_{k+l} is the kernel of the morphism of vector bundles

$$p_l(P): J_{k+l}(E) \to J_l(F)$$

which sends $j_{k+l}(s)(x)$ into $j_l(Ps)(x)$, if s is a section of E over a neighborhood of $x \in X$. Moreover, R_{k+l} depends only on R_k, and if the dimension of the fibers of R_{k+l} is constant, then R_{k+l} is a vector bundle. It is clear that $R_{k+l,x}$ is the space of solutions of $Pu = 0$ of order $k + l$ at x. Hence π_{k+l} maps R_{k+l+1} into R_{k+l}, so we can set $R_\infty = \text{proj lim } R_{k+l}$; this is the space of formal solutions of the equation $Pu = 0$.

DEFINITION. We say that a differential equation $R_k \subset J_k(E)$ is formally integrable if R_{k+l} is a vector bundle and if the projection $\pi_{k+l}: R_{k+l+1} \to R_{k+l}$ is surjective, for all $l \geqq 0$.

We recall the following result:

THEOREM 1. *Let $R_k \subset J_k(E)$ be a formally integrable differential equation of order k on E and let $P: \mathscr{E} \to \mathscr{F}$ be a differential operator of order k associated to R_k. Then there exist a vector bundle G and a differential operator $Q: \mathscr{F} \to \mathscr{G}$ of order l associated to a formally integrable differential equation of order l on F such that*

$$\mathscr{E} \xrightarrow{P} \mathscr{F} \xrightarrow{Q} \mathscr{G} \tag{1}$$

is a complex and is formally exact in the sense that the sequences

$$J_{k+l+m}(E) \xrightarrow{p_{l+m}(P)} J_{l+m}(F) \xrightarrow{p_m(Q)} J_m(G)$$

are exact for $m \geqq 0$. Furthermore the cohomology $H^1(P)$ of the sequence (1) depends only on R_k and is independent of the choice of the operators P and Q satisfying the above properties.

We therefore define the Spencer cohomology groups of R_k by

$$H^1(R_k) = H^1(P)$$

and inductively by

$$H^j(R_k) = H^{j-1}(Q) \quad \text{for } j > 1.$$

The differential operator Q is the compatibility condition for P and expresses all the formal obstructions to solving the inhomogeneous equation $Pu = f$ if $f \in \mathscr{F}$. Local solvability holds for R_k or for P if and only if $H^1(R_k) = 0$.

One of the main problems in the theory of overdetermined partial differential equations is to find sufficient conditions for the vanishing of the Spencer cohomology groups of a differential equation. One condition one may impose on R_k is ellipticity. We shall say that a differential equation $R_k \subset J_k(E)$ is elliptic if, for some differential operator $P: \mathscr{E} \to \mathscr{F}$ associated to R_k, the symbol $\sigma_\xi(P): E \to F$ is injective for all nonzero cotangent vectors ξ to X (see [**1**] or [**11**]). If $R_k \subset J_k(E)$ is elliptic and analytic relative to an analytic structure on E and X, then $H^j(R_k) = 0$, for $j > 0$ (see [**11**]).

2. **Lie equations.** We shall now consider certain differential equations on the tangent bundle T of X. We shall impose the condition on our differential equation $R_k \subset J_k(T)$ that, if vector fields ξ, η are solutions of R_k, then the bracket $[\xi, \eta]$ also be a solution of R_k.

Let us now describe the precise restriction on R_k we wish to consider. If we set

$$[j_k(\xi)(x), j_k(\eta)(x)] = j_{k-1}[\xi, \eta](x),$$

where ξ, η are vector fields defined on a neighborhood of $x \in X$, we obtain a bracket

$$J_k(T)_x \otimes J_k(T)_x \to J_{k-1}(T)_x$$

making $J_\infty(T)_x = \text{proj lim } J_m(T)_x$ into a Lie algebra.

DEFINITION. A formally integrable equation $R_k \subset J_k(T)$ of order k on T is said to be a *Lie equation* if $[R_{k+1}, R_{k+1}] \subset R_k$. A Lie equation is formally transitive if $\pi_0: R_k \to J_0(T)$ is surjective.

If R_k is a Lie equation, then for $x \in X$, the subspace $R_{\infty,x} = \text{proj lim } R_{k+l,x}$ of $J_\infty(T)_x$ is a subalgebra (see [**8**]). In [**3**], we consider Lie equations which are not necessarily formally integrable; as in the above definition, we only impose conditions on R_k itself, although a priori it might seem more natural to impose conditions on differential operators associated to R_k.

The infinitesimal transformations of a continuous pseudogroup in the sense of Lie and Cartan are solutions of a Lie equation. If the pseudogroup is transitive, then this Lie equation is formally transitive.

EXAMPLES. Suppose that $X = \boldsymbol{R}^n$ and let $\mathfrak{g} \subset \mathfrak{gl}(n, \boldsymbol{R})$ be a Lie subalgebra. If $(x^1, \ldots, x^n)$ are the standard coordinates on $\boldsymbol{R}^n$, we consider the differential equation, for the vector field $\xi = \sum_{j=1}^n \xi^j \partial/\partial x^j$ on $U \subset \mathbf{R}^n$,

$$\partial \xi^j(x)/\partial x^i \in \mathfrak{g}$$

for all $x \in U$. This is a formally transitive Lie equation $R_1(\mathfrak{g})$ of order 1 on $\mathbf{R}^n$.

(a) If $n = 2m$, for some integer m, and $\mathfrak{g} = \mathfrak{gl}(m, \mathbf{C}) \subset \mathfrak{gl}(n, \mathbf{R})$, then $R_1(\mathfrak{g})$ is the equation for holomorphic vector fields on $\mathbf{C}^m$.

(b) If $(\; , \;)$ is a nondegenerate bilinear symmetric form on $\mathbf{R}^n$ and $\mathfrak{g}$ is the Lie algebra of all endomorphisms A of $\mathbf{R}^n$ verifying

$$(Au, v) + (u, Av) = 0, \tag{2}$$

then $R_1(\mathfrak{g})$ is the equation for Killing vector fields.

(c) If $(\; , \;)$ is a nondegenerate bilinear skew-symmetric form on $\mathbf{R}^n$ and $\mathfrak{g}$ is the Lie algebra of all endomorphisms A of $\mathbf{R}^n$ verifying (2), then $R_1(\mathfrak{g})$ is the equation for symplectic vector fields.

If $R_k \subset J_k(T)$ is a formally transitive Lie equation and if $x \in X$, we endow $R_{k+l,x}$ with the discrete topology and $R_{\infty,x} = \text{proj lim } R_{k+l,x}$ with the projective limit topology. Then $R_{\infty,x}$ is a topological Lie algebra and a closed subalgebra of $J_\infty(T)_x$

(i) whose underlying topological vector space is a linearly compact topological vector space (that is, the topological dual of a discrete vector space),

(ii) possesses a neighborhood of 0 containing no ideals other than $\{0\}$.

In fact, the kernel $R^0_{\infty,x}$ of $\pi_0 : R_{\infty,x} \to J_0(T)_x$ contains no nontrivial ideals of $R_{\infty,x}$. We shall call a topological Lie algebra satisfying conditions (i) and (ii) *transitive* (see [**4**]). If $x, x' \in X$, then $R_{\infty,x}$ and $R_{\infty,x'}$ are isomorphic as topological Lie algebras, so that R_k determines an isomorphism class of transitive Lie algebras.

THEOREM 2 (GUILLEMIN-STERNBERG [**5**]). *Let L be a transitive Lie algebra. Then there exist a manifold X and $x \in X$ such that L is isomorphic to a closed subalgebra L' of $J_\infty(T)_x$ for which $\pi_0 : L' \to J_0(T)_x$ is surjective.*

Using the above theorem and existence theorems for analytic nonlinear differential equations, in [**3**] we prove:

THEOREM 3 (THIRD FUNDAMENTAL THEOREM). *Let L be a transitive Lie algebra. Then there exists an analytic formally transitive Lie equation R_k of order k on an analytic manifold X such that, for $x \in X$, the transitive Lie algebra $R_{\infty,x}$ is isomorphic to L.*

This analogue of Lie's third fundamental theorem for Lie groups shows that all transitive Lie algebras arise from formally transitive Lie equations. The following theorem shows that closed ideals of a transitive Lie algebra also correspond to Lie equations.

THEOREM 4. *Let R_k be a formally transitive Lie equation of order k on X; let*

$x \in X$ and $I \subset R_{\infty,x}$ a closed ideal. Assume that X is simply connected. There exist an integer $m \geqq k$ and a unique Lie equation $R'_m \subset R_m$ of order m on X such that

$$R'_{\infty,x} = I \quad \text{and} \quad [R_{m+1}, R'_{m+1}] \subset R'_m.$$

3. **The Spencer cohomology of Lie equations.** To a Lie equation $R_k \subset J_k(T)$ one can associate a nonlinear Spencer cohomology group $\tilde{H}^1(R_k)$ (see [**8**] and [**9**]). We shall not attempt to define this set here but only give a description of it, in particular in the case of Example (b).

An almost-complex structure on an open set $X \subset \boldsymbol{R}^{2m}$ is a section J of $T \otimes T^*$ over X satisfying $J^2 = -\mathrm{Id}$. Recall that any complex structure on X defines such a section J: a choice of complex coordinates on a neighborhood of a point $x \in X$ induces a complex vector space structure on T_x which is easily seen to be independent of the choice of local coordinates. An almost-complex structure is said to be *integrable* if it comes from a complex structure. For such a structure J on X, one easily sees that

$$[\xi, J\eta] + [J\xi, \eta] - J[\xi, \eta] - J[J\xi, J\eta] = 0 \tag{3}$$

for all vector fields ξ, η on X. An almost-complex structure J on X satisfying (3) will be called *formally integrable*. Two germs at $x \in \boldsymbol{R}^{2m}$ of almost-complex structures J, J' will be said to be equivalent if there exists a germ of a diffeomorphism $\varphi: (\boldsymbol{R}^{2m}, x) \to (\boldsymbol{R}^{2m}, x)$ such that $\varphi_* \circ J \circ \varphi_*^{-1} = J'$. It is clear that J is formally integrable if and only if J' is, and that J is integrable if and only if it is equivalent to the germ at x of the standard complex structure on $\boldsymbol{R}^{2m}$. If $R_1(\mathfrak{g})$ is the Lie equation of Example (a), we define $\tilde{H}^1(R_1(\mathfrak{g}))_x$ to be the set of equivalence classes of germs at $x \in \boldsymbol{R}^{2m}$ of formally integrable almost-complex structures. The Newlander-Nirenberg theorem says precisely that a formally integrable almost-complex structure is integrable, or equivalently that $\tilde{H}^1(R_1(\mathfrak{g}))$ is trivial. i.e. for all $x \in \boldsymbol{R}^{2m}$, the cohomology group $\tilde{H}^1(R_1(\mathfrak{g}))_x$ has only one element.

In general, one can define the notion of a structure associated to a Lie equation $R_k \subset J_k(T)$, the equivalence, the integrability and formal integrability of such R_k-structures. Then $\tilde{H}^1(R_k)_x$ is the set of equivalence classes of germs at $x \in X$ of formally integrable R_k-structures. The second fundamental theorem for R_k states that $\tilde{H}^1(R_k)$ is trivial, or equivalently that any formally integrable R_k-structure is integrable. Although this theorem is false in general, it is known to hold in many cases. We now describe most of the known results about the Spencer cohomology groups of Lie equations.

(I) If R_k is analytic with respect to a real-analytic structure on X, then, if we restrict our attention to analytic maps and analytic sections, then $H^j(R_k) = 0$ for $j > 0$ and $\tilde{H}^1(R_k)$ is trivial. This follows from existence theorems for analytic differential equations.

(II) If R_k is elliptic and is either analytic with respect to a real-analytic structure on X or formally transitive, then $H^j(R_k) = 0$ for $j > 0$ and $\tilde{H}^1(R_k)$ is trivial. The triviality of $\tilde{H}^1(R_k)$ for equations R_k which are elliptic and analytic was recently proved by Malgrange [**8**] and [**9**], generalizing the Newlander-Nirenberg theorem. In [**3**], using the Third Fundamental Theorem (Theorem 3), it is shown that Malgrange's result implies that a formally transitive elliptic Lie equation is analytic with respect to a real-analytic structure on X.

(III) For the examples $R_1(\mathfrak{g})$ of §2, $H^j(R_1(\mathfrak{g})) = 0$ for $j > 0$ by the Ehrenpreis-Malgrange theorem. The Second Fundamental Theorem is not known to be true in this case. However, if $\mathfrak{g}$ is the Lie algebra of Examples (a), (b) or (c), then $\tilde{H}^1(R_1(\mathfrak{g}))$ is trivial. For Example (b), this follows from Darboux's theorem and for Example (c) from Frobenius' theorem.

The remainder of this section is devoted to the relation between the vanishing of the Spencer cohomology groups and the validity of the Second Fundamental Theorem for a Lie equation $R_k \subset J_k(T)$ and the transitive Lie algebras determined by R_k. In the following, we shall identify two cohomology groups if they are isomorphic as vector spaces.

To any transitive Lie algebra L, we wish to associate Spencer cohomology groups $H^j(L)$ such that:

(i) the Spencer cohomology groups $H^j(L)$ depend only on the isomorphism class of L as a topological Lie algebra;

(ii) one can extend the definition of these cohomology groups to closed ideals of L to obtain Spencer cohomology groups $H^j(I;L)$ of a closed ideal I of L depending only on the isomorphism class of (L, I) as a pair of topological Lie algebras;

(iii) for any exact sequence

$$0 \to I \to L \xrightarrow{\varphi} L' \to 0$$

where I is a closed ideal of L and $\varphi: L \to L'$ is a continuous homomorphism of transitive Lie algebras, one obtains an exact sequence of cohomology groups

$$\cdots \to H^j(I;L) \to H^j(L) \to H^j(L') \to H^{j+1}(I;L) \to \cdots.$$

This can be done as follows. Let L be a transitive Lie algebra. Then by the Third Fundamental Theorem, there exists an analytic formally transitive Lie equation $R_k \subset J_k(T)$ on a real-analytic manifold X such that for $x \in X$, the transitive Lie algebra $R_{\infty,x}$ is isomorphic to L as a topological Lie algebra. Then we define

$$H^j(L) = H^j(R_k)_x$$

for some $x \in X$. If I is a closed ideal of L, by Theorem 4, we obtain an analytic Lie equation $R'_m \subset R_m$ of order m on some neighborhood of $x \in X$ such that $[R_{m+1}, R'_{m+1}] \subset R'_m$ and such that (L, I) is isomorphic to $(R_{\infty,x}, R'_{\infty,x})$ as pairs of topological Lie algebras; we define

$$H^j(I;L) = H^j(R'_m)_x.$$

In fact, it is not obvious that these cohomology groups are well defined, but, if they are, it is clear that they satisfy properties (i) and (ii). We shall prove in a future publication that they are well defined and satisfy (iii). It follows then that the vanishing of the Spencer cohomology groups in positive degrees of an analytic formally transitive Lie equation $R_k \subset J_k(T)$ depends only on the isomorphism class of topological Lie algebras determined by R_k. An analogous result holds for the triviality of $\tilde{H}^1(R_k)$ for any formally transitive Lie equation R_k; in fact, one can also associate a nonlinear Spencer cohomology group $\tilde{H}^1(L)$ to a transitive Lie algebra L so that analogues of conditions (i), (ii) and (iii) hold.

REFERENCES

1. H. Goldschmidt, *Existence theorems for analytic linear partial differential equations*, Ann. of Math. (2) **86** (1967), 246–270. MR **36** #2933.

2. ———, *Formal theory of overdetermined linear partial differential equations*, Proc. Sympos. Pure Math., vol. 16, Amer. Math. Soc., Providence, R.I., 1970, pp. 187–194. MR **41** #9307.

3. ———, *Sur la structure des équations de Lie:* I. *Le troisième théorème fondamental*, J. Differential Geometry **6** (1972); II. *Équations formellement transitives*, ibid. (to appear).

4. V. W. Guillemin, *A Jordan-Hölder decomposition for a certain class of infinite dimensional Lie algebras*, J. Differential Geometry **2** (1968), 313–345. MR **41** #8481.

5. V. W. Guillemin and S. Sternberg, *An algebraic model of transitive differential geometry*, Bull. Amer. Math. Soc. **70** (1964), 16–47. MR **30** #533.

6. ———, *Deformation theory of pseudogroup structures*, Mem. Amer. Math. Soc. No. 64 (1966). MR **35** #2302.

7. A. Kumpera and D. C. Spencer, *Lie equations*, Ann. of Math. Studies, Princeton Univ. Press, Princeton, N.J. (to appear).

8. B. Malgrange, *Pseudo-groupes de Lie elliptiques*, Séminaire Leray, Collège de France, 1969–1970.

9. ———, *Equations de Lie*, J. Differential Geometry (to appear).

10. D. C. Spencer, *Deformation of structures on manifolds defined by transitive, continuous pseudogroups*. I, II, Ann. of Math. (2) **76** (1962), 306–445. MR **27** #6287a, b.

11. ———, *Overdetermined systems of linear partial differential equations*, Bull. Amer. Math. Soc. **75** (1969), 179–239. MR **39** #3533.

UNIVERSITÉ SCIENTIFIQUE ET MÉDICALE DE GRENOBLE, FRANCE

CURVATURE FUNCTIONS FOR 2-MANIFOLDS

JERRY L. KAZDAN[1] AND F. W. WARNER[1]

1. **Two geometric problems.** Let M be a compact connected 2-dimensional manifold (not necessarily orientable). If a smooth function K is the Gaussian curvature of some metric $\hat{g}$ on M, then the Gauss-Bonnet theorem tells us that

$$\int_M K\, dA = 2\pi\chi(M), \tag{1}$$

where dA is the element of area with respect to $\hat{g}$ and $\chi(M)$ is the Euler characteristic of M. This clearly imposes a sign condition on K depending on $\chi(M)$. We were naturally led to ask the following:

Question 1. Is a smooth function K on M the Gaussian curvature of some metric on M if and only if

(a) $\chi(M) > 0$: K is positive somewhere,

(b) $\chi(M) = 0$: K changes sign (unless $K \equiv 0$),

(c) $\chi(M) < 0$: K is negative somewhere?

One way to attack this problem is to prescribe some metric g on M and let k be its Gaussian curvature, and then to try to realize K (or $K \circ \varphi$, where φ is an arbitrary diffeomorphism of M) as the curvature of a metric $\hat{g}$ pointwise conformal to g, i.e. of the form $\hat{g} = e^{2u}g$ for some $u \in C^\infty(M)$. (Note that this notion of pointwise conformal should not be confused with a conformal diffeomorphism F, where metrics g and $\hat{g}$ are conformally equivalent if $F^*(g) = e^{2u}\hat{g}$ for some diffeomorphism F of M and some $u \in C^\infty(M)$. It is this notion that customarily appears in complex analysis.) This approach has the advantage that, in principle at least, it is applicable to any genus. Moreover, it leads directly to the specific partial differential equation

AMS 1970 *subject classifications*. Primary 35J20, 35J60, 53A30, 53C20; Secondary 53C45.

[1] Supported in part by National Science Foundation Grants GP-28976X and GP-29258.

$$\Delta u = k - Ke^{2u}, \tag{2}$$

where Δ is the Laplacian with respect to the metric g.

Question 2. Let g be a given metric on M with curvature k. If K satisfies the sign condition of Question 1 (depending on $\chi(M)$), is K the curvature of some metric $\hat{g}$ that is pointwise conformal to g?

In other words, can one solve (2)? Of course, if one cannot solve (2), one asks for necessary and sufficient conditions. This question is related to a geometric question of L. Nirenberg; his question reduces to the existence of a solution of (2) on S^2, where g is the standard metric (with $k \equiv 1$) and where one assumes $K > 0$.

2. A summary of geometric results.

THEOREM 1. *If $\chi(M) = 0$, that is, if M is a torus or a Klein bottle, the answer to Question 2 is NO. In fact, excluding the trivial case $K \equiv 0$, K is the curvature of a metric $\hat{g}$ pointwise conformal to g if and only if K changes sign and $\int_M Ke^{2v}\,dA < 0$, where v is a solution of $\Delta v = k$ and dA is the element of area for g.*

This result enables us to conclude

THEOREM 2. *The answer to Question* 1(b) *is YES.*

A more subtle consequence of Theorem 1 is the following result, which one might expect since there is no Gauss-Bonnet theorem for the plane $\boldsymbol{R}^2$.

THEOREM 3. *Every $K \in C^\infty(\boldsymbol{R}^2)$ is the Gaussian curvature of some Riemannian metric on $\boldsymbol{R}^2$.*

We can also prove similar theorems for a wide class of open manifolds. Note that one cannot hope to get a complete metric, since, for example by Bonnet's theorem, completeness and $K \geqq \text{const} > 0$ implies compactness [**11**, p. 254]. We can sharpen this and prove the following precise result.

THEOREM 4. *$K \in C^\infty(\boldsymbol{R}^2)$ is the curvature of some complete Riemannian metric on $\boldsymbol{R}^2$ if and only if*

$$\lim_{r\to\infty} \inf_{|p|\geqq r} K(p) \leqq 0.$$

In the case of S^2 we have an incomplete result which, however, appears to be considerably deeper than those above. Here Δ will denote the Laplacian of the standard metric on S^2.

THEOREM 5. *The answer to Question 2 for S^2 is NO. A necessary condition on K for there to exist a solution of $\Delta u = 1 - Ke^{2u}$ on S^2 is that*

$$\int_{S^2} e^{2u}\nabla K \cdot \nabla F\, dA = 0, \tag{3}$$

for all spherical harmonics F of degree 1 *(here* ∇ *denotes the gradient on* S^2).

For example, if K is a nontrivial spherical harmonic of degree 1, then the integral is positive for $F = K$. Since the integral (3) is unchanged if one adds constants to K, one sees that there are positive functions K on S^2 for which one cannot solve $\Delta u = 1 - Ke^{2u}$; an example is $K = 2 + \cos\varphi$ (we use spherical coordinates with $z = \cos\varphi$). This shows that the answer to L. Nirenberg's question is NO. Some motivation for this necessary condition is the rotationally symmetric case done in [**4**].

If $\chi(M) < 0$, we have a result similar to the necessary condition of Theorem 1.

THEOREM 6. *For* $\chi(M) < 0$, *the answer to Question* 2 *is NO. If* K *is the curvature of a metric* $\hat{g}$ *pointwise conformal to* g *(with curvature* k*) then*

$$\int Ke^{2v}\,dA < 0,$$

where v *is a solution of* $\Delta v = k - c$ *and* $c = (\int k\,dA)/\text{area}\,(M)$, *and where the area is with respect to* g.

3. **Table of known results.** It is convenient to summarise the above and comment on related work of others by tabulating the known results ("n and s" means that necessary and sufficient conditions are known).

	Question 1	*Question* 2
S^2	? (yes if $K > 0$ or if $K(x) = K(-x)$)	no
P^2	yes	yes (if g is the standard metric)
T^2	yes	no, "n and s"
Klein bottle	yes	no, "n and s"
$\chi(M) < 0$	? (yes if $K < 0$)	no
$\boldsymbol{R}^2$	yes	no

S^2: The assertion "yes" if $K > 0$ on S^2 was established by H. Gluck [**3**] who reduced it to the Minkowski problem. The "yes" for antipodally symmetric functions, $K(-x) = K(x)$, was proved by J. Moser [**8**]. The special case of antipodally symmetric functions sufficiently close to 1 had previously been done by D. Koutroufiotis [**5**]. The "no" to Question 2 is our Theorem 5.

P^2: These follow from the results of J. Moser on antipodally symmetric functions on S^2.

T^2 *and Klein bottle*: These follow from Theorems 1 and 2 above. (We mentioned our question for S^2 to Melvyn Berger who subsequently obtained a partial solution for the torus and for $\chi(M) < 0$ below [**1**].)

$\chi(M) < 0$: The "yes" if $K < 0$ is due to Berger [**1**] for orientable surfaces. His proof can easily be extended to cover the nonorientable cases also. It is pertinent to observe in this case that the solution to (2) is actually unique. The "no" to Question 2 is Theorem 6 above.

$\boldsymbol{R}^2$: These follow from Theorems 3 and 4 above. The "no" to Question 2 follows from a result of H. Wittich [**13**], see also [**10**], which asserts, for example, that there are no solutions of $\Delta u = e^{2u}$ defined on all of $\boldsymbol{R}^2$.

Before leaving the geometric results, we should mention that one can rephrase Question 1 in terms of curvature forms, $\Omega = K\,dA$. In this case, it turns out [**12**] that the Gauss-Bonnet condition (1) is a necessary and sufficient condition for a two form Ω to be the curvature form of a Riemannian metric on a compact orientable M.

4. **Supplementary remarks on differential equations.** The general equation we have been looking at is

$$\Delta u = f + he^{\alpha u}, \tag{4}$$

where f and h are prescribed smooth functions on a compact 2-dimensional Riemannian manifold M and $\alpha > 0$ is a constant. Let

$$c = \frac{1}{\text{area }(M)} \int_M f\,dA$$

and exclude the trivial case $h \equiv 0$ (which of course is consistent with (4) only when $c = 0$). Then we can generalize Theorem 1 to any 2-dimensional compact manifold M.

THEOREM 1′. *For $c = 0$, equation* (4) *has a solution if and only if the two conditions below both hold*:

(a) *h changes sign*;

(b) *$\int he^{\alpha v}\,dA > 0$, where v is a solution of $\Delta v = f$.*

For $c > 0$, we have the following theorem, which has been observed by both Berger and Moser in a special case.

THEOREM 7. *There exists a constant $\beta > 0$ such that for $0 < c\alpha < \beta$, equation* (4) *has a solution if and only if h is negative somewhere on M.*

The proofs of these use the calculus of variations. Using his sharp version of the Trudinger inequality for S^2, Moser [7] has shown that for $M = S^2$ with the standard metric, one can take $\beta = 2$. Our Theorem 5 shows that $\beta = 2$ is the best possible constant on S^2. One of the striking phenomena is the fact that there are entirely different constraints on h in the three cases $c\alpha = 0$, $0 < c\alpha < 2$ and $c\alpha = \beta = 2$ on S^2. For $c\alpha = 0$ we have the peculiar constraint (b) in Theorem 1′, while for $c\alpha = 2$, we have the even more peculiar constraint of Theorem 5 (where other notation was used). This appears to be closely related to the fact that $\lambda = 0$ and $\lambda = 2$ are the first two eigenvalues of $-\Delta$ on S^2 (with the standard metric). We have no information on S^2 for the range $2 < c\alpha < 6$, but again at $c\alpha = 6$, which is the next eigenvalue of $-\Delta$, we have a constraint analogous to (3) which shows that there are rotationally symmetric K, for example $K = 3\cos^2\varphi - 1$, for which $\Delta u = 1 - Ke^{6u}$ has no rotationally symmetric solutions.

Our equation is of the form

$$Lu = f(x, u), \qquad x \in M,$$

where L is a linear elliptic operator. One can use the theory found, for example in [**2**, p. 369–370] if $Lu = 0$ has no solutions except $u = 0$ and if $f(x, u)$ is bounded. If there are nontrivial solutions of $Lu = 0$, and if f is bounded, then the recent interesting results of Landesman and Lazer [**6**] (see also the simplification and generalization by Nirenberg [**9**]) are applicable.

Our equation $\Delta u = k - Ke^{2u}$ does not yield to these techniques for the two reasons

(1) Δ has a nontrivial kernel on compact M, and

(2) $f(x, u) = k(x) - K(x)e^{2u}$ is unbounded as a function of u.

Detailed proofs of all of the theorems stated above will appear in [**14**], [**15**].

ADDED IN PROOF. In recent work on this problem we have been able to prove that the answer to Question 1 if $\chi(M) < 0$ is "yes" [**16**]. Theorem 3 has been extended to show that for many open 2-manifolds any K is a Gaussian curvature [**16**]. Analogously, if M is compact and $\dim M \geqq 3$, then we can prove that any function K that is negative somewhere is the *scalar curvature* of some metric [**17**]. This negativity condition is known to be necessary for certain spin manifolds if $K \not\equiv 0$. For many open manifolds, $\dim M \geqq 3$, we can prove that any K is a scalar curvature [**17**]. Concerning the paragraph after Theorem 7, we now have a constraint similar to (3) obstructing the solvability on S^2 (with the standard metric) for any value of $c\alpha \geqq 2$. The proofs of these assertions have led us to some new existence and nonexistence theorems for a class of nonlinear elliptic equations, $Lu = f(x, u)$, including some where L has a nontrivial kernel and f is unbounded as a function of u [**18**].

BIBLIOGRAPHY

1. Melvyn Berger, *On Riemannian structures of prescribed Gauss curvature for compact two-dimensional manifolds*, J. Differential Geometry **5** (1971), 325–332.

2. R. Courant and D. Hilbert, *Methods of mathematical physics*. Vol. II: *Partial differential equations*, Interscience, New York, 1962. MR **25** #4216.

3. H. Gluck, *Deformations of normal vector fields and the generalized Minkowski problem*, Bull. Amer. Math. Soc. **76** (1971), 1106–1110.

4. Jerry L. Kazdan and F. W. Warner, *Surfaces of revolution with monotonic increasing curvature and an application to the equation $\Delta u = 1 - Ke^{2u}$ on S^2*, Proc. Amer. Math. Soc. **32** (1972), 139–141.

5. D. Koutroufiotis, *On Gaussian curvature and conformal mapping* (to appear).

6. E. M. Landesman and A. C. Lazer, *Nonlinear perturbations of linear elliptic boundary value problems at resonance*, J. Math. Mech. **19** (1970), 609–623. MR **42** #2171.

7. J. Moser, *A sharp form of an inequality by N. Trudinger*, Indiana Univ. Math. J. **20** (1971), 1077–1092.

8. ———, *On a nonlinear problem in differential geometry* (to appear).

9. L. Nirenberg, *An application of generalized degree to a class of nonlinear problem*, in Contributions to nonlinear functional analysis, Academic Press. New York, 1971.

10. R. Osserman, *On the inequality* $\Delta u \geqq f(u)$, Pacific J. Math. **7** (1957), 1641–1647. MR **20** #4701.

11. J. J. Stoker, *Differential geometry*, Pure and Appl. Math., vol. 20, Interscience, New York, 1969. MR **39** #2072.

12. N. Wallach and F. W. Warner, *Curvature forms for* 2*-manifolds*, Proc. Amer. Math. Soc. **25** (1970), 712–713.

13. H. Wittich, *Ganze Lösungen der Differentialgleichung* $\Delta u = e^u$, Math. Z. **49** (1944), 579–582. MR **6**, 228.

14. Jerry L. Kazdan and F. W. Warner, *Curvature functions for* 2*-manifolds*. I (to appear).

15. ———, *Curvature functions for* 2*-manifolds*. II (to appear).

16. ———, *Curvature functions for* 2*-manifolds*. III. *The negative Euler characteristic case* (to appear).

17. ———, *Scalar curvature and conformal deformation of Riemannian structure* (to appear).

18. ———, *Remarks on some nonlinear elliptic equations* (to appear).

UNIVERSITY OF PENNSYLVANIA

SCATTERING WITH LONG RANGE POTENTIALS[1]

P. ALSHOLM AND TOSIO KATO

1. **Introduction.** We consider the Schrödinger operators

$$H_0 = -\Delta/2 = -|p|^2/2,$$
$$H = H_0 + V(x),$$

in $\boldsymbol{H} = L^2(R^n)$, $n = 1, 2, \ldots$. Here $V(x)$ is the operator of multiplication by a real-valued measurable function (potential) $V(x)$. For simplicity we assume $V(x)$ to be bounded, so that H is selfadjoint. $p = -i \operatorname{grad}_x$ is the vector differential operator, but we regard it also as the multiplication by the variable $p \in R^n$ in the momentum (Fourier) representation of $\boldsymbol{H}$. Similarly we often interpret $x \in R^n$ as the differential operator $x = i \operatorname{grad}_p$ in the momentum representation.

If $V(x) \to 0$ sufficiently fast as $|x| \to \infty$ (short range potential), it is known that the wave operators

$$W_\pm = \operatorname*{s-lim}_{t \to \pm\infty} e^{itH} e^{-itH_0}$$

exist and are complete. Roughly, this is the case if $V(x) = O(|x|^{-1-\varepsilon})$, $\varepsilon > 0$ (see e.g. [**1**]).

For long range potentials $V(x)$ which decay more slowly, $W_\pm$ as defined above need not exist. This was proved by Dollard [**2**],[**3**] for the Coulomb potential $V(x) = c|x|^{-1}$. In this case he was able to show that there exist the modified wave operators

$$W_{D,\pm} = \operatorname*{s-lim}_{t \to \pm\infty} e^{itH} e^{-itH_0 - iX_t}$$

AMS 1970 *subject classifications*. Primary 47A40, 81A45; Secondary 35J10.

[1] An improved version of a part of the lecture given by the second-named author at the American Mathematical Society Summer Institute on Partial Differential Equations, Berkeley, 1971. This work was partly supported by NSF grant GP-29369X.

which have all the desired properties of $W_\pm$. Here X_t is a certain selfadjoint operator, depending on t and commuting with H_0. His result suggests that in a more general case, X_t should be chosen in such a way that $(d/dt)X_t = V(tp)$. Amrein, Martin, and Misra [**4**] proved that $W_{D,\pm}$ indeed exist with such an X_t if $V(x) = c|x|^{-\alpha}$ with $3/4 \leqq \alpha \leqq 1$ and $n = 3$.

Recently, Buslaev and Matveev [**5**] proved that the same is true if

$$(1.1) \qquad |D^k V(x)| \leqq c(1 + |x|)^{-k-\alpha}, \qquad \alpha > 1/2, \qquad k \leqq [n/2] + 2,$$

where D^k denotes an arbitrary derivative of order k. They give a similar result for the case $0 < \alpha \leqq 1/2$ with a different choice of X_t, assuming more differentiability of $V(x)$.

The main object of the present paper is to show that it suffices to assume (1.1) for $k \leqq 2$, by a method quite different from that of Buslaev-Matveev. More precisely, we shall prove

THEOREM 1. *Let* $V(x) = V_S(x) + V_L(x)$, *where*

$$(\mathrm{S}) \qquad |V_S(x)| \leqq c(1 + |x|)^{-1-\varepsilon},$$

$$(\mathrm{L1}) \qquad |DV_L(x)| \leqq c(1 + |x|)^{-1-\beta},$$

$$(\mathrm{L2}) \qquad |D^2 V_L(x)| \leqq c(1 + |x|)^{-2-\gamma},$$

with

$$(1.2) \qquad \varepsilon > 0, \quad 1/2 < \beta \leqq 1, \quad (1 - \beta)^2/\beta < \gamma.$$

Then $W_{D,\pm}$ *exist with the choice of*

$$(1.3) \qquad X_t = X_t(p) = \int_0^t V_L(sp)\, ds.$$

REMARKS. (1) V_S and V_L are the short range and long range parts of V, respectively. (2) The derivatives such as DV_L are to be taken in the distribution sense. (3) As a special case we may choose $\beta = \gamma > 1/2$. Thus the theorem contains the result of Buslaev-Matveev mentioned above. (4) We do not assume explicitly any condition about the decay rate of $V_L(x)$ itself. But (L1) implies that $\lim_{|x|\to\infty} V_L(x) = V_0$ exists if $n \geqq 2$ and that $\lim_{x\to\pm\infty} V_L(x) = V_\pm$ exist if $n = 1$. One may assume $V_0 = 0$ without essentially changing the problem; then (L1) implies that $V_L(x) = O(|x|^{-\beta})$. But there is in general no such simplification when $n = 1$. (5) Condition (S) can be weakened to allow $V(x)$ to have certain local singularities.

THEOREM 2. *Under the assumption of Theorem* 1, $W_{D,\pm}$ *are isometric and have the following intertwining properties*:

$$HW_{D,\pm} = W_{D,\pm}(H_0 + V_0) \qquad (n \geqq 2),$$

$$HW_{D,\pm} = W_{D,\pm}(H_0 + V_{\pm \mathrm{sign}(p)}) \qquad (n = 1),$$

where the constants V_0, $V_{\pm}$ *are defined in Remark* (4) *to Theorem* 1.

REMARK. The last result for the case $n = 1$ is of some interest, for it shows that (a part of) H is unitarily equivalent, not to H_0 itself but to a slightly modified operator.

2. **Proof of Theorem 1.** I. First we show that we may assume, without loss of generality, that V_L is C^∞ and satisfies

$$\text{(L3)} \qquad |D^3 V_L(x)| \leqq c(1 + |x|)^{-2-\gamma}$$

in addition to (L1) and (L2). To see this, we mollify V_L to $V'_L = \omega * V_L$, where $\omega \in C_0^\infty(R^n)$, $\omega \geqq 0$, and $\|\omega\|_{L^1} = 1$. V'_L also satisfies (L1) and (L2). It follows further from (L1) that $V'_L - V_L = O(|x|^{-1-\beta})$. If we construct X'_t from V'_L by (1.3), therefore, we see that $\lim_{t\to\pm\infty}(X'_t(p) - X_t(p))$ exist pointwise. Hence

$$\operatorname*{s-lim}_{t\to\pm\infty} e^{-i(X_t - X'_t)}$$

exist, so that $W_{D,\pm}$ exist if and only if they exist when X_t is replaced by X'_t. Since $V = (V_S + V_L - V'_L) + V'_L$ in which the first summand satisfies (S) by the remark given above, it suffices to prove the theorem with V_L replaced by V'_L. But V'_L is C^∞ and satisfies (L3) by $D^3 V'_L = D\omega * D^2 V_L$ and (L2).

II. Let us now consider $W_{D,+}$. As is well known, it suffices to show that

$$\phi(t) = \|(d/dt)e^{itH}e^{-itH_0 - iX_t}u\| = \|(V(x) - V_L(tp))e^{-itH_0 - iX_t}u\|$$

is integrable on $[1, \infty)$ for each $u \in \boldsymbol{S}$, where $\boldsymbol{S}$ is the set of all $u \in \boldsymbol{H}$ with its Fourier transform in $C_0^\infty(R^n - \{0\})$.

Write $\phi = \phi_S + \phi_L$, where

$$\phi_S(t) = \|V_S(x)e^{-itH_0 - iX_t}u\|,$$
$$\phi_L(t) = \|(V_L(x) - V_L(tp))e^{-itH_0 - iX_t}u\|.$$

III. First we estimate ϕ_L. Using the identity

$$\text{(2.1)} \qquad e^{it|p|^2/2}F(x)e^{-it|p|^2/2} = e^{-i|x|^2/2t}F(tp)e^{i|x|^2/2t}$$

(both sides being equal to $F(x + tp)$), we can write

$$\begin{aligned}
\phi_L(t) &= \|\{e^{-i|x|^2/2t}V_L(tp)e^{i|x|^2/2t} - V_L(tp)\}e^{-iX_t}u\| \\
&\leqq \int_0^1 \left\| \frac{d}{d\lambda}(e^{-i\lambda|x|^2/2t}V_L(tp)e^{i\lambda|x|^2/2t})e^{-iX_t}u \right\| d\lambda \\
&= \frac{1}{2t}\int_0^1 \|[V_L(tp), |x|^2]e^{i\lambda|x|^2/2t}e^{-iX_t}u\| \, d\lambda,
\end{aligned}$$

where [,] denotes the commutator. Computing the commutator and using (2.1) again, we obtain

$$\phi_L(t) \leqq \int_0^1 \Big\{ \|(\text{grad } V_L)(\lambda x)\cdot e^{-iY}(x + \text{grad } X_t)(p))u\| \tag{2.2}$$
$$+ \frac{t}{2}\|(\Delta V)(\lambda x)e^{-iY}u\| \Big\}\, d\lambda,$$

where

$$Y = Y_{t,\lambda}(p) = t|p|^2/2\lambda + X_t(p).$$

In view of (L1) and (L2), estimating (2.2) is reduced to estimating the following quantities:

$$\begin{gathered} a_j(t) = \|(1 + |\lambda x|)^{-\rho_j}e^{-iY}v_j\|, \qquad j = 1, 2, 3, \\ \rho_1 = \rho_2 = 1 + \beta, \qquad \rho_3 = 2 + \gamma, \\ v_1 = xu, \qquad v_2 = (\text{grad } X_t)(p)u, \qquad v_3 = u. \end{gathered} \tag{2.3}$$

In what follows we shall show that

$$\begin{gathered} a_1(t) \leqq c_u t^{-1-\beta}, \qquad a_2(t) \leqq c_u t^{-1+(1-\beta)2-\beta\gamma}, \\ a_3(t) \leqq c_u t^{-2-\gamma^2}, \qquad \text{for } t \geqq t_u, \end{gathered} \tag{2.4}$$

uniformly for $\lambda \in [0, 1]$, where the c_u and t_u are positive constants depending only on $u \in \boldsymbol{S}$. Since (1.2) implies that $a_1(t)$, $a_2(t)$ and $ta_3(t)$ are integrable in t, (2.2) shows that $\phi_L(t)$ is integrable.

IV. Before estimating the $a_j(t)$, we prepare some inequalities satisfied by the derivatives of $X_t(p)$. We use the symbol D_r for the radial derivative: $D_rF(p) = (\partial/\partial|p|)F(p) = |p|^{-1}p\cdot\text{grad } F(p)$. Then it follows easily from (1.3) and (L1) to (L3) that

$$\begin{gathered} |D_rX_t(p)| \leqq c_K t^{1-\beta}, \qquad |D_r^2X_t(p)| \leqq c_K t^{1-\beta}, \\ |D_r^3X_t(p)| \leqq c_K t^{1-\gamma}, \qquad |D_r^4X_t(p)| \leqq c_K t^{2-\gamma}, \\ |\text{grad } X_t(p)| \leqq c_K t^{1-\beta}, \qquad |D_r \text{ grad } X_t(p)| \leqq c_K t^{1-\beta}, \\ |D_r^2 \text{ grad } X_t(p)| \leqq c_K t^{1-\gamma}, \qquad \text{for } p \in K, t \geqq t_K, \end{gathered} \tag{2.5}$$

where K is any compact subset of $R^n - \{0\}$ and c_K, t_K are positive constants depending only on K. Here we have assumed that $\gamma \leqq \beta$, which one can do without loss of generality as is easily seen from (1.2).

V. To estimate the $a_j(t)$, we start from the identity

$$\lambda x e^{-iY} = e^{-iY}\{tp + \lambda(\text{grad } X_t)(p) + \lambda x\}. \tag{2.6}$$

Scalar multiplication by p from the right gives

$$\lambda x\cdot e^{-iY}p = e^{-iY}(tQ(p) + \lambda x\cdot p), \tag{2.7}$$

where

$$Q(p) = Q_{t,\lambda}(p) = |p|^2 + \frac{\lambda}{t}|p|D_r X_t(p) \tag{2.8}$$

and D_r is the radial derivative introduced in the preceding paragraph. The second term on the right of (2.8) is of the order $t^{-\beta}$ by virtue of (2.5), so that $Q(p)^{-1}$ exists if $p \in K$ and t is sufficiently large. Thus (2.7) gives for any $v \in \boldsymbol{S}$ or $\boldsymbol{S}^n$

$$e^{-iY}v = \frac{\lambda}{t}x \cdot e^{-iY}pQ(p)^{-1}v - \frac{\lambda}{t}e^{-iY}x \cdot pQ(p)^{-1}v, \qquad t \geqq t_v. \tag{2.9}$$

Now (2.9) gives

$$\begin{aligned} \|(1 + |\lambda x|)^{-1}e^{-iY}v\| &\leqq t^{-1}(\|pQ(p)^{-1}v\| + \|x \cdot pQ(p)^{-1}v\|) \\ &\leqq c_K t^{-1}(\|v\| + \|x \cdot pv\|), \qquad t \geqq t_K, \end{aligned} \tag{2.10}$$

where $K = \operatorname{supp} \hat{v}$ ($\hat{v}$ is the Fourier transform of v). In deducing the last inequality of (2.10), we have used the fact that $|D_r Q(p)^{-1}| \leqq c_K$ for $p \in K$ and t sufficiently large. In fact, it follows from (2.5) that

$$\begin{aligned} &|D_r Q(p)^{-1}| \leqq c_K, \qquad |D_r^2 Q(p)^{-1}| \leqq c_K, \\ &|D_r^3 Q(p)^{-1}| \leqq c_K t^{1-\gamma} \quad \text{for } p \in K,\ t \geqq t_K. \end{aligned} \tag{2.11}$$

From (2.9) we further obtain

$$\begin{aligned} \|(1 + |\lambda x|)^{-2}e^{-iY}v\| \leqq\ & t^{-1}\|(1 + |\lambda x|)^{-1}e^{-iY}pQ(p)^{-1}v\| \\ &+ t^{-1}\|(1 + |\lambda x|)^{-1}e^{-iY}x \cdot pQ(p)^{-1}v\|. \end{aligned}$$

Substitution from (2.10) gives

$$\begin{aligned} \|(1 + |\lambda x|)^{-2}e^{-iY}v\| &\leqq c_K t^{-2}(\|pQ(p)^{-1}v\| + \|x \cdot ppQ(p)^{-1}v\| \\ &\qquad + \|x \cdot pQ(p)^{-1}v\| + \|(x \cdot p)^2 Q(p)^{-1}v\|) \\ &\leqq c_K t^{-2}(\|v\| + \|x \cdot pv\| + \|(x \cdot p)^2 v\|), \qquad t \geqq t_K, \end{aligned} \tag{2.12}$$

where we have used the first two inequalities of (2.11).

In the same way we obtain

$$\begin{aligned} \|(1 + |\lambda x|)^{-3}e^{-iY}v\| \leqq\ & c_K t^{-3}(t^{1-\gamma}\|v\| + \|x \cdot pv\| \\ &+ \|(x \cdot p)^2 v\| + \|(x \cdot p)^3 v\|), \qquad t \geqq t_K, \end{aligned} \tag{2.13}$$

the factor $t^{1-\gamma}$ coming from the use of the last inequality of (2.11).

VI. Now it is easy to deduce the required estimates (2.4). To estimate $a_1(t)$, we set $v = v_1 = xu$ in (2.10) and (2.12), obtaining

$$\|(1 + |\lambda x|)^{-1}e^{-iY}v_1\| \leqq c_u t^{-1}, \qquad \|(1 + |\lambda x|)^{-2}e^{-iY}v_1\| \leqq c_u t^{-2}.$$

A simple interpolation, based on the inequality $\|A^\beta w\| \leqq \|w\|^{1-\beta}\|Aw\|^\beta$ for a selfadjoint operator $A \geqq 0$, then gives

$$a_1(t) = \|(1 + |\lambda x|)^{-1-\beta}e^{-iY}v_1\| \leqq c_u t^{-1-\beta},$$

as required.

Estimating $a_2(t)$ is somewhat more complicated because $v_2 = (\text{grad } X_t(p))u$ involves t. Again we set $v = v_2$ in (2.10) and (2.12). Since

$$x \cdot pv_2 = in(\text{grad } X_t(p))u + p(\text{grad } X_t(p)) \cdot xu + i|p|(D_r \text{ grad } X_t(p))u$$

is at most of the order $t^{1-\beta}$ by (2.5), (2.10) gives

$$\|(1 + |\lambda x|)^{-1}e^{-iY}v_2\| \leqq c_u t^{-\beta}, \qquad t \geqq t_u. \tag{2.14}$$

Treating $(x \cdot p)^2 v_2$ in a similar manner, we notice that the largest contribution comes from a term involving $(D_r^2 \text{ grad } X_t(p))u$, which is of the order $t^{1-\gamma}$ by (2.5) [we assume $\gamma \leqq \beta$ as remarked above]. Thus (2.12) gives

$$\|(1 + |\lambda x|)^{-2}e^{-iY}v_2\| \leqq c_u t^{-1-\gamma}, \qquad t \geqq t_u. \tag{2.15}$$

By interpolation, we obtain, from (2.14) and (2.15),

$$a_2(t) = \|(1 + |\lambda x|)^{-1-\beta}e^{-iY}v_2\| \leqq c_u t^{-(1-\beta)\beta - \beta(1+\gamma)} = c_u t^{-1+(1-\beta)^2 - \beta\gamma},$$

as required.

$a_3(t)$ is again easy to estimate since $v_3 = u$ does not depend on t. Setting $v = v_3 = u$ in (2.12) and (2.13) gives

$$\|(1 + |\lambda x|)^{-2}e^{-iY}v_3\| \leqq c_u t^{-2}, \qquad \|(1 + |\lambda x|)^{-3}e^{-iY}v_3\| \leqq c_u t^{-2-\gamma},$$

and the interpolation leads to the desired result $a_3(t) \leqq c_u t^{-2-\gamma^2}$.

VII. Estimating $\phi_S(t)$ is the same as for $a_1(t)$, with $\beta = \varepsilon$, $\lambda = 1$ and $v_1 = u$. Thus

$$\phi_S(t) \leqq c_u t^{-1-\varepsilon}, \qquad t \geqq t_u,$$

and $\phi_S(t)$ is integrable. This completes the proof of Theorem 1.

3. **Proof of Theorem 2.** It is obvious that $W_{D,\pm}$ are isometric To prove the intertwining properties, we note that

$$\begin{aligned} e^{isH}W_{D,\pm}e^{-isH_0} &= \operatorname*{s-lim}_{t \to \pm\infty} e^{i(t+s)H}e^{-i(t+s)H_0 - iX_t} \\ &= W_{D,\pm} \text{ s-lim } e^{i(X_{t+s} - X_t)}. \end{aligned} \tag{3.1}$$

If $n \geqq 2, V(x) \to V_0$ as $|x| \to \infty$. Hence

$$X_{t+s}(p) - X_t(p) = \int_t^{t+s} V(rp)\,dr \to sV_0, \qquad t \to \pm\infty,$$

provided $p \neq 0$, and (3.1) gives $e^{isH}W_{D,\pm} = W_{D,\pm}e^{is(H_0+V_0)}$, from which the desired result follows.

If $n = 1$, $V(x) \to V_\pm$ as $x \to \pm\infty$. Hence $X_{t+s}(p) - X_t(p) \to sV_{\pm\text{sign}(p)}$ as $t \to \pm\infty$, from which the desired result follows in the same way.

BIBLIOGRAPHY

1. T. Kato, *Some results on potential scattering*, Proc. Internat. Conference on Functional Analysis and Related Topics (Tokyo, 1969), Univ. of Tokyo Press, Tokyo, 1970. MR **42** #3610.

2. J. D. Dollard, *Asymptotic convergence and the Coulomb interaction*, J. Mathematical Phys. **5** (1964), 729–738. MR **29** #921.

3. ———, *Quantum-mechanical scattering theory for short-range and Coulomb interactions*, Rocky Mountain J. Math. **1** (1971), 5–88. MR **42** #5561.

4. W. O. Amrein, Ph. A. Martin and B. Misra, *On the asymptotic condition of scattering theory*, Helv. Phys. Acta **43** (1970), 313–344.

5. V. S. Buslaev and V. B. Matveev, *Wave operators for the Schrödinger equation with a slowly decreasing potential*, Theor. Math. Phys. **1** (1970), 367–376.

UNIVERSITY OF CALIFORNIA, BERKELEY

WHAT IS RENORMALIZATION?

JAMES GLIMM[1] AND ARTHUR JAFFE[2]

1. **Introduction and notation.** For some 30 years, physicists have assumed that elementary particles and their interactions can be described by quantum mechanics. The natural approach is to study quantum fields, and time-honored calculations in electrodynamics yield perfect agreement with experiment. Surprisingly, then, this vast area of physics has received relatively little mathematical attention. Theorems that establish the existence of interacting fields are only now being proved, and results have been obtained only in the case of one or two (not yet three) space dimensions. Furthermore, while most of the Wightman axioms [**19**] for fields have been verified in two dimensions, little has been proved about other properties of the solutions; nor have any numerical calculations been carried out with the known models.

One reason that field theory has discouraged study is the need for "renormalization." Renormalization has both a mathematical and a physical basis; both aspects must be considered to deal with the problems effectively. Roughly, the physics of renormalization deals with the fact that the interaction of particles may alter their energy, mass, etc. We therefore may adjust certain parameters to chosen values, so that masses and charges of the interacting particles take their experimental values. These choices are connected with the mathematical problem of defining the nonlinear terms in the field equations. When our definitions are physically sound, they also yield a mathematically well behaved theory.

In what follows, we will develop these ideas more precisely and in detail. Since, however, the physics differs for each quantum field, renormalization for each prob-

AMS 1970 *subject classifications*. Primary 35-02, 35R20, 81-02, 81A06, 81A09, 81A17, 81A18, 81A19; Secondary 46L05.

[1] Supported in part by the National Science Foundation Grant GP-24003.

[2] Alfred P. Sloan Fellow. Supported in part by the Air Force Office of Scientific Research, Contract AF44620-70-C0030, and by the National Science Foundation Grant GP-31239X.

lem will be different. There is a whole hierarchy of physical and mathematical pitfalls. While renormalization has some unifying ideas, the deepest results concern particular problems and the estimates that lead to their solution. After some basic introduction, we review some proved theorems about renormalizations of one theory, in order to convey the flavor of the subject. Finally, we will review some progress more broadly, and mention several open problems.

Let us begin with a rapid review of quantum mechanics: A pure physical state is described by a unit vector in a Hilbert space $\mathscr{H}$. An observable A is a selfadjoint operator on $\mathscr{H}$. It is often convenient to consider a C^* algebra $\mathfrak{A}$ of bounded operators on $\mathscr{H}$, with all bounded observables contained in $\mathfrak{A}$. This algebra $\mathfrak{A}$ is the closure of the union of algebras $\mathfrak{A}(B)$, where B is a bounded region of space-time. The operators in $\mathfrak{A}(B)$ are called local observables.

A basic observable is the energy or Hamiltonian H. By definition, the Hamiltonian is the infinitesimal generator of time translations. Hence we require the existence of a one-parameter unitary group $U(t) = \exp(itH)$ on $\mathscr{H}$, and a one-parameter automorphism group σ_t of $\mathfrak{A}$, with σ_t implemented by $U(t)$,

$$\sigma_t(A) = U(t)AU(t)^*, \qquad A \in \mathfrak{A}.$$

We assume that our physical system is stable. Mathematically, we assume that there is a state Ω of minimum energy, and for convenience we let Ω have energy zero. Thus

$$0 \leqq H, \qquad H\Omega = 0.$$

The momentum P of the state Ω is also zero, $P\Omega = 0$, and Ω is called the *vacuum state*.

Particles enter our theory as states with nonzero energy and momentum. If, for simplicity, we consider a single type of particle of mass m, the energy-momentum spectrum would consist of a hyperboloid corresponding to single particle states with the energy

$$\mu(p) = (p^2 + m^2)^{1/2}$$

and a continuum arising from two or more particles $H \geqq (P^2 + 4m^2)^{1/2}$.

The field $\phi(x, t)$ provides a description of particles and their interaction. We require that if $f(x, t) \in \mathscr{S}$ has a Fourier transform $f^\sim(p, E)$ supported in $0 < E < (p^2 + 4m^2)^{1/2}$ and intersects $E = \mu(p)$, then

$$\phi(f)\Omega = \int \phi(x, t) f(x, t)\, dx\, dt\Omega \neq 0.$$

The physical interpretation of $\phi(f)\Omega$ is a single-particle state with a wave packet $f^\sim(p, \mu(p))$. Assuming the existence of such one-particle states, n-particle scattering states exist as strong limits as $t \to \pm\infty$ of the vectors $U(t)\phi(f_{-t})^n\Omega$, where $(f_{-t})^\sim = \exp(-it\mu(p))f^\sim$. This standard construction is the Haag-Ruelle scattering

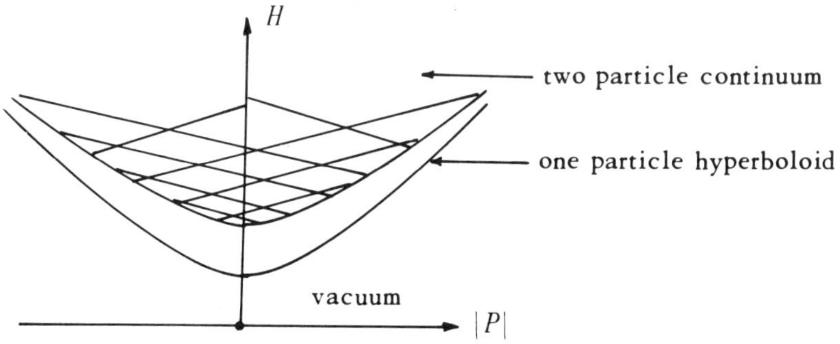

THE ENERGY-MOMENTUM SPECTRUM

theory [**16**]. The scattering states behave as free particles, and the canonical linear transformation of incoming to outgoing states is the scattering matrix S.

Another requirement (and the one which causes the most difficulty) is the requirement of Lorentz covariance. There should exist a strongly continuous unitary representation of the Poincaré group on $\mathscr{H}$, and the field $\phi(x, t)$ should transform according to one of the standard representations.

2. **Quantum fields and differential equations.** The above requirements are quite general, and do not determine fields; in fact, it is somewhat difficult to obtain a Lorentz covariant field. One productive approach has been to study covariant field equations. One then expects to obtain Lorentz covariant fields as solutions. The linear case describes free particles, so we are mainly interested in nonlinear wave equations. In the problems studied up to now, it has been convenient to specify the $t = 0$ Cauchy data to agree with time zero free fields.

In classical problems, wave equations such as $(\Box + m^2)\phi = F(\phi)$ have been studied for a large class of functions F. For quantum fields, only certain polynomials have been studied (with minor exception for selfinteracting bosons in one-space dimension). The renormalization difficulties for general functions F have not been understood. However, all the interactions considered by physicists as describing nature are polynomials of degree three or four.

The simplest quantum field problem concerns the equation

$$(\Box + m^2)\phi(x, t) = -\mathscr{P}'(\phi(x, t)) \tag{2.1}$$

where $\mathscr{P}'$ is the derivative of a positive polynomial $\mathscr{P}$. We call this a $\mathscr{P}(\phi)$-theory, since $\mathscr{P}$ enters the energy integral for the equation. Often it is convenient to use a subscript to denote the number of space-time dimensions. Thus the ϕ_4^4 theory means the wave equation in 3-space dimensions, $(\Box + m^2)\phi = -4\phi^3$.

Later we also consider the "Yukawa interaction" of a boson field ϕ and a fermion field ψ. The corresponding classical equations are

$$(\square + m^2)\phi(x,t) = -\bar{\psi}\psi,$$

(2.1′)

$$\left\{\gamma_0 \frac{\partial}{\partial t} + \gamma \cdot \nabla + M\right\}\psi(x,t) = \phi\psi.$$

Here ψ is a spinor and the γ_μ are Dirac matrices. Also, $\bar{\psi} = -i\psi^*\gamma_0$. The Yukawa interaction is so named since it gives rise to a Yukawa binding force between the fermions.

A formal solution to the field equations (2.1) is given by the automorphism σ_t implemented by a Hamiltonian H.

$$\phi(x,t) = \sigma_t(\phi(x,0)) = e^{itH}\phi(x,0)e^{-itH}. \tag{2.2}$$

Here the Hamiltonian H is written in terms of the time zero fields $\phi(x,0)$ and $\dot{\phi}(x,0)$. In the linear case ($\mathscr{P} = 0$),

$$H_0 = \int \tfrac{1}{2}\{\dot{\phi}^2 + (\nabla\phi)^2 + m^2\phi^2\}\, dx, \tag{2.3}$$

and the relations $[\dot{\phi}(x,0), \phi(y,0)] = -i\delta(x-y)$ ensure that $(\square + m^2)\phi(x,t) = 0$. Note that the mass of the field is determined by the coefficient of $\int \phi^2\, dx$ in the Hamiltonian. Thus adding an interaction $\frac{1}{2}\delta m^2 \int \phi^2\, dx$ to H_0 replaces m^2 by $m^2 + \delta m^2$.

For a nonlinear selfinteraction ($\deg \mathscr{P} \geqq 4$), we have formally

$$H = H_0 + \int \mathscr{P}(\phi(x,0))\, dx = H_0 + H_I. \tag{2.4}$$

Then (2.2) satisfies (2.1).

Thus we reduce the problem of solving (2.1) to the problem of finding a selfadjoint linear operator (2.4). The nonlinear wave equation (2.1) is reduced to a question in linear functional analysis. We shall find this equivalence very useful, since most of the progress made on field theory interactions has resulted from using linear methods. Nonlinear equation techniques have never been successful in direct application to (2.1).

The problems in field theory are problems with an infinite number of degrees of freedom. In fact, the H of (2.4) can be represented as a (elliptic) partial differential operator in an infinite number of variables; see [**15**], [**10**]. With this representation, H_0 equals

$$\sum_{i,j=1}^{\infty} a_{ij}\left\{-\frac{\partial^2}{\partial q_i \partial q_j} + q_i q_j - \delta_{ij}\right\},$$

where a_{ij} is positive definite. In the same representation, the interaction Hamiltonian H_I is multiplication by a function.

Since we have an infinite number of degrees of freedom, we must be careful in defining a Hamiltonian. In fact, (2.4) as it stands is not a densely-defined operator, and the Schrödinger equation $i\, d\theta/dt = H\theta$ has no meaning. In order to be precise, we replace (2.4) by a regularized Hamiltonian $H_{\kappa,V}$, and then *define* H as the limit $\kappa, V \to \infty$. For $\kappa < \infty$ or $V < \infty$, our fields can no longer be Lorentz covariant, so we must re-establish this property for the limiting fields. The regularized theory has the advantage of being mathematically well-behaved.

We modify (2.4) in two ways to arrive at $H_{\kappa,V}$. First, we replace $\phi(x, 0)$ in $\mathscr{P}(\phi(x, 0))$ by a regularized field

$$\phi_\kappa(x, 0) = (\chi_\kappa * \phi)(x, 0).$$

Here $\chi_\kappa(x) = \kappa\chi(\kappa x)$ is a C_0^∞ approximation to the Dirac measure δ. Hence

$$\lim_{\kappa\to\infty} \phi_\kappa(x, 0) = \phi(x, 0).$$

We refer to κ as an "ultraviolet cutoff," since it damps the large-frequency Fourier components of ϕ. The regularized field $\phi_\kappa(x, 0)$ is a selfadjoint operator, and the function $\mathscr{P}(\phi_\kappa(x, 0))$ is defined by the spectral theorem.

The second modification to (2.4) is to replace the integration over all space (which may lead to an infinite multiple of the identity) by integration over $|x| \leqq V$. This is called a "volume cutoff." With these two modifications, we define the approximate Hamiltonian

$$H_{\kappa,V} = H_0 + \int_{|x|\leqq V} \mathscr{P}(\phi_\kappa(x, 0))\, dx. \tag{2.5}$$

In the following section, we will discuss the limits $\kappa \to \infty$ and $V \to \infty$, and we shall see how to modify (2.5) so these limits yield physically meaningful, selfadjoint Hamiltonians H, which define the time translation of nonlinear, covariant fields.

In summary, we are studying nonlinear, operator wave equations, or, equivalently, partial differential operators in an infinite number of variables. We approximate by a finite number of degrees of freedom (partial differential operator), and then pass to a limit.

3. **Renormalization.** Let us begin with two simple examples of mass renormalization, one physical and one mathematical. A simple example in classical physics was studied by Green in the 19th century [**13**]. He considered how the motion of a solid spheroid of density ρ would be changed by placing it in a uniform, incompressible fluid of density ρ'; he concluded the object moves as in a vacuum but with an effective density $\rho + \lambda\rho'$. He computed that the geometrical "renormalization constant" λ equals $\frac{1}{2}$ in the case of a sphere. (Of course, the buoyancy must be added to the force.)

A simple mathematical example is obtained by perturbing a linear field of mass m, which satisfies

$$(\Box + m^2)\phi(x, t) = 0.$$

With a perturbation $\frac{1}{2}\lambda\phi^2$, the new equation

$$(\Box + m^2)\phi = -\lambda\phi$$

describes particles with the "renormalized mass" $(m^2 + \lambda)^{1/2}$. The shift in the square of the mass is just $\delta m^2 = \lambda$. In §2, we already discussed the interaction Hamiltonians $\frac{1}{2}\lambda\int\phi^2\,dx$ for this mass shift.

We now proceed with nonlinear quantum fields.

3.1. *Physical interpretation.* In the nonlinear case, just as in the linear equation above, the mass of particles described by the field can be altered by the interaction (nonlinear terms). As explained in §1, the mass equals the size of the gap in the spectrum of the Hamiltonian H, so we call this gap the mass gap. The point of the mass renormalization of the $\mathscr{P}(\phi)$-theory consists of adding a quadratic term $-\frac{1}{2}\delta m^2\phi^2$ to $\mathscr{P}(\phi)$, with δm^2 chosen to adjust the mass gap to a pre-assigned "physical mass." It is usual to assume that physical mass is the m in $(\Box + m^2)\phi$, so the full linear term in the equation of motion is $\{m^2 - \delta m^2\}\phi$. Sometimes $\{m^2 - \delta m^2\}^{1/2}$ is called the "bare mass."

It is not possible to carry out this procedure exactly, since no useful formula exists for δm^2. However, we can compute δm^2 (or the mass gap) in any order of perturbation theory. The importance of writing the equations in renormalized form lies in the fact that in most theories, $-\delta m^2$ tends to ∞ with κ. Thus only the renormalized equations have a physical interpretation in the limit $\kappa \to \infty$. It is heartening to note that if the renormalization is not performed as described below, the Hamiltonian $H_{\kappa,V}$ has no limit as a selfadjoint operator (although it may have a limit as a physically irrelevant, sesquilinear form). Also, only when $H_{\kappa,V}$ is renormalized do the fields have a limit as the cutoffs are removed.

Let us now return to perturbation theory for the spectrum of $H_{\kappa,V}$. To be concrete, we describe the Yukawa$_2$ Hamiltonian. Let $E_{\kappa,V}$ be the minimum eigenvalue of $H_{\kappa,V}$. (It is easy to prove by Kato perturbation theory that such an eigenvalue exists.) We call $E_{\kappa,V}$ the vacuum energy, and compute E_2, its expansion to second order: We assume Ω_0 is the unperturbed vacuum, $H_0\Omega_0 = 0$, and $\langle\Omega_0, H_{\kappa,V}\Omega_0\rangle = 0$. Then abbreviating $H_{I,\kappa,V}$ by H_I,

$$E_2 = \langle\Omega_0, H_0\Omega_0\rangle + \langle\Omega_0, H_I\Omega_0\rangle - \langle\Omega_0, H_I H_0^{-1} H_I\Omega_0\rangle = -\langle\Omega_0, H_I H_0^{-1} H_I\Omega_0\rangle.$$

The first disturbing fact is that for any interaction, when κ is fixed,

$$E_2 = -O(V). \tag{3.1}$$

Furthermore, for V fixed,

$$\lim_{\kappa\to\infty} E_2 = -\infty \tag{3.2}$$

(except for the $\mathscr{P}(\phi)_2$-Hamiltonian). We interpret (3.1) as saying that the interaction causes a shift in the energy density, and hence it causes an infinite shift in the energy when the volume tends to infinity. (It is for this reason we have the volume cutoff in $H_{\kappa,V}$.) Likewise, (3.2) says that some interactions are so singular that the energy density may be shifted an infinite amount unless the interaction is regularized by an ultraviolet cutoff. These two effects have different physical origins and are dealt with separately.

We can expect a well-behaved limit for $H_{\kappa,V}$ as $\kappa \to \infty$ or as $V \to \infty$ only if we add a constant (multiple of the identity) to $H_{\kappa,V}$ in order to re-adjust the spectrum of $H_{\kappa,V}$ to have a convergent lower bound. Since only energy differences, but not energies, can be measured, adding a constant to $H_{\kappa,V}$ does not change any physics. In fact, the automorphism $\sigma_{t,\kappa,V}$ implemented by $H_{\kappa,V}$ is completely unchanged.

In a similar fashion, we can compute the mass gap in perturbation theory by computing the next-to-lowest point in the spectrum of $H_{\kappa,V}$, and subtracting. In the Yukawa$_2$ theory, the second order mass gap M_2 is proportional to $\log \kappa$. We therefore also add a mass term to $H_{\kappa,V}$. (Recall that a mass term is proportional to $\int \phi_\kappa^2\, dx$.) The renormalized Hamiltonian is defined by

$$H_{\kappa,V,\mathrm{ren}} = H_{\kappa,V} - \delta E_{\kappa,V} - \tfrac{1}{2}\delta m_\kappa^2 \int_V \phi_\kappa^2\, dx. \tag{3.3}$$

It turns out that for the Yukawa$_2$ Hamiltonian, only the second order contributions to δE and δm^2 diverge as $\kappa \to \infty$. We therefore choose for δE and δm^2 precisely those numbers given by second order perturbation theory. This renormalized Hamiltonian will have an operator limit as $\kappa \to \infty$. The remaining finite renormalization of the energy and the mass need not be included in $H_{\kappa,V}$ until after taking the $\kappa \to \infty$ limit.

The constants δE and δm^2 are called "renormalization constants." When perturbation theory gives closed expressions for the divergent parts of the renormalization constants, the Hamiltonian is called "super-renormalizable."

We note that the renormalization of $H_{\kappa,V}$ changes our Yukawa$_2$ equations. Instead of (2.1′), we have the renormalized equations

$$\begin{gathered}(\Box + m^2)\phi(x,t) = -\bar{\psi}\psi + \delta m^2\phi = j(x,t),\\ \left\{\gamma_0 \frac{\partial}{\partial t} + \gamma_1 \frac{\partial}{\partial x} + M\right\}\psi(x,t) = \phi\psi,\end{gathered} \tag{3.4}$$

where the total current $j(x,t)$, but not its components $\bar{\psi}\psi$ or $\delta m^2\phi$, have limits as $\kappa \to \infty$.

The limit $V \to \infty$ for the Yukawa$_2$ theory, as with other interactions, requires the renormalization of the vacuum energy. Perturbation theory contributes a volume divergence in every order (arising from the translation invariance of the Hamiltonian). Partial results have been obtained for this limit, using the abstract definition

$$\inf \text{spectrum } H_{V,\text{ren}} = 0.$$

These results are described in §4; see [**10**], [**7**], [**18**], [**11**].

3.2. *Mathematical problems.* In addition to supplying the correct renormalization constants, lowest order perturbation theory provides the correct guide to what mathematical problems occur as the cutoffs are removed. Let us discuss three Hamiltonians: ϕ_2^4, Yukawa$_2$, and ϕ_3^4. We treat $\kappa \to \infty$ and $V \to \infty$ separately.

In ϕ_2^4, there are no divergences in δE or δm^2 as $\kappa \to \infty$. The interpretation of this result is that the domain of H_V includes $D(H_0) \cap D(H_{I,V})$. In fact, $H_0 + H_{I,V}$ is selfadjoint on this domain.

In a model with ultraviolet divergences, however, the limit will be more singular. The first difficulty we encounter is a change of domain. Perturbation theory indicates for Yukawa$_2$ that $D(H_V) \cap D(H_0) = \{\varnothing\}$, and this appears true. In more singular theories, however, the ultraviolet divergence may force a change in the representation of $\mathfrak{A}$ to a unitarily inequivalent representation for $\kappa = \infty$. This is the case in ϕ_3^4 ([**4**], [**3**], [**14**]), and is indicated in perturbation theory: The vacuum vector $\Omega_{\kappa,V}$ has an inner production $\langle \Omega_0, \Omega_{\kappa,V} \rangle$ that vanishes as $\kappa \to \infty$ (in perturbation theory), suggesting that the vacuum $\Omega_{\kappa,V}$ is conveying to a vector outside the space.

In the limit $V \to \infty$, the perturbation theory vacuum converges weakly to zero in all theories. In fact, the representation changes in this limit [**7**]. These problems of change in representation are outlined further in the charts of §4.

To conclude this section, we mention two other types of renormalization which occur in three-space dimensions; e.g., ϕ_4^4. There are no mathematical results on *field strength* or *charge* renormalization, but there exist extensive computations in the physics literature.

Field strength renormalization simply multiplies the field ϕ by a constant $Z^{-1/2}$ to normalize the length of the one particle states:

$$\|Z^{-1/2}\phi(f)\Omega\|^2 = \langle f, f \rangle$$

for a suitable scalar product on the hyperboloid $E^2 = p^2 + m^2$.

Charge renormalization is required to adjust the long range force between particles to reflect their classical charge. For instance, charge renormalization is required in electrodynamics to ensure that two electrons repel each other at large distances with an e^2/r^2 force law, where e is the classical electric charge.

Charge renormalization causes serious difficulties, since when it diverges as $\kappa \to \infty$, it diverges in *every* order of perturbation theory. It is unclear whether Hamiltonian methods are applicable with such nonsuper-renormalizable field theories. In perturbation theory of such interactions, the sharp time field appears not to be an operator, but rather a sesquilinear form (for $\kappa = \infty$). Thus the C^*-algebra $\mathfrak{A}$ should be defined in terms of the space-time averaged fields, and so we expect that it is necessary not only to change representations of $\mathfrak{A}$ as $\kappa \to \infty$, but to

Model	*Perturbation Theory*	*Status of H_V*
$\mathscr{P}(\phi)_2$ [5], [17]	$\delta E, \delta m^2$ both finite	$H_V = H_V^*$. Inclusion of domain: $$H_V \supset H_0 + H_{I,V}.$$ Norm convergence of $(H_{\kappa,V} - \zeta)^{-1}$ as $\kappa \to \infty$. *Open*. $H_V = H_0 + H_{I,V}$? (Known if deg $\mathscr{P} = 4$.)
Yukawa$_2$ [8], [10]	$\delta E = O(\mathrm{Log}^2 \kappa)$ $\delta m^2 = O(\mathrm{Log}\, \kappa)$	$0 \leqq H_V = H_V^*$ with proper counterterms. Counterterms yield change in *domain*: $$D(H_V) \cap D(H_0) = \{\varnothing\}.$$ Norm convergence of $(H_{\kappa,V} - \zeta)^{-1}$ as $\kappa \to \infty$. Definitions of $\bar{\psi}\psi - \delta m^2 \phi$, etc. [1]. *Open*. Uniqueness of vacuum Ω_V?
ϕ_3^4	$\delta E = O(\kappa)$ $\delta m^2 = O(\mathrm{Log}^2 \kappa)$	H_V densely defined [4]. Infinite counterterms yield change in representation of $\mathfrak{A}$ to $\pi(\mathfrak{A})$ on $\mathscr{F}_{\mathrm{ren}}$ as $\kappa \to \infty$ [3], [14]. *Work in progress*. $0 \leqq H_{\kappa,V}$ uniformly in κ [12].
ϕ_4^4	$\delta E = O(\kappa^4)$ $\delta m^2 = O(\kappa^2)$ $\delta\lambda = O(\mathrm{Log}\, \kappa)$	*Open*. Sharp time yields bilinear forms but not operators? This presumably yields change of *C^*-algebra* $\mathfrak{A}$ to $\mathfrak{A}_{\mathrm{ren}}$ as $\kappa \to \infty$.

The $\kappa \to \infty$ Limit

actually change the algebra itself. We note that mass and energy renormalizations occur as lower degree polynomials in $H_{\kappa,V}$. Charge renormalization, however, affects the term in $\mathscr{P}$ of maximum degree. We might expect it to be unusually difficult.

4. **Some results.** In order to give a feeling for what can or does occur as κ, $V \to \infty$, we first state some theorems about the Yukawa$_2$ theory. We then illustrate with two charts some problems of increasing complexity for models of increasing difficulty.

THEOREM 1 [8]. *As $\kappa \to \infty$, the resolvents of $H_{\kappa,V,\mathrm{ren}}$ defined in* (3.3) *converge in norm to the resolvent of a selfadjoint operator H_V.*

THEOREM 2 [9]. *The Hamiltonians H_V define a V-independent automorphism of $\mathfrak{A}$, with finite propagation speed.*

Model	*Status of H*
$\mathscr{P}(\phi)_2$ Yukawa$_2$ [7], [**18**]	Change in representation of $\mathfrak{A}$ to $\pi(\mathfrak{A})$ on $\mathscr{F}_{\text{ren}}$ only in limit $V = \infty$. Positive selfadjoint H on $\mathscr{F}_{\text{ren}}$. Convergent subsequence of states $\omega_V(\cdot) = \langle \Omega_V, \cdot \Omega_V \rangle$ on $\mathfrak{A}$ as $V \to \infty$, norm convergent on each $\mathfrak{A}(B)$. *Open*. Convergences of the full sequence ω_V? Uniqueness of the vacuum Ω. Gap in the spectrum of H (mass gap).
ϕ_3^4	Inequivalent representations of $\mathfrak{A}$ for each change in volume [3], [**14**]. However, unitary equivalence of representations of $\mathfrak{A}(B)$ for B fixed and V sufficiently large [**2**]. *Open*. Any question about $V = \infty$ representation.
ϕ_4^4	*Open*. No change of representation of $\mathfrak{A}(B)_{\text{ren}}$ for B fixed and V sufficiently large?

The $V \to \infty$ Limit

We note that, given B and a range of $|t| \leqq T$, for V sufficiently large,

$$\sigma_t(A) = \exp(itH_V)A\exp(-itH_V), \qquad A \in \mathfrak{A}(B),$$

is independent of V. This follows from the finite propagation speed property for propagation with H_V. Thus we may use locally correct Hamiltonians H_V to implement an automorphism σ_t on the dense subalgebra $\bigcup_B \mathfrak{A}(B)$ of $\mathfrak{A}$.

THEOREM 3 [5], [**18**]. *There exists a cyclic representation π of $\mathfrak{A}$ by bounded operators $\pi(\mathfrak{A})$ on a Hilbert space $\mathscr{F}_{\text{ren}}$, and a selfadjoint operator H such that, for $A \in \mathfrak{A}$,*

$$e^{itH}\pi(A)e^{-itH} = \pi(\sigma_t(A)).$$

The Hamiltonian H is positive, and $H\Omega = 0$ where Ω is the cyclic vector in $\mathscr{F}_{\text{ren}}$.

THEOREM 4 [1]. *Each side of the field equations* (3.4) *can be defined, and these equations hold as sesquilinear forms on $\mathscr{F}_{\text{ren}}$. On a dense domain, the equations are continuous functions of* (x, t).

To conclude, let us summarize: The infamous divergences of quantum field theory are present; in fact, they are necessary both for a physically reasonable and for a mathematically well-behaved quantum field theory. Our fields are obtained as limits of approximate, regularized fields, and the hardest work goes into establishing uniform estimates that ensure the existence of these limits.

In the above charts, we summarize the status of Hamiltonians $H_V = \lim_\kappa H_{\kappa,V}$ and $H = \lim_V H_V$ for some models.

REFERENCES

Note. A systematic treatment and extensive references can be found in [**10**].

1. J. Dimock, *Estimates, renormalized currents and field equations for the Yukawa$_2$ field theory*, Thesis, Harvard University, Cambridge, Mass., 1971.

2. J.-P. Eckmann (to appear).

3. J. Fabrey, *Exponential representations of the canonical commutation relations*, Comm. Math. Phys. **19** (1970), 1–30.

4. J. Glimm, *Boson fields with the ϕ^4 interaction in three dimensions*, Comm. Math. Phys. **10** (1968), 1–47. MR **37** #7154.

5. J. Glimm and A. Jaffe, *A $\lambda\phi^4$ quantum field theory without cutoffs*. I, Phys. Rev. (2) **176** (1968), 1945–1951. MR **40** #1106.

6. ———, *The $\lambda(\phi^4)_2$ quantum field theory without cutoffs*. II. *The field operators and the approximate vacuum*, Ann. of Math. (2) **91** (1970), 362–401. MR **41** #1333.

7. ———, *The $\lambda(\phi^4)_2$ quantum field theory without cutoffs*. III. *The physical vacuum*, Acta. Math. **125** (1970), 203–267. MR **42** #4130.

8. ———, *Self-adjointness of the Yukawa$_2$ Hamiltonian*, Ann. Phys. **60** (1970), 321–383.

9. ———, *The Yukawa$_2$ quantum field theory without cutoffs*, J. Functional Analysis **7** (1971), 323–357.

10. ———, *Field theory models*, 1970 Les Houches Lectures, C. DeWitt and R. Stora (Editors), Gordon and Breach, New York, 1971.

11. ———, *The energy momentum spectrum and vacuum expectation values in quantum field theory*. II, Comm. Math. Phys. **22** (1971), 1–22.

12. ——— (work in progress).

13. G. Green, *Researches on the vibration of pendulums in fluid media*, Mathematical papers of George Green, Chelsea, New York, 1970.

14. K. Hepp, *Théorie de la renormalization*, Springer-Verlag, Heidelberg, 1969.

15. A. Jaffe, *Existence theorems for a cutoff $\lambda\phi^4$ field theory*, Proc. Conf. Mathematical Theory of Elementary Particles (Dedham, Mass., 1965), M.I.T. Press, Cambridge, Mass., 1966, pp. 45–58. MR **36** #1167.

16. R. Jost, *The general theory of quantized fields*, Lectures in Appl. Math., vol. 4, Amer. Math. Soc., Providence, R.I., 1965. MR **31** #1929.

17. L. Rosen, *A $\lambda\phi^{2n}$ field theory without cutoffs*, Comm. Math. Phys. **16** (1970), 127–183.

18. R. Schrader, *A Yukawa interaction without cutoffs in space time of two dimensions*, Ann. Physics (to appear).

19. R. Streater and A. Wightman, *PCT, spin and statistics, and all that*, Benjamin, New York, 1964. MR **28** #4807.

COURANT INSTITUTE OF MATHEMATICAL SCIENCES, NEW YORK UNIVERSITY

HARVARD UNIVERSITY

QUANTUM FIELDS AND MARKOFF FIELDS

EDWARD NELSON[1]

Wightman [7] showed how the requirements of relativistic quantum theory lead one from partial differential equations to the rather elaborate constructs of quantum field theory, and Jaffe [**8**] showed how the equations can be solved if one can construct the appropriate Hamiltonian. What I wish to discuss is a new method for constructing the Hamiltonian. I say new, but of course it is strongly influenced by previous work of others, in particular of Symanzik [**6**].

The method I am advocating is probabilistic. In recent years, probabilistic methods, including path space integrals, have been used in quantum field theory, particularly by Glimm, Jaffe, Rosen, and myself. The usual reaction of workers in the field is to recoil in horror and to attempt to find alternate or partially alternate methods. I want to begin by attempting to show that probability is a natural tool in the subject.

1. **Second quantization.** To say that something is natural is to assert the existence of a functor from one category to another. Let $\mathscr{h}$ be the category whose objects are Hilbert spaces and whose morphisms are contractions (that is, linear operators of norm $\leqq 1$). Let $\mathscr{p}$ be the category whose objects are $\mathscr{L}^2$ spaces of probability spaces; a morphism

$$P: \mathscr{L}^2(\Omega_1, d\mu_1) \to \mathscr{L}^2(\Omega_2, d\mu_2)$$

from one such object to another is a positivity-preserving linear operator such that $P1 = 1$ and $\int Pf\, d\mu_2 = \int f\, d\mu_1$ for all f in $L^2(\Omega_1, d\mu_1)$. It is easy to see that the isomorphisms in $\mathscr{h}$ are the unitary operators, and quantum theory is so constructed as to be covariant under unitary operators. The isomorphisms of $\mathscr{p}$ are measure-preserving transformations. Consequently, if we construct a covariant functor from

AMS 1970 *subject classifications*. Primary 81A18, 60J99.

[1] This work was supported by the NSF. A complete account will appear elsewhere.

$\mathscr{h}$ to $\mathscr{p}$, as we shall, this means that any probabilistic object constructed by means of this functor is an intrinsic aspect of quantum theory.

Let ϕ be the Gaussian stochastic process indexed by the Hilbert space $\mathscr{H}$, with mean 0 and covariance given by the inner product, and let $(\Omega, d\mu)$ be the underlying probability space of this process. Thus for each u in $\mathscr{H}$, $\phi(u)$ is a Gaussian random variable on $(\Omega, d\mu)$ with expectation (mean)

$$E\phi(u) = \int_\Omega \phi(u)(w)\, d\mu(w) = 0,$$

and we have the covariance

$$E\overline{\phi(v)}\phi(u) = \int_\Omega \overline{\phi(v)(w)}\phi(u)(w)\, d\mu(w) = \langle v, u\rangle.$$

(The inner product on $\mathscr{H}$ is assumed to be conjugate linear in the first variable.) We let $\Gamma(\mathscr{H}) = \mathscr{L}^2(\Omega, d\mu)$.

Since ϕ is Gaussian, $\phi(u_1)\cdots\phi(u_n)$ is in $\mathscr{L}^2(\Omega, d\mu)$ if $u_1, \ldots, u_n$ are in $\mathscr{H}$. Let $\Gamma(\mathscr{H})^{(\leqq m)}$ be the closed linear span of such elements with $n \leqq m$, and let $\Gamma(\mathscr{H})^{(n)}$ be the orthogonal complement of $\Gamma(\mathscr{H})^{(\leqq(n-1))}$ in $\Gamma(\mathscr{H})^{(\leqq n)}$. Thus we have $\Gamma(\mathscr{H})$ represented as the direct sum

$$\Gamma(\mathscr{H}) = \bigoplus_{n=0}^{\infty} \Gamma(\mathscr{H})^{(n)},$$

where $\Gamma(\mathscr{H})^{(0)} = \mathbf{C}$. We let $:\phi(u_1)\cdots\phi(u_n):$ be the orthogonal projection of $\phi(u_1)\cdots\phi(u_n)$ onto $\Gamma(\mathscr{H})^{(n)}$. We have

$$\langle :\phi(v_1)\cdots\phi(v_n):, :\phi(u_1)\cdots\phi(u_n):\rangle = \sum_\pi \langle v_{\pi_1}, u_1\rangle \cdots \langle v_{\pi_n}, u_n\rangle,$$

where the sum is over all permutations π of $1, \ldots, n$. From this it follows that there is a unique unitary equivalence of the symmetric n-fold tensor product $\mathscr{H}^{\otimes}_{\text{sym}}$ onto $\Gamma(\mathscr{H})^{(n)}$ in which

$$u \otimes \cdots \otimes u \mapsto :\phi(u)^{(n)}:,$$

provided that we give $\mathscr{H}^{\otimes n}_{\text{sym}}$ the norm in which

$$\|u \otimes \cdots \otimes u\|^2 = n!\|u\|^{2n}.$$

Thus $\Gamma(\mathscr{H})$ is equivalent to the Fock space for bosons (that is, symmetric statistics) that Wightman discussed in [7], and $\Gamma(\mathscr{H})^{(n)}$ is the n-particle space. This construction of Fock space is due to Segal.

Now let $A: \mathscr{H}_1 \to \mathscr{H}_2$ be a morphism of $\mathscr{h}$. We define $\Gamma(A): \Gamma(\mathscr{H}_1) \to \Gamma(\mathscr{H}_2)$ by letting $\Gamma(A)$ be the direct sum of the $\Gamma(A)^{(n)}: \Gamma(\mathscr{H}_1)^{(n)} \to \Gamma(\mathscr{H}_2)^{(n)}$ with $\Gamma(A) = A \otimes \cdots \otimes A$ (n factors). It is not hard to verify that Γ is a morphism in $\mathscr{p}$, and Γ is a covariant functor from $\mathscr{h}$ to $\mathscr{p}$.

LEMMA 1. *Let* F *be in* $\Gamma(\mathscr{H})^{(n)}$. *Then, for* $j = 1, 2, \ldots, F$ *is in* $\mathscr{L}^{2j}(\Omega, d\mu)$, *where* $\Gamma(\mathscr{H}) = \mathscr{L}^2(\Omega, d\mu)$, *and*

$$\|F\|_{2j} \leqq (2j)^{n/2}\|F\|_2.$$

This lemma can be proved quickly using stochastic integrals. The constant $(2j)^{n/2}$ is not the best possible, and if $2j = 2$ it can of course be replaced by 1. Then by interpolation we have the following:

LEMMA 2. *Let* F *be in* $\Gamma(\mathscr{H})^{(n)}$. *Then, for* $2 \leqq p \leqq 4$,

$$\|F\|_p \leqq 4^{(1-(2/p))n}\|F\|_2.$$

In recent years, $\mathscr{L}^p$ estimates have proved quite useful in constructive quantum field theory. Using these two lemmas we can give a trivial proof of the basic property which Barry Simon calls hypercontractivity.

PROPOSITION. *Let* $A: \mathscr{H}_1 \to \mathscr{H}_2$ *with* $\|A\| < 1$. *Let* $\Gamma(\mathscr{H}_1) = \mathscr{L}^2(\Omega_1, d\mu_1)$ *and* $\Gamma(\mathscr{H}_2) = \mathscr{L}^2(\Omega_2, d\mu_2)$. *Then for some* $p > 2$, $\Gamma(A)$ *is a bounded operator from* $\mathscr{L}^2(\Omega_1, d\mu_1)$ *to* $\mathscr{L}^p(\Omega_2, d\mu_2)$.

PROOF. Let F be in $\Gamma(\mathscr{H}_1) = \mathscr{L}^2(\Omega_1, d\mu_1)$, and let $F^{(n)}$ be its component in $\Gamma(\mathscr{H}_1)^{(n)}$. Then

$$\begin{aligned}\|\Gamma(A)F\|_p &\leqq \sum_n \|\Gamma(A)F^{(n)}\|_p \\ &\leqq \sum_n 4^{(1-(2/p))n}\|\Gamma(A)F^{(n)}\|_2 \\ &\leqq \sum_n (4^{(1-(2/p))}\|A\|)^n\|F\|_2,\end{aligned}$$

where we have used the definition of $\Gamma(A)$ on $\Gamma(\mathscr{H}_1)^{(n)}$ as $A \otimes \cdots \otimes A$. Since $\|A\| < 1$, and $4^{(1-(2/p))}$ can be made arbitrarily close to 1 for $p > 2$, we have that this is $\leqq$ constant $\|F\|_2$, concluding the proof.

We will return to this proposition later.

2. **Markoff fields.** Let us consider a particular case of this construction in which more structure is present.

Let $\mathscr{H}^{-1}(\mathbf{R}^d)$ be the Sobolev space of index -1 over d-dimensional Euclidean space $\mathbf{R}^d$. The inner product is given by $\langle g, f\rangle_{-1} = \langle g, (-\Delta + m^2)^{-1}f\rangle$ where $m > 0$. (For $d \geqq 3$ we may also consider the case $m = 0$.)

Consider the Gaussian process ϕ as above. For $E \subset \mathbf{R}^d$, let $\mathscr{O}(E)$ be the family of all random variables on $(\Omega, d\mu)$ generated by the $\phi(f)$ with f in $\mathscr{H}^{-1}(\mathbf{R}^d)$ and $\operatorname{supp} f \subset E$. (The family is closed under sums, products, and all Borel functions.) If U is open in $\mathbf{R}^d$, let ∂U be its boundary and U' its complement.

THEOREM. *If* U *is open in* $\mathbf{R}^d$ *and* A *is in* $\mathscr{L}^1(\Omega, d\mu) \cap \mathscr{O}(U)$ *then*

$$E\{A|\mathscr{O}(U')\} = E\{A|\mathscr{O}(\partial U)\}. \tag{1}$$

This says that the conditional expectation of A given $\mathcal{O}(U')$ is the same as the conditional expectation of A given just $\mathcal{O}(\partial U)$. In other words, if we want to know something about the behavior of the field ϕ inside U (namely, A) and we know all about the behavior of the field outside U (namely, $\mathcal{O}(U')$), this gives us no more information than just the knowledge of the field on the boundary of U (namely, $\mathcal{O}(\partial U)$).

Using simple properties of Gaussian random variables of mean 0 (the fact that orthogonal such random variables are independent) and the fact that ϕ is isometric from $\mathscr{H}^{-1}(\mathbf{R}^d)$ into $\mathscr{L}^2(\Omega, d\mu)$, the proof of the theorem reduces to showing the following:

Let f be in $\mathscr{H}^{-1}(\mathbf{R}^d)$ with $\operatorname{supp} f \subset U$, let

$$\mathscr{M} = \{g \in \mathscr{H}^{-1}(\mathbf{R}^d) : \operatorname{supp} g \subset U'\},$$

$$\mathscr{N} = \{g \in \mathscr{H}^{-1}(\mathbf{R}^d) : \operatorname{supp} g \subset \partial U\},$$

and let h be the orthogonal projection of f onto $\mathscr{M}$. We need to show that h is in the smaller space $\mathscr{N}$.

We have

$$\langle g, (-\Delta + m^2)^{-1} h\rangle = \langle g, (-\Delta + m^2)^{-1} f\rangle$$

for all g in $\mathscr{M}$, in particular for all g in $C^\infty_{\text{com}}(U'^\circ)$ (where U'° is the interior of U'). That is,

$$(-\Delta + m^2)^{-1} h = (-\Delta + m^2)^{-1} f$$

as distributions on U'°. Since $(-\Delta + m^2)$ is a local operator, $h = f$ on U'°. But $f = 0$ on U'°, so $\operatorname{supp} h \subset \partial U$, concluding the proof.

The property (1) will be called the *Markoff property*, since it is a generalization to higher dimensions of the notion of a Markoff process indexed by $\mathbf{R}$. Any process satisfying (1) and certain mild technical conditions will be called a *Markoff field*, and if it is covariant under the action of the Euclidean group of $\mathbf{R}^d$ (as in the free Markoff field constructed above) then it will be called a *Euclidean field*.

3. **Quantum fields.** Let ϕ be a Euclidean field, on the probability space $(\Omega, d\mu)$, over $\mathbf{R}^d$, where d is the number of space-time dimensions. We want to construct from it a corresponding quantum field θ. I wish to emphasize that a novel aspect of this construction is that d is the number of space *and time* dimensions. The Hilbert space $\mathscr{K} = \mathscr{L}^2(\Omega, d\mu)$ is not the Hilbert space of state vectors of the quantum field (the space which in the case of the free field is called Fock space)—it is a much bigger Hilbert space. The first task is to construct the Hilbert space $\mathscr{H}$ of the quantum theory.

Fix the hyperplane $\mathbf{R}^{d-1}$ in $\mathbf{R}^d$. Let $\mathscr{H} = \mathscr{K} \cap \mathcal{O}(\mathbf{R}^{d-1})$ and let E be the orthogonal projection operator of $\mathscr{K}$ onto $\mathscr{H}$. Let T^t, for $-\infty < t < \infty$, be the action of the one-parameter translation subgroup in the direction orthogonal to $\mathbf{R}^{d-1}$. Since ϕ is a Euclidean field, this is a one-parameter group of unitary operators on $\mathscr{K}$ (and it is strongly continuous because of a continuity assumption on the field which I have not made explicit here). Now define

$$P^t u = ET^t u, \qquad u \in \mathscr{H}, t \geqq 0.$$

THEOREM. *There is a unique positive selfadjoint operator H on $\mathscr{H}$ such that* $P^t = e^{-tH}, t \geqq 0$.

To prove this, we need to show that $P^t P^s = P^{t+s}$ for $t, s \geqq 0$, and this follows in a standard way from the Markoff property (applied to open sets U in $\mathbf{R}^d$ which are half-spaces), that the P^t are contractions (which is trivial since E is an orthogonal projection and T^t is unitary) and that the P^t are selfadjoint. The selfadjointness of P^t is seen by considering the action of reflection in a hyperplane. The point is that if an element of $\mathscr{H}^{-1}(\mathbf{R}^d)$ with support in a hyperplane is reflected in the hyperplane it is left unchanged. This is not true for the Sobolev spaces of higher negative index, for an element with support in a hyperplane may involve differentiation in the normal direction. Thus it is essential to assume that our process ϕ is indexed by $\mathscr{H}^{-1}(\mathbf{R}^d)$, as we have done, or by some similar space.

The operator H constructed in this way is the Hamiltonian. Once we have the Hamiltonian H and the field $\phi(\mathbf{x}, 0)$ at time 0, we may define the quantum field θ by

$$\theta(\mathbf{x}, t) = e^{itH}\phi(\mathbf{x}, 0)e^{-itH}.$$

A rather mild technical assumption ensures the existence of the vacuum expectation values $\langle \Omega, \theta(f_1) \cdots \theta(f_n)\Omega \rangle$, where Ω is the function identically 1 on $(\Omega, d\mu)$, each f_i is in the Schwartz space $\mathscr{S}(\mathbf{R}^d)$, and

$$\theta(f_i) = \int_{\mathbf{R}^d} f_i(x)\theta(x)\, dx$$

(we use the notation $x = (\mathbf{x}, t)$ for d-vectors). These vacuum expectation values are given by tempered distributions $\mathscr{W}^{(n)}$ on $\mathbf{R}^{nd}$, the Wightman distributions, so that

$$\langle \Omega, \theta(f_1) \cdots \theta(f_n)\Omega \rangle = \int \cdots \int f_1(x_1) \ldots f_n(x_n) \mathscr{W}^{(n)}(x_1, \ldots, x_n)\, dx_1 \ldots dx_n.$$

The main result is that the $\mathscr{W}^{(n)}$ have the principal properties required in order to reconstruct a quantum field theory from them. These properties are:

(a) relativistic covariance,
(b) spectrum condition,
(c) hermiticity,
(d) local commutativity,
(e) positive definiteness,

and they are explicitly described in Chapter 3 of Streater-Wightman [**5**].

Properties (c) and (e) are trivial consequences of the definition of the vacuum expectation values in terms of an inner product. Property (a) is reasonable, because we have taken a Euclidean covariant theory and replaced t by it. However, the proof uses a partial differential equation and only yields the invariance of $\mathscr{W}^{(n)}$ under transformations in the connected component of the identity of the inhomogeneous Lorentz group. We know that the energy operator H has positive spectrum, and this together with (a) ensures that the energy-momentum operators have spectrum in the forward cone, which is (b). Property (d) should correctly be formulated as follows:

$$\mathscr{W}^{(n)}(x_1, \ldots, x_j, x_{j+1}, \ldots, x_n) = \mathscr{W}^{(n)}(x_1, \ldots, x_{j+1}, x_j, \ldots, x_n),$$

$$(x_j - x_{j+1})^2 < 0.$$

That is, they are equal as distributions on the open set in which x_j and x_{j+1} are space-like separated in Minkowski space. This expresses the commutativity of the quantum field operators at space-like separated points, an essential feature of quantum field theory which was discussed by Wightman [7]. From a formal point of view, it is easy to make this result plausible. Suppose that x_j and x_{j+1} are space-like separated. Then, using (a), we may make an inhomogeneous Lorentz transformation to ensure that the two points lie in the hyperplane $\mathbf{R}^{d-1}$ given by $t = 0$. But for $t = 0$, the quantum field θ is by definition equal to the Markoff field ϕ, and the values of ϕ are random variables which of course commute. The trouble with this argument is that the fields at individual points are not defined, and θ must be smeared over Minkowski space, whereas ϕ must be smeared over Euclidean space.

From a technical point of view, these aspects are the most interesting part of the theory. The very existence of the vacuum expectation values requires the construction of a scale of Hilbert spaces in $\mathscr{H}$ analogous to the classical Sobolev spaces, and a general study of smoothing properties of operators associated with the scale. The proof of local commutativity involves a study of functions of several complex variables, the Wightman functions, and their distribution boundary values, and the proof is clinched by a theorem of Jost [**2**, p. 83]. I will say no more about this here.

4. **Multiplicative functionals.** I have described in outline how to attach a quantum field theory to certain Markoff fields. The only example of a Markoff field which I have given is the free Markoff field (the Gaussian field indexed by $\mathscr{H}^{-1}(\mathbf{R}^d)$), and it gives rise of course to the well-known free quantum field of mass m. We need a method for constructing new Markoff fields from old ones. One approach is by means of multiplicative functionals.

Let ϕ be a Markoff field on $\mathbf{R}^d$ with probability space $(\Omega, d\mu)$. For example, ϕ might be the free Markoff field. An *additive functional* is a random variable A on $(\Omega, d\mu)$ such that for every finite open cover $\{U_i\}$ of $\mathbf{R}^d$, there exist A_i in $\mathscr{O}(U_i)$ with

$A = \sum A_i$. A *multiplicative functional* is a strictly positive random variable M on $(\Omega, d\mu)$ such that for every finite open cover $\{U_i\}$ of $\mathbf{R}^d$, there exist strictly positive M_i in $\mathcal{O}(U_i)$ with $M = \prod M_i$. Clearly, $M = e^A$ establishes a bijection between the two classes.

Now suppose that M is an integrable multiplicative functional, and assume it normalized so that $EM = 1$, where E now denotes the expectation (integral with respect to $d\mu$). Then $(\Omega, M\, d\mu)$ is again a probability space, and one can prove the theorem that ϕ is again a Markoff field, a new one, on $(\Omega, M\, d\mu)$.

Let me conclude by sketching an application of this idea without going into any detail. If ϕ is the free Markoff field for $d = 2$, one can make sense of the expression

$$A_g = \int_{\mathbf{R}^2} g(x){:}\phi(x)^n{:}\, dx$$

for g in $\mathcal{L}^1(\mathbf{R}^2) \cap \mathcal{L}^2(\mathbf{R}^2)$. For $n = 4$, this is a Euclidean version of the ϕ_2^4 theory studies by Glimm and Jaffe; the general case has been studied by Rosen. If g is positive and n is even, then e^{-A_g} is integrable. Now A_g is an additive functional (because it is a limit of sums of local quantities) so that

$$M_g = (e^{-A_g}/Ee^{-A_g})$$

is a multiplicative functional of expectation 1. Thus ϕ is a Markoff field on the new probability space $(\Omega, M_g\, d\mu)$.

It is not a Euclidean field because of the volume cut-off g. In fact, it is hopeless to construct a Euclidean field directly by means of a multiplicative functional because the Euclidean group acts ergodically on the free field, and the only Euclidean invariant multiplicative functional is a constant. I hope that it is true that, for each bounded region U, $\lim_{g\to 1} E\{M_g|\mathcal{O}(U)\}$ exists and gives rise to a Euclidean field, but I have no result to report.

Suppose now that g is the characteristic function of a rectangle in $\mathbf{R}^2$ with base in the hyperplane $\mathbf{R}^1$. Let u be in $\mathcal{H}$. Since T^t, where t is the height of the rectangle, is unitary, $T^t u$ is also in $\mathcal{L}^2(\Omega, d\mu)$, so that uT^tu is in $\mathcal{L}^1$. Using the Proposition of §1, it may be shown that uT^tu is actually in $\mathcal{L}^q$ for some $q > 1$. The same proof that shows that e^{-A_g} is in $\mathcal{L}^1$ shows that it is in $\mathcal{L}^p$ for all $p < \infty$. Thus $uT^tue^{-A_g}$ is in $\mathcal{L}^1$. In this way we have that the Hamiltonian is bounded below on Fock space, corresponding to the fact that there is no infinite energy renormalization in this theory. as Jaffe explained. In particular, we may choose $u = \Omega$, and by the spectral theorem we get an exponential bound (exponential in the height t) on Ee^{-A_g}. Now turn the rectangle over on its side. The free field is Euclidean, so we have an exponential bound, exponential in the width of the rectangle, on Ee^{-A_g}. Now repeat the above argument. In this way we obtain the fact that the Hamiltonian has a lower bound proportional to the volume of the support of the cut-off g. This is the result established by Glimm and Jaffe in §5 of [**1**]. This is one application of Markoff fields to constructive quantum field theory, and I hope there will be others.

References

1. James Glimm and Arthur Jaffe, *The $\lambda(\phi^4)_2$ quantum field theory without cutoffs*. III. *The physical vacuum*, Acta Math. **125** (1970). 203–267. MR **42** #4130.

2. Res Jost, *The general theory of quantized fields*, Lectures in Appl. Math., vol. 4, Amer. Math. Soc., Providence, R.I., 1965. MR **31** #1929.

3. Edward Nelson, *A quartic interaction in two dimensions*, Proc. Conference Mathematical Theory of Elementary Particles (Dedham, Mass., 1965), M.I.T. Press, Cambridge, Mass., 1966, pp. 69–73. MR **35** #1309.

4. L. Rosen, *A $\lambda\phi^{2n}$ field theory without cutoffs*, Comm. Math. Phys. **16** (1970), 157–183. MR **42** #5559.

5. R. F. Streater and A. S. Wightman, *PCT, spin and statistics, and all that*, Benjamin, New York, 1968.

6. K. Symanzik, *Euclidean quantum field theory*, Rendiconti della Scuola Internazionale di Fisica "E. Fermi," XLV Corso. [This paper has an appendix by S. R. S. Varadhan which establishes the most interesting property of the Wiener process which has been discovered in recent years.]

7. Arthur Wightman, *Relativistic wave equations as singular hyperbolic systems*, Proc. Sympos. Pure Math., vol. 23, Amer. Math. Soc., Providence, R.I., 1972.

8. James Glimm and Arthur Jaffe, *What is renormalization*?, Proc. Sympos. Pure Math., vol. 23, Amer. Math. Soc., Providence, R.I., 1972.

Princeton University

ON THE STEADY FALL OF A BODY IN A NAVIER-STOKES FLUID[1]

H. F. WEINBERGER

1. **Introduction.** We say that a body undergoes a steady falling motion in an infinite viscous fluid if the motion of the fluid as seen by an observer attached to the body is independent of time. Such a motion is of interest because it may be a limit of a class of unsteady falling motions.

We formulate the problem of finding a steady falling motion of a given body in two different ways in §2.

In Problem I we prescribe both the shape and a downward orientation of the body. We think of the body as a hollow one inside which we are free to move masses about, and we seek a position of the center of mass which will result in a steady falling motion with the given downward orientation. In general the body must undergo a rotation about the vertical axis as well as a translation in this motion.

In Problem II the mass distribution in the body is prescribed, and we look for a steady falling motion with the orientation to be determined. Again the body will in general undergo both translation and a rotation about the vertical axis.

The linearized (Stokes flow) versions of Problems I and II have been considered in another paper [**11**]. We have shown that they both can be solved for any body with positive capacity. We have been able to characterize the solution of the linearized Problem I by several variational principles, which lead to approximations and bounds for the speed of fall.

Our purpose is not only to prove the existence of solutions of Problems I and II, but also to show that these solutions are approximated by those of the corresponding linear problems when the net mass is sufficiently small.

In §3 we show that Problem I always has a solution. Our method of proof is the Leray-Schauder method [**10**], which was used for the first boundary value problem

AMS 1970 *subject classifications*. Primary 76D05, 35Q10; Secondary 55C25.

[1] This work was supported by the National Science Foundation through grant GP-27275.

by J. Leray [**8**], [**9**]. However, we prefer to use the formulation given by O. A. Ladyženskaja [**6**], [**7**].

In §4 we show that the solution for the Stokes flow version of Problem I can be obtained as a limit of solutions of the Navier-Stokes problem as a certain dimensionless parameter $m\rho|\mathbf{g}|/\mu^2$ approaches zero. Furthermore, we show that the falling speed in the Navier-Stokes fluid is no greater than that for a body of the same mass falling in a fluid of the same viscosity in the Stokes approximation. That is, the nonlinear terms tend to increase the drag.

The situation is not as clean for Problem II. The existence proof in the linear case in [**11**] used not only the existence but also the uniqueness of the solution of Problem I. The only uniqueness proof, even for small masses, for Navier-Stokes problems in unbounded domains is that of R. Finn [**2**]. Finn's proof, however, makes strong use of the fundamental solution of the Oseen equation, which allows translation but not rotation.

For this reason we have to confine ourselves to positions of the center of mass such that the linearized problem has an isolated direction of fall. Under this condition we establish in §5 the existence of a steady fall when $m\rho|\mathbf{g}|/\mu^2$ is sufficiently small. We do show that for a given body the condition on the linear problem is violated for at most two positions of the center of mass.

In §6 we show that under some symmetry conditions Problem II may be solved even when the conditions of Theorem 2 are violated.

This paper represents a first step in treating a rather large class of problems of physical interest. We do not examine the question of whether any of the steady solutions that we obtain is stable or whether it is a limit of unsteady solutions.

Related problems which we do not discuss include the steady fall of a body in a vertical cylindrical tube and the steady falling motion of several bodies.

2. **Formulation of the problems.** We consider the fall under its own weight of a bounded connected rigid body in an infinite Navier-Stokes fluid which is at rest at infinity. We shall say that the falling motion is steady if the velocity and pressure of the fluid in a coordinate system which is attached to the body are independent of time. The velocity at infinity in this moving coordinate frame is $-[\mathbf{U} + \boldsymbol{\omega} \times \mathbf{r}]$, where $\mathbf{U}$ and $\boldsymbol{\omega}$ are the linear and angular velocities of the body relative to a Galilean frame. Since the flow is to be independent of time, $\mathbf{U}$ and $\boldsymbol{\omega}$ must be constant. These two vectors are to be determined by equating the viscous force and torque on the body to the force and torque due to gravity. The viscous force must be constant. The direction of the gravitational field $\mathbf{g}$ in the moving coordinates will rotate about the $\boldsymbol{\omega}$-axis, so that the force is not constant, unless $\boldsymbol{\omega}$ is in the direction of $\mathbf{g}$. That is, we must have

$$\boldsymbol{\omega} = \lambda \mathbf{g}, \tag{2.1}$$

where λ is a constant scalar and $\mathbf{g}$ is a vector which is fixed in the body.

Since $\mathbf{U}$ and $\boldsymbol{\omega}$ are constant, the difference between the velocity and its limiting value $-[\mathbf{U} + \boldsymbol{\omega} \times \mathbf{r}]$ will also be independent of time in the moving coordinates. We call this difference $\mathbf{u}(x)$. Its components u_i are the components of the velocity in a Galilean coordinate system resolved along the coordinate directions of the moving coordinate frame.

A short computation shows that the steady-state Navier-Stokes equation becomes

$$\rho[(\mathbf{u} - \mathbf{U} - \lambda\mathbf{g} \times \mathbf{r}) \cdot \operatorname{grad}]\mathbf{u} + \lambda\rho\mathbf{g} \times \mathbf{u} + \operatorname{grad} p - \mu\Delta\mathbf{u} = \rho\mathbf{g}, \qquad \operatorname{div} \mathbf{u} = 0. \tag{2.2}$$

The viscous stress is given by the tensor

$$\sigma_{ij} = \mu(u_{i,j} + u_{j,i}) - p\delta_{ij}. \tag{2.3}$$

(We shall use the notation $_{,i}$ for $\partial/\partial x_i$.)

The equations (2.2) are to hold in a fixed domain D, which is the exterior of a closed bounded connected set (the body) B. The constant vectors $\mathbf{U}$ and $\lambda\mathbf{g}$ are to be determined from the equilibrium conditions

$$\oint_{\partial B} \mathbf{f}\, dS = m'\mathbf{g}, \qquad \oint_{\partial B} \mathbf{f} \times \mathbf{r}\, dS = m'\mathbf{g} \times \mathbf{r}'. \tag{2.4}$$

Here m' and $\mathbf{r}'$ are the mass and the position of the center of mass of the mass distribution on B, and $f_i = \sigma_{ij}n_j$ is the surface force per unit area on B. (We shall use the summation convention throughout this paper.)

We shall study two different boundary value problems:

Problem I. Given a vector $\mathbf{g}$ fixed in B and the total mass m', find a position $\mathbf{r}'$ of the center of mass such that there is a solution $\mathbf{u}$ of the equation (2.2) which vanishes at infinity and has boundary values of the form $\mathbf{U} + \lambda\mathbf{g} \times \mathbf{r}$, and for which the equilibrium conditions (2.4) hold.

Problem II. Given the mass distribution on B (that is, given m' and $\mathbf{r}'$), find a direction $\mathbf{g}$ (that is, a vertical orientation of the body) such that there is a solution $\mathbf{u}$ of the equation (2.2) which vanishes at infinity and has boundary values of the form $\mathbf{U} + \lambda\mathbf{g} \times \mathbf{r}$ and for which the equilibrium conditions (2.4) hold.

In Problem I we notice that we can choose $\mathbf{r}'$ so that the second condition in (2.4) is satisfied if and only if

$$\mathbf{g} \cdot \oint_{\partial B} \mathbf{f} \times \mathbf{r}\, dS = 0. \tag{2.5}$$

In fact, once this condition is satisfied it is only necessary to choose

$$\mathbf{r}' = -\frac{1}{m|\mathbf{g}|^2}\mathbf{g} \times \oint_{\partial B} \mathbf{f} \times \mathbf{r}\, dS + \alpha \mathbf{g} \tag{2.6}$$

for any α, so that the center of mass may lie anywhere on the line (2.6) which is parallel to $\mathbf{g}$. The boundary conditions for Problem I are thus

$$\begin{aligned} \oint_{\partial B} \mathbf{f}\, dS &= m\mathbf{g}, \\ \mathbf{g}\cdot\oint_{\partial B} \mathbf{f} \times \mathbf{r}\, dS &= 0. \end{aligned} \tag{2.7}$$

These four conditions are to be used to determine the vector $\mathbf{U}$ and the scalar λ.

We wish to formulate weak versions of these problems. We denote by J' the linear space of vector fields $\boldsymbol{\varphi}$ each of which satisfies the conditions

$$\begin{aligned} \operatorname{div} \boldsymbol{\varphi} &= 0 && \text{in } D, \\ \boldsymbol{\varphi} &= \mathbf{V} + \mathbf{W} \times \mathbf{r} && \text{in some neighborhood of } B, \\ \boldsymbol{\varphi} &= 0 && \text{outside a bounded set,} \end{aligned} \tag{2.8}$$

where $\mathbf{V}$ and $\mathbf{W}$ are constant vectors which may depend on $\boldsymbol{\varphi}$. (Note that the subspace of J' of codimension six which is obtained by making $\mathbf{V} = \mathbf{W} = 0$ is the space $\dot{J}$ introduced by O. A. Ladyženskaja [7].)

In view of the definition (2.3) and the fact that $\operatorname{div} \mathbf{u} = 0$, we can rewrite the first equation in (2.2) as

$$-\sigma_{ij,j} + \rho[(\mathbf{u} - \mathbf{U} - \lambda\mathbf{g} \times \mathbf{r}) \cdot \operatorname{grad}]u_i + \lambda\rho(\mathbf{g} \times \mathbf{u})_i = \rho g_i.$$

We take the scalar product of an arbitrary $\boldsymbol{\varphi}$ in J' with both sides of this equation and integrate by parts to obtain the equation

$$\begin{aligned} \int_D \sigma_{ij}\varphi_{i,j} - \rho \int_D \{\varphi_{i,j}u_i(u_j - U_j - \lambda\varepsilon_{jkl}g_k x_l) - \lambda\boldsymbol{\varphi}\cdot\mathbf{g} \times \mathbf{u}\}\, dx \\ - \int_{\partial B} \varphi_i \sigma_{ij} n_j\, dS = \rho \int_D \boldsymbol{\varphi}\cdot\mathbf{g}\, dx. \end{aligned} \tag{2.9}$$

Since $\boldsymbol{\varphi} = \mathbf{V} + \mathbf{W} \times \mathbf{r}$ on ∂B, the boundary term can be evaluated by means of the conditions (2.4). Because $\mathbf{g} = \operatorname{grad}(\mathbf{g}\cdot\mathbf{r})$, we find that

$$\begin{aligned} \int_D \boldsymbol{\varphi}\cdot\mathbf{g}\, dx &= \oint_{\partial B} \mathbf{g}\cdot\mathbf{r}\boldsymbol{\varphi}\cdot\mathbf{n}\, dS \\ &= -|B|(\mathbf{V}\cdot\mathbf{g} + \mathbf{W} \times \mathbf{r}''\cdot\mathbf{g}) \end{aligned}$$

where $\mathbf{r}''$ is the position of the centroid of B and $|B|$ is the volume of B.

Finally, we see that since $\operatorname{div} \boldsymbol{\varphi} = 0$,

$$\sigma_{ij}\varphi_{i,j} = \tfrac{1}{2}\mu(u_{i,j} + u_{j,i})(\varphi_{i,j} + \varphi_{j,i}).$$

Thus, if we define the symmetric bilinear form

$$E(\boldsymbol{\varphi}, \boldsymbol{\psi}) = \tfrac{1}{2}\mu \int_D (\varphi_{i,j} + \varphi_{j,i})(\psi_{i,j} + \psi_{j,i})\, dx, \tag{2.10}$$

we can rewrite (2.9) as

$$\begin{aligned} E(\boldsymbol{\varphi}, \mathbf{u}) - \rho \int_D \{\varphi_{i,j}u_i(u_j - U_j - \lambda\varepsilon_{jkl}g_k x_l) - \lambda\boldsymbol{\varphi}\cdot\mathbf{g}\times\mathbf{u}\}\, dx \\ -(m' - \rho|B|)\mathbf{V}\cdot\mathbf{g} - \mathbf{W}\cdot(m'\mathbf{r}' - \rho|B|\mathbf{r}'')\times\mathbf{g} = 0. \end{aligned} \tag{2.11}$$

We observe that if $\boldsymbol{\varphi} = \mathbf{V} + \mathbf{W}\times\mathbf{r}$ and $\boldsymbol{\psi} = \mathbf{P} + \mathbf{Q}\times\mathbf{r}$ near B,

$$\int_D \varphi_{j,i}\psi_{i,j}\, dx = \oint_{\partial B} \varphi_{j,i}\psi_i n_j\, dS = 2\mathbf{W}\cdot\mathbf{Q}|B|$$

so that

$$E(\boldsymbol{\varphi}, \boldsymbol{\psi}) = \mu\left\{\int_D \varphi_{i,j}\psi_{i,j}\, dx + 2\mathbf{W}\cdot\mathbf{Q}|B|\right\}. \tag{2.12}$$

Therefore $E(\boldsymbol{\varphi}, \boldsymbol{\varphi})$ is positive definite on J'.

We complete the space J' with respect to the norm

$$\|\boldsymbol{\varphi}\| = \{E(\boldsymbol{\varphi}, \boldsymbol{\varphi})\}^{1/2}, \tag{2.13}$$

and call the resulting Hilbert space H'.

We now assume that the capacity C of B is positive, and we prove the following lemma, which shows that the notion of boundary values $\mathbf{V} + \mathbf{W}\times\mathbf{r}$ is preserved in the completion H'.

LEMMA 1. *If the capacity C of B is positive, there exists a constant α depending only on B such that for any $\boldsymbol{\varphi}$ in J' with boundary values $\mathbf{V} + \mathbf{W}\times\mathbf{r}$ the inequalities*

$$\begin{gathered} |\mathbf{V}|^2 \leqq \alpha^2\|\boldsymbol{\varphi}\|^2/\mu, \qquad |\mathbf{W}|^2 \leqq \alpha^2\|\boldsymbol{\varphi}\|^2/\mu, \\ (\mathbf{V}\cdot\mathbf{W})^2/\mathbf{W}^2 \leqq \|\boldsymbol{\varphi}\|^2/4\pi\mu C \end{gathered} \tag{2.14}$$

hold.

PROOF. There exist generalized solutions of the problems

$$\begin{aligned} \Delta q &= 0 \quad \text{in } D, \\ q &= 1 \quad \text{on } \partial B,\ u \to 0 \text{ at } \infty; \\ \Delta q_i &= 0 \quad \text{in } D, \\ q_i &= x_i \quad \text{on } \partial B,\ q_i \to 0 \text{ at } \infty,\ i = 1, 2, 3. \end{aligned} \tag{2.15}$$

Moreover,

$$\int_D |\mathrm{grad}\, q|^2\, dx = 4\pi C, \tag{2.16}$$

and there are constants c_i so that

$$q_i = c_i q + O((r)^{-2}) \quad \text{as } r \to \infty.$$

Then

$$\int_D \mathrm{grad}\, q \cdot \mathrm{grad}(q_i - c_i q)\, dx = 0. \tag{2.17}$$

The polarization matrix is defined by the equation

$$P_{ij} = \int_\Omega \mathrm{grad}(q_i - c_i q) \cdot \mathrm{grad}(q_j - c_j q)\, dx. \tag{2.18}$$

Now for $\boldsymbol{\varphi} \in J'$,

$$\varphi_i = V_i q + \varepsilon_{ijk} W_j q_k \quad \text{near } \partial B.$$

Then by Dirichlet's principle

$$\begin{aligned} \int_\Omega \varphi_{i,j}\varphi_{i,j}\, dx &\geqq \sum_i \int_D |\mathrm{grad}(V_i q + \varepsilon_{ijk} W_j q_k)|^2\, dx \\ &= 4\pi C|\mathbf{V} + \mathbf{W} \times \mathbf{c}|^2 + (P_{kk}\delta_{ij} - P_{ij})W_i W_j \end{aligned} \tag{2.19}$$

so that, by (2.12),

$$E(\boldsymbol{\varphi}, \boldsymbol{\varphi}) \geqq 4\pi\mu C|\mathbf{V} + \mathbf{W} \times \mathbf{c}|^2 + \mu[(P_{kk} + 2|B|)\delta_{ij} - P_{ij}]W_i W_j. \tag{2.20}$$

It is easily seen that the matrix P_{ij} is positive semidefinite. Let us assume that it is in diagonal form. If $P_{11} = 0$, then the function $q_1 - c_1 q$, which has the boundary values $x_1 - c_1$, has Dirichlet integral zero. It follows that B must lie in the plane $x_1 = c_1$ except for a set of capacity zero. If P_{22} were also zero, then B would have to lie on the line $x_1 = c_1$, $x_2 = c_2$ except for a set of capacity zero. But a line segment has capacity zero. Since B has positive capacity, we conclude that P_{ij} can have at most one zero eigenvalue. When P_{ij} is in diagonal form, the matrix $P_{kk}\delta_{ij} - P_{ij}$ is the diagonal matrix with elements $(P_{22} + P_{33}, P_{33} + P_{11}, P_{11} + P_{22})$, all of which are positive. Therefore this matrix is positive definite.

It is an easy consequence of the maximum principle that $\mathbf{c}$ must lie in the convex hull of B. Therefore $|\mathbf{c}|$ is bounded by the maximum distance from the origin of points of ∂B.

It then follows from the inequality (2.20) that $\mu|\mathbf{W}|^2$ and then also $\mu|\mathbf{V}|^2$ can be bounded by a constant times $E(\boldsymbol{\varphi}, \boldsymbol{\varphi}) = \|\boldsymbol{\varphi}\|^2$. The last inequality in (2.14) follows from the fact that $|\mathbf{V} + \mathbf{W} \times \mathbf{c}|^2 \geqq (\mathbf{V} \cdot \mathbf{W})^2/|\mathbf{W}|^2$. Q.E.D.

Lemma 1 shows that V_i and W_i are bounded linear functionals, and are therefore also defined on the closure H'. In terms of these, we can then speak of the boundary values $\mathbf{V} + \mathbf{W} \times \mathbf{r}$ of a function in H'.

We can now formulate weak versions of Problems I and II. In order to simplify the formulas, we define the net mass

$$m = m' - \rho|B| \tag{2.21}$$

and the effective center of mass

$$\mathbf{r}_c = (m'\mathbf{r}' - \rho|B|\mathbf{r}'')/m. \tag{2.22}$$

Problem I. Given a vector $\mathbf{g}$, find a function $\mathbf{u} \in H'$ with boundary values of the form $\mathbf{U} + \lambda\mathbf{g} \times \mathbf{r}$ such that for every $\boldsymbol{\varphi} \in J'$ with boundary values of the form $\mathbf{V} + \beta\mathbf{g} \times \mathbf{r}$ the equation

$$E(\boldsymbol{\varphi}, \mathbf{u}) - \rho \int [\varphi_{i,j}u_i(u_j - U_j - \lambda\varepsilon_{jkl}g_k x_l) - \lambda\boldsymbol{\varphi}\cdot\mathbf{g} \times \mathbf{u}]\, dx - m\mathbf{V}\cdot\mathbf{g} = 0 \tag{2.23}$$

is satisfied.

Problem II. Find vectors $\mathbf{U}$ and $\mathbf{g}$ with $\mathbf{g}$ of given length $|\mathbf{g}|$ and a function $\mathbf{u} \in H'$ with boundary values $\mathbf{U} + \lambda\mathbf{g} \times \mathbf{r}$ such that for every $\boldsymbol{\varphi} \in J'$ with boundary values $\mathbf{V} + \mathbf{W} \times \mathbf{r}$ the equation

$$\begin{aligned} E(\boldsymbol{\varphi}, \mathbf{u}) - \rho\int \{\varphi_{i,j}u_i(u_j - U_j - \lambda\varepsilon_{jkl}g_k x_l) - \lambda\boldsymbol{\varphi}\cdot\mathbf{g} \times \mathbf{u}\}\, dx \\ - m(\mathbf{V} + \mathbf{W} \times \mathbf{r}_c)\cdot\mathbf{g} = 0 \end{aligned} \tag{2.24}$$

is satisfied.

Note that $\mathbf{r}_c$ only occurs in the expression $\mathbf{W} \times \mathbf{r}_c \cdot \mathbf{g}$. It follows that once we have a solution $\mathbf{u}$ of Problem II with a downward direction $\mathbf{g}$ for a given effective center of mass $\mathbf{r}_c$, then this same $\mathbf{u}$ is still a solution if the effective center of mass is moved to a point $\mathbf{r}_c + \alpha\mathbf{g}$ for any constant α.

REMARK. We note that, by introducing the new velocity variable

$$\bar{\mathbf{u}} = \mu\mathbf{u}/m|\mathbf{g}|, \tag{2.25}$$

Problems I and II can be reduced to problems where $\mu = m = |\mathbf{g}| = 1$ while the density ρ is replaced by the dimensionless constant $m^* \equiv \rho m|\mathbf{g}|/\mu^2$.

3. **Existence of a solution of Problem** I. We shall now show that the weak form of Problem I can always be solved.

THEOREM 1. *For any closed connected bounded set B with positive capacity, any positive ρ, m, and μ, and any given vector* $\mathbf{g}$ *there is a weak solution of Problem* I.

PROOF. Let H''_N be the subspace of those vector fields of H' which vanish for $|\mathbf{r}| > N$

and which have boundary values of the form $\mathbf{U} + \lambda\mathbf{g} \times \mathbf{r}$. The integer N is so large that B lies in the ball $|\mathbf{r}| < N$, but is otherwise arbitrary.

We observe that the left-hand side of (2.23) is a bounded linear functional on φ in H''_N and define the operator $A_N : H''_N \to H''_N$ by the equation

$$(3.1) \quad E(\boldsymbol{\varphi}, A_N[\mathbf{u}]) = \rho \int \{\varphi_{i,j} u_i(u_j - U_j - \lambda\varepsilon_{jkl} g_k x_l) - \lambda\boldsymbol{\varphi} \cdot \mathbf{g} \times \mathbf{u}\}\, dx + m\mathbf{V} \cdot \mathbf{g}.$$

We first solve the problem of finding a $\mathbf{u} \in H''_N$ such that

$$E(\boldsymbol{\varphi}, \mathbf{u}) - E(\boldsymbol{\varphi}, A_N[\mathbf{u}]) = 0$$

for all $\boldsymbol{\varphi} \in H''_N$. The existence of such a $\mathbf{u}$ is established by observing that an argument used by Ladyženskaja [6], [7, p. 117] shows that A_N is a completely continuous operator. The family of equations

$$(3.2) \qquad \mathbf{u} - \kappa A_N[\mathbf{u}] = 0, \qquad 0 \leqq \kappa \leqq 1,$$

has the unique solution $\mathbf{u} = 0$ when $\kappa = 0$. An integration by parts shows that any solution $\mathbf{u}$ in H''_N of (3.2) must satisfy

$$(3.3) \qquad E(\mathbf{u}, \mathbf{u}) = \kappa E(\mathbf{u}, A_N[\mathbf{u}]) = \kappa m \mathbf{g} \cdot \mathbf{U}.$$

It follows from this identity and the last inequality in (2.14) that

$$(3.4) \qquad \|\mathbf{u}\|^2 \leqq \frac{\kappa^2 m^2}{4\pi\mu C} |\mathbf{g}|^2 \leqq \frac{m^2}{4\pi\mu C} |\mathbf{g}|^2.$$

Thus, the Leray-Schauder theorem can be applied, and it follows that for each N there is a $\mathbf{u}_N \in H''_N$ for which

$$E(\boldsymbol{\varphi}, \mathbf{u}_N) = E(\boldsymbol{\varphi}, A_N[\mathbf{u}_N])$$

for all $\boldsymbol{\varphi} \in H''_N$. Moreover, each $\mathbf{u}_N$ satisfies the inequality (3.4). It then follows in the usual way that there is a sequence $N_k \to \infty$ such that $\mathbf{u}_{N_k}$ converges weakly to a function $\mathbf{u}$ in H' and the boundary values converge to the boundary values $\mathbf{U} + \lambda\mathbf{g} \times \mathbf{r}$ of $\mathbf{u}$. By a result of O. A. Ladyženskaja [6], [7], $\mathbf{u}_{N_k}$ converges to $\mathbf{u}$ strongly in L_4 on any bounded set. It follows that for each $\boldsymbol{\varphi}$ in H' with bounded support and boundary values of the form $\mathbf{V} + \beta\mathbf{g} \times \mathbf{r}$ the equation (2.23) is satisfied. Thus $\mathbf{u}$ is the required solution. It again satisfies the inequality (3.4) and, since each $\mathbf{u}_N$ satisfies (3.3) with $\kappa = 1$, we have

$$(3.5) \qquad E(\mathbf{u}, \mathbf{u}) \leqq m\mathbf{U} \cdot \mathbf{g} \leqq \frac{m^2}{4\pi\mu C} |\mathbf{g}|^2.$$

In particular, the vertical component $\mathbf{U} \cdot \mathbf{g}/|\mathbf{g}|$ is bounded by $m|\mathbf{g}|/4\pi\mu C$, so that the effective drag coefficient is at least $4\pi\mu C$.

We remark that once $\mathbf{U}$ and λ have been found, $\mathbf{u}$ is the solution of a Dirichlet problem with smooth data, so that the usual regularity theorems hold.

4. **Relations with Stokes flow.** Let $\mathbf{u}$ be a solution of equation (2.23), the weak form of Problem I. Let m be the corresponding net mass. We see from the inequality (3.4) that the quantity $\mathbf{u}/m$ is bounded in norm. Hence, out of any sequence $\mathbf{u}^{(n)}$ of solutions with corresponding masses $m^{(n)}$ which approach zero, we can select a subsequence $\mathbf{u}^{(n)'}$ so that $\mathbf{u}^{(n)'}/m^{(n)'}$ converges weakly to an element $\tilde{\mathbf{u}}$ of H'. Also, the boundary values $(\mathbf{U}^{(n)'} + \lambda^{(n)'}\mathbf{g} \times \mathbf{r})/m^{(n)'}$ converge to a limiting value $\tilde{\mathbf{U}} + \tilde{\lambda}\mathbf{g} \times \mathbf{r}$. Moreover, as has been pointed out by O. A. Ladyženskaja [**6**], [**7**], $\mathbf{u}^{(n)'}/m^{(n)'}$ then converges strongly to $\tilde{\mathbf{u}}$ in L_4 on any bounded set.

Dividing (2.23) with $\mathbf{u} = \mathbf{u}^{(n)'}$ and $m = m^{(n)'}$ by $m^{(n)'}$ and taking limits, we see that $\tilde{\mathbf{u}}$ satisfies the equation

$$\mu\left(\int \varphi_{i,j}\tilde{u}_{i,j} + 2|B|\beta\tilde{\lambda}|\mathbf{g}|^2\right) - \mathbf{V}\cdot\mathbf{g} = 0$$

for any $\boldsymbol{\varphi}$ in J' with boundary values $\mathbf{V} + \beta\mathbf{g} \times \mathbf{r}$. This implies that $\tilde{\mathbf{u}}$ is a weak solution of the Stokes flow problem

$$\begin{aligned} -\mu\Delta\tilde{\mathbf{u}} + \operatorname{grad}\tilde{p} &= \rho\mathbf{g}, \\ \operatorname{div}\tilde{\mathbf{u}} &= 0 \quad \text{in } D, \\ \tilde{\mathbf{u}} &= \tilde{\mathbf{U}} + \tilde{\lambda}\mathbf{g} \times \mathbf{r} \quad \text{on } \partial B, \\ \oint_{\partial B} \tilde{\mathbf{f}}\, dS &= (\rho|B| + 1)\mathbf{g}, \\ \mathbf{g}\cdot\oint_{\partial B} \tilde{\mathbf{f}} \times \mathbf{r}\, dS &= 0, \end{aligned} \tag{4.1}$$

where

$$\tilde{f}_i = \mu(\tilde{u}_{i,j} + \tilde{u}_{j,i})n_j - \tilde{p}n_i \quad \text{on } \partial B.$$

This is the Stokes flow analogue of Problem I for net mass one. We have shown in [**11**] that this problem has a unique solution. It follows that as $m \to 0$ the ratio $\mathbf{u}/m$ converges to $\tilde{\mathbf{u}}$ weakly in H'. In particular, the ratio $\mathbf{U}\cdot\mathbf{g}/m|\mathbf{g}|$ of terminal speed to mass has the limit $\tilde{\mathbf{U}}\cdot\mathbf{g}/|\mathbf{g}|$.

We have also shown in [**11**] that the solution $\tilde{\mathbf{u}}$ of the problem (4.1) can be characterized by the variational principle

$$\tilde{\mathbf{U}}\cdot\mathbf{g} = \max\frac{(\mathbf{V}\cdot\mathbf{g})^2}{E(\boldsymbol{\varphi}, \boldsymbol{\varphi})} \tag{4.2}$$

where the maximum is taken with respect to $\boldsymbol{\varphi}$ in H' with boundary values $\mathbf{V} + \beta\mathbf{g} \times \mathbf{r}$.

We now choose for $\boldsymbol{\varphi}$ the solution $\mathbf{u}$ of the problem (2.23) with a given net mass m, which has been constructed in §3. We then see from the inequality (3.5) and (4.2) that

$$\tilde{\mathbf{U}} \cdot \mathbf{g} \geqq \mathbf{U} \cdot \mathbf{g}/m$$

or

$$(\mathbf{U} \cdot \mathbf{g})/|\mathbf{g}| \leqq m(\tilde{\mathbf{U}} \cdot \mathbf{g})/|\mathbf{g}|. \tag{4.3}$$

Thus, if we graph the terminal speed

$$t(m, \mathbf{g}) \equiv (\mathbf{U} \cdot \mathbf{g})/|\mathbf{g}| \tag{4.4}$$

against the net mass m, we see that the curve has a tangent with slope $\tilde{\mathbf{U}} \cdot \mathbf{g}/|\mathbf{g}|$ at $m = 0$ and that for $m > 0$, $t(m, \mathbf{g})$ lies below this tangent line. Physically, this means that the fall in a Navier-Stokes fluid is slower than the fall with the same orientation and the same net mass in the Stokes approximation.[2] (The fact that the orientation is to remain fixed means that, in general, the center of mass may have to be moved. Moreover, our statement only applies to the class of flows which we have constructed; there could conceivably be others.)

A negative value of m means that the body tends to float rather than sink. The resulting problem is equivalent to that which is obtained by replacing m by $|m| = -m$ and $\mathbf{g}$ by $-\mathbf{g}$. This corresponds to the body of mass $-m$ falling in an upside down orientation. By the above considerations, we see that, for $m < 0$, $t(m, \mathbf{g})$ lies above the tangent line $m\tilde{\mathbf{U}} \cdot \mathbf{g}/|\mathbf{g}|$. Thus the graph of $t(m, \mathbf{g})$ not only has a tangent at $m = 0$, but also has an inflection there. We can therefore expect that the Stokes approximation $m\tilde{\mathbf{U}} \cdot \mathbf{g}/|\mathbf{g}|$ is a rather good approximation to the terminal speed.

REMARK. Since Problem I may have more than one solution, the function $t(m, \mathbf{g})$ need not be single-valued except at $m = 0$. However, the graph of the multiple valued function $t(m, \mathbf{g})$ has the properties outlined above.

The inequality (4.3) allows us to transfer the upper bounds for terminal speeds in Stokes flow found in **[11]** to the solution of Problem I. For example, if $\mathbf{g}_1, \mathbf{g}_2, \mathbf{g}_3$ are any three mutually perpendicular vectors of length $|\mathbf{g}|$ and we consider the speeds of the body falling with these downward orientations, we have

$$t(m, \mathbf{g}_1) + t(m, \mathbf{g}_2) + t(m, \mathbf{g}_3) \leqq m|\mathbf{g}|/2\pi\mu C. \tag{4.5}$$

We note that by (3.5) and (4.3), and the fact that $\tilde{\mathbf{u}}$ is well-behaved at infinity

$$E(\mathbf{u}/m, \mathbf{u}/m) \leqq \mathbf{U} \cdot \mathbf{g}/m \leqq \tilde{\mathbf{U}} \cdot \mathbf{g} = E(\tilde{\mathbf{u}}, \tilde{\mathbf{u}}).$$

It follows that $\mathbf{u}/m$ converges to $\tilde{\mathbf{u}}$ strongly as $m \to 0$.

Finally, we recall the remark at the end of §2. According to this remark, we may write $\mathbf{u} = m|\mathbf{g}|\bar{\mathbf{u}}/\mu$, where $\bar{\mathbf{u}}$ depends only on the dimensionless parameter $\rho m|\mathbf{g}|/\mu^2$. Thus, we may state our results in the form:

[2] The result that the nonlinear terms in the Navier-Stokes equations increase the drag has been found in other situations by Kearsley **[4]** and by Keller, Rubenfeld, and Molyneux **[5]** by applying different variational principles for Stokes flow.

The quantity $\mu\mathbf{u}/m|\mathbf{g}|$ converges strongly to the solution $\tilde{\mathbf{u}}$ of the problem (4.1) *with $\rho = \mu = 1$ and $\mathbf{g}$ replaced by $\mathbf{g}/|\mathbf{g}|$ as the dimensionless parameter $\rho m|\mathbf{g}|/\mu^2$ approaches zero.*

5. **The existence of solutions of Problem** II. In view of the Remark at the end of §2, we may, without loss of generality, take $\mu = m = |\mathbf{g}| = 1$ in Problem II. Accordingly, we rewrite Problem II in the form

(5.1)
$$E(\varphi, \mathbf{u}) = m^* \int \{\varphi_{i,j}u_i(u_j - U_j - \varepsilon_{jkl}\omega_k x_l) - \varphi \cdot \omega \times \mathbf{u}\}\, dx + (\mathbf{V} + \mathbf{W} \times \mathbf{r}_c)\cdot \mathbf{g},$$
$$\mathbf{u} \in H', \qquad \omega = \lambda \mathbf{g}, \qquad |\mathbf{g}| = 1,$$

for all φ in H' with bounded support, where $\mathbf{V} + \mathbf{W} \times \mathbf{r}$ are the boundary values of φ. The parameter μ is to be set equal to one in the definition (2.10) of E.

We are to find not only the vector field $\mathbf{u}$ but also the vector $\mathbf{g}$ on the sphere $|\mathbf{g}| = 1$ and the real scalar λ. Thus our unknown is the triplet $\{\mathbf{u}, \mathbf{g}, \lambda\}$ on $H' \times S_2 \times E_1$.

As in §3, we first look at the restriction of this problem to the space H'_N of elements of H' which vanish outside the sphere $|\mathbf{r}| = N$. We define the operator $B_N: H'_N \to H'_N$ by

(5.2)
$$E(\varphi, B_N\mathbf{u}) = \int_D \{\varphi_{i,j}u_i(u_j - U_j - \varepsilon_{jkl}\omega_j x_l) + \varphi \cdot \omega \times \mathbf{u}\}\, dx$$

for all $\varphi \in H'_N$. It again follows from the arguments of Ladyženskaja [**6**], [7] that B_N is compact. We define the linear operator $R: E_3 \to H'$ by

(5.3)
$$E(\varphi, R\mathbf{g}) = (\mathbf{V} + \mathbf{W} \times \mathbf{r}_c)\cdot \mathbf{g}$$

for all $\varphi \in H'$ with boundary values $\mathbf{V} + \mathbf{W} \times \mathbf{r}$. The function $\tilde{\mathbf{u}} = R\mathbf{g}$ is the weak solution of the linear (Stokes flow) problem

(5.4)
$$\Delta\tilde{\mathbf{u}} - \operatorname{grad} p = 0 \quad \text{in } D,$$
$$\operatorname{div}\tilde{\mathbf{u}} = 0,$$
$$\tilde{\mathbf{u}} = \tilde{\mathbf{U}} + \tilde{\omega} \times \mathbf{r} \quad \text{on } \partial B \text{ for some } \tilde{\mathbf{U}}, \tilde{\omega},$$
$$\oint_{\partial B} \mathbf{f}\, dS = \mathbf{g},$$
$$\oint_{\partial B} \mathbf{f} \times \mathbf{r}\, dS = \mathbf{g} \times \mathbf{r}_c,$$

where

$$f_i = (\tilde{u}_{i,j} + \tilde{u}_{j,i})n_j - pn_i.$$

Let P_N be the orthogonal projection operator from H' onto H'_N. Then the restriction of the problem (5.1) to H'_N can be written in the form

$$\begin{aligned} &\mathbf{u} - m^* B_N \mathbf{u} - P_N R\mathbf{g} = 0, \\ &\mathbf{u} \in H'_N, \qquad \omega = \lambda \mathbf{g}, \qquad |\mathbf{g}| = 1. \end{aligned} \tag{5.5}$$

It follows from the inequality (2.14) that there is a bounded linear operator $T: H' \to E_3$ which gives the angular velocity on the boundary of a vector field in H'. Thus, the conditions $\omega = \lambda \mathbf{g}$, becomes

$$\lambda \mathbf{g} - m^* T B_N \mathbf{u} - T P_N R \mathbf{g} = 0.$$

We note that the operator

$$Q_N \equiv T P_N R$$

is a linear mapping from E_3 to E_3, and can therefore be represented by a 3×3 matrix. Moreover,

$$\lim_{N \to \infty} Q_N = TR \equiv Q,$$

which is again a 3×3 matrix.

The equation (5.5) becomes

$$\begin{aligned} &\mathbf{u} - m^* B_N \mathbf{u} - P_N R\mathbf{g} = 0, \\ &Q_N \mathbf{g} - \lambda \mathbf{g} = -m^* T B_N \mathbf{u}, \qquad |\mathbf{g}| = 1. \end{aligned} \tag{5.6}$$

We now consider the one parameter family of mappings

$$\{\mathbf{u}, \mathbf{g}, \lambda\} \to \{\mathbf{u} - \kappa B_N \mathbf{u} - P_N R\mathbf{g}, \lambda \mathbf{g} - \kappa T B_N \mathbf{u} - Q_N \mathbf{g}\} \tag{5.7}$$

from $H'_N \times S_2 \times E_1$ to $H'_N \times E_3$.

A zero of this mapping when $\kappa = m^*$ gives a solution of the restriction (5.5) of (5.1) to H'_N. In order to prove the existence of such a zero by Leray-Schauder theory, we first compute the degree of the mapping at $\kappa = 0$.

For fixed $\mathbf{g}$ the mapping $\mathbf{u} \to \mathbf{u} - P_N R\mathbf{g}$ is clearly one-to-one and onto. Therefore, the degree of the mapping when $\kappa = 0$ is just that of the mapping $\{\mathbf{g}, \lambda\} \to \lambda \mathbf{g} - Q_N \mathbf{g}$ of $S_2 \times E_1$ to E_3. We must thus look at the equations

$$\lambda \mathbf{g} - Q_N \mathbf{g} = \mathbf{h}, \qquad |\mathbf{g}| = 1,$$

which constitute a system of four equations for λ and $\mathbf{g}$.

The solutions of these equations when $\mathbf{h} = 0$ are just the eigenvectors and eigenvalues of Q_N. We note that if $\mathbf{g}$ is an eigenvector with $|\mathbf{g}| = 1$, the same is true of $-\mathbf{g}$. An easy computation shows that the contributions of $\mathbf{g}$ and $-\mathbf{g}$ to the degree cancel, so that the degree of the mapping (5.7) at zero on $H'_N \times S_2 \times E_1$ is always zero. Thus, in order to obtain existence by means of Leray-Schauder theory, we must restrict $\mathbf{g}$ to a subset of S_2 on which the degree is not zero. Such a restriction is possible when we deal with an isolated eigenvector.

We note that the linearized (Stokes flow) version of Problem II is obtained from (5.1) by setting $m^* = 0$. This problem is equivalent to the equations

$$\tilde{\mathbf{u}} = R\tilde{\mathbf{g}}, \qquad \tilde{\lambda}\tilde{\mathbf{g}} = \mathrm{Q}\tilde{\mathbf{g}}, \qquad |\tilde{\mathbf{g}}| = 1, \tag{5.8}$$

where $\mathrm{Q} = TR$ is the limit of the Q_N as $N \to \infty$. The problem (5.8) is the weak form of the Stokes flow version of Problem II:

$$\begin{gathered} \Delta\tilde{\mathbf{u}} - \operatorname{grad}\tilde{p} = 0, \qquad \operatorname{div}\tilde{\mathbf{u}} = 0, \\ \oint_{\partial B} \tilde{\mathbf{f}}\, dS = \tilde{\mathbf{g}}, \qquad \oint \tilde{\mathbf{f}} \times \mathbf{r}\, dS = \tilde{\mathbf{g}} \times \mathbf{r}_c, \\ \tilde{\omega} = \tilde{\lambda}\tilde{\mathbf{g}}, \qquad |\tilde{\mathbf{g}}| = 1. \end{gathered} \tag{5.9}$$

The orientations of the body in steady falling motion in the Stokes flow are just the real eigenvectors of unit length of Q. Since Q is a real 3×3 matrix, it always has at least one real eigenvector.

We shall say that $\mathbf{g}^*$ is an *isolated direction of fall* of the Stokes flow problem (5.9) if it is an eigenvector of Q and if the eigenvector space of the corresponding eigenvalue λ^* is one-dimensional. That is, the geometric multiplicity of λ^* is one.

We shall prove the following lemma.

LEMMA 2. *If $\mathbf{g}^*$ is an isolated direction of fall, then there exist a positive constant η and a positive function $M(\varepsilon)$ defined for $0 < \varepsilon \leqq \eta$ such that if $0 < \varepsilon \leqq \eta$ and $m^* \leqq M(\varepsilon)$, the problem* (5.1) *has a solution $\{\mathbf{u}, \mathbf{g}, \lambda\}$ with $|\mathbf{g} - \mathbf{g}^*| < \varepsilon$.*

PROOF. Since $\mathbf{g}^*$ is an isolated direction of fall, there is an $\eta > 0$ such that all eigenvectors g of Q of length one different from $\mathbf{g}^*$ satisfy $|\mathbf{g} - \mathbf{g}^*| \geqq 2\eta$, and hence $|\mathbf{g} - \mathbf{g}^*| \geqq 2\varepsilon$ whenever $0 < \varepsilon \leqq \eta$.

Since Q_N converges to Q, we can for any such ε find an N_0 so large that for $N \geqq N_0$ the pencils $\beta\mathrm{Q} + (1 - \beta)\mathrm{Q}_N$, $0 \leqq \beta \leqq 1$, have no eigenvector $\mathbf{g}$ of length one for which $|\mathbf{g} - \mathbf{g}^*| = \varepsilon$.

An elementary computation shows that the degree of the mapping $\{\mathbf{g}, \lambda\} \to \lambda\mathbf{g} - \mathrm{Q}\mathbf{g}$ on the set $|\mathbf{g}| = 1$, $|\mathbf{g} - \mathbf{g}^*| \leqq \varepsilon$ is equal in absolute value to the algebraic multiplicity of the eigenvalue λ^* which corresponds to $\mathbf{g}^*$. Since this degree is preserved on the pencil $\beta\mathrm{Q} + (1 - \beta)\mathrm{Q}_N$, we conclude that the degree of the mapping $\{\mathbf{g}, \lambda\} \to \lambda\mathbf{g} - \mathrm{Q}_N\mathbf{g}$ on the set $|\mathbf{g}| = 1$, $|\mathbf{g} - \mathbf{g}^*| \leqq \varepsilon$ is not zero for $N \geqq N_0$.

We now consider any solution of the problem

$$\begin{gathered} \mathbf{u} - \kappa B_N\mathbf{u} - P_N R\mathbf{g} = 0, \qquad \lambda\mathbf{g} - \mathrm{Q}_N\mathbf{g} = \kappa T B_N\mathbf{u}, \\ |\mathbf{g}| = 1, \qquad \mathbf{u} \in H'_N. \end{gathered} \tag{5.10}$$

Taking the scalar product with $\mathbf{u}$ and observing that, by the definition (5.2), $E(\mathbf{u}, B_N\mathbf{u}) = 0$, we find that

$$E(\mathbf{u}, \mathbf{u}) = E(\mathbf{u}, R\mathbf{g}) = \mathbf{U} \cdot \mathbf{g}.$$

Therefore, by (2.14),

$$(5.11) \qquad E(\mathbf{u}, \mathbf{u}) \leqq \alpha^2, \qquad |\lambda| \leqq \alpha^2.$$

Suppose there were a sequence of positive κ_ν converging to zero and integers $N_\nu \geqq N_0$ for which the problem (5.10) has solutions $\{\mathbf{u}_\nu, \mathbf{g}_\nu, \lambda_\nu\}$ with $|\mathbf{g}_\nu - \mathbf{g}^*| = \varepsilon$. By (5.11) there would be a subsequence along which either N_ν would be constant, say $N_\nu = \bar{N} \geqq N_0$, or N_ν would converge to infinity. Moreover, λ_ν would converge to some λ, and $\mathbf{g}_\nu$ would converge to a $\mathbf{g}$ with

$$|\mathbf{g}| = 1, \qquad |\mathbf{g} - \mathbf{g}^*| = \varepsilon.$$

Taking limits in (5.10) and recalling that $\kappa_\nu \to 0$, we would find that either

$$\lambda \mathbf{g} - \mathrm{Q}_{\bar{N}} \mathbf{g} = 0, \qquad |\mathbf{g}| = 1, \qquad |\mathbf{g} - \mathbf{g}^*| = \varepsilon,$$

contrary to the definition of N_0 or

$$\lambda \mathbf{g} - \mathrm{Q} \mathbf{g} = 0, \qquad |\mathbf{g}| = 1, \qquad |\mathbf{g} - \mathbf{g}^*| = \varepsilon,$$

contrary to the definition of η.

We conclude that there is a positive constant $M(\varepsilon)$ so that for $\kappa \leqq M(\varepsilon)$ the equation (5.10) has no solution on the boundary of the set $E(\mathbf{u}, \mathbf{u}) \leqq 2\alpha^2$, $|\mathbf{g} - \mathbf{g}^*| \leqq \varepsilon$, $|\mathbf{g}| = 1$, $|\lambda| \leqq 2\alpha^2$. Therefore, the degree of the mapping (5.7) at the origin remains constant. Since this degree is not zero at $\kappa = 0$, it is not zero at $\kappa = m^* \leqq M(\varepsilon)$, and the existence of a solution $\{\mathbf{u}_N, \mathbf{g}_N, \lambda_N\}$ of (5.6) follows from the Leray-Schauder theorem.

As in §3, we find that a subsequence of the sequence $\mathbf{u}_N$ has a weak limit $\mathbf{u}$, while the corresponding $\mathbf{g}_N$ have a limit $\mathbf{g}$ which satisfies $|\mathbf{g} - \mathbf{g}^*| < \varepsilon$ and λ_N converges to a real λ, so that $\{\mathbf{u}, \mathbf{g}, \lambda\}$ is a solution of (5.1).

We can now derive the principal result of this section.

THEOREM 2. *Suppose that the linearized problem* (5.9) *has* $2l$ *isolated directions of fall. Then there is a positive constant* M *depending only on* B *and* $\mathbf{r}_c$ *such that if* $m\rho|\mathbf{g}|/\mu^2 \leqq M$, *Problem* II *has at least* $2l$ *solutions. Moreover, every solution* $\{\tilde{\mathbf{u}}, \tilde{\mathbf{g}}, \tilde{\lambda}\}$ *of* (5.9) *for which* $\tilde{\mathbf{g}}$ *is an isolated direction of fall is the strong limit as* $\rho m|\mathbf{g}|/\mu^2$ *approaches zero of* $\{\mu\mathbf{u}/m|\mathbf{g}|, \mathbf{g}, \lambda\}$ *where* $\{\mathbf{u}, \mathbf{g}, \lambda\}$ *is a family of solutions of Problem* II.

PROOF. Given Problem II, we make the transformation (2.25) to obtain a new problem with $m = |\mathbf{g}| = \mu = 1$ and ρ replaced by $m^* = \rho m|\mathbf{g}|/\mu^2$ for the new variable $\bar{\mathbf{u}} = \mu\mathbf{u}/m|\mathbf{g}|$. We apply Lemma 2 to this new problem, taking for M the smallest $M(\eta)$. When $\mathbf{g}^*$ is an isolated direction of fall, so is $-\mathbf{g}^*$, so that the total number of such directions is an even number $2l$, $1 \leqq l \leqq 3$.

The weak convergence of $\{\bar{\mathbf{u}}, \mathbf{g}, \lambda\}$ to $\{\tilde{\mathbf{u}}, \tilde{\mathbf{g}}, \tilde{\lambda}\}$ follows from Lemma 2 and the bound (5.11).

Since $\bar{\mathbf{u}}$ is a weak limit of functions $\bar{\mathbf{u}}_N$ which satisfy $E(\bar{\mathbf{u}}_N, \bar{\mathbf{u}}_N) = \bar{\mathbf{U}}_N \cdot \mathbf{g}_N$ where

$\bar{\mathbf{U}}_N$ is the boundary velocity of $\bar{\mathbf{u}}_N$, we have $E(\bar{\mathbf{u}}, \bar{\mathbf{u}}) \leqq \bar{\mathbf{U}} \cdot \mathbf{g}$. Moreover, since $\bar{\mathbf{u}}$ converges to $\tilde{\mathbf{u}}$ weakly, $\bar{\mathbf{U}}$ converges to the boundary velocity $\tilde{\mathbf{U}}$. Thus,

$$E(\bar{\mathbf{u}}, \bar{\mathbf{u}}) \leqq \bar{\mathbf{U}} \cdot \mathbf{g} \to \tilde{\mathbf{U}} \cdot \mathbf{g} = E(\tilde{\mathbf{u}}, \tilde{\mathbf{u}}).$$

Therefore, $E(\tilde{\mathbf{u}}, \tilde{\mathbf{u}})$ is the limit of $E(\bar{\mathbf{u}}, \bar{\mathbf{u}})$, and the convergence is strong.

In view of Theorem 2 it is useful to investigate the possibility that the problem (5.9) has no isolated directions of fall.

H. Brenner [**1**], [**3**, p. 173] has shown that there is a positive definite symmetric 6×6 matrix

$$\begin{pmatrix} \mathrm{K} & \mathrm{C}^* \\ \mathrm{C} & \Omega \end{pmatrix} \tag{5.12}$$

depending only on B with the properties that the force and torque exerted by a Stokes flow around B with boundary values $\tilde{\mathbf{U}} + \tilde{\omega} \times \mathbf{r}$ are given by $\mathrm{K}\tilde{\mathbf{U}} + \mathrm{C}^*\tilde{\omega}$ and $\mathrm{C}\tilde{\mathbf{U}} + \Omega\tilde{\omega}$, respectively. Thus, if the solution $\tilde{\mathbf{u}}$ of the problem (5.4) has boundary values $\tilde{\mathbf{U}} + \tilde{\omega} \times \mathbf{r}$, we must have

$$\mathrm{K}\tilde{\mathbf{U}} + \mathrm{C}^*\tilde{\omega} = \tilde{\mathbf{g}}, \qquad \mathrm{C}\tilde{\mathbf{U}} + \Omega\tilde{\omega} = \tilde{\mathbf{g}} \times \mathbf{r}_c.$$

Solving for $\tilde{\omega}$, we find that

$$\tilde{\omega} = [\Omega - \mathrm{CK}^{-1}\mathrm{C}^*]^{-1}[\tilde{\mathbf{g}} \times \mathbf{r}_c - \mathrm{CK}^{-1}\tilde{\mathbf{g}}].$$

But by definition $\tilde{\omega} = \mathrm{Q}\tilde{\mathbf{g}}$. Thus, we can express the matrix Q in terms of the matrix (5.12) by the relation

$$\mathrm{Q}\mathbf{g} = [\Omega - \mathrm{CK}^{-1}\mathrm{C}^*]^{-1}[\mathbf{g} \times \mathbf{r}_c - \mathrm{CK}^{-1}\mathbf{g}] \tag{5.13}$$

for all $\mathbf{g}$.

We now observe that if Q has two distinct eigenvalues, at least one of them is real and simple so that an isolated direction of fall exists. Therefore, there is no isolated direction of fall if and only if Q has an eigenvalue τ of algebraic multiplicity three and the rank of $\mathrm{Q} - \tau\mathrm{I}$ is either zero or one.

Suppose τ is an eigenvalue of Q of algebraic multiplicity three. Then clearly $\tau = \frac{1}{3}\operatorname{tr} \mathrm{Q}$ or, by (5.13),

$$\tau = -\tfrac{1}{3}\operatorname{tr}\{[\Omega - \mathrm{CK}^{-1}\mathrm{C}^*]^{-1}\mathrm{CK}^{-1}\}, \tag{5.14}$$

which is independent of $\mathbf{r}_c$.

Thus the problem (5.9) has no isolated direction of fall if and only if the matrix $\mathrm{Q} - \tau\mathrm{I}$ with τ given by (5.14) has rank at most one and square zero. We are now in a position to prove the following theorem.

Theorem 3. *For a given body B there are at most two positions of the effective center of mass $\mathbf{r}_c$ for which the problem* (5.9) *has no isolated direction of fall.*

Proof. Suppose that there is a point where (5.9) has no isolated direction of fall.

We choose the origin at this point, so that

$$(5.15)\quad -([\Omega - CK^{-1}C^*]^{-1}CK^{-1})_{ij} + \tfrac{1}{3}\mathrm{tr}\{[\Omega - CK^{-1}C^*]^{-1}CK^{-1}\}\delta_{ij} = \alpha_i\beta_j$$

where

$$(5.16)\qquad \boldsymbol{\alpha}\cdot\boldsymbol{\beta} = 0.$$

We now seek other values of $\mathbf{r}_c$ with the same property, so that

$$[\Omega - CK^{-1}C^*]^{-1}_{ip}\varepsilon_{pjq}x_q^{(c)} + \alpha_i\beta_j = \gamma_i\delta_j$$

with

$$(5.17)\qquad \boldsymbol{\gamma}\cdot\boldsymbol{\delta} = 0.$$

Multiplying through by $\Omega - CK^{-1}C^*$, we have

$$(5.18)\qquad \varepsilon_{ijq}x_q^{(c)} = \bar{\gamma}_i\delta_j - \bar{\alpha}_i\beta_j$$

where we have defined

$$(5.19)\qquad \bar{\boldsymbol{\alpha}} = [\Omega - CK^{-1}C^*]\boldsymbol{\alpha}, \qquad \bar{\boldsymbol{\gamma}} = [\Omega - CK^{-1}C^*]\boldsymbol{\gamma}.$$

Taking the symmetric part of (5.18), we see that

$$(5.20)\qquad \bar{\gamma}_i\delta_j + \bar{\gamma}_j\delta_i = \bar{\alpha}_i\beta_j + \bar{\alpha}_j\beta_i.$$

We take the scalar product of this identity with $\boldsymbol{\gamma}$ and use the condition (5.17) to find that

$$(5.21)\qquad \boldsymbol{\gamma}\cdot[\Omega - CK^{-1}C^*]\boldsymbol{\gamma}\boldsymbol{\delta} = \boldsymbol{\gamma}\cdot\bar{\boldsymbol{\alpha}}\boldsymbol{\beta} + \boldsymbol{\gamma}\cdot\boldsymbol{\beta}\bar{\boldsymbol{\alpha}}.$$

In particular, this shows that if either $\boldsymbol{\alpha}$ or $\boldsymbol{\beta}$ is zero, that is, if Q is scalar, then either $\boldsymbol{\gamma}$ or $\boldsymbol{\delta}$ is zero. It then follows from (5.18) that $\mathbf{r}_c$ must be zero. That is, *if* Q *is a scalar matrix for some value of* $\mathbf{r}_c$, *then* Q *has isolated directions of fall for all other* $\mathbf{r}_c$.

Conversely, we see by reversing the roles of $\{\boldsymbol{\alpha}, \boldsymbol{\beta}\}$ and $\{\boldsymbol{\gamma}, \boldsymbol{\delta}\}$ in (5.20) that if neither $\boldsymbol{\alpha}$ nor $\boldsymbol{\beta}$ is zero then neither $\boldsymbol{\gamma}$ nor $\boldsymbol{\delta}$ can be zero.

If we take the scalar product of (5.20) with $[\Omega - CK^{-1}C^*]^{-1}\boldsymbol{\delta}$ and use (5.17), we see that

$$(5.22)\qquad \boldsymbol{\delta}\cdot[\Omega - CK^{-1}C^*]^{-1}\boldsymbol{\delta}\bar{\boldsymbol{\gamma}} = \boldsymbol{\alpha}\cdot\boldsymbol{\delta}\boldsymbol{\beta} + \boldsymbol{\beta}\cdot[\Omega - CK^{-1}C^*]^{-1}\boldsymbol{\delta}\bar{\boldsymbol{\alpha}}.$$

It is easily seen that if the equation (5.18) has a solution $\mathbf{r}_c$, it is given by

$$(5.23)\qquad \mathbf{r}_c = \tfrac{1}{2}(\bar{\boldsymbol{\gamma}}\times\boldsymbol{\delta} - \bar{\boldsymbol{\alpha}}\times\boldsymbol{\beta}).$$

The equations (5.21) and (5.22) show that $\bar{\boldsymbol{\gamma}}$ and $\boldsymbol{\delta}$ are linear combinations of $\bar{\boldsymbol{\alpha}}$ and $\boldsymbol{\beta}$. Therefore, there is a constant c so that $\bar{\boldsymbol{\gamma}}\times\boldsymbol{\delta} = c\bar{\boldsymbol{\alpha}}\times\boldsymbol{\beta}$ and hence

$$(5.24)\qquad \mathbf{r}_c = \tfrac{1}{2}(c-1)\bar{\boldsymbol{\alpha}}\times\boldsymbol{\beta}.$$

Then (5.18) becomes

(5.25) $$\tfrac{1}{2}(c-1)(\bar{\alpha}_i\beta_j - \bar{\alpha}_j\beta_i) + \bar{\alpha}_i\beta_j = \bar{\gamma}_i\delta_j.$$

Applying the matrix $[\Omega - \mathrm{C}\mathrm{K}^{-1}\mathrm{C}^*]^{-1}$ gives

$$\tfrac{1}{2}(c+1)\alpha_i\beta_j - \tfrac{1}{2}(c-1)([\Omega - \mathrm{C}\mathrm{K}^{-1}\mathrm{C}^*]^{-1}\boldsymbol{\beta})_i([\Omega - \mathrm{C}\mathrm{K}^{-1}\mathrm{C}^*]\boldsymbol{\alpha})_j = \gamma_i\delta_j.$$

We now square both matrices and recall the conditions (5.16) and (5.17) to find

$$-\tfrac{1}{4}(c^2-1)\{\boldsymbol{\beta}\cdot[\Omega - \mathrm{C}\mathrm{K}^{-1}\mathrm{C}^*]^{-1}\boldsymbol{\beta}\alpha_i([\Omega - \mathrm{C}\mathrm{K}^{-1}\mathrm{C}^*]\boldsymbol{\alpha})_j$$
$$+ \boldsymbol{\alpha}\cdot[\Omega - \mathrm{C}\mathrm{K}^{-1}\mathrm{C}^*]\boldsymbol{\alpha}([\Omega - \mathrm{C}\mathrm{K}^{-1}\mathrm{C}^*]^{-1}\boldsymbol{\beta})_i\beta_j\} = 0.$$

Since $[\Omega - \mathrm{C}\mathrm{K}^{-1}\mathrm{C}^*]$ times the matrix in braces is positive definite, this equation implies $c = \pm 1$. We see from (5.24) that $c = 1$ corresponds to $\mathbf{r}_c = 0$, and hence $\boldsymbol{\gamma} = \boldsymbol{\alpha}$, $\boldsymbol{\delta} = \boldsymbol{\beta}$. If $c = -1$, we find

(5.26) $$\mathbf{r}_c = -(\bar{\boldsymbol{\alpha}} \times \boldsymbol{\beta}),$$

and (5.25) shows that $\bar{\boldsymbol{\gamma}} = \boldsymbol{\beta}$, $\boldsymbol{\delta} = \bar{\boldsymbol{\alpha}}$. Thus, if the matrix Q has a quadratic elementary divisor for some value of $\mathbf{r}_c$, then there is a second value of $\mathbf{r}_c$ where the same is true, and at all other points the problem (5.9) has at least one isolated direction of fall.

Thus our theorem is proved.

We note that if $\mathrm{C} = 0$, then the matrix Q is zero when $\mathbf{r}_c = 0$. This, then, is an example of the case in which the matrix becomes scalar at one point so that the problem (5.9) has a solution $\{\tilde{\mathbf{u}}, \tilde{\mathbf{g}}, \tilde{\lambda}\}$ with any prescribed $\tilde{\mathbf{g}}$. Thus, there is no isolated direction of fall when $\mathbf{r}_c = 0$, but there are always isolated directions of fall when $\mathbf{r}_c \neq 0$.

H. Brenner [**1**] has defined B to be a *nonskew body* if $\mathrm{C} = 0$ for a suitable choice of origin (the center of stress). The class of nonskew bodies includes all axially symmetric and orthotropic bodies (see [**1**], [**3**, p. 187]) and all bodies which are symmetric about a point (see [**11**]).

The above considerations show that if B is a nonskew body, if the effective center of mass is not at the center of stress, and if $\rho m|\mathbf{g}|/\mu^2$ is sufficiently small, then there exist at least two solutions of Problem II. Taking $\mathrm{C} = 0$ and $\mathbf{r}_c \neq 0$, one easily sees from (5.13) that Q has the eigenvalue zero and two imaginary eigenvalues, so that the problem (5.9) has exactly two directions of fall, $\pm\tilde{\mathbf{g}}^*$ with the corresponding $\tilde{\lambda} = 0$.

Since all explicitly known Stokes flows seem to be about nonskew bodies, it is not known whether the case of a triple eigenvalue with a quadratic elementary divisor actually occurs.

6. **Some symmetry considerations.** Theorem 2 leaves unanswered the question of whether Problem II for such a symmetric body as a sphere with the center of mass at its center has a solution. In this section we shall show that sufficient symmetry can replace the existence of an isolated direction of fall for the linearized problem and can, in fact, provide existence of the motion for all values of the parameters.

We shall say that a body B has rotational symmetry of order $k \geqq 2$ about a line L if it is invariant under a rotation by an angle $2\pi/k$ about L. Our principal result is the following:

THEOREM 4. *Let B be rotationally symmetric of order $k \geqq 2$ about the x_3-axis. If the center of mass of B lies on the x_3-axis, and if $\mathbf{g}$ is any vector parallel to the x_3-axis, then for any given values of m, ρ, and μ there exists a solution $\{\mathbf{u}, \mathbf{g}, \lambda\}$ of Problem* II. *Moreover, if $\mathrm{R}: E_3 \to E_3$ represents rotation about the x_3-axis through angle $2\pi/k$, then*

$$\mathbf{u}(\mathrm{R}\mathbf{r}) = \mathrm{R}\mathbf{u}(\mathbf{r})$$

and in particular $\mathbf{u}$ is parallel to $\mathbf{g}$ along the x_3-axis.

PROOF. We define the vector field mapping

$$\mathscr{T}\mathbf{u}(\mathbf{r}) \equiv \mathrm{R}^{-1}\mathbf{u}(\mathrm{R}\mathbf{r}).$$

It is easily verified that $\operatorname{div} \mathscr{T}\mathbf{u}(\mathbf{r}) = \operatorname{div} \mathbf{u}(\mathrm{R}\mathbf{r})$. Hence $\mathscr{T}$ maps J' into J' and H' into H'. Moreover, one can check that $\mathscr{T}$ is a unitary operator on H'.

We define $\tilde{J}$ to be the subspace of J' which consists of vector fields which are invariant under $\mathscr{T}$. That is, $\mathbf{u} \in \tilde{J}$ if and only if $\mathbf{u} \in J'$ and its components in cylindrical coordinates are periodic of period $2\pi/k$ in θ. The boundary values of such a $\mathbf{u}$ must be of the form $\alpha\mathbf{g} + \beta\mathbf{g} \times \mathbf{r}$. Let $\tilde{H}$ be the closure of $\tilde{J}$ in H'.

By an argument exactly like the proof of Theorem 1 we show that there is a $\mathbf{u} \in \tilde{H}$ so that

$$E(\varphi, \mathbf{u}) - \rho \int_D [\varphi_{i,j}u_i(u_j - U_j - \lambda\varepsilon_{jkl}g_k x_l) - \lambda\varphi \cdot \mathbf{g} \times \mathbf{u}]\, dx = m\mathbf{V} \cdot \mathbf{g} \tag{6.1}$$

for all $\varphi \in \tilde{J}$. Consider now an arbitrary $\varphi \in J'$. Since $\mathscr{T}$ is unitary and $\mathbf{u} \in \tilde{H}$,

$$E(\mathscr{T}\varphi, \mathbf{u}) = E(\mathscr{T}\varphi, \mathscr{T}\mathbf{u}) = E(\varphi, \mathbf{u}).$$

In a similar manner we see that the second term on the left of (6.1) is also not changed if φ is replaced by $\mathscr{T}\varphi$. Let φ have the boundary values $\mathbf{V} + \mathbf{W} \times \mathbf{r}$. Then since $\mathscr{T}^k\varphi = \varphi$, we see that the vector field

$$\psi \equiv k^{-1} \sum_{l=0}^{k-1} \mathscr{T}^l\varphi \tag{6.2}$$

lies in $\tilde{J}$ and that it has the boundary values $|\mathbf{g}|^{-2}\{\mathbf{V} \cdot \mathbf{g}\mathbf{g} + \mathbf{W} \cdot \mathbf{g}\mathbf{g} \times \mathbf{r}\}$. Hence, for any $\varphi \in J'$,

$$E(\varphi, \mathbf{u}) - \rho \int_D [\varphi_{i,j}u_i(u_j - U_j - \lambda\varepsilon_{jkl}g_k x_l) - \lambda\varphi \cdot \mathbf{g} \times \mathbf{u}]\, dx$$

$$= E(\psi, \mathbf{u}) - \rho \int_D [\psi_{i,j}u_i(u_j - U_j - \lambda\varepsilon_{jkl}g_k x_l) - \lambda\psi \cdot \mathbf{g} \times \mathbf{u}]\, dx = \mathbf{V} \cdot \mathbf{g}$$

by (6.1). Thus, (6.1) holds for all φ in J'.

Since both the center of mass and the centroid lie on the x_3-axis, the same must be true of $\mathbf{r}_c$. Thus, $\mathbf{W} \times \mathbf{r}_c \cdot \mathbf{g} = 0$ for all $\mathbf{W}$, and we see that $\{\mathbf{u}, \mathbf{g}, \lambda\}$ is a solution of Problem II as formulated in (2.24).

Theorem 4 gives the existence of a steady fall for an object like a k-bladed propeller with its axis vertical.

If in addition to having rotational symmetry B is invariant under reflection in the x_2-x_3 plane, we can show that there is a solution $\{\mathbf{u}, \mathbf{g}, \lambda\}$ in which $\mathbf{u}$ is invariant not only under $\mathcal{T}$ but also under the transformation $S^{-1}\mathbf{u}(S\mathbf{r})$ where S represents reflection in the x_2-x_3 plane. If $\mathbf{g}$ lies in this plane, it then follows that $\lambda = 0$, so that there is a fall with no spin.

If B is axially symmetric, we can use the same arguments to obtain the following result.

THEOREM 5. *Let B be a body of revolution, and let its center of mass lie on the axis of symmetry. Then for arbitrary values of m, ρ, and μ there is a solution $\{\mathbf{u}, \mathbf{g}, 0\}$ of Problem* II *without spin, where* $\mathbf{g}$ *is any vector directed along the axis of symmetry. Moreover,* $\mathbf{u}$ *is axially symmetric and the body moves vertically.*

BIBLIOGRAPHY

1. H. Brenner, *The Stokes resistance of an arbitrary particle.* II, Chem. Eng. Sci. **19** (1964), 599–624.

2. R. Finn, *On the Stokes paradox and related questions*, Proc. Sympos. Nonlinear Problems (Madison, Wis., 1962), Univ. of Wisconsin Press, Madison, Wis., 1963, pp. 99–115. MR **27** #2739.

3. J. Happel and H. Brenner, *Low Reynolds number hydrodynamics with special applications to particulate media*, Prentice-Hall, Englewood Cliffs, N.J., 1965. MR **33** #3562.

4. E. A. Kearsley, *Bounds on the dissipation of energy in steady flow of a viscous incompressible fluid around a body rotating within a finite region*, Arch. Rational Mech. Anal. **5** (1960), 347–354. MR **22** #10431.

5. J. B. Keller, L. A. Rubenfeld and J. E. Molyneux, *Extremum principles for slow viscous flows with applications to suspension*, J. Fluid Mech. **30** (1967), 97–125.

6. O. A. Ladyzenskaja, *Investigation of the Navier-Stokes equation for a stationary flow of an incompressible fluid*, Uspehi Mat. Nauk **14** (1959), no. 3 (87), 75–97; English transl., Amer. Math. Soc. Transl. (2) **25** (1963), 173–197. MR **22** #10437.

7. ———, *The mathematical theory of viscous incompressible flow*, Fizmatgiz, Moscow, 1961; English transl.; 2nd English ed., Gordon and Breach, New York, 1969. MR **27** #5034a; MR **40** #7610.

8. J. Leray, *Étude de diverses équations intégrales non linéaires et de quelques problèmes que pose l'hydrodynamique*, J. Math. Pures Appl. (9) **12** (1933), 1–82.

9. ———, *Les problèmes non linéaires*, Enseignment Math. **35** (1936), 139–151.

10. J. Leray and J. Schauder, *Topologie et équations fonctionnelles*, Ann. Sci. Ecole Norm. Sup. (3) **51** (1934), 45–78.

11. H. F. Weinberger, *Variational properties of steady fall in Stokes flow*, J. Fluid Mech. **52** (1972), 321–344.

UNIVERSITY OF MINNESOTA

RELATIVISTIC WAVE EQUATIONS AS SINGULAR HYPERBOLIC SYSTEMS

A. S. WIGHTMAN

1. **Introduction.** The family of mathematical problems discussed here has emerged in recent years as a result of efforts to put a small chapter of quantum field theory, the so-called external field problem, on a sound mathematical footing. The external field problem is special because the partial differential equation for the unknown field is *linear*, but the coefficients are allowed to vary in space and time and that gives rise to some surprises, which seem to be of general interest. There is a vast and in large part turgid mathematical physics literature on the subject. To make the general wisdom which has accumulated there more readily available to a mathematical audience I have, in the following, tried to place the problems in their physical context, and still to bring out the essential mathematical questions many of which remain to be answered. The exposition assumes only a vague general knowledge of quantum mechanics [**1**].

The relativistic wave equations to be considered mostly arose in connection with quantum theories of elementary particles, although several examples e.g. the wave equation

$$\Box\varphi(x) = 0, \qquad \Box = \frac{1}{c^2}\frac{\partial^2}{\partial t^2} - \Delta, \tag{1.1}$$

for the real valued function, $\varphi(x)$, on space-time and Maxwell's equations

$$\begin{aligned} &\nabla \cdot \boldsymbol{E}(x) = 0, \qquad \nabla \cdot \boldsymbol{B}(x) = 0, \\ &\nabla \times \boldsymbol{E}(x) = -\frac{1}{c}\frac{\partial \boldsymbol{B}(x)}{\partial t}, \qquad \nabla \times \boldsymbol{B}(x) = \frac{1}{c}\frac{\partial \boldsymbol{E}(x)}{\partial t}, \end{aligned} \tag{1.2}$$

for the three real-dimensional vector valued functions, $\boldsymbol{E}(x)$ (electric field strength)

AMS 1970 *subject classifications*. Primary 35L40, 81A18.

and $\boldsymbol{B}(x)$ (magnetic induction) antedate quantum theory itself. (Here and throughout the following $x = (x^0, \boldsymbol{x})$, $x^0 = ct$. Also $(\nabla \times \boldsymbol{E})^j = (\partial/\partial x^k)E^l - (\partial/\partial x^l)E^k$, j, k, l a cyclic permutation of 1, 2, 3 and $\nabla \cdot \boldsymbol{E} = \sum_{j=1}^{3}(\partial/\partial x^j)E^j$.) Two early examples of such wave equations were the so-called *Klein-Gordon equation* (1926)

$$[\Box + m^2]\psi(x) = 0 \tag{1.3}$$

for a complex valued function ψ and the *Dirac equation* (1928)

$$[\gamma^\mu D_\mu + m]\psi(x) = 0 \tag{1.4}$$

which is a system of four equations for a four-dimensional complex vector valued function $\psi(x)$. Here the γ^μ, $\mu = 0, 1, 2, 3$, are generators of the Clifford algebra in space-time, i.e. they are 4×4 complex matrices satisfying

$$\gamma^\mu\gamma^\nu + \gamma^\nu\gamma^\mu = 2g^{\mu\nu}. \tag{1.5}$$

The Minkowski metric $g^{\mu\nu}$ is given by

$$\begin{aligned} g^{\mu\nu} &= 0, \qquad \mu \neq \nu, \\ &= 1, \qquad \mu = \nu = 0, \\ &= -1, \qquad \mu = \nu = 1, 2, 3, \end{aligned} \tag{1.6}$$

$D_\mu = (1/i)(\partial/\partial x^\mu)$ and m is a positive constant.

For the application of such equations to the relativistic description of a single elementary particle, what is essential is a *Hilbert space structure on a family of solutions, together with a unitary representation of the Poincaré group* ($\equiv$ inhomogeneous Lorentz group) or more generally the universal covering group of its connected component, $ISL(2, \boldsymbol{C})$, in that Hilbert space. This construction will be described in the next section. At this point I want only to introduce some notation and to describe the physical interpretation.

An element of $ISL(2, \boldsymbol{C})$ is written $\{a, A\}$ where a is a space-time translation and A is an element of the 2×2 complex unimodular group, $SL(2, \boldsymbol{C})$. The group multiplication law in $ISL(2, \boldsymbol{C})$ is

$$\{a, A\}\{b, B\} = \{a + \Lambda(A)b, AB\} \tag{1.7}$$

where $\Lambda(A)$ is the Lorentz transformation that corresponds to A under the homomorphism of $SL(2, \boldsymbol{C})$ onto $L_+^\uparrow$, the component of the identity in the Lorentz group.

If ψ is a solution of some wave equation, $U_1(a, A)\psi$ stands for the corresponding solution Lorentz transformed by A and translated by a. The physical meaning of $U_1(a, A)$ derives in part from the interpretation of the infinitesimal space-time translation operators as observables of energy-momentum. Explicitly,

$$U_1(a, \mathbf{1}) = \int \exp(ip \cdot a)\, dE(p) \tag{1.8}$$

where the integration runs over $\boldsymbol{R}^4$ and E is a projection valued measure on $\boldsymbol{R}^4$.

For the differential equations (1.3) and (1.4), E has its support on the hyperboloid $p^2 = m^2$ which has two components, the positive energy, $p^0 = (m^2 + \boldsymbol{p}^2)^{1/2}$, and the negative energy, $p^0 = -(m^2 + \boldsymbol{p}^2)^{1/2}$. For the more general equations considered later, we will get a *mass spectrum*; E will have its support on a family of hyperboloids $p^2 = m_1^2, p^2 = m_2^2, \ldots, p^2 = m_n^2$. This set of momenta p will be referred to as the *mass shell*. Since particles of imaginary mass have never been seen, it is a physical requirement that $m_1^2 \geqq 0, \ldots, m_n^2 \geqq 0$. Complex masses, with non-vanishing real and imaginary parts, cannot occur as long as U_1 is unitary.

For strictly positive masses, a second part of the physical interpretation of U_1 is obtained by considering the transformation law of solutions that have $\boldsymbol{p} = 0$, $p^0 = m_i$, under the subgroup of $SL(2, \boldsymbol{C})$, that leaves this p fixed: $\Lambda(A)p = p$; it is $SU(2)$, the group of 2×2 unitary unimodular matrices. (Such solutions are not normalizable but what follows is the essence of a rigorous discussion which considers only normalizable solutions.) The irreducible unitary representations of $SU(2)$ are labeled by the spin angular momentum quantum number, j, which takes the values $0, \frac{1}{2}, 1, \frac{3}{2}, 2, \ldots$. The multiplicities of these irreducible representations in the restriction of U_1 to the solutions with $\boldsymbol{p} = 0$ and $p^0 = m_i$ give the *spin spectrum*.

For $m = 0$, the analogue of $SU(2)$ is the isotropy group of a light-like vector. It turns out to be isomorphic to the two sheeted covering group of the Euclidean group of the plane. When the "translation" subgroup (which has nothing to do with space-time translations, being an abelian subgroup of $SL(2, \boldsymbol{C})$) is trivially represented, the isotropy group is effectively just a one parameter subgroup of $SU(2)$ so the irreducible representations are one dimensional: $e^{i\theta/2} \to e^{ih\theta}$ where h is the *helicity* quantum number which takes the values $h = 0, \pm\frac{1}{2}, \pm 1, \ldots$. Thus, for mass zero, the spin spectrum is replaced by a *helicity spectrum*. We will ignore the case in which the "translation" subgroup is nontrivially represented; it turns out to be of little physical interest.

All the relativistic wave equations we will consider have the property that if there is a nontrivial solution of energy momentum p, there is also one of energy momentum $-p$. Since negative energy particles have never been observed, we use only the positive energy solutions of the wave equations for the description of elementary particles. The analysis of the representations of $ISL(2, \boldsymbol{C})$ shows that the subspace spanned by the positive energy solutions is left invariant by U_1 so the restriction to positive energies is compatible with the transformation law under $ISL(2, \boldsymbol{C})$.

The restriction to positive energy solutions has another advantage. The scalar product that is the natural candidate to define the Hilbert space structure in the chosen family of solutions is often positive on the positive energy solutions and negative on the negative energy solutions. For example, the Klein-Gordon equation (1.3) has a scalar product

$$(\psi, \chi) = i \int_\Sigma d\Sigma_\mu(x) [\overline{\psi(x)} \partial^\mu \chi(x) - \overline{\partial^\mu \psi(x)} \chi(x)] \tag{1.9}$$

with this property. Here the integral is over a space-like hyperplane Σ.

If the representation U_1 is irreducible, the mass spectrum contains a single mass and the spin (or helicity) spectrum a single spin (or helicity). It is natural to call a wave equation equipped with such a U_1, *irreducible.* A final filip was added to this theory in 1939 by Wigner. He showed that up to unitary equivalence the irreducible unitary representations of $ISL(2, C)$ are determined by the mass and the representation of the isotropy group of an occurring momentum, and hence by mass and spin, or for $m = 0$, helicity. This means that in the description of an elementary particle the choice of a particular irreducible wave equation does not matter, provided that it yields the desired mass and spin (or helicity). Apparently what is essential is the unitary equivalence of the representation of $ISL(2, C)$. In fact, this statement is true for free, i.e. noninteracting, particles but false for interacting particles. In the historical development of the subject (1926–1939), the group-theoretical point of view played relatively little role because physicists were interested from the beginning in the extension of the theory to describe interaction.

The simplest theory of this kind is provided by a coupling to an *external field.* The adjective external here refers to the fact that the perturbation is described by a given fixed function or family of functions defined on space-time. For example, one replaces the Klein-Gordon equation (1.3) by

$$\{[\partial^\mu - eiA^\mu(x)][\partial_\mu - eiA_\mu(x)] + m^2\}\psi(x) = 0 \tag{1.10}$$

and the Dirac equation (1.4) by

$$[\gamma^\mu/i[\partial_\mu - eiA_\mu(x)] + m]\psi(x) = 0, \tag{1.11}$$

where $A_\mu(x)$ is a vector potential describing a given electromagnetic field, and $-e$ is the charge of the particle described by the wave function $\psi(x)$. This kind of coupling, induced by the replacement $\partial^\mu \to \partial^\mu - eiA^\mu(x)$, is called *minimal electromagnetic* coupling in the physics literature.

One physical phenomenon for which these external field couplings provide a quantitative description is the binding of particles to an external potential. For example, suppose that $\boldsymbol{A}(x) = 0$ and $A^0(x)$ is a function of $\boldsymbol{x}$ independent of x^0. Then equations like (1.10) and (1.11) may have normalizable stationary solutions, i.e. normalizable solutions of the form,

$$\psi(x) = \psi_1(\boldsymbol{x})\exp[-iEt]. \tag{1.12}$$

Such a solution is called a *bound state of energy E.* If one takes, for $A^0(\boldsymbol{x})$ the Coulomb potential of a point charge, Ze,

$$A^0(\boldsymbol{x}) = Ze/|\boldsymbol{x}|, \tag{1.13}$$

one gets an approximate description of a hydrogen-like atom in which the nucleus is idealized as an external field and the elementary particle in question plays the role of the electron. One of the first triumphs of Dirac's theory of the electron was the resulting formula for the bound state energies (the so-called Sommerfeld formula).

The solutions of the external field problem contain a description of a second physical phenomenon: *scattering by an external potential.* But it is just here that a crisis arises for the physical interpretation. To bring out the difficulty in as clear a way as possible, consider an external field which is C^∞ and of compact support in space-time. Then for $|t|$ sufficiently large the wave equation for an external field reduces to a wave equation for a free particle so the solution can be written for such x as the Fourier transform of a function whose support is on the hyperboloid $p^2 = m^2$. If it is assumed that for large negative t the solution written in this way has momenta concentrated around some fixed p, then the solution evolves in such a way that for large positive t the corresponding momenta will be spread over the hyperboloid $p^2 = m^2$. That is precisely the phenomenon of scattering. The difficulty arises because if the incoming wave is of positive energy, the outgoing wave will, in general, contain both positive and negative energies; the theory predicts that positive energy particles are scattered into unphysical negative energy states. This circumstance is generally known as the *Klein paradox* after Oscar Klein who discovered it for the Dirac equation in a slightly different context in 1928. Of course, there were valiant attempts to outlaw the transitions to negative energy states, but, to make a long story short, none of them worked. *There is no adequate resolution of the Klein paradox within the framework of single particle theory.*

The period 1928–1936 saw a fundamental conceptual transformation in the external field problem. What was formerly a single particle theory was transformed to a many particle theory. Whereas in the single particle interpretation the wave equation was a differential equation for the particle wave function, in the many particle theory it became an equation for the field operator. What formerly was a nonvanishing probability for a transition from positive to negative energies turned out to be related to a probability for the creation of particle antiparticle pairs by the external field. I will now sketch briefly the constructions that realize all these statements.

First, one has to introduce Hilbert spaces to describe the states of arbitrarily many particles. These are the so-called *Фок spaces.* If $\mathscr{H}_1$ is the space of one particle states, they are the two Hilbert spaces

$$\mathscr{F}_\varepsilon(\mathscr{H}_1) = \bigoplus_{n=0}^{\infty} (\mathscr{H}_1^{\otimes n})_\varepsilon. \tag{1.14}$$

Here the summand for $n = 0$ is by definition the one-dimensional Hilbert space. The subscript ε takes the two values s and a referring respectively to the symmetrized and antisymmetrized tensor product. The symmetric case is said to have *Bose-Einstein statistics*; both possibilities occur in nature, the former for mesons, the latter for electrons, for example. A vector, Φ, in $\mathscr{F}_\varepsilon(\mathscr{H}_1)$ is given by a sequence

$$\{\Phi^{(0)}, \Phi^{(1)}, \dots\} \tag{1.15}$$

where $\Phi^{(n)}$ belongs to $(\mathscr{H}_1^{\otimes n})_\varepsilon$. The probability of finding exactly n particles in the state Φ is

$$\|\Phi^{(n)}\|_n^2 \left[\sum_{k=0}^{\infty} \|\Phi^{(k)}\|_k^2 \right]^{-1} \tag{1.16}$$

where $\| \ \|_n$ is the norm in $(\mathscr{H}^{\otimes n})_\varepsilon$. The expectation value of the number of particles is given by

$$\frac{(\Phi, N\Phi)}{\|\Phi\|^2} = \frac{\sum_{n=0}^{\infty} n\|\Phi^{(n)}\|_n^2}{\sum_{n=0}^{\infty} \|\Phi^{(n)}\|_n^2}, \tag{1.17}$$

where N is the operator defined by

$$(N\Phi)^{(n)} = n\Phi^{(n)}. \tag{1.18}$$

The state $\Psi_0 = \{1, 0, 0, 0, \ldots\}$ has no particles. It is appropriately interpreted as the *vacuum*.

The representation U_1 of $ISL(2, C)$ in $\mathscr{H}_1$ induces a representation U in the corresponding Фок spaces

$$U(a, A) = \bigoplus_{n=0}^{\infty} U_1(a, A)^{\otimes n} \tag{1.19}$$

where by definition, for $n = 0$, the direct summand is the trivial representation. Consequently, the vacuum state is invariant under U:

$$U(a, A)\Psi_0 = \Psi_0. \tag{1.20}$$

U describes the transformation law of the system of arbitrarily many particles under $ISL(2, C)$.

Having derived the state spaces of the many particle theory, we turn to the construction of some operators in them. Here the so-called annihilation and creation operators play a special role. To obtain these quickly notice that the $\mathscr{F}_\varepsilon(\mathscr{H}_1)$ have dense subsets which are the symmetric and antisymmetric tensor algebras over $\mathscr{H}_1$ for $\varepsilon = s$ and a respectively. They are just those vectors $\{\Phi^{(0)}, \Phi^{(1)}, \ldots\}$ of $\mathscr{F}_\varepsilon(\mathscr{H}_1)$ for which $\Phi^{(n)} = 0$ for all sufficiently large n. The product operation that makes them algebras is respectively $\vee$ and $\wedge$, i.e. the operation of tensor product followed by symmetrization and antisymmetrization respectively. The operation of left multiplication by an element of the tensor algebra can be extended from the elements of the tensor algebra itself to all of $\mathscr{F}_\varepsilon(\mathscr{H}_1)$. Thus, it makes sense to define for any $\Psi \in \mathscr{H}_1$ the creation operator

$$a^*(\Psi) = N^{1/2}\Psi \overset{\vee}{\wedge}. \tag{1.21}$$

It takes $\{\Phi^{(0)}, \Phi^{(1)}, \ldots\}$ into $\{0, \Phi^{(0)}\Psi, 2^{1/2}\Psi \overset{\vee}{\wedge} \Phi^{(1)}, 3^{1/2}\Psi \overset{\vee}{\wedge} \Phi^{(2)}, \ldots\}$ and therefore clearly "creates" a particle in the single particle state Ψ. The corresponding annihilation operator $a(\chi)$ is defined for every χ in the dual (complex linear dual) of

$\mathscr{H}_1$ as

$$a(\chi)\Phi = (N + 1)^{1/2}\langle \chi, \Phi \rangle \tag{1.22}$$

where the symbol $\langle \chi, \Phi \rangle$ stands for

$$\{\langle \chi, \Phi^{(1)} \rangle, \langle \chi, \Phi^{(2)} \rangle, \ldots\} \tag{1.23}$$

and $\langle \chi, \Phi^{(n)} \rangle$ is the element $(\mathscr{H}^{\otimes(n-1)})_\varepsilon$ obtained by evaluating χ on the first argument of $\Phi^{(n)}$. Elementary calculations show that for the symmetric case a and a^* satisfy the *CCR* (*canonical commutation relations*):

$$\begin{aligned} a(\chi_1)a(\chi_2) - a(\chi_2)a(\chi_1) &= 0, \\ a(\chi)a^*(\Psi) - a^*(\Psi)a(\chi) &= \langle \chi, \Psi \rangle \mathbf{1}, \end{aligned} \tag{1.24}$$

while for the antisymmetric case they satisfy the *CAR* (*canonical anticommutation relations*): (Then (1.25) is a pair of equations.)

$$\begin{aligned} a(\chi_1)a(\chi_2) + a(\chi_2)a(\chi_1) &= 0, \\ a(\chi)a^*(\Psi) + a^*(\Psi)a(\chi) &= \langle \chi, \Psi \rangle \mathbf{1}. \end{aligned} \tag{1.25}$$

To make these definitions and statements precise, a word should be said about the domains on which $a(\chi)$ and $a^*(\Phi)$ are defined. We choose to regard (1.21) as defining $a^*(\Psi)$ and (1.22) as defining $a(\chi)$ on vectors in the corresponding tensor algebras. Then we extend the domain of definition by passing to the closures. It turns out that in the antisymmetric case $a(\chi)$ and $a^*(\Phi)$ are bounded and everywhere defined so (1.25) is a relation among bounded operators. On the other hand, in the symmetric case $a(\chi)$ and $a^*(\Psi)$ are unbounded for $\Psi \neq 0$, $\chi \neq 0$ and can be defined only on the domain of $(N + 1)^{1/2}$; the CCR (1.24) hold on the domain of N.

The creation and annihilation operators have simple transformation laws under $ISL(2, \boldsymbol{C})$. For the creation operators

$$U(a, A)a^*(\Psi)U(a, A)^{-1} = a^*(U_1(a, A)\Psi) \tag{1.26}$$

for all $\Psi \in \mathscr{H}_1$. To state the transformation law for the annihilation operators it is convenient to introduce the canonical antilinear bijection J, which maps linear functionals on $\mathscr{H}_1$ onto $\mathscr{H}_1$,

$$\langle \chi, \Phi \rangle = (J\chi, \Phi) \tag{1.27}$$

for all $\Psi \in \mathscr{H}_1$ and χ in the dual of $\mathscr{H}_1$. ((,) is the sequilinear scalar product in $\mathscr{H}_1$.) Then we can write

$$U(a, A)a(J^{-1}\Psi)U(a, A)^{-1} = a(J^{-1}U_1(a, A)\Psi). \tag{1.28}$$

The mapping J also appears when one computes the adjoint of creation operators

$$a^*(\Phi)^* = a(J^{-1}\Phi). \tag{1.29}$$

Thus the star on $a^*(\Phi)$ makes it almost but not quite the adjoint of $a(\Phi)$.

This completes the necessary machinery of the many particle theory. Our next step is to relate it to the notion of quantized field.

There exist numerous elaborate inductive discussions of the properties to be required of a quantized field (see for example [**2**]). To avoid further lengthening this already lengthy introduction, I will skip all that, and merely state that a *quantized field* is an operator valued distribution where the operators act in the quantum mechanical state space and the distribution means distribution on space-time. The test functions are taken to have as many components as the field so that, formally, if $\psi_\alpha(x)$, $\alpha = 1, \ldots, N$, is an operator valued distribution

$$\psi(f) = \sum_{\alpha=1}^{N} \int d^4x f_\alpha(x)\psi_\alpha(x), \tag{1.30}$$

where the f_α are typically complex-valued C^∞ functions of fast decrease on space-time. The continuity properties appropriate to a distribution are imposed by requiring that matrix elements $(\Phi, \psi(f)\Psi)$ be distributions in f for each fixed pair of vectors Φ, Ψ in some appropriate domain in $\mathscr{H}$, the Hilbert space of states about which nothing more will be said here. We will not commit ourselves to any hermiticity properties of $\psi(f)$. Both fields which are hermitian

$$\psi(f)^* = \psi(\bar{f}), \tag{1.31}$$

where $\bar{f}$ is the complex conjugate of f and those which are not turn out to be useful. Somewhat more general than hermitian fields are self charge conjugate fields for which there is a charge conjugation matrix C acting on the components of the test functions such that

$$\psi(f)^* = \psi(C\bar{f}). \tag{1.32}$$

These will also appear in the following. In the absence of external fields a quantized field is required to have a transformation law, under $ISL(2, \boldsymbol{C})$,

$$U(a, A)\psi(f)U(a, A)^{-1} = \psi(\{a, A\}f) \tag{1.33}$$

where

$$(\{a, A\}f)_\alpha(x) = \sum_{\beta=1}^{N} S(A^{-1})_{\beta\alpha} f_\beta(\Lambda(A^{-1})(x - a)) \tag{1.34}$$

as it ought to be to make $\psi_\alpha(x)$ formally transform as

$$U(a, A)\psi_\alpha(x)U(a, A)^{-1} = \sum_{\beta=1}^{N} S(A^{-1})_{\alpha\beta}\psi_\beta(\Lambda(A)x + a). \tag{1.35}$$

Here $A \to S(A)$ is some finite-dimensional representation of $SL(2, \boldsymbol{C})$. Finally, quantized fields are required to be *local*. That means that if f and g are test functions of compact support and every point of the support of f is space-like with respect to every point of the support of g, one of the conditions

$$[\psi(f), \psi(g)]_{\pm} = 0 \tag{1.36}$$

and one of the conditions

$$[\psi(f), \psi(g)^*]_{\pm} = 0 \tag{1.37}$$

hold. Here $[A, B]_-$ stands for the ordinary commutator $AB - BA$ while $[A, B]_+$ stands for the anticommutator $AB + BA$. It turns out that the signs have to be the same in (1.36) and (1.37). The wisdom of including the anticommutator as well as the commutator in (1.36) and (1.37) was born in the discovery of the significance of the CAR, equation (1.25), for the description of electrons and other particles having Fermi-Dirac statistics. The locality of quantized fields is one of the key general properties that give quantum field theory its special flavor. It is an expression of the relativistic propagation of influence. That is why it will be such a shock when we find that the solutions of certain equations with external fields necessarily violate locality.

Apart from the general requirements just described the fields considered here will satisfy linear differential equations. In the absence of external fields these will be assumed to take the form

$$[\beta^\mu \partial_\mu + m]\psi(x) = 0, \tag{1.38}$$

or expressed in the language of distributions

$$\psi(f) = 0 \tag{1.39}$$

for all f of the form

$$f(x) = -\beta^{\mu T}\partial_\mu h + mh,$$

where T denotes transpose and $h \in \mathscr{S}$. The Klein-Gordon equation is rewritten in this form in §2.

How can one construct quantized fields satisfying a given wave equation (still without external fields) in terms of the annihilation and creation operators of many particle theory? Two distinct answers were given to this question in the period 1926–1936. Both were based on the idea of using both positive and negative energy solutions of the single particle wave equation, but associating the positive energy solutions with annihilation operators for particles and the negative energies with creation operators for antiparticles. The two solutions differed in this: the first assumes a particle is identical with its antiparticle; the second assumes that the two are distinct although the antiparticle has the same mass and spin. In the former case, the state space is $\mathscr{F}_\varepsilon(\mathscr{H}_1)$ while in the latter it is taken as $\mathscr{F}_\varepsilon(\mathscr{H}_1) \otimes \mathscr{F}_\varepsilon(\mathscr{H}_{\bar{1}})$, $\mathscr{H}_{\bar{1}}$ being the state space for a single antiparticle. The Ansatz for the field in the former case is

$$\psi(f) = a(\Pi_+ f) + a^*(\Pi_- f), \tag{1.40}$$

while in the latter it is

$$\psi(f) = a(\Pi_+ f) \otimes \mathbf{1} + \begin{Bmatrix} \mathbf{1} & \text{if } \varepsilon = s \\ (-1)^N & \text{if } \varepsilon = a \end{Bmatrix} \otimes a^*(\Pi_- f). \tag{1.41}$$

Here N is the total number of particles and the effect of the operator $(-1)^N$ is to make the second term anticommute with the first for $\varepsilon = a$. The symbol Π_+ stands for a linear map from the test function space to the dual of the one particle space. $\Pi_+ f = 0$ if the Fourier transform of f vanishes for positive energies. Similarly Π_- is a linear map from the test function space to the one antiparticle space such that $\Pi_- f = 0$ if the Fourier transform of f vanishes for negative energies.

The explicit form of Π_+ and Π_- will be given in §3 and it will be evident that they have sufficient continuity properties that the fields as defined by (1.40) and (1.41) are tempered operator valued distributions. Let us see what conditions are imposed on Π_+ and Π_- by the transformation law under $ISL(2, \boldsymbol{C})$. Consider, for simplicity, the case of (1.40).

$$\begin{aligned} U(a, A)\psi(f)U(a, A)^{-1} &= a(J^{-1}U_1(a, A)J\Pi_+ f) + a^*(U_1(a, A)\Pi_- f) \\ &= \psi(\{a, A\} f) = a(\Pi_+\{a, A\} f) + a^*(\Pi_-\{a, A\} f). \end{aligned} \tag{1.42}$$

Clearly, what is needed is

$$\begin{aligned} J^{-1}U_1(a, A)J\Pi_+ f &= \Pi_+\{a, A\} f, \\ U_1(a, A)\Pi_- f &= \Pi_-\{a, A\} f. \end{aligned} \tag{1.43}$$

A similar calculation for the theory with distinct antiparticles yields these same conditions. We leave their verification to §3.

To check the local property consider first the theory with distinct antiparticles. There evidently

$$[\psi(f), \psi(g)]_\pm = 0 \tag{1.44}$$

with the minus sign holding for the symmetric case and the plus sign for the antisymmetric case. The nonvanishing commutator or anticommutator is between the field and the adjoint of the field. It is convenient to introduce a notation $\psi^+(g)$ to denote

$$\psi^+(g) = \psi(\eta g)^*, \tag{1.45}$$

(this is the field analogue of the definition $\psi^+(x) = \overline{\psi(x)}\eta$ of the single particle theory, which appears in §2, to which the reader is referred for the definition of η). The advantages of using ψ^+ will appear shortly. The commutation relation between ψ and ψ^+ is

$$\begin{aligned} [\psi(f), \psi^+(g)]_\pm &= [a(\Pi_+ f), a(\Pi_+\eta g)^*]_\pm + [b^*(\Pi_- f), (b^*(\Pi_-\eta g))^*]_\pm \\ &= \{\langle \Pi_+ f, J\Pi_+\eta g\rangle \pm \langle J^{-1}\Pi_-\eta g, \Pi_- f\rangle\}\mathbf{1}. \end{aligned} \tag{1.46}$$

The success of the Ansatz is a result of the fact that for the theories we will consider the right-hand side of (1.46) is

$$\iint f(x)\frac{1}{i}S(x-y)g(y)\,d^4x d^4y \tag{1.47}$$

where

$$S(x) = S_R(x) - S_A(x) \tag{1.48}$$

is an $N \times N$ matrix solution of the homogeneous equation

$$(\beta^\mu \partial_\mu + m)S(x) = 0 \tag{1.49}$$

and S_R and S_A are respectively the retarded and advanced fundamental solutions associated with (1.49):

$$(\beta^\mu \partial_\mu + m)S_{R;A}(x) = \delta(x)\mathbf{1} \tag{1.50}$$

and

$$S_R(x) = 0, \qquad x^0 < 0 \quad \text{and} \quad x^2 < 0, \tag{1.51}$$

$$S_A(x) = 0, \qquad x^0 > 0 \quad \text{and} \quad x^2 < 0. \tag{1.52}$$

Because it turns out that S_R and S_A have supports on the future and past light cones respectively, $S(x-y) = 0$ for space-like $x - y$; that guarantees the local property. (There is a regrettable coincidence in our notation between $S(A)$, the representative of $SL(2, \mathbf{C})$, and $S(x)$, the $N \times N$ matrix solution of the homogeneous wave equation which reaches a climax in (1.54) below. The reader will have to tell the difference between the two from their indicated arguments and context.)

The reason for introducing the field ψ^+ is that the commutator or anticommutator (1.46) is then Lorentz-invariant

$$U(a, A)[\psi(f), \psi^+(g)]_\pm U(a, A)^{-1} = [\psi(f), \psi^+(g)]_\pm \tag{1.53}$$

by virtue of the simple Lorentz transformation property of $S(x)$:

$$S(A^{-1})S(\Lambda(A)x)S(A) = S(x). \tag{1.54}$$

In the theory (1.40) in which particle and antiparticle are identical, the commutation relation for the field ψ and ψ^+ is again (1.46). (In the deduction, there are again two terms, the second one now having b replaced by a; the result is the same.) However ψ^+ is now ψ acted on by a linear transformation of components so (1.44) is replaced by

$$[\psi(f), \psi(g)] = \iint d^4x\, d^4y f(x)\frac{1}{i}S(x-y)(C^T\eta)^{-1}g(y). \tag{1.55}$$

This completes the sketch of the theory of free fields. It is the standard stuff of quantum field theory and has been combed over in a mathematically rigorous way by a number of authors. It is in great shape no matter how you look at it [**3**]. Now I turn to the external field problem in quantum field theory where things are in a much less settled state.

In accordance with the ideas described above, we want to replace the field equation (1.38) by a perturbed equation which in the case of minimal coupling is

$$[\beta^\mu(\partial_\mu - eiA_\mu(x)) + m]\psi(x) = 0, \tag{1.56}$$

or more generally

$$[\beta^\mu\partial_\mu + m + \mathscr{B}(x)]\psi(x) = 0, \tag{1.57}$$

where $\mathscr{B}(x)$ is an $N \times N$ matrix function on space-time. Two questions occur immediately: what is it reasonable to require of $\mathscr{B}(x)$ in the way of (a) local smoothness, (b) behavior at infinity? Since we will be looking for operator-valued distribution solutions the only completely safe answer to question (a) is C^∞, otherwise we cannot be sure a priori that it makes sense to multiply $\psi(x)$ by $\mathscr{B}(x)$. As far as (b) is concerned, there are several interesting possibilities. We have already discussed in the context of single particle theory the case of a static potential. What was not remarked there, however, is that the theory is rather unstable against this kind of perturbation even if the $\mathscr{B}(x)$ is of compact support in the space coordinates; the theory will fail to exist in any respectable sense if the $\mathscr{B}(x)$ is too large. Viewed from the point of view of many particle theory, the explanation of this is simple. If $\mathscr{B}(x)$ is strong enough it will become capable of creating pairs at a nonzero rate. Because $\mathscr{B}(x)$ has been assumed static, this process has been going on for an infinite time and should therefore have created an infinite number of pairs. Since this formulation is not set up to handle such a possibility, it should break down and does so. This explanation makes it plausible that for perturbations which are of rapid decrease in all directions in space-time, or even better of compact support, one might hope for the existence of well-behaved solutions in which only a finite number of pairs are produced. It is this case that will be discussed throughout the following.

There is one last requirement, which will be imposed on $\mathscr{B}(x)$,

$$\mathscr{B}(x)^* = \eta\mathscr{B}(x)\eta^{-1}. \tag{1.58}$$

Then the form $\psi^+(x)\mathscr{B}(x)\psi(x)$ is hermitian and ψ^+ satisfies the equation

$$-\partial^\mu\psi^+(x)\beta_\mu + \psi^+(x)(m + \mathscr{B}(x)) = 0 \tag{1.59}$$

which can be formally derived along with (1.57) from the Lagrangian density

$$\mathscr{L}(x) = \psi^+(\beta^\mu\partial_\mu + m)\psi + \psi^+\mathscr{B}\psi. \tag{1.60}$$

The condition (1.58) guarantees the conservation of the current

$$j^\mu(x) = \psi^+(x)i\beta^\mu\psi(x), \tag{1.61}$$

and that is the main reason for imposing it.

The fact that $\mathscr{B}(x)$ is of rapid decrease suggests that there might be some use in replacing the differential equation (1.56) by an integral equation

$$\psi(x) = \psi^{\text{in}}(x) - \int S_R(x - y)\mathscr{B}(y)\psi(g)\, d^4y, \tag{1.62}$$

where in some sense $\psi \to \psi^{\text{in}}$ as $x^0 \to -\infty$, or alternatively

$$\psi(x) = \psi^{\text{out}}(x) - \int S_A(x - y)\mathscr{B}(y)\psi(y)\, dy^4, \tag{1.63}$$

where in some sense $\psi \to \psi^{\text{out}}$ as $x^0 \to +\infty$. The equation (1.62) for ψ has extra content in quantum field theory as compared with standard scattering theory if we require not only that $\psi^{\text{in}}(x)$ satisfy the free wave equation (1.38), but also the free field commutation relations. The idea that this should be used as a basis for a formal theory of scattering was first proposed by Yang and Feldman and therefore equations (1.62) and (1.63) are sometimes called *Yang-Feldman equations.*

The formal usefulness of the Yang-Feldman equations is shown by the following reduction of the problem of solving (1.57) for fields to that of solving it for functions. (This reduction is sometimes called, among physicists, "the reduction of the q-number to the c-number problem"; q-number being Dirac's original notation for operators, and c-number his shorthand for ordinary complex numbers.) Smear (1.62) with a test function to get

$$\psi(f) = \psi^{\text{in}}(f) - \int f(x)\, dx^4 S_R(x - y)\mathscr{B}(y)\psi(y)\, d^4y. \tag{1.64}$$

This may be written as

$$\psi(T_R f) = \psi^{\text{in}}(f), \tag{1.65}$$

where

$$(T_R f)(x) = f(x) + \int d^4y f(y) S_R(y - x)\mathscr{B}(x). \tag{1.66}$$

Analogously

$$\begin{aligned} \psi(T_A f) &= \psi^{\text{out}}(f), \\ (T_A f)(x) &= f(x) + \int d^4y f(y) S_A(y - x)\mathscr{B}(x). \end{aligned} \tag{1.67}$$

Now, if ψ^{in} is regarded as given, it is natural to try and "solve" (1.65).

$$\psi(f) = \psi^{\text{in}}(T_R^{-1} f). \tag{1.68}$$

A sufficient condition for the right-hand side of (1.68) to define a field is that T_R^{-1}

be a continuous mapping of the test function space $\mathscr{S}$ onto $\mathscr{S}$, $\mathscr{S}$ being the space of C^∞ functions of fast decrease. Then it follows that

$$\psi^{\text{out}}(f) = \psi(T_A f) = \psi^{\text{in}}(T_R^{-1} T_A f). \tag{1.69}$$

It is easy to see that T_R defined by (1.66) and T_A by (1.67) define continuous mappings of $\mathscr{S}$ into itself. Since a continuous bijection of a Frechet space onto a Frechet space has a continuous inverse and $\mathscr{S}$ is a Frechet space it suffices to prove that T_R and T_A are bijections of $\mathscr{S}$ onto $\mathscr{S}$.

It turns out (the precise statement is in §4) that proving T_R and T_A are one-to-one is roughly equivalent to proving that weakly retarded and weakly advanced fundamental solutions of (1.57) exist, i.e. $N \times N$ matrix distributions satisfying

$$\left[\beta^\mu \frac{\partial}{\partial x^\mu} + m + \mathscr{B}(x)\right] S_{R;A}(x, y; \mathscr{B}) = \delta(x - y)\mathbf{1}. \tag{1.70}$$

Weak retardedness replaces the requirement of retardness

$$S_R(x, y; \mathscr{B}) = 0 \quad \text{for } (x^0 - y^0) < 0 \text{ and } (x - y)^2 < 0$$

with the requirement that it approach zero faster than power as $x - y$ approaches infinite along any direction outside the future light cone. More precisely, for each positive integer n, and each $f, g \in \mathscr{S}$ and l not in the future light cone

$$\left| \iint f(x) S_R(x + \tau l, y; \mathscr{B}) g(y)\, d^4x\, d^4y \right| \leq \frac{C(f, g, n, l)}{1 + |\tau|^l} \tag{1.71}$$

for some constant $C(f, g, n, l)$. The field then satisfies

$$[\psi(f), \psi^+(g)]_\pm = \iint f(x) \frac{1}{i} S(x, y; \mathscr{B}) g(y)\, d^4x\, d^4y \tag{1.72}$$

where

$$S(x, y; \mathscr{B}) = S_R(x, y; \mathscr{B}) - S_A(x, y; \mathscr{B}). \tag{1.73}$$

If $S_{R;A}(x, y; \mathscr{B})$ are only weakly retarded and advanced then ψ is only quasi-local; i.e. the commutator distribution $S(x, y; \mathscr{B})$ does not vanish for space-like separations, but only approaches zero rapidly for $x - y$ approaching infinity in a space-like direction.

The notion of weak retardness was introduced in this connection in response to an important discovery of Velo and Zwanziger [**4**] who showed that for certain equations and couplings $\mathscr{B}(x)$ *if* a weakly retarded fundamental solution exists at all, it is *not* retarded. This is a kind of instability of relativistic wave equations not anticipated by earlier work by physicists: *the particles described by the wave equation move faster than light in the region where the external perturbation is nonvanishing*. From the mathematical point of view the situation is still quite obscure because the known existence theorems for fundamental solutions do not, in general, cover the case at hand [**5**]. Thus it could be that there are equations of the class under consideration and perturbations $\mathscr{B}(x)$ for which the Lewy phenomenon takes place; no fundamental solution of the required character exists at all. Whatever the situation may be, it seems to me that these equations provide an interesting class of hyperbolic systems unstable under perturbation by "lower order terms."

There are a number of interesting auxiliary questions that have been looked into under the assumption of the existence of weakly retarded and advanced fundamental solutions. An example is the existence of a reasonable scattering theory. Here the main result proved under some slightly restrictive hypotheses is that the standard scattering theory goes through whether or not the weakly retarded fundamental solution is retarded [**6**]. Since unlike the scattering theory for a single particle theory, this scattering theory with pair production seems not to be widely known, I will sketch its formal foundations.

The solutions (1.68) and (1.69) express ψ and ψ^{out} in terms of ψ^{in}. Elementary calculations show that, so defined, ψ satisfies the differential equation (1.57) with the external field present, while ψ^{out} satisfies the free wave equation (1.38). It is also easy to show that ψ^{out} satisfies the free field commutation relations. The out field annihilation and creation operators can then be defined by writing a general test function f as

$$f = f_+ + f_-$$

where the Fourier transform, $\hat{f}_+$, of f_+ vanishes for p in the physical spectrum and of negative energy while $\hat{f}_-$ vanishes for p in the physical spectrum and of positive energy, and splitting

$$(1.74)\qquad \begin{aligned} a^{\text{out}}(\Pi_+ f) &= \psi^{\text{out}}(f_+) = \psi^{\text{in}}(T_R^{-1}T_A f_+),\\ b^{\text{out}}(\Pi_- f) &= \psi^{\text{out}}(f_-) = \psi^{\text{in}}(T_R^{-1}T_A f_-). \end{aligned}$$

What is not so clear from this definition is that there is an out vacuum, i.e. a vector Ψ_0^{out} satisfying

$$(1.75)\qquad \begin{aligned} a^{\text{out}}(\Pi_+ f)\Psi_0^{\text{out}} &= 0,\\ b^{\text{out}}(\Pi_+ f)\Psi_0^{\text{out}} &= 0. \end{aligned}$$

Now a^{out} and b^{out} can be written directly, in terms of a^{in} and b^{in}, as

$$(1.76)\qquad \begin{aligned} a^{\text{out}}(\Pi_+ f) &= a^{\text{in}}(\Pi_+ T_R^{-1}T_A f_+) + b^{\text{in}*}(\Pi_- T_R^{-1}T_A f_+),\\ b^{\text{out}}(\Pi_+ f) &= b^{\text{in}}(\Pi_+ {}^c(T_R^{-1}T_A({}^cf)_-)) + a^{\text{in}*}(\Pi_- {}^c(T_R^{-1}T_A({}^cf)_-)) \end{aligned}$$

where the charge conjugation operation $(\ ) \to {}^c(\ )$ is defined on the test functions by

$$(1.77)\qquad ({}^cf)(x) = C^T\overline{f(x)}.$$

Furthermore, conditions under which operators of the form (1.75) have a unique (up to a phase factor) vacuum vector satisfying (1.75) are known **[9]** and are discussed in §4. For the symmetric case they are always satisfied, while for the antisymmetric case an additional assumption is needed. When a unique Ψ_0^{out} exists there is a unitary operation S (again S!) called variously the *S-matrix*, the *S-operator*, the *collision operator*, etc. such that

$$\psi^{\text{out}}(f) = S^{-1}\psi^{\text{in}}(f)S, \qquad \Psi_0^{\text{out}} = S^{-1}\Psi_0^{\text{in}}. \tag{1.78}$$

It is terms of this unitary operator that the probability amplitudes for scattering and pair creation in the external field are expressed. For example,

$$(\Psi_0^{\text{out}}, \Psi_0^{\text{in}}) = (\Psi_0^{\text{in}}, S\Psi_0^{\text{in}}) \tag{1.79}$$

is the probability amplitude that if no particle is present initially, none will be present finally. If the external field can create pairs we will have

$$|(\Psi_0^{\text{out}}, \Psi_0^{\text{in}})|^2 < 1. \tag{1.80}$$

(We have normalized $\|\Psi_0^{\text{out}}\| = \|\Psi_0^{\text{in}}\| = 1$.) The probability amplitude for the creation of exactly one pair with particle in a state Φ_1 and antiparticle in a state Φ_2 is

$$(a^{\text{out}*}(\Phi_1)b^{\text{out}*}(\Phi_2)\Psi_0^{\text{out}}, \Psi_0^{\text{in}}) = (a^{\text{in}*}(\Phi_1)b^{\text{in}*}(\Phi_2)\Psi_0^{\text{in}}, S\Psi_0^{\text{in}}). \tag{1.81}$$

When this expression is written in terms of the scattering matrix for the one particle theory, one has the resolution of the Klein paradox. What comes out is this: (1.81) is a bilinear functional in Φ_1 and Φ_2. It has a kernel that depends on two four-momenta of positive energy lying on mass shell. If S_{++} is the scattering operator from positive energy to positive energy in the one particle theory and S_{-+} is that from positive to negative then $S_{++}^{-1}S_{-+}$ has a kernel which coincides with that of (1.81).

Further elaboration of the scattering theory shows that it contains only three independent probability amplitudes: (1.81), which is the amplitude for the production of a pair, the amplitude for scattering of one particle

$$(a^{\text{out}*}(\Phi_1)\Psi_0^{\text{out}}, a^{\text{in}*}(\Phi_2)\Psi_0^{\text{in}}) \tag{1.82}$$

and the amplitude for scattering of an antiparticle

$$(b^{\text{out}*}(\Phi_1)\Psi_0^{\text{out}}, b^{\text{in}*}(\Phi_2)\Psi_0^{\text{in}}). \tag{1.83}$$

The full scattering operator can be built up using just these three amplitudes. Thus, the scattering theory of the external field problem in quantum field theory is only a slight extension of the Schrödinger theory of scattering by a potential.

This completes our attempt to provide a motivation and orientation for the external field problem. Before turning to the detailed discussion, I want to remark that the methods discussed here do *not* work for coupled quantized fields. For an introduction to that subject I refer you to the paper by Glimm and Jaffe in these Proceedings.

2. **Single particle equations; no external field.** The most general determined first order system is of the form

$$[\beta^\mu \partial_\mu + \rho]\psi(x) = 0, \tag{2.1}$$

where $\psi(x)$ is an N-component complex-valued function and the β^μ, $\mu = 0, 1, 2, 3$, and ρ are complex $N \times N$ matrices independent of x. The usual formulation of the requirement of relativistic invariance for the equation is that there exists an $N \times N$ matrix representation of $SL(2, C)$: $A \mapsto S(A)$ such that

$$S(A)^{-1}\beta^\mu S(A) = \Lambda(A)^\mu{}_\nu \beta^\nu, \tag{2.2}$$

$$S(A)^{-1}\rho S(A) = \rho. \tag{2.3}$$

Then if ψ is a solution so is $U(a, A)\psi$ where

$$(U(a, A)\psi)(x) = S(A)\psi(\Lambda(A^{-1})(x - a)), \tag{2.4}$$

and it has the obvious meaning of the solution ψ transformed by a transformation of $ISL(2, C)$,

If we look for tempered distribution solutions of (2.1) we may Fourier transform it to get

$$(-i\beta^\mu p_\mu + \rho)\hat{\psi}(p) = 0, \tag{2.5}$$

which says that $\hat{\psi}(\rho)$ vanishes except where the matrix $(-i\beta^\mu p_\mu + \rho)$ is singular, i.e. except where

$$\mathscr{P}(p) = \det(-i\beta^\mu p_\mu + \rho) = 0. \tag{2.6}$$

By virtue of the transformation properties (2.2) and (2.3) the polynomial $\mathscr{P}(p)$ is invariant under Lorentz transformations lying in the component of the identity in the Lorentz group

$$\mathscr{P}(\Lambda p) = \mathscr{P}(p). \tag{2.7}$$

Hence by the fundamental theorem on vector invariants, $\mathscr{P}$ is actually a polynomial in $p^2 = (p^0)^2 - \boldsymbol{p}^2$. Its roots $p^2 = m_1^2, \ldots, m_n^2$ give the candidates for the mass spectrum of a theory of a single particle. We will assume $m_1^2 \geqq 0, \ldots, m_n^2 \geqq 0$. That is to be regarded as a restriction on β^μ and ρ and will not be the last we impose.

Under what circumstances will the equation have zero mass solutions? For this there is a simple criterion.

LEMMA (D. KWOH [**10**]). *A necessary condition that* (2.1) *have a zero mass solution is that*

$$\det(-i\beta^\mu p_\mu + \lambda\rho) = 0 \tag{2.8}$$

for all real λ *and all light-like* p, *i.e. all* p *such that* $p^2 = 0$.

PROOF. There exist Lorentz transformations called *boosts* which transform a given light-like p into $\lambda^{-1}p$ for any $\lambda > 0$. If A is a corresponding element of $SL(2, \boldsymbol{C})$ and u is a nontrivial solution of

$$(-i\beta^\mu p_\mu + \rho)u = 0, \tag{2.9}$$

then $S(A)u$ satisfies

$$(-i\beta^\mu p_\mu \lambda^{-1} + \rho)S(A)u = 0$$

and does not vanish. Hence (2.8) for the given p. Since $SL(2, \boldsymbol{C})$ acts transitively on the light cone, the same holds for all light-like p.

Although the case of ρ singular is interesting, we will primarily be concerned in the following with equations for which $m_j^2 > 0, j = 1, \ldots, n$. There is then no loss in generality in assuming ρ a constant multiple of the identity $\rho = m\mathbf{1}$.

To make the invariant system of equations (2.1) with the transformation law (2.4) into a single particle theory, we need an invariant sesquilinear form on the positive energy solutions. The traditional method for obtaining it is to assume that there exists a nonsingular hermitian matrix η such that

$$-\beta^{\mu *} = \eta\beta^\mu\eta^{-1}, \qquad \rho^* = \eta\rho\eta^{-1}, \tag{2.10}$$

$$S(A)^* = \eta S(A^{-1})\eta^{-1}, \tag{2.11}$$

where the * is hermitian adjoint.

Then, writing

$$\psi^+(x) = \overline{\psi(x)}\eta, \tag{2.12}$$

where the bar is complex conjugation,

$$-\partial^\mu\psi^+(x)\beta_\mu + \psi^+(x)\rho = 0, \tag{2.13}$$

and, if ψ and χ are any two solutions of (2.1) smooth enough to multiply,

$$\partial^\mu[\psi^+(x)i\beta_\mu\chi(x)] = 0. \tag{2.14}$$

Thus, integrating over a space-like hyperplane Σ we have

$$\int_\Sigma d\Sigma_\mu\psi^+(x)i\beta^\mu\chi(x) = (\psi, \chi) \tag{2.15}$$

as a candidate for a hermitian scalar product. (The i is included because from (2.10) it is $\eta i\beta^\mu$ which is hermitian.) To guarantee that this form is positive it is necessary and sufficient to require that

$$u^+ i\beta^0 u \geqq 0 \tag{2.16}$$

for all positive energy solutions of (2.9).

The final requirement on the β's and ρ arises from the need for an invariant connection between positive and negative energy solutions. The traditional operation connecting them is known in physics as *charge conjugation* or the *particle antiparticle* transformation. It is

(2.17) $$\psi(x) \to C^{-1}\overline{\psi(x)}.$$

To guarantee that the latter is again a solution of (2.1) it suffices to require

(2.18) $$\overline{\beta^\mu} = C\beta^\mu C^{-1}, \qquad \bar{\rho} = C\rho C^{-1}.$$

It then turns out that it is natural to assume

(2.19) $$\overline{S(A)} = CS(A)C^{-1}$$

and

(2.20) $$\eta = (-1)^\sigma C^T \eta C^{-1},$$

where $(-1)^\sigma$ is $+1$ on the subspace where $S(A)$ is single valued as a representation of the Lorentz group and -1 on that where it is double valued. The physical significance of charge conjugation becomes clear if we note that, if u satisfies (2.9), $C^{-1}\bar{u}$ satisfies (2.9) with p replaced by $-p$.

The equations (2.2), (2.3), (2.10), (2.11), (2.16), (2.18), (2.19) and (2.20) exhaust the list of general requirements on β^μ and ρ. Let us describe some elementary consequences, restricting attention for simplicity to the special case $\rho = m\mathbf{1}, m > 0$.

If u is a solution of (2.9) for a time like vector p, i.e. $p^2 > 0$, then, by a suitable Lorentz transformation $\Lambda(A)$ we can bring p into $\Lambda(A)p = (E, 0, 0, 0)$. Then $v = S(A)u$ satisfies

(2.21) $$(-iE\beta^0 + m)v = 0.$$

Conversely, knowing the solutions of (2.21) one can recover the solutions of (2.9). Thus, there is no loss of generality in considering (2.21). (This is called going to the rest system of the particle.) The eigenvalues λ_j of $i\beta^0$ evidently are m/E. Thus the mass spectrum is given by

(2.22) $$m_j^2 = m^2/\lambda_j^2$$

for $\lambda_j \neq 0$. If there is an eigenvalue zero of $i\beta^0$, it does not contribute to the mass spectrum. The nature of the nonzero eigenvalues of $i\beta^0$ is controlled by the following:

LEMMA (SPEER [**11**]). *If the form $u^+ i\beta^0 u$ is strictly positive on the positive energy solutions of*

(2.23) $$(-i\beta^0 E + m)u = 0$$

then the nonzero eigenvalues of $i\beta^0$ are semisimple.

PROOF. Let the distinct eigenvalues of $i\beta^0$ be $\lambda_1, \ldots$ with eigenvectors and generalized eigenvectors

$$u_{jk}^{\lambda_1}, \quad j = 1, \ldots, j(\lambda_1),\ k = 0, \ldots, k(j, \lambda_1),$$
$$u_{jk}^{\lambda_2}, \quad j = 1, \ldots, j(\lambda_2),\ k = 0, \ldots, k(j, \lambda_2),$$
$$\vdots$$

such that

$$\begin{aligned} i\beta^0 u_{jk}^{\lambda_l} &= \lambda_l u_{jk}^{\lambda_l} && \text{if } k = 0, \\ &= \lambda_l u_{jk}^{\lambda_l} + u_{jk-1}^{\lambda_l} && \text{if } k > 0. \end{aligned} \tag{2.24}$$

Take $k = 1$ and multiply (2.24) by $u_{j0}^{\lambda_l +}$,

$$u_{j0}^{\lambda_l +} i\beta^0 j_{j1}^{\lambda_l} = \lambda_l u_{j0}^{\lambda_l +} u_{j1}^{\lambda_l} + u_{j0}^{\lambda_l +} u_{j0}^{\lambda_l}. \tag{2.25}$$

From the hermiticity of $\eta i\beta^0$ and the reality of the mass spectrum, the left-hand side cancels the first term on the right-hand side. Thus, for all λ_l and j such that $k(j, \lambda_l) > 0$,

$$u_{j0}^{\lambda_l +} u_{j0}^{\lambda_l} = 0. \tag{2.26}$$

Now for $\lambda_l \neq 0$, this equation is equivalent to

$$\frac{1}{\lambda_l} u_{j0}^{\lambda_l +} i\beta^0 u_{j0}^{\lambda_l} = 0. \tag{2.27}$$

This is a contradiction for $\lambda_l > 0$, since the second factor is strictly positive by hypothesis. If $\lambda_l < 0$, the charge conjugate solution $v = C^{-1} u_{j0}^{\lambda_l}$ is of positive energy and satisfies

$$\frac{1}{|\lambda_l|} v^+ i\beta^0 v = 0, \tag{2.28}$$

also a contradiction. Therefore $k(j, \lambda_l) = 0$ for all $\lambda_l \neq 0$, and the lemma is proved.

As a consequence of this lemma and the charge conjugation invariance of the spectrum of $i\beta^0$, the minimal equation of $i\beta^0$ is

$$\prod_j [(i\beta^0)^2 - \lambda_j^2](i\beta^0)^q = 0, \tag{2.29}$$

where the product is over all distinct λ_j^2 and q is $\sup_j k(j, 0) + 1$. If we use the Lorentz transformation law of the β^μ, (2.2), (2.29) can be rewritten as

$$\left[\prod_j \left((\not{\xi})^2 - \left(\frac{m^2}{m_j^2}\right)\xi^2\right)\right](\not{\xi})^q = 0, \tag{2.30}$$

where ξ is any complex four component vector and the notation $\not{\xi} = i\beta^\mu \xi_\mu$ has been introduced. This relation is a generalization of one derived for the case of a single mass by Harish-Chandra in 1947 [**12**].

If we take for granted that the spin spectrum of the equation is simple, i.e. that for the mass m_j the representation of $ISL(2, C)$ is irreducible, say of spin s_j, then the nonzero λ_j have multiplicity $(2s_j + 1)$ and we can immediately evaluate

$$\det(-\not{\xi} + m) = \prod_j \left(\frac{m^2}{m_j^2}\right)^{2s_j+1} (-\xi^2 + m_j^2)^{2s_j+1} m^{N-\Sigma_j 2(2s_j+1)} \tag{2.31}$$

where, consistent with our previous convention, products and sums over j are over distinct values of m_j^2. If the spin spectrum is degenerate, (2.31) has an easy generalization which we will not write down.

At first sight, one might think that cases in which $N - \sum_j 2(2s_j + 1) > 0$ are in some sense unnatural, because one has introduced many "unnecessary" components to produce a given mass and spin spectrum. In fact most of the equations discussed in the physics literature have this character. From the point of view of the theory of partial differential equations, such wave equations are degenerate because the determinant is not of degree N in ξ. Thus, from a general point of view it is not surprising that small perturbations can change the characteristics of such equations.

Knowing the minimal equation for $\not{\xi}$, one can derive a number of useful relations. For example, there is a so-called Klein-Gordon divisor. It is so named because for a single mass, $m_j = m$, it is an $N \times N$ matrix, the solution $d(\xi)$ of

$$d(\xi)(-\not{\xi} + m) = (-\xi^2 + m^2)\mathbf{1}. \tag{2.32}$$

In this case, (see [**12**, Appendix C])

$$d(\xi) = (\not{\xi} + m)\left(\frac{\not{\xi}}{m}\right)^q - \left(\frac{\xi^2 - m^2}{m}\right)\sum_{r=0}^{q-1}\left(\frac{\not{\xi}}{m}\right)^r. \tag{2.33}$$

More generally, with a mass spectrum $m_1^2 \cdots m_n^2$, $d(\xi)$ is defined as the matrix satisfying

$$d(\xi)(-\not{\xi} + m) = \prod_{j=1}^{n} (-\xi^2 + m_j^2)\mathbf{1}. \tag{2.34}$$

For $n = 2$,

$$d(\xi) = \frac{1}{m}(-\xi^2 + m_1^2)(-\xi^2 + m_2^2)\sum_{r=0}^{q-1}\left(\frac{\not{\xi}}{m}\right)^r + (m + \not{\xi})\left(\frac{\not{\xi}}{m}\right)^q\left[\left(\frac{m_1^2}{m^2}\right)\left(\frac{m_2^2}{m^2}\right)\not{\xi}^2 - \xi^2\left(\frac{m_1^2}{m^2} + \frac{m_2^2}{m^2}\right) + \frac{m_1^2 m_2^2}{m^2}\right].$$

For $n = 3$,

$$d(\xi) = \frac{}{m} \prod_{j=1} (-\xi^2 - m_j^2) \sum_{r=0}^{q-1} \left(\frac{\not{\xi}}{m}\right)^r + (m + \not{\xi}) \left(\frac{\not{\xi}}{m}\right)^q \left(\prod_{j=1}^{3} \frac{m_j^2}{m^2}\right)$$

(2.35)
$$\cdot \left\{ (\not{\xi})^4 + (\not{\xi})^2 \left[m^2 - \xi^2 \sum_{j=1}^{3} \left(\frac{m^2}{m_j^2}\right) \right] + \left[(m^2)^2 - m^2 \xi^2 \sum_{j=1}^{3} \left(\frac{m^2}{m_j^2}\right) + (\xi^2)^2 \sum_{j<k} \left(\frac{m^4}{m_j^2 m_k^2}\right) \right] \right\}.$$

$$\vdots$$

Clearly, the formula for the Klein-Gordon divisor makes it possible to write a simple expression for the fundamental solutions of the wave equation in terms of a fundamental solution of a scalar ($N = 1$) wave equation

$$(\beta^\mu \partial_\mu + m) S_R(x) = \delta(x)\mathbf{1}, \tag{2.36}$$

where

$$S_R(x) = d(i\partial) \Delta_R(m_1^2, \ldots, m_n^2, x) \tag{2.37}$$

and

$$\Delta_R(m_1^2, \ldots, m_n^2, x) = (2\pi)^{-4} \int d^4p \left[\prod_{j=1}^{n} (-p^2 + m_j^2) \right]^{-1} \exp(-ipx) \tag{2.38}$$

is the retarded fundamental solution satisfying

$$\prod_{j=1}^{n} (\Box + m_j^2) \Delta_R(x) = \delta(x). \tag{2.39}$$

A second set of useful formulae can be derived from the knowledge of the minimal equation for $i\beta^0$. Let $u_j^{\lambda_l}, j = 1, \ldots, 2s_l + 1$ denote the solutions of

$$i\beta^0 u_j^{\lambda_l} = \lambda_l u_j^{\lambda_l} \tag{2.40}$$

orthogonalized for distinct j and normalized so that

$$u_j^{\lambda_l +} u_k^{\lambda_l} = \delta_{jk} \operatorname{sgn} \lambda_l (-1)^q. \tag{2.41}$$

(They are already orthogonal for distinct l:

$$u_j^{\lambda_l +} u_k^{\lambda_{l'}} = 0, \qquad l \neq l'.) \tag{2.42}$$

Then, for $\lambda_l > 0$,

$$\sum_j u_j^{\lambda_l} \otimes u_j^{\lambda_l +} = \prod_{j \neq l} \frac{(i\beta^0 - \lambda_j)}{\lambda_l - \lambda_j} \left(\frac{i\beta^0}{\lambda_l}\right)^q \tag{2.43}$$

while, for $\lambda_l < 0$,

$$\sum_j u_j^{\lambda_l} \otimes u_j^{\lambda_l +} = (-1)^q \prod_{j \neq l} \frac{(i\beta^0 - \lambda_j)}{\lambda_l - \lambda_j}\left(\frac{i\beta^0}{\lambda_l}\right)^q. \tag{2.44}$$

Here the products on the right side run over all distinct λ_j, positive as well as negative. Equation (2.43) is true because both sides are projection operators annihilating all eigenvectors and generalized eigenvectors of eigenvalues different from λ_l and equal to one on the eigenvectors of λ_l. The same holds for (2.44) when $(-1)^q = 1$, i.e. when the representation of $SL(2, \mathbf{C})$ associated with the equation is double valued as a representation of the Lorentz group. When $(-1)^q = -1$, each side is the negative of a projection operator and the argument goes through with this slight change.

By multiplying (2.43) and (2.44) on the left by $S(A)$ and the right by $S(A^{-1})$ and choosing A appropriately we convert them into formulae for the projection onto solutions of mass $m_l = m/|\lambda_l|$ and sign of energy sgn λ_l.

$$\begin{aligned} E_{m_l,+}(p) &= \sum_j u_j^{\lambda_l}(p) \otimes u_j^{\lambda_l}(p)^+ \\ &= \prod_{j \neq l} \frac{(p\!\!\!/ /(p^2)^{1/2})^2 - \lambda_j^2}{\lambda_l^2 - \lambda_j^2} \frac{(p\!\!\!/ /p^2 + \lambda_l)}{2\lambda_l}\left(\frac{p\!\!\!/}{(p^2)^{1/2}\lambda_l}\right)^q, \end{aligned} \tag{2.45}$$

$$\begin{aligned} E_{m_l,-}(p) &= (-1)^q \sum_j u_j^{\lambda_l}(p) \otimes u_j^{\lambda_l}(p)^+ \\ &= \prod_{j \neq l} \frac{(p\!\!\!/ /(p^2)^{1/2})^2 - \lambda_j^2}{\lambda_l^2 - \lambda_j^2} \frac{(p\!\!\!/ /(p^2)^{1/2} - |\lambda_l|)}{-2|\lambda_l|}\left(\frac{p\!\!\!/}{(p^2)^{1/2}|\lambda_l|}\right)^q. \end{aligned} \tag{2.46}$$

Here the products are over $\lambda_j^2 \neq \lambda_l^2$. The charge conjugation matrix C establishes a connection between these two projections

$$\begin{aligned} C^{-1}\overline{E_{m_j,+}(p)}C &= (-1)^q \sum_j C^{-1}\overline{u_j^{\lambda_l}(p)} \otimes [C^{-1}\overline{u_j^{\lambda_l}(p)}]^+ \\ &= E_{m_m,-}(-p). \end{aligned} \tag{2.47}$$

In the following section, use will be made of the projection operators onto the space spanned by all positive energy solutions of momentum p, $p^0 > 0$, and the projection onto the space spanned by all negative energy solutions of momenta $(-p)$. The notation which will be used for these is

$$\Pi(p) = \sum_l E_{m_l,+}(p), \tag{2.48}$$

$$\Pi(-p) = \sum_l E_{m_l,-}(-p). \tag{2.49}$$

The relation between Π and its adjoint is

$$\Pi(p)^* = \eta\Pi(p)\eta^{-1},$$

$$\Pi(-p)^* = \eta\Pi(-p)\eta^{-1}.$$

To complete this section we give some examples of wave equations satisfying the assumptions of this section.

P(etiau)-D(uffin)-K(emmer) Spin 0 [**13**]. Here ψ has $N = 5$ components, a scalar φ, and four vector components φ^μ. The equations are just (1.3) scaled to fit the general first order formalism

$$\begin{aligned} \partial^\mu \varphi_\mu + m\varphi &= 0, \\ -\partial^\mu \varphi + m\varphi^\mu &= 0, \end{aligned} \tag{2.50}$$

so

$$\beta^0 = \left\{\begin{array}{c|cccc} 0 & 1 & 0 & 0 & 0 \\ \hline -1 & 0 & 0 & 0 & 0 \\ 0 & 0 & 0 & 0 & 0 \\ 0 & 0 & 0 & 0 & 0 \\ 0 & 0 & 0 & 0 & 0 \end{array}\right\}, \text{ etc.} \tag{2.51}$$

The representation of the $ISL(2, \boldsymbol{C})$ is

$$S = \mathscr{D}^{(0,0)} \oplus \mathscr{D}^{(1/2,1/2)}, \tag{2.52}$$

i.e.,

$$S(A) = \left\{\begin{array}{c|cccc} 1 & 0 & 0 & 0 & 0 \\ \hline 0 & & & & \\ 0 & & \Lambda(A) & & \\ 0 & & & & \\ 0 & & & & \end{array}\right\}. \tag{2.53}$$

(The irreducible representations of $SL(2, \boldsymbol{C})$ are labeled $\mathscr{D}^{(j_1,j_2)}$ where j_1 and j_2 take the values $0, \frac{1}{2}, \frac{3}{2}, 2, \ldots$ and $\overline{\mathscr{D}^{(j_1,j_2)}}$ is equivalent to $\mathscr{D}^{(j_2,j_1)}$.) Here

$$\det(-\not\xi + m) = m^3(-\xi^2 + m^2). \tag{2.54}$$

Dirac Spin $\frac{1}{2}$. Here the equation is (1.4) with $\gamma^\mu = i\beta^\mu$. The representation of $ISL(2, \boldsymbol{C})$ is

$$S = \mathscr{D}^{(1/2,0)} \oplus \mathscr{D}^{(0,1/2)} \tag{2.55}$$

and

$$\det(-\not\xi + m) = (-\xi^2 + m^2)^2. \tag{2.56}$$

PDK Spin 1 [**13**]. Here the equations arise from the following equations of Proca:

$$\begin{aligned} \partial^\kappa F_{\kappa\lambda} + m\varphi_\lambda &= 0, \\ -(\partial_\kappa \varphi_\lambda - \partial_\lambda \varphi_\kappa) + mF_{\kappa\lambda} &= 0. \end{aligned} \tag{2.57}$$

If we introduce the ten component wave function

$$\psi = \{\varphi^0, \varphi^1, \varphi^2, \varphi^3, F^{01}, F^{02}, F^{03}, F^{23}, F^{31}, F^{12}\}$$

then the β's are 10×10 matrices, e.g.

$$\beta^0 = \left\{ \begin{array}{cccc|cccccc} 0 & 0 & 0 & 0 & 0 & 0 & 0 & 0 & 0 & 0 \\ 0 & 0 & 0 & 0 & 1 & 0 & 0 & 0 & 0 & 0 \\ 0 & 0 & 0 & 0 & 0 & 1 & 0 & 0 & 0 & 0 \\ 0 & 0 & 0 & 0 & 0 & 0 & 1 & 0 & 0 & 0 \\ \hline 0 & -1 & 0 & 0 & & & & & & \\ 0 & 0 & -1 & 0 & & & & & & \\ 0 & 0 & 0 & -1 & & & & 0 & & \\ 0 & 0 & 0 & 0 & & & & & & \\ 0 & 0 & 0 & 0 & & & & & & \\ 0 & 0 & 0 & 0 & & & & & & \end{array} \right\}. \tag{2.58}$$

The representation of $SL(2, \boldsymbol{C})$ is

$$S = \mathscr{D}^{(1/2,1/2)} \oplus \mathscr{D}^{(0,1)} \oplus \mathscr{D}^{(1,0)}, \tag{2.59}$$

and

$$\det(-\not\xi + m) = m^4(-\xi^2 + m^2)^3. \tag{2.60}$$

Fierz-Pauli Spin $\frac{3}{2}$ [**14**]. Here ψ has sixteen components, $\psi_{\alpha\mu}$, $\alpha = 1, 2, 3, 4$, $\mu = 0, 1, 2, 3$, and transforms like the tensor product of a Dirac spinor and a vector

$$\begin{aligned} S &= (\mathscr{D}^{(0,1/2)} \oplus \mathscr{D}^{(1/2,0)}) \otimes \mathscr{D}^{(1/2,1/2)} \\ &= \mathscr{D}^{(1/2,0)} \oplus \mathscr{D}^{(0,1/2)} \oplus \mathscr{D}^{(1,1/2)} \oplus \mathscr{D}^{(1/2,1)}. \end{aligned} \tag{2.61}$$

Here the β^μ matrices are

$$(\beta^\mu)^\kappa{}_\lambda = i[\varepsilon^{\kappa\mu}{}_{\lambda\nu}\gamma^5\gamma^\nu - \tfrac{1}{3}\varepsilon^{\kappa\mu}{}_{\rho\nu}\gamma^5\gamma^\nu\gamma^\rho\gamma_\lambda] \tag{2.62}$$

and the hermitizing matrix η is

$$(\eta)^\kappa{}_\lambda = \eta_{\text{Dirac}}(g^\kappa{}_\lambda - \tfrac{1}{3}\gamma^\kappa\gamma_\lambda), \tag{2.63}$$

γ^5 is defined as $\gamma^5 = i\gamma^0\gamma^1\gamma^2\gamma^3$.

For spin zero and one $i\beta^0$ has no elementary divisors so the minimal equation is

$$(-\not\xi^2 + \xi^2)\not\xi = 0. \tag{2.64}$$

For spin $\frac{3}{2}$, on the other hand, $i\beta^0$ has an eigensubspace of dimension 8 in which there are four eigenvectors of zero eigenvalue each accompanied by a generalized eigenvector so $q = 2$ and the minimal equation is

$$(-\not\xi^2 + \xi^2)(\not\xi)^2 = 0. \tag{2.65}$$

The determinant of the equation is given by

$$\det(-\not\xi + m) = m^8(-\xi^2 + m^2)^4. \tag{2.66}$$

For further examples and references see [**15**].

3. **Free fields for the single particle equations.** With the machinery of the single particle theory in hand, it is easy to construct the mappings $\Pi_\pm$ which appear in definitions (1.40) and (1.41) of the free field. For this purpose it is convenient to take the momentum space form of $\mathscr{H}_1$. Then $\Phi \in \mathscr{H}_1$ is an N-component function, defined on the positive energy mass shell satisfying

$$\Pi(p)\Phi(p) = \Phi(p) \tag{3.1}$$

for almost all p on the mass shell, and square integrable with respect to the scalar product

$$\sum_l \int d\Omega_{m_l}(p)\Phi(p)^+\Psi(p) = (\Phi, \Psi) \tag{3.2}$$

where $d\Omega_m(p)$ is the Lorentz invariant volume element on $p^2 = m^2$,

$$d\Omega_m(p) = d^3p/(m^2 + \mathbf{p}^2)^{1/2}. \tag{3.3}$$

The dual of $\mathscr{H}_1$ is composed of all χ satisfying the same conditions except that (3.1) is replaced by

$$\Pi(p)^T\chi(p) = \chi(p). \tag{3.4}$$

The linear functional χ is expressed as

$$\langle \chi, \Phi \rangle = \sum_l \int d\Omega_{m_l}(p)\chi(p)\Phi(p), \tag{3.5}$$

and therefore the antilinear mapping J from the linear functionals on $\mathscr{H}_1$ to vectors of $\mathscr{H}_1$ is given by

$$(J\chi)(p) = \eta^{-1}\overline{\chi(p)} \tag{3.6}$$

and its inverse by

$$(J^{-1}\Phi)(p) = \bar{\eta}\overline{\Phi(p)}. \tag{3.7}$$

The mapping Π_+ from the test function space to the dual of $\mathscr{H}_1$ is

$$(3.8)\qquad (\Pi_+ f)(p) = \pi^{1/2}\Pi(p)^T f(p),$$

i.e. it is Fourier transformation of f followed by restriction to the positive energy mass shell followed by multiplication by the transpose of the matrix $\Pi(p)$ defined in (2.48). The constant in the formula has been fixed so as to produce the conventional results for the commutation relations as will be seen shortly.

The verification of the transformation law (1.43) for Π_+ is a straightforward computation:

$$(3.9)\qquad \begin{aligned}(J^{-1}U_1(a, A)J\Pi_+ f)(p) &= \bar{\eta}\overline{(U_1(a, A)J\Pi_+ f)(p)} \\ &= \bar{\eta}\exp(-ip\cdot a)\overline{S(A)}\bar{\eta}^{-1}(\Pi_+ f)(\Lambda(A^{-1})p).\end{aligned}$$

But (2.11) implies

$$(3.10)\qquad \overline{\eta S(A)\eta^{-1}} = S(A^{-1})^T$$

and we have also

$$(3.11)\qquad S(A)\Pi(p)S(A^{-1}) = \Pi(\Lambda(A)p).$$

So (3.9) is

$$(3.12)\qquad \begin{aligned}&= \pi^{1/2}S(A^{-1})^T\exp(-ip\cdot a)\Pi(\Lambda(A^{-1})p)^T\hat{f}(\Lambda(A^{-1})p) \\ &= \pi^{1/2}\Pi(p)^T\exp(-ip\cdot a)S(A^{-1})^T\hat{f}(\Lambda(A^{-1})p) = (\Pi_+\{a, A\}f)(p).\end{aligned}$$

The definition of Π_- as a map from the test function space to $\mathscr{H}_{\bar{1}}$ is

$$(3.13)\qquad (\Pi_- f)(p) = \pi^{1/2}\Pi(p)(C^T\eta)^{-1}\hat{f}(-p).$$

The verification of the transformation law is in this case

$$(3.14)\qquad (U_1(a, A)\Pi_- f)(p) = \pi^{1/2}\exp(ip\cdot a)S(A)\Pi(\Lambda(A^{-1})p)(C^T\eta)^{-1}\hat{f}(-\Lambda(A^{-1})p)$$

$$(3.15)\qquad \begin{aligned}&= \pi^{1/2}\Pi(p)\exp(-ip\cdot a)S(A)(C^T\eta)^{-1}\hat{f}(-(A^{-1})p) \\ &= \pi^{1/2}\Pi(p)(C^T\eta)^{-1}\exp(-ip\cdot a)S(A^{-1})^T\hat{f}(-(A^{-1})p) = (\Pi_-\{a, A\}f)(p).\end{aligned}$$

It remains to be shown that the commutation relations (1.46) for the free fields have, in fact, the local form expressed in (1.47) and (1.48). To see this we compute

$$(3.16)\qquad \begin{aligned}\langle\Pi_+ f, J\Pi_+\bar{\eta}g\rangle &= \sum_l\int d\Omega_{m_l}(p)(\Pi_+ f)(p)(J\Pi_+\bar{\eta}g)(p) \\ &= \pi\sum_l\int d\Omega_{m_l}(p)\hat{f}(p)\Pi(p)\eta^{-1}\Pi(p)^*\eta\hat{g}(-p) \\ &= \pi\sum_l\int d\Omega_{m_l}(p)\hat{f}(p)\Pi(p)\hat{g}(-p) = \iint d^4x\, d^4y f(x)\frac{1}{i}S^{(+)}(x-y)g(y),\end{aligned}$$

where

$$S^{(+)}(x) = \frac{i}{2(2\pi)^3} \int \sum_l d\Omega_{m_l}(p)\Pi(p)\exp(-ip \cdot x) \tag{3.17}$$

and $\Pi(p)^2 = \Pi(p)$, $\eta^{-1}\Pi(p)^*\eta = \Pi(p)$ have been used.

Similarly, the second term on the right-hand side of (1.46) is

$$\begin{aligned} &\langle J^{-1}\Pi_-\bar{\eta}g, \Pi_- f\rangle \\ &\quad = \pi \sum_l \int d\Omega_{m_l}(p)[\overline{\bar{\eta}\Pi(p)}\,\overline{(C^T\eta)^{-1}}\eta\hat{g}(p)]\Pi(p)(C^T\eta)^{-1}\hat{f}(-p). \end{aligned} \tag{3.18}$$

Since

$$\bar{\eta}\overline{\Pi(p)}\eta^{-1} = \Pi(p)^T, \tag{3.19}$$

this is

$$\begin{aligned} &\pi \sum_l \int d\Omega_{m_l}(p)\hat{g}(p)\eta^T C\Pi(-p)\eta^{-1}(C^T)^{-1}\hat{f}(-p) \\ &\quad = \pi(-1)^\sigma \sum_l \int d\Omega_{m_l}(p)\hat{f}(-p)C^{-1}\overline{\Pi(p)}C\hat{g}(p) \\ &\quad = \pi(-1)^\sigma \sum_l \int d\Omega_{m_l}(p)\hat{f}(-p)\Pi(-p)\hat{g}(p) \\ &\quad = -(-1)^\sigma \iint d^4x\, d^4y f(x)\frac{1}{i}S^{(-)}(x-y)g(y), \end{aligned} \tag{3.20}$$

where

$$S^{(-)}(x) = \frac{-i}{2(2\pi)^3} \sum_l \int d\Omega_{m_l}(p)\Pi(-p)\exp(+ip \cdot x). \tag{3.21}$$

Consequently, we have (1.47) with

$$S(x) = S^{(+)}(x) + S^{(-)}(x). \tag{3.22}$$

It is not obvious that the $S(x)$ defined in this way coincides with the definition (1.48) as the difference of the advanced and retarded functions. However, it is true as a consequence of identities connecting the Klein-Gordon divisor with $\Pi(p)$ which will not be spelled out here.

One last remark on the free fields. With the one particle and one antiparticle subspaces identified, the field operator becomes equal to its own charge conjugate

$$\begin{aligned} \psi^c(f) &= \psi(C^T\bar{f})^* = a(\Pi_+C^T\bar{f})^* + a^*(\Pi_-C^T\bar{f})^* \\ &= a^*(J\Pi_+C^T\bar{f}) + a(J^{-1}\Pi_-C^T\bar{f}) = \psi(f), \end{aligned} \tag{3.23}$$

because

(3.24) $$J\Pi_+ C^T \bar{f} = \Pi_- f \quad \text{and} \quad J^{-1}\Pi_- C^T \bar{f} = \Pi_+ f$$

as one easily verifies using (3.6), (3.7) and (2.47).

4. **External field problem for quantized fields.** Our discussion will be based on the formula (1.68) expressing ψ in terms of ψ^{in} via the transformation of the test function space, T_R^{-1}. The essential result relating the mapping T_R to the existence of fundamental solutions is

THEOREM [6]. *A necessary and sufficient condition that there exist unique tempered distributions* $S_R(x, y; \mathscr{B})$ *and* $S_A(x, y; \mathscr{B})$

(a) *satisfying the differential equations*

$$[\beta^\mu \partial_\mu + m + \mathscr{B}(x)]S_{R;A}(x, y; \mathscr{B}) = \delta(x - y)\mathbf{1},$$

$$-\frac{\partial}{\partial y^\mu} S_{R;A}(x, y; \mathscr{B})\beta^\mu + S_{R;A}(x, y; \mathscr{B})(m + \mathscr{B}(y)) = \delta(x - y)\mathbf{1},$$

(b) *weakly retarded and advanced respectively,*

(c) *such that*

$$f \to \int f(x) S_{R;A}(x, y; \mathscr{B})\, d^4x$$

and

$$g \to \int S_{R;A}(x, y; \mathscr{B}) g(y)\, d^4y$$

define continuous mappings of $\mathscr{S}$ *into* $\mathscr{O}_M$
is that the transformations T_R *and* T_A *be bijections of* $\mathscr{S}$ *onto* $\mathscr{S}$.

Here the notation $\mathscr{O}_M$ stands for the C^∞ functions all of whose derivatives are polynomially bounded, and equipped with the usual topology.

Thus, by the argument described in the Introduction $\psi(f) = \psi^{\text{in}}(T_R^{-1}f)$ and $\psi^{\text{out}}(T_R^{-1}T_A f)$ are well-defined operator-valued tempered distributions provided that one has obtained unique weakly retarded and advanced fundamental solutions satisfying (c).

Using these fundamental solutions we can easily obtain explicit formulae for the inverse operators T_R^{-1} and T_A^{-1}. For example,

(4.1) $$(T_R^{-1}f)(x) = f(x) - \int d^4y f(y) S_R(y, x; \mathscr{B})\mathscr{B}(x).$$

To verify this formula one needs the identity

(4.2) $$S_R(x, y; \mathscr{B}) - \int S_R(x - z) d^4z \mathscr{B}(z) S_R(z, y; \mathscr{B}) = S_R(x - y),$$

which follows if one notes that the left-hand side satisfies the free differential equations and is weakly retarded. (A weakly retarded solution of the free equations is retarded, and the retarded solution is unique.)

Next we derive a formula for $T_R^{-1}T_A$ which is useful in scattering theory. Note that, directly from the definitions (1.66) and (1.67),

$$((T_A - T_R)f)(x) = -\int f(y)d^4yS(y-x)\mathscr{B}(x)$$

so

$$(T_R^{-1}T_Af)(x) = f(x) - T_R^{-1}\left[\int f(y)d^4yS(y-x)\mathscr{B}(x)\right]$$

$$= f(x) - \int d^4yf(y)S(y-x)\mathscr{B}(x) \tag{4.3}$$

$$+ \int d^4yf(y)S(y-z)\mathscr{B}(z)\,d^4zS_R(z,x;\mathscr{B})\mathscr{B}(x).$$

From this formula it is evident that ψ^{out} satisfies the free differential equation

$$\psi^{\text{out}}(f) = 0 \quad \text{when } f = (-\beta^{\mu T}\partial_\mu + m)h,\ h \in \mathscr{S}, \tag{4.4}$$

because the insertion of (4.3) in $\psi^{\text{in}}(T_R^{-1}T_Af)$ yields three terms of which the last two vanish, because $S(x)$ is a solution of the free equation while the first, $\psi^{\text{in}}(f)$ vanishes, and because ψ^{in} has been constructed to satisfy the free equation.

The formula also displays explicitly the fact that $T_R^{-1}T_Af$ is f plus terms that depend only on the value of $\hat{f}(p)$ restricted to the mass shell and projected onto the subspace satisfying the transposed Dirac equation of the appropriate momentum. This establishes the legitimacy of the definition (1.74) of a^{out} and b^{out} and yields an explicit form for the transformations appearing in (1.76).

$$\begin{aligned}(\Pi_+T_R^{-1}T_Af_\pm)(p) = \pi^{1/2}\Bigg[&\hat{f}_\pm(p)\Pi(p)\\ &\mp \frac{i}{4\pi}\sum_l \int d\Omega_{m_l}(q)\hat{f}_\pm(q)\Pi(q)\hat{\mathscr{B}}(q-p)\Pi(p)\\ &\pm \frac{i}{2(2\pi)^3}\sum_l \int d\Omega_{m_l}(q)\hat{f}_\pm(q)\Pi(q)\hat{\mathscr{B}}(q-r)\,d^4r\\ &\cdot \hat{S}_R(r,s;\mathscr{B})\,d^4s\hat{\mathscr{B}}(s-p)\Pi(p)\Bigg],\end{aligned} \tag{4.5}$$

$$(\Pi_- T_R^{-1} T_A f_\pm)(p) = \pi^{1/2}\Big[\Pi(p)(C^T\eta)^{-1}\hat{f}_\pm(-p)$$

$$\pm \frac{i}{4\pi}\sum_l \int d\Omega_{m_l}(q)\hat{f}_\pm(q)\Pi(q)\hat{\mathscr{B}}(q+p)(\eta^T C)^{-1}\Pi(p)^T \tag{4.6}$$

$$\mp \frac{i}{2(2\pi)^3}\sum_l \int d\Omega_{m_l}(q)\hat{f}_\pm(q)\Pi(q)\hat{\mathscr{B}}(q-r)\,d^4r$$

$$\cdot \hat{S}_R(r, s; \mathscr{B})\,d^4s\hat{\mathscr{B}}(s+p)(\eta^T C)^{-1}\Pi(p)^T\Big],$$

where for the upper sign the integrals run over the positive energy mass shell and over the negative energy mass shell for the lower sign. The last two terms in each of these expressions are integral operators with C^∞ kernels. There are analogous expressions for the transformations $\Pi_\pm{}^c(T_R^{-1}T_A{}^c(f)_-)$ appearing in the second equation of (1.76) giving b^{out} in terms of b^{in} and $b^{\text{in}*}$. Rather than writing them out we note that

$$({}^cf)_- = {}^c(f_+) \tag{4.7}$$

and that

$${}^c(T_R^{-1}T_A{}^c(f_+)) = (T_R^c)^{-1}(T_A^c)f_+ \tag{4.8}$$

where T_R^c and T_A^c are T_R and T_A but computed with the charge conjugate $B: B^c(x) = C^{-1}\overline{B(x)}C$. In passing, note that this remark has the consequence that for an external field invariant under charge conjugation, i.e. $B(x) = B^c(x)$, the antiparticle scattering amplitude (1.83) and (1.82) are equal.

We have already remarked in the Introduction that the manipulation necessary to establish the free field commutation relations

$$[\psi^{\text{out}}(x), \psi^{\text{out}+}(y)]_\pm = \frac{1}{i}S(x-y) \tag{4.9}$$

are elementary. (Furthermore, they are carried out in [**1**].) There is only one essential property of ψ^{out} to be established, the existence and uniqueness of Ψ_0^{out}. Let us sketch some of the steps involved. The direct study of the defining equations

$$[a^{\text{in}}(\Pi_+ T_R^{-1}T_A f_+) + b^{\text{in}*}(\Pi_- T_R^{-1}T_A f_+)]\Psi_0^{\text{out}} = 0, \tag{4.10}$$

$$[b^{\text{in}}(\Pi_+(T_R^c)^{-1}T_A^c f_+) + b^{\text{in}*}(\Pi_-(T_R^c)^{-1}T_A^c f_+)]\Psi_0^{\text{out}} = 0 \tag{4.11}$$

proceeds by projecting the left-hand sides onto the subspace with n_1 particles and n_2 antiparticles. Then the following lemma appears immediately.

Lemma [**9**]. *A necessary and sufficient condition for the uniqueness (up to a phase factor) of the solution of* (4.10) *and* (4.11) *is that*

$$\{\Pi_+ T_R^{-1}T_A f_+ ; f \in \mathscr{S}\} \tag{4.12}$$

and

(4.13) $$\{\Pi_+ T_R^{c-1} T_A^c f_+ ; f \in \mathscr{S}\}$$

be dense in $\mathscr{H}_1$.

As far as existence is concerned

LEMMA [9]. *For the existence of amplitudes* $\Psi_0^{\text{out}(n_1,n_2)}$ *satisfying* (4.10) *and* (4.11) *it suffices that a pair amplitude* $\Psi_0^{\text{out}(1,1)}$ *exist satisfying the equations for* $n_1 = 0, n_2 = 1$ *and for* $n_1 = 1, n_2 = 0$.

In fact, the general amplitudes for $n_1 = n_2 > 0$ may then be defined as determinants (antisymmetric case) or permanents constructed from the pair amplitude, while $\Psi_0^{\text{out}(0,0)} = 1$ and $\Psi_0^{\text{out}(n_1,n_2)} = 0$ for $n_1 \neq n_2$.

The final step is less elementary and involves applying the criteria of [9], to the integral operators defined by (4.5).

LEMMA [9]. *If the uniqueness conditions* (4.12) *and* (4.13) *are satisfied it is necessary and sufficient for the existence of* Ψ_0^{out} *and of a unitary S-operator that the integral operators on the space of single particle and antiparticle states defined by*

$$f_+ \to \Pi_- T_R^{-1} T_A f_+ \quad \textit{and} \quad f_+ \to \Pi_-(T_R^c)^{-1} T_A^c f_+$$

be Hilbert Schmidt.

That the criteria of these lemmas are in fact satisfied for the second terms of (4.5) and (4.6) is a simple consequence of relativistic kinematics and the fact that the kernel contains a factor $\mathscr{B}(p + q)$ where *both* p and q are of positive energy [5]. The analogous $\mathscr{B}(q - p)$ appearing $\Pi_+ T_R^{-1} T_A f_+$ does not lead to a Hilbert Schmidt kernel in general. For the last terms the restriction of the Lemma appears in general to be a further condition on the fundamental solutions.

It remains to verify the denseness of the sets (4.12) and (4.13). In the Bose-Einstein case (representation of $SL(2, C)$ single-valued as a representation of the Lorentz group) it turns out that that can be obtained from the commutation relations plus a little general argument [6]. On the other hand, in the Fermi-Dirac case, the change in sign in the commutation relations is crucial; the possibility remains that there is a finite-dimensional subspace of single particle states orthogonal to (4.12) and (4.13). Up to this point I have neither been able to show that this possibility is realized nor that it is excluded. Thus, there remains one point in the scattering theory to be elucidated.

5. **General remarks; open problems.** The results described in the preceding show how, using elementary quantum field theory, a satisfactory quantum mechanical interpretation of the solutions of the external field problem can be given in general, provided that weakly retarded and advanced fundamental solutions of the wave

equation exist with a certain regularity property. What is probably more interesting from the point of view of the theory of partial differential operators is the noncausality phenomenon itself. Let me describe first how we know that it actually occurs.

Consider the behavior of the fundamental solutions of (1.57) in the special case in which $\mathscr{B}$ is constant in an open set. Then the theory of equations with constant coefficients can be used in the open set to give a simple complete description of the characteristics. The result is a *sufficient* condition for noncausality: if the light cone opens up to a larger cone for such an external field there is noncausality.

Let me state precisely the results needed from the theory of PDO with constant coefficients. If $P(D)$ is an $N \times N$ matrix of partial differential operators, it is *hyperbolic with respect to a half-space* $n^\mu x_\mu \geqq 0$ if there exists an $N \times N$ matrix, E, of distributions in $\mathscr{D}'$ such that

$$P(D)E(x) = \delta(x)\mathbf{1} \tag{5.1}$$

and the support of E is in a cone lying, except for $x = 0$, entirely in $n^\mu x_\mu > 0$. The set of all such $P(D)$ we call $\mathrm{Hyp}_N(n)$ where the subscript refers to the number of rows and columns and the argument n to the half-space. The first step in the theory is to relate hyperbolicity for $N \times N$ systems of equations to hyperbolicity for a single equation. The key is the folk-lemma (which I learned from L. Gårding).

LEMMA.

$$P(D) \in \mathrm{Hyp}_N(n) \Leftrightarrow \det P(D) \in \mathrm{Hyp}_1(n). \tag{5.2}$$

This also works in a connected open set containing the origin. For a single equation a necessary and sufficient condition for hyperbolicity has been known for two decades. (See for example [**16**].)

THEOREM. $P(D) \in \mathrm{Hyp}_1(n) \Leftrightarrow$

(a) $P(n) \neq 0$.

(b) *There exists a* τ_0 *such that, for all real* ξ *and all* $\tau < \tau_0$,

$$P(\xi + i\tau n) \neq 0.$$

If $P(D)$ is homogeneous the τ_0 in (b) may be replaced by 0; for all real ξ,

$$P(\xi + \tau\eta) = 0$$

has only real roots in τ.

The last result which will be needed describes what lower order terms can be added to a hyperbolic principal part without destroying hyperbolicity.

THEOREM [**16**]. *If the principal part* $P_k(D)$ *of* $P(D)$ *is hyperbolic, then*

$$P(D) \in \mathrm{Hyp}_1(n) \Leftrightarrow P(D) \prec P_k(D).$$

If $P(D)$ *is written as a sum of homogeneous contributions*

$$P(D) = P_k(D) + P_{k-1}(D) + \cdots,$$

then

$$P(D) \prec P_k(D) \Leftrightarrow P_l(D) \prec P_k(D), \qquad l = 1, 2, \ldots, k-1.$$

If the roots of $P_k(\xi)$ are simple, then $P_l(D) \prec P_k(D)$ for all P_l with $l < k$. ($\prec$ is Hörmander's weaker than relation: if $P(\xi)$ and $Q(\xi)$ are polynomials Q is weaker than P, $Q \prec P$, if there exists a constant C such that

$$\tilde{Q}(\xi) \leqq C\tilde{P}(\xi) \tag{5.4}$$

for all ξ. Here

$$\tilde{Q}(\xi) = \left[\sum_{\alpha} |D^\alpha Q(\xi)|^2\right]^{1/2} \tag{5.5}$$

and analogously for $\tilde{P}$.)

The criterion of this theorem long known to be sufficient has recently been proved necessary by Svensson. With these results in hand, the test for noncausality is reduced to the classification of the external fields $\mathscr{B}$ for a given equation and the computation of $\det(-\not{p} + m + \mathscr{B})$.

Notice that of the examples in (2.54), (2.56), (2.60) and (2.66) only the Dirac equation yields a nonvanishing determinant when $m = \mathscr{B} = 0$, and only the PDK spin 0 has a simple root for p^2 where $\mathscr{B} = 0$, $m > 0$. Thus in all these cases except PDK spin 0, lower order terms could, a priori, destroy hyperbolicity.

Next consider the problem of classifying the $\mathscr{B}$. In the examples listed, the β^μ are irreducible so $\mathscr{B}$ may be expanded uniquely in a set of irreducible tensor quantities built from the β^μ. For example, for PDK spin 0 the general interaction is

$$\begin{aligned}\mathscr{B}(x) = {} & \sigma(x) - \tfrac{1}{3}(1 + \beta^\rho\beta_\rho)\rho(x) - eiA_\mu(x)\beta^\mu \\ & + B^\mu(x)[-\tfrac{1}{3}(1 + \beta^\rho\beta_\rho), \beta_\mu] + \frac{i}{4}F_{\mu\nu}(\beta^\mu\beta^\nu - \beta^\nu\beta^\mu) \\ & + \tfrac{1}{2}G_{\mu\nu}[\beta^\mu\beta^\nu + \beta^\nu\beta^\mu - \tfrac{1}{2}g^{\mu\nu}\beta^\rho\beta_\rho],\end{aligned} \tag{5.6}$$

where σ, ρ, A_μ, $\mathscr{B}_\mu$, $F_{\mu\nu}$, $G_{\mu\nu}$ are real, $F_{\mu\nu}$ is antisymmetric and $G_{\mu\nu}$ is symmetric and of trace 0. Thus there are 25 different kinds of coupling classified into 6 different tensor classes. Similarly, for the Dirac equation there are 16 different kinds of five different tensor types.

$$\begin{aligned}\mathscr{B}(x) = {} & \sigma(x) + i\gamma^5\rho(x) - eA^\mu(x)\gamma_\mu \\ & + B^\mu(x)\gamma_\mu\gamma^5 + \frac{i}{4}F^{\mu\nu}(\gamma_\mu\gamma_\nu - \gamma_\nu\gamma_\mu).\end{aligned} \tag{5.7}$$

I will not give the 100 kinds of 15 tensor types for PDK spin 1 nor the 256 kinds of 31 tensor types for Fierz Pauli spin $\frac{3}{2}$. They have been classified and listed explicitly by A. Glass [**17**].

For spin zero with the external fields given by (5.6),

$$\det(-\not p + m + \mathscr{B}) = (m + \rho)\det\,[mg^{\mu}{}_{\nu} - (G^{\mu}{}_{\nu} + iF^{\mu}{}_{\nu})] \tag{5.8}$$

$$+ \det\left\{\begin{array}{c|c} 0 & i(p_{\mu} + eA_{\mu}) + B_{\mu} \\ \hline i(p^{\mu} + eA^{\mu}) + B^{\mu} & [mg^{\mu}_{\nu} - G^{\mu}_{\nu} - iF^{\mu}_{\nu}] \end{array}\right\}.$$

The principal part comes from the second term and is

$$\begin{aligned} \mathscr{P}_2(p) = p_{\rho}[(m^2 - \tfrac{1}{2}(G^{\kappa\lambda}G_{\kappa\lambda} + F^{\kappa\lambda}F_{\kappa\lambda}))(-mg^{\rho\sigma} - G^{\rho\sigma}) \\ - m(G^{\rho}{}_{\kappa}G^{\kappa\sigma} - F^{\rho}{}_{\kappa}F^{\kappa\sigma}) - G^{\rho}{}_{\kappa}G^{\kappa}{}_{\mu}G^{\mu\sigma} \\ + G^{\rho}{}_{\kappa}F^{\kappa}{}_{\lambda}F^{\lambda\sigma} + F^{\rho}{}_{\kappa}G^{\kappa}{}_{\lambda}F^{\lambda\sigma} + F^{\rho}{}_{\kappa}F^{\kappa}{}_{\lambda}G^{\lambda\sigma}]p_{\sigma}. \end{aligned} \tag{5.9}$$

For some F and G, no matter how small their matrix elements, $\mathscr{P}_2(p) \geqq 0$ will include p's not satisfying $p^2 \geqq 0$, so the theory will in general be noncausal unless $F = G = 0$. On the other hand, with $F = G = 0$, the remaining coupling terms yield contributions of lower degree in p which are harmless because the characteristics are simple.

For the Dirac equation the characteristics are not simple but all the occurring lower order terms are weaker than the principal part, so no noncausality occurs in constant fields [**15**].

For PDK spin 1 it turns out that for minimal and dipole coupling

$$B(x) = -eiA_{\mu}(x)\beta^{\mu} + \frac{i\mu}{4}F_{\mu\nu}(x)(2 + \beta^{\rho}\beta_{\rho})(\beta^{\mu}\beta^{\nu} - \beta^{\nu}\beta^{\mu}) \tag{5.10}$$

there is no noncausality, but with quadrupole coupling

$$\mathscr{B}(x) = -iq(\partial_{\lambda}F_{\mu\nu}(x))(5 + 2\beta^{\rho}\beta_{\rho})\beta^{\mu}\beta^{\nu} \tag{5.11}$$

there is. From the physical point of view this result (due to Velo and Zwanziger) is more shocking than the corresponding statement for F and G coupling. The reason is that the F and G coupling when looked at in detail turn out to be rather strange velocity-dependent couplings. On the other hand, the electric quadrupole coupling, although idealized, is not unreasonable physically; there are several particles in nature one would have been pleased to describe causally using the coupling. Should we be worrying that this disease is really widespread and might also have infected the theory of coupled fields?

It is difficult to answer this question at this stage of development because we do not know whether the noncausal theory really exists, in general. Here is an open mathematical problem:

Do weakly retarded and advanced fundamental solutions exist for these systems even when the external fields destroy the hyperbolic character according to the criteria discussed above? Or possibly does the Levy phenomenon occur here: the lack of causality could be telling us that for "general" $\mathscr{B}(x) \in \mathscr{S}$ no tempered weakly retarded or advanced fundamental solution exists at all?

When this mathematical problem is solved, it will be time to worry about the implications of the answer for elementary particle theory.

In closing, I want to mention another ancient family of questions associated with these theories which may be related to the above questions. I say may be related because I am not sure. In fact, so far I have found this family completely opaque.

The questions originate in a remark of Fierz and Pauli in [**14**]. They consider equations for a vector field derivable from a Lagrangean density

$$\mathscr{L}(x) = \partial^\lambda \bar{\varphi}^\mu \partial_\lambda \varphi_\mu - m^2 \bar{\varphi}^\lambda \varphi_\lambda - (1 + b)\, \partial^\lambda \bar{\varphi}^\mu \partial_\mu \varphi_\lambda, \tag{5.12}$$

where b is some external field. The Euler equations are

$$\partial^\lambda(\partial_\lambda \varphi_\mu - \partial_\mu \varphi_\lambda) - \partial^\lambda(b\, \partial_\mu \varphi_\lambda) + m^2 \varphi_\mu = 0, \tag{5.13}$$

and they have the property that when $b = 0$, the identity

$$\partial^\mu \varphi_\mu = 0 \tag{5.14}$$

follows. (Take the divergence of (5.13).) Thus in the absence of the external field the time derivative of φ^0 is fixed in terms of the space derivatives of $\varphi^1, \varphi^2, \varphi^3$. Cauchy data for the initial value problem have to be given subject to this constraint. In the presence of an external field an analogous calculation yields an equation from which $\partial^0 \varphi_0$ is not determinable in terms of the space derivatives of $\varphi^1, \varphi^2, \varphi^3$. The Cauchy problem shows a discontinuity at zero external field. Later on P. Federbush [**18**] showed that a similar phenomenon takes place with minimal coupling and a certain form of spin 2 equation. There is vast untidy physics literature devoted to this matter of constraints in the presence of an external field. On the other hand, the weakly retarded fundamental solution solves the inhomogeneous Cauchy problem with null initial data at $-\infty$ and therefore might not say anything at all about such constraints at finite times. My second open problem is:

Are discontinuities in the Cauchy problem at finite times related to the eyistence of weakly retarded (advanced) fundamental solutions which are not retarded (advanced)?

References

For a previous attempt to provide entry into this subject for a mathematically minded reader see:

1. A. S. Wightman, *Partial differential equations and relativistic quantum field theory*, Lectures in Differential Equations, vol. II, A. K. Aziz (Editor), Van Nostrand, Princeton, N.J., 1969, pp. 1–52. Some calculations omitted in the present notes are carried out in detail there.

2. A. S. Wightman and L. Gårding, *Fields as operator-valued distributions in relativistic quantum theory*, Ark. Fys. **28** (1964), 129–184.

3. I. E. Segal, *Mathematical problems of relativistic physics*, Lectures in Appl. Math., vol. 2, Amer. Math. Soc., Providence, R.I., 1963. MR **26** #1774.

4. G. Velo and D. Zwanziger, *Propagation and quantization of Rarita-Schwinger waves in an external electromagnetic potential*, Phys. Rev. **186** (1969), 1337–1341. *Noncausality and other defects of interaction*

Lagrangians, for particles with spin one and higher, Phys. Rev. **188** (1969), 2218–2222.

5. See for a special case, however, P. Minkowski and R. Seiler, *Massive vector meson in external fields*, Phys. Rev. **D4** (1971), 359.

6. A. S. Wightman, *Qausi-local and local solutions of the external field problem and their scattering theories* (in preparation). Earlier published results for special cases are contained in [**7**] and [**8**].

7. A. Capri, Thesis, Princeton University, Princeton, N.J., 1967 (unpublished). *Electron in a given time-dependent electromagnetic field*, J. Mathematical Phys. **10** (1969), 575–580.

8. B. Schroer, R. Seiler and J. A. Swieca, *Problems of stability for quantum fields in external time dependent potentials*, Phys. Rev. **D2** (1970), 2927.

9. P. Kristensen, L. Mejlbo and E. T. Poulsen, *Tempered distributions in infinitely many dimensions*. I. *Canonical field operators*, Comm. Math. Phys. **1** (1965), 175–214; II. *Displacement operators*, Math. Scand. **14** (1964), 129–150; III. *Linear transformations of field operators*, Comm. Math. Phys. **6** (1967), 29–48. MR **31** #5085; #5086; **36** #1170.

10. D. Kwoh, Senior Thesis, Princeton University, Princeton, N.J., 1970 (unpublished).

11. E. R. Speer, *Generalized Feynman amplitudes*, Ann. of Math. Studies, no. 62, Princeton Univ. Press, Princeton, N.J.; Univ. of Tokyo Press, Tokyo, 1969. MR **39** #2422.

12. Harish-Chandra, *On relativistic wave equations*, Phys. Rev. (2) **71** (1947), 793–805. MR **8**, 554.

13a. G. Petiau, *Contribution à la théorie des équations d'ondes corpusculaires*, Acad. Roy. Belg. **16** (1936), fasc. 2.

13b. R. J. Duffin, *On the characteristic matrices of covariant systems*, Phys. Rev. **54** (1938), 1114.

13c. N. Kemmer, *The particle aspect of meson theory*, Proc. Roy. Soc. London **173A** (1939), 91–116.

14. M. Fierz and W. Pauli, *On relativistic wave equations for particles of arbitrary spin in an electromagnetic field*, Proc. Soc. London **173A** (1939), 211–232.

15. A. S. Wightman, *The stability of representations of the Poincaré group*, Proc. Fifth Coral Gables Conference, 1968, pp. 291–312.

16. L. Hörmander, *Linear partial differential operators*, Die Grundlehren der math. Wissenschaften, Band 116, Academic Press, New York; Springer-Verlag, Berlin, 1963. MR **28** #4221.

17. A. Glass, Thesis, Princeton University, Princeton, N.J., 1971 (unpublished); Comm. Math. Phys. **23** (1971), 176–184.

18. P. Federbush, *Minimal electromagnetic coupling for spin two particles*, Nuovo Cimento (10) **19** (1961), 572–573. MR **22** #11902.

PRINCETON UUNIVERSITY

L^p-L^q-ESTIMATES FOR SINGULAR INTEGRAL OPERATORS ARISING FROM HYPERBOLIC EQUATIONS

WALTER LITTMAN[1]

1. Recently R. Strichartz has obtained some L^p-estimates for the wave equation. In [**9**] he estimates the L^q-norm at time $t > 0$ in terms of the L^p-norm of the Cauchy data at $t = 0$. These results are applied in [**10**] to obtain analogous estimates for the nonhomogeneous equation, and these in turn are used to obtain a number of applications to existence theorems for nonlinear equations, the existence of wave operators for nonlinear equations, and other problems. The basic method used in [**9**] consists in expressing the solution operator in terms of Bessel functions and interpolating with respect to the index of the Bessel function. It is the use of Bessel functions which restricts this method to equations having spherical symmetry.

We shall restrict ourselves to estimates of the L^p-$L^{p'}$-type, where the $L^{p'}$-norm of the image is estimated in terms of the L^p-norm of the initial function $(1 \leqq p \leqq 2)$ $1/p + 1/p = 1$. L^p-L^p-estimates can be obtained similarly as well as other means (see for example [**5**], [**11**]) and combined with our present results to yield more general L^p-L^q-estimates.

2. **Kernels with singularities on hypersurfaces.** Let S be a smooth $n - 1$ surface smoothly embedded in R^n; let $s(x)$ be a real valued function defined in a neighborhood $\mathcal{N}$ of S such that grad $s(x) \neq 0$ on S; and let $\varphi(x)$ be a C^∞ (or "sufficiently smooth") function with compact support contained in $\mathcal{N}$. We consider the distribution $\beta(s(x))\varphi(x)$ defined on R^n induced by the one-dimensional distribution $\beta(\cdot)$. Specifically, we shall take for $\beta(\cdot)$ the one-dimensional distributions

$$\frac{1}{\Gamma(1-\alpha)}(x_1^{-\alpha})_{x_1>0} \equiv H_+^{(\alpha)}(x_1), \qquad \frac{1}{\Gamma(1-\alpha)}(x_1^{-\alpha})_{x_1<0} \equiv H_-^{(\alpha)}(x_1).$$

AMS 1970 *subject classifications*. Primary 44A25, 35L30; Secondary 42A18.

[1] This work was supported by Air Force Grant 883–67.

The Γ represents the gamma function and the distributions $x_1^{-\alpha}$ are defined by analytic continuation as in [**3**]. $H_+^{(\alpha)}$ is the analytic continuation of the αth derivative of the Heaviside step function, while $H_-(x_1) \equiv H_+(-x_1)$. Suppose k of the $n-1$ principal curvatures of S are bounded away from zero. Then we have

THEOREM 1. *Let the operator $T_+^{(\alpha)}$ be defined as convolution with the kernel $H_\pm^{(\alpha)}(s(x))\varphi(x)$. Then $T_+^{(\alpha)}: L^p(R^n) \to L^{p'}(R^n)$ is a bounded operator for $p' \leqq p_0' \equiv (2+k)/\alpha$, $1/p + 1/p' = 1$.*

OUTLINE OF PROOF (FOR $p' = p_0'$). For $\operatorname{Re}\alpha = 0$ the kernel is bounded and has bounded support, yielding $T_\pm^{(\alpha)}: L^1 \to L^\infty$. In general (lemma) the Fourier transform of the kernel appearing in the statement of the theorem is $O(|\xi|^{\operatorname{Re}\alpha - 1 - (k/2)})$. (For a special case of this lemma see [**5**].) Applying E. Stein's interpolation theorem ("with a complex parameter") [**11**] gives us the theorem.

REMARK. Let $\log^{(\alpha)}(x_1)$ denote the analytic continuation of the αth derivative of $\log(x_1)$. Then the above theorem holds for the kernel $\log^{(\alpha)}(s(x))\varphi(x)$ for $2 \leqq p' < p_0'$.

3. **Fourier multipliers**

THEOREM 2. *Let $\varphi(\xi)$, $\psi(\xi)$ be C^∞ (or "sufficiently smooth") positively homogeneous functions of degree zero. Suppose moreover that $|\xi|\psi(\xi)$ is the support function of a smooth convex body with positive gaussian curvature. Then*

$$M(\xi) \equiv |\xi|^{-a}\varphi(\xi)e^{i|\xi|\psi(\xi)}, \qquad a \geqq 0.$$

is a Fourier multiplier from $L^p(R^n) \to L^{p'}_{\mathrm{loc}}(R^n)$ for $p > p_0$ where $1/p_0 = 1/2 + a/(n+1)$. (Equality holds in some cases; see below.)

OUTLINE OF PROOF. One first obtains (lemma) asymptotic expansions (near infinity) for the Fourier transforms of the distributions discussed in §2 (as was done in [**1**] for the case $H_+^{(1)}(x_1) = \delta(x)$ or $\log^{(1)}(x_1) = 1/x_1$). One then represents $M(\xi)$ as a sum of two functions, each of which is the first (or "worst") term of the asymptotic expansion of one of the three types of kernels described in §2. If a kernel of "log type" has to be used one gets only the strict inequality $p > p_0$ in the statement of the theorem. Otherwise the theorem is valid for $p \geqq p_0$. Specifically, equality holds if $a + (n/2)$ is not an integer.

THEOREM 2′. *Suppose we have two functions*

$$M_j(\xi) \equiv |\xi|^{-a}\varphi_j(\xi)e^{i|\xi|\psi_j(\xi)}, \qquad j = 1, 2,$$

satisfying the conditions of the previous theorem. If $\psi_1(\xi) = -\psi_2(-\xi)$, $\varphi_1(\xi) = -\varphi_2(-\xi)$, then $M(\xi) = M_1(\xi) + M_2(\xi)$ is a Fourier multiplier $L^p \to L^{p'}_{\mathrm{loc}}$ for $p \geqq p_0$.

4. **Applications to hyperbolic equations.** The results of either §2 or §3 can be

used to give L^p-$L^{p'}$-estimates for the solutions of strictly hyperbolic equations with constant coefficients and no lower order terms. If the (even) order of the equation is m, suppose we prescribe $D_t^k u(x, 0) = 0$ for $0 \leqq k \leqq m - 2$ and $D_t^{m-1} u(x, 0) = \varphi$. Then the solution operator $\varphi(\cdot) \to u(\cdot, t)$ can be expressed in terms of a Fourier multiplier of the type mentioned in Theorem 2′ (see [**2**, p. 75]). Alternately one may decompose the kernel R of the solution operator as a sum of terms of the type occurring in Theorem 1 plus a bounded remainder with compact support, and apply Theorem 1. If the normal surfaces are convex and have positive curvature, the kernel R has the form

$$R = \sum_{j=0}^{N} H_{\pm}^{((n+3)/2-m-j)}(s(x))\varphi_j(x) + \text{harmless},$$

as can be deduced, for example, from [**7**], [**8**] (see also [**4**]).

The case of variable coefficients is the same in principle, although some of the technical details differ. One combines the known energy estimates for the solution with a representation for R, similar to the one above, valid for small t. Interpolation is affected by introducing a differentiation operator (in x) of complex order α and interpolating with respect to α.

Bibliography

1. O. Arena and W. Littman, *Far field behavior of solution to partial differential equations: asymptotic expansions and maximal rates of decay along a ray*, Ann. Scuola Norm. Sup. Pisa (to appear).

2. L. Bers, F. John and M. Schechter, *Partial differential equations*, Lectures in Appl. Math., vol. 3, Interscience, New York, 1964. MR **29** #346.

3. I. M. Gel′fand and Z. Ja. Šapiro, *Homogeneous functions and their extensions*, Uspehi Mat. Nauk **10** (1955), no. 3 (65), 3–70; English transl., Amer. Math. Soc. Transl. (2) **8** (1958), 21–85. MR **17**, 371; MR **20** #1061.

4. J. Leray, *Un prolongement de la transformation de la place qui transforme la solution unitaire d'un opérateur hyperbolique en sa solution élémentaire* (*Problème de Cauchy*. IV), Bull. Soc. Math. France **90** (1962), 39–156. MR **26** #1625.

5. W. Littman, *Fourier transforms of surface-carried measures and differentiability of surface averages*, Bull. Amer. Math. Soc. **69** (1963), 766–770. MR **27** #5086.

6. ———, *Decay at infinity of solutions to partial differential equations with constant coefficients*, Trans. Amer. Math. Soc. **123** (1966), 449–459. MR **33** #6110.

7. D. A. Ludwig, *Exact and asymptotic solutions of the Cauchy problem*, Comm. Pure Appl. Math. **13** (1960), 473–508. MR **22** #5816.

8. ———, *Singularities of superpositions of distributions*, Pacific J. Math. **15** (1965), 215–239. MR **31** #1553.

9. R. S. Strichartz, *Convolutions with kernels having singularities on a sphere*, Trans. Amer. Math. Soc. **148** (1970), 461–471. MR **41** #876.

10. ———, *A priori estimates for the wave equation and some applications*, J. Functional Analysis **5** (1970), 218–235. MR **41** #2231.

11. E. M. Stein, *L^p boundedness of certain convolution operators*, Bull. Amer. Math. Soc. **77** (1971), 404–405.

12. A. Zygmund, *Trigonometrical series*, 2nd rev. ed., Cambridge Univ. Press, New York, 1959. MR **21** #6498.

University of Minnesota

ONE-SIDED CONDITIONS FOR FUNCTIONS HARMONIC IN THE UNIT DISC

VICTOR L. SHAPIRO[1]

1. **Introduction.** Let $\mathscr{D}$ designate the open unit disc, and let $u(r, \theta)$ be a function harmonic in $\mathscr{D}$. We define $u_*(\theta) = \liminf_{r \to 1} u(r, \theta)$. Also for (r, θ) in $\mathscr{D}$, we shall set $P(r, \theta) = \frac{1}{2} + \sum_{n=1}^{\infty} r^n \cos n\theta$ and observe that $P(r, \theta)$ is the familiar Poisson kernel. We intend to establish the following result.

THEOREM. *Let $u(r, \theta)$ be harmonic in $\mathscr{D}$ and let $f(\theta)$ be a finite-valued function in $L^1[-\pi, \pi)$. Suppose*
(i) *there are positive constants M and ε with $0 < \varepsilon < 2$ such that $u(r, \theta) \geqq -M(1 - r)^{\varepsilon - 2}$ for (r, θ) in $\mathscr{D}$;*
(ii) *$u_*(\theta) = f(\theta)$ for $-\pi \leqq \theta < \pi$.*
Then for (r, θ) in $\mathscr{D}$,

$$u(r, \theta) = \pi^{-1} \int_{-\pi}^{\pi} P(r, \theta - \varphi) f(\varphi)\, d\varphi.$$

Since there is a positive constant M' such that

$$\sup_{-\pi \leqq \theta < \pi} |\partial P(r, \theta)/\partial \theta| \leqq M'(1 - r)^{-2} \quad \text{for } 0 \leqq r < 1$$

and since

$$\lim_{r \to 1} \partial P(r, \theta)/\partial \theta = 0 \quad \text{for } -\pi \leqq \theta < \pi,$$

we see that the conclusion to the theorem is false if we set $\varepsilon = 0$ in (i).

AMS 1970 *subject classifications.* Primary 31A05, 31A25, 35C15, 42A48; Secondary 31A20, 42A32, 42A24.

[1] Research sponsored by the Air Force Office of Scientific Research, Office of Aerospace Research, USAF, under Grant No. AFOSR 69-1689. The United States Government is authorized to reproduce and distribute reprints for governmental purposes notwithstanding any copyright notation hereon.

Likewise, since $P(r, \theta) \geqq 0$ for (r, θ) in $\mathscr{D}$, and $\lim_{r \to 1} P(r, \theta) = 0$ for θ in $[-\pi, \pi) - 0$, we see that the conclusion to the theorem is false if we replace (ii) by the weaker condition "$u_*(\theta) = f(\theta)$ in $[-\pi, \pi)$ except possibly for one point."

This paper is to a certain extent motivated by our previous paper [**3**].

2. **Proof of the theorem.** With no loss in generality we can suppose from the start that $M = 1$, and $(2\pi)^{-1} \int_{-\pi}^{\pi} u(r, \theta)\, d\theta = 0$, i.e., u evaluated at the origin is zero. Also, as is well-known, every function harmonic in $\mathscr{D}$ is the real part of a function holomorphic in $\mathscr{D}$. Therefore, u can be represented by the Abelian means of a trigonometric series which is the real part of a corresponding power series. Consequently, using the periodicity of u in θ, we see that we are assuming the following (where we now have replaced θ by x):

(1) $$u(r, x) \geqq -(1 - r)^{\varepsilon - 2} \quad \text{for } 0 \leqq r < 1 \text{ and } -\infty < x < \infty.$$

(2) $u(r, x) = \sum_{n=1}^{\infty} (a_n \cos nx + b_n \sin nx) r^n$ for $0 \leqq r < 1$ and $-\infty < x < \infty$. Also the series converges uniformly in x for $0 \leqq r \leqq r_0$ when $0 < r_0 < 1$.

(3) $\liminf_{r \to 1} u(r, x) = f(x)$ for $-\infty < x < \infty$ where $f(x)$ is a finite-valued periodic function in $L^1[-\pi, \pi)$.

(In this paper, when we refer to a function defined on $(-\infty, \infty)$ as being periodic we shall always mean periodic of period 2π.)

We first establish the following:

(4) $$a_n = O(n^{2-\varepsilon}) \quad \text{and} \quad b_n = O(n^{2-\varepsilon}) \quad \text{as } n \to \infty.$$

We shall establish (4) for a_n. A similar proof will establish the corresponding result for b_n.

It follows from (2) that for $n \geqq 2$ and for $0 \leqq r < 1$,

$$a_n r^n = \pi^{-1} \int_{-\pi}^{\pi} \cos nx\, u(r, x)\, dx = \pi^{-1} \int_{-\pi}^{\pi} \cos nx [u(r, x) + (1 - r)^{\varepsilon - 2}]\, dx.$$

Consequently, from (1), (2), and this last fact,

(5) $$|a_n r^n| \leqq \pi^{-1} \int_{-\pi}^{\pi} [u(r, x) + (1 - r)^{\varepsilon - 2}]\, dx = 2(1 - r)^{\varepsilon - 2}.$$

Taking $r = 1 - n^{-1}$, we conclude from (5) that

(6) $$n^{\varepsilon - 2} |a_n| \leqq 2(1 - n^{-1})^{-n} \quad \text{for } n \geqq 2.$$

From (6), it follows immediately that $a_n = O(n^{2-\varepsilon})$ as $n \to \infty$ and the proof of (4) is complete.

Next, using (4), we set, for $0 \leqq r < 1$,

(7) $$U(r, x) = -\sum_{n=1}^{\infty} (a_n \cos nx + b_n \sin nx) r^n n^{-2}$$

and observe that

(8) $U(r, x)$ is uniformly continuous in (r, x) for $0 \leqq r \leqq r_0$ and $-\infty < x < \infty$ for each r_0 such that $0 < r_0 < 1$.

Also we observe that

$$U(r, x) = -\int_0^r s^{-1}\,ds \int_0^s u(\rho, x)\rho^{-1}\,d\rho, \qquad 0 \leqq r < 1. \tag{9}$$

Next we set

$$U_1(r, x) = -\int_{1/2}^r s^{-1}\,ds\left[\int_{1/2}^s u(\rho, x)\rho^{-1}\,d\rho\right], \qquad \tfrac{1}{2} \leqq r < 1, \tag{10}$$

and observe that, for $\frac{1}{2} \leqq r < 1$,

$$U(r, x) = U_1(r, x) + U(\tfrac{1}{2}, x) - \log 2r \int_0^{1/2} u(\rho, x)\rho^{-1}\,d\rho. \tag{11}$$

Next, we set, for $\frac{1}{2} \leqq r < 1$,

$$V(r, x) = -\int_{1/2}^r s^{-1}\,ds \int_{1/2}^s [u(\rho, x) + (1-\rho)^{\varepsilon-2}]\rho^{-1}\,d\rho. \tag{12}$$

It follows from (1) and (12) that, for each x in $(-\infty, \infty)$,

(13) $V(r, x)$ is nonincreasing in the interval $\frac{1}{2} \leqq r < 1$.

Consequently,

(14) $\lim_{r\to 1} V(r, x) = V(x)$ exists for $-\infty < x < \infty$,

(In particular, we note that $-\infty \leqq V(x) < +\infty$.) Also, we see from (12) that for each r with $\frac{1}{2} \leqq r < 1$, $V(r, x)$ is a continuous periodic function of x. We consequently conclude from (13) (see [**1**, pp. 40–41]) that

(15) $V(x)$ is an extended real-valued function which is upper semicontinuous for $-\infty < x < \infty$.

Now, from (10) and (12), we have that

$$V(r, x) = U_1(r, x) - \int_{1/2}^r s^{-1}\,ds \int_{1/2}^s (1-\rho)^{\varepsilon-2}\rho^{-1}\,d\rho. \tag{16}$$

Since $\varepsilon > 0$, the double integral on the right-hand side of the equal sign in (16) tends to a finite limit as $r \to 1$. We consequently conclude from this fact and (11), (4), (15), and (16) that the following two facts hold:

(17) $\lim_{r\to 1} U(r, x) = U(x)$ exists for $-\infty < x < \infty$.

(18) $U(x)$ is an extended real-valued periodic function which is upper semi-continuous in the interval $-\infty < x < \infty$.

It follows from (18) and [**1**, p. 41] that

(19) there exists a finite constant M_1 such that $U(x) \leqq M_1$ for $-\infty < x < \infty$.

Next, we use (4) and set

(20) $$H(x) = \sum_{n=1}^{\infty} (a_n \cos nx + b_n \sin nx)/n^4.$$

We observe in particular from (4) and (20) that

(21) $H(x)$ is a continuous periodic function.

Next, as in [**5**, p. 23], we set

$$D_{\#}^2 H(x) = \liminf_{t\to 0}[H(x + t) + H(x - t) - 2H(x)]/t^2.$$

It follows from [**5**, p. 353] and (7), (17), (19), and (20) that

(22) $$D_{\#}^2 H(x) \leqq M_1 \quad \text{for } -\infty < x < \infty.$$

Consequently, we have that $D_{\#}^2[H(x) - M_1 x^2/2] \leqq 0$ for $-\infty < x < \infty$. But then it follows from (21) and [**5**, Lemma 10.7, p. 23] that $H(x) - M_1 x^2/2$ is a concave function on the interval $(-\infty, \infty)$. In turn this implies by [**5**, Lemma 3.16, p. 328] that $\lim_{t\to 0} [H(x + t) + H(x - t) - 2H(x)]/t^2$ exists and is finite almost everywhere and represents a function which is in L^1 on every finite interval. We conclude from (7), (17), and (2) and [**5**, Lemma 8.2, p. 356] that

(23) $U(x)$ is finite almost everywhere and in L^1 on every finite interval.

Next, we see from (11), (17), and (23) that $\lim_{r\to 1} U_1(r, x) = U_1(x)$ exists for $-\infty < x < \infty$ and that $U_1(x)$ is in L^1 on compact subsets of $(-\infty, \infty)$. Consequently it follows from (14) and (16) that $V(x)$ is in L^1 on compact subsets of $(-\infty, \infty)$, and therefore from (12), (13), (14), and the Lebesgue monotone convergence theorem that for every finite interval $[\alpha, \beta]$, $\int_\alpha^\beta |V(r, x) - V(x)|\, dx \to 0$ as $r \to 1$. But then it follows from (11) and (16) that in particular

(24) $$\int_{-\pi}^{\pi} |U(r, x) - U(x)|\, dx \to 0 \quad \text{as } r \to 1.$$

Letting $S[U]$ designate the Fourier series of U, we conclude immediately from (7) and (24) that

(25) $$S[U] = -\sum_{n=1}^{\infty} [a_n \cos nx + b_n \sin nx] n^{-2}.$$

Next, we define a more general generalized second derivative than before as follows:

$$(26) \qquad D^{*2}U(x) = \limsup_{t\to 0}\left[(2t)^{-1}\int_{-t}^{t} U(x+s)\,ds - U(x)\right]6t^{-2}.$$

We note that if $U(x) = -\infty$, then $D^{*2}U(x)$ is well-defined and is equal to $+\infty$. Also using the proof of [**5**, p. 353] (or setting $k = 1$ in [**4**, Lemma 7, p. 66]) we see from (2), (3), (7), and (17) that wherever $U(x)$ is finite-valued, $D^{*2}U(x) \geqq f(x)$. Since from (3), $f(x)$ is always finite-valued and from (9) for every x, $U(x) < +\infty$, we conclude that

$$(27) \qquad f(x) \leqq D^{*2}U(x) \quad \text{for } -\infty < x < +\infty.$$

Since $f(x)$ is a finite-valued function in L^1 on compact subsets of $(-\infty, \infty)$, given a finite interval (α, β), we invoke the Vitali-Caratheodory theorem [**2**, p. 75] and obtain a sequence of functions $\{f_n\}_{n=1}^{\infty}$ with the following properties:

$$(28) \qquad f_n(x) \leqq f_{n+1}(x) \leqq f(x) \text{ for } \alpha < x < \beta \text{ and } n = 1, 2, \dots;$$

$$(29) \qquad f_n(x) \text{ is upper semicontinuous on } (\alpha, \beta) \text{ for } n = 1, 2, \dots;$$

$$(30) \qquad f_n(x) \text{ is in } L^1(\alpha, \beta) \text{ for } n = 1, 2, \dots;$$

$$(31) \qquad \lim_{n\to\infty} f_n(x) = f(x) \text{ almost everywhere in } (\alpha, \beta).$$

Using (30), we set

$$(32) \qquad F_n(x) = \int_0^x dy\left[\int_0^y f_n(s)\,ds\right] \quad \text{for } \alpha < x < \beta \text{ and } n = 1, 2, \dots$$

and observe from (26) and (29) that

$$(33) \qquad D^{*2}F_n(x) \leqq f_n(x) \quad \text{for } \alpha < x < \beta \text{ and } n = 1, 2, \dots .$$

From (27), (28), and (33), we have that $D^{*2}F_n(x) < +\infty$ and $D^{*2}U(x) > -\infty$. Consequently $D^{*2}[U(x) - F_n(x)] \geqq D^{*2}U(x) - D^{*2}F_n(x)$. We conclude from this fact and from (27), (28), and (33) that

$$(34) \qquad D^{*2}[U(x) - F_n(x)] \geqq 0 \quad \text{for } \alpha < x < \beta \text{ and } n = 1, 2, \dots .$$

Now with n fixed, if $\alpha < x_1 < x_2 < \beta$ and $g(x)$ is a linear function with the property that $g(x_j) \geqq U(x_j) - F_n(x_j)$, for $j = 1, 2$, then using (34) and the upper semicontinuity of $U(x) - F_n(x)$, it is easy to show that $g(x) \geqq U(x) - F_n(x)$ for $x_1 \leqq x \leqq x_2$. (I.e., observe that if $U(x) - F_n(x) - g(x)$ is positive at a point in the interior of (x_1, x_2), so is $U(x) - F_n(x) - g(x) + \delta(x - x_1)(x - x_2)$ for δ sufficiently small and positive. Consequently there exists an x_0 with $x_1 < x_0 < x_2$ such that $D^{*2}[U(x) - F_n(x) - g(x) + \delta(x - x_1)(x - x_2)]$ evaluated at x_0 is $\leqq 0$, which is a contradiction to (34).)

As a consequence of this last fact and the upper semicontinuity of $U(x) - F_n(x)$ we have first that, for each n, $U(x) - F_n(x)$ is finite-valued in the interval (x_1, x_2), and next (from the actual definition of a convex function [5, p. 21]) that

$$(35) \qquad U(x) - F_n(x) \text{ is convex in } (\alpha, \beta) \text{ for } n = 1, 2, \ldots .$$

Continuing, we next set

$$(36) \qquad F(x) = \int_\alpha^x dy \left[\int_\alpha^y f(s)\, ds \right] \quad \text{for } \alpha < x < \beta,$$

and observe from (28), (31), and (32) that

$$(37) \qquad \lim_{n\to\infty} [U(x) - F_n(x)] = U(x) - F(x) \quad \text{for } \alpha < x < \beta.$$

Since the finite limit of a sequence of convex functions is convex, we conclude from (35) and (37) that

$$(38) \qquad U(x) - F(x) \text{ is convex in } (\alpha, \beta).$$

Next, we adapt the argument in [5, pp. 358–359] to our purposes here and observe from (38) and [5, p. 22] (with D^+ designating the right-hand derivative) that $D^+[U(x) - F(x)]$ exists and is finite in (α, β) and

(39) $D^+U(x) - \int_\alpha^x f(y)\, dy$ is a nondecreasing function of x in the interval $\alpha < x < \beta$.

From (38), (39), and [5, p. 24], we in turn obtain that

(40) $D^+U(x)$ is in L^1 on compact subsets of (α, β) and $U(x_2) - U(x_1) = \int_{x_1}^{x_2} D^+U(x)\, dx$ for $\alpha < x_1 < x_2 < \beta$.

Since (α, β) was an arbitrary finite subinterval of $(-\infty, \infty)$, we set

$$(41) \qquad D^+U(x) = \Phi(x) \quad \text{for } -\infty < x < \infty$$

and observe from (39) and (40) that

(42) $\Phi(x)$ is continuous on $(-\infty, \infty)$ except possibly at a countable number of points. Also for each x, $\Phi(x^+)$ and $\Phi(x^-)$ exist and are finite.

Furthermore,

(43) $\Phi(x)$ is a periodic function in $L^1[-\pi, \pi)$ and $U(x_2) - U(x_1) = \int_{x_1}^{x_2} \Phi(x)\, dx$ for $-\infty < x_1 < x_2 < \infty$.

We conclude from (25) and (43) that

$$(44) \qquad S[\Phi] = \sum_{n=1}^{\infty} [a_n \sin nx - b_n \cos nx]/n.$$

Next, we shall establish the fact that

(45) $\Phi(x)$ is a continuous function for $-\infty < x < \infty$.

Suppose that (45) is false. Then it follows from (42) that there exists an x_0 such that $\Phi(x_0^+) \neq \Phi(x_0^-)$. From Fatou's theorem [**5**, p. 100] and from (2), (3), and (44), we have, for $-\infty < x < \infty$,

(46)
$$\text{(a)}\ \liminf_{t\to 0+}[\Phi(x+t) - \Phi(x-t)]/2t \leqq f(x),$$
$$\text{(b)}\ f(x) \leqq \limsup_{t\to 0+}[\Phi(x+t) - \Phi(x-t)]/2t.$$

If $\Phi(x_0^-) < \Phi(x_0^+)$, (46)(a) gives $f(x_0) = +\infty$; if $\Phi(x_0^-) > \Phi(x_0^+)$, (46)(b) gives $f(x_0) = -\infty$. But by (3), $f(x_0)$ is finite. Therefore no x_0 exists such that $\Phi(x_0^-) \neq \Phi(x_0^+)$, and (45) is established.

Once again let (α, β) be an arbitrary finite interval. For (α, β), we now invoke the other part of the Vitali-Caratheodory theorem [**2**, p. 75] and choose a sequence of functions $\{g_n(x)\}_{n=1}^{\infty}$ with the following properties:

(47) $g_n(x) \geqq g_{n+1}(x) \geqq f(x)$ for $\alpha < x < \beta$ and $n = 1, 2, \ldots$.

(48) $g_n(x)$ is lower semicontinuous on (α, β) for $n = 1, 2, \ldots$.

(49) $g_n(x)$ is in $L^1(\alpha, \beta)$ for $n = 1, 2, \ldots$.

(50) $\lim_{n\to\infty} g_n(x) = f(x)$ almost everywhere in (α, β).

Using (49) we set

(51)
$$G_n(x) = \int_\alpha^x g_n(y)\,dy \quad \text{for } \alpha < x < \beta,$$

and with $D_1 G_n(x) = \liminf_{t\to 0+} [G_n(x+h) - G(x-h)]/2t$, we see from (48) that

(52)
$$D_1 G_n(x) \geqq g_n(x) \quad \text{for } \alpha < x < \beta.$$

Noting that in (α, β), $D_1[\Phi(x) - G_n(x)] \leqq D_1\Phi(x) - D_1 G_n(x)$, we conclude from (46)(a), (47), and (52) that

(53)
$$D_1[\Phi(x) - G_n(x)] \leqq 0 \quad \text{for } \alpha < x < \beta \text{ and } n = 1, 2, \ldots.$$

From (39), (41), and (45), we have that $\Phi(x)$ is a continuous function of bounded variation on each closed subinterval in the interior of (α, β). Consequently, we have from (51), (53), and [**5**, Lemma 8.19, p. 359] that

(54) $\Phi(x) - G_n(x)$ is nonincreasing for $\alpha < x < \beta$ and $n = 1, 2, \ldots$.

But the Lebesgue dominated convergence theorem, in conjunction with (47), (49), (50), (51) and (54) gives us that

(55)
$$\Phi(x) - \int_\alpha^x f(y)\,dy \text{ is nonincreasing for } \alpha < x < \beta.$$

From (39), (41), (45), and (55), we conclude that there is a constant c such that $\Phi(x) - \int_\alpha^x f(y)\,dy = c$ for $\alpha < x < \beta$. Since (α, β) was an arbitrary finite interval, we have that

$$\Phi(x_2) - \Phi(x_1) = \int_{x_1}^{x_2} f(x)\,dx \quad \text{for } -\infty < x_1 < x_2 < \infty. \tag{56}$$

Observing from (43) and (45) that $\Phi(x)$ is a continuous periodic function, we obtain from (56) that $\int_{-\pi}^{\pi} f(x)\,dx = 0$ and that for $n \geqq 1$, $\int_{-\pi}^{\pi} f(x)\cos nx\,dx = n\int_{-\pi}^{\pi}\Phi(x)\sin nx\,dx$ and $\int_{-\pi}^{\pi} f(x)\sin nx\,dx = -n\int_{-\pi}^{\pi}\Phi(x)\cos nx\,dx$. We conclude from (44) that

$$S[f] = \sum_{n=1}^{\infty} a_n\cos nx + b_n \sin nx. \tag{57}$$

With $P(r, x)$ designating the Poisson kernel as before (see §1), we obtain from (2) and (57) that, for every x and for $0 \leqq r < 1$,

$$u(r, x) = \pi^{-1}\int_{-\pi}^{\pi} f(y)P(r, x - y)\,dy,$$

and the proof of the theorem is complete.

BIBLIOGRAPHY

1. H. L. Royden, *Real analysis*, Macmillan, New York, 1965.
2. S. Saks, *Théorie de l'intégrale*, Monografie Mat., vol. 7, PWN, Warsaw, 1937.
3. V. L. Shapiro, *The uniqueness of functions harmonic in the interior of the unit disk*, Proc. London Math. Soc. (3) **13** (1963), 639–652. MR **27** #5916.
4. ———, *Fourier series in several variables*, Bull. Amer. Math. Soc. **70** (1964), 48–93. MR **28** #1448.
5. A. Zygmund, *Trigonometrical series*. Vol. I, 2nd ed., Cambridge Univ. Press, New York, 1959. MR **21** #6498.

UNIVERSITY OF CALIFORNIA, RIVERSIDE

AUTHOR INDEX

Italic numbers refer to pages on which a complete reference to a work by the author is given.

Roman numbers refer to pages on which a reference is made to a work of the author. For example, under Shiffman would be the page on which a statement like the following occurs: "The following corollary generalizes a result of Shiffman. . . ."

Boldface numbers indicate the first page of the articles in this volume.

Subject Index